The Catskill Delta

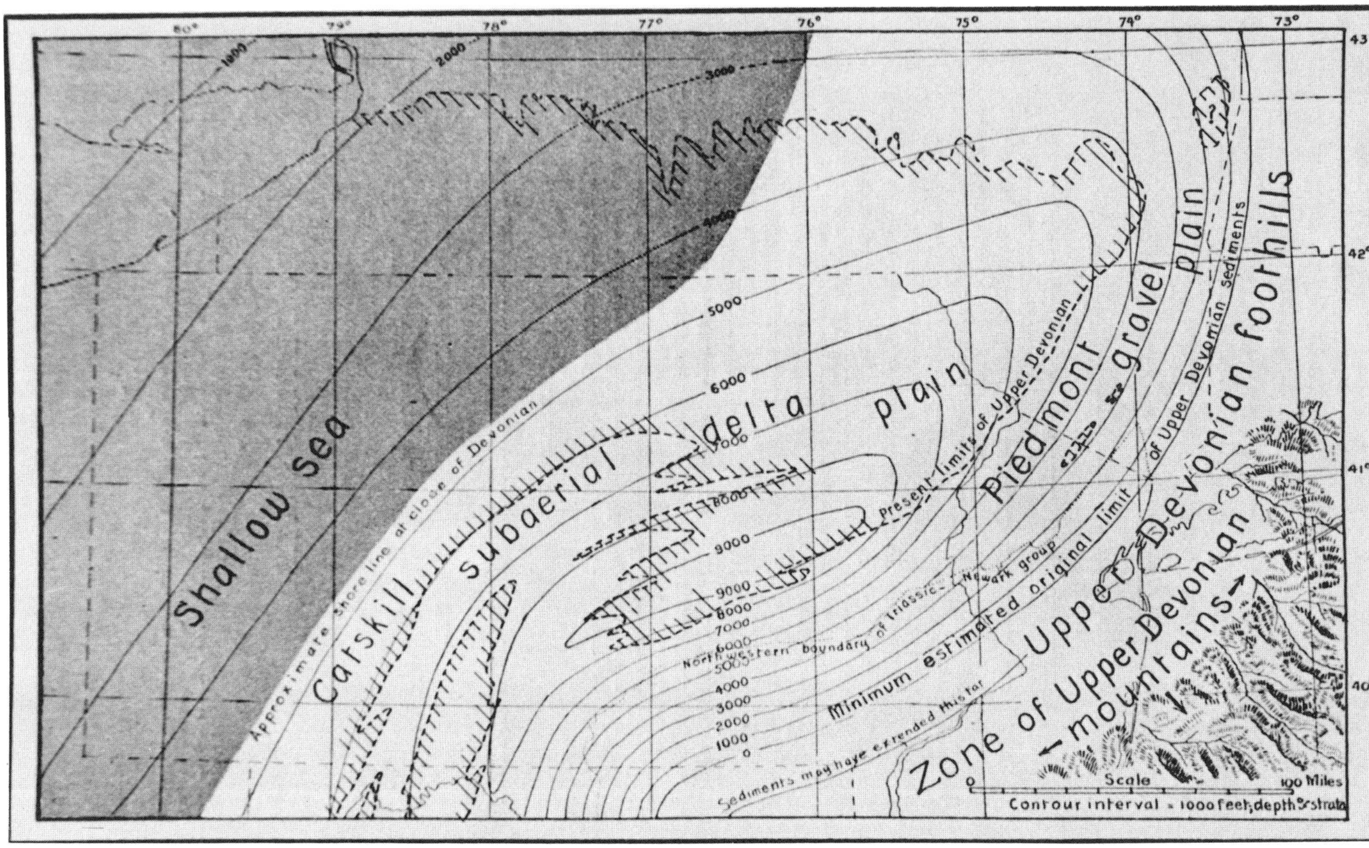

An Isopach Map and aerial reconstruction of the Catskill Delta prepared by Joseph Barrell (Barrell, J., 1913, The Upper Devonian Delta of the Appalachian Geosyncline: American Journal of Science, 4th ser., vol. 36, p. 429–472. Reproduced here with permission.)

The Catskill Delta

Edited by

Donald L. Woodrow
Department of Geoscience
Hobart and William Smith Colleges
Geneva, New York 14456

William D. Sevon
Pennsylvania Geological Society
P.O. Box 2357
Harrisburg, Pennsylvania 17120

THE GEOLOGICAL SOCIETY OF AMERICA · 1888 ·

SPECIAL PAPER

201

Published by The Geological Society of America, Inc.
3300 Penrose Place, P.O. Box 9140, Boulder, Colorado 80301

Printed in U.S.A.

Library of Congress Cataloging in Publication Data
Main entry under title:

The Catskill Delta.

(Special paper / Geological Society of America ; 201)
Bibliography: p.
1. Geology, Stratigraphic—Devonian—Congresses.
2. Geology—Appalachian Region—Congresses. I. Woodrow,
Donald L. II. Sevon, William D. III. Series: Special
paper (Geological Society of America); 201.
QE665.C38 1985 551.7'4 85-946
ISBN 8137-2201-2

Cover concept by Carole Woodrow.

Contents

Preface

Catskill Delta rocks have long had an attraction for geologists and the decade of the 80's is no exception. Since January, 1980, thousands of oil and gas wells have been completed in the Delta's rocks, thus perpetuating an industry started by Edwin Drake in 1859. Enormous tonnages of sandstone, siltstone, and shale have been extracted for a variety of construction purposes over the same period. Even the great flagstone industry that flourished in Pennsylvania and New York during the latter part of the 19th century continues to be active, though at a greatly diminished pace. And, for a brief period prior to 1980, Catskill rocks were being actively investigated as a source of uranium.

The attraction and value of these rocks goes far beyond the market place. Rarely do historical geology, stratigraphy, or sedimentology texts fail to mention the Catskill Delta and since geology was introduced into academic curricula in the nineteenth century, thousands of students have learned about sedimentary rocks, paleontology, paleoenvironments, and structural geology by examining these strata. Nor is interest limited to students. Scholars of every rank, industrial geologists, and geologists from the surveys, both American and foreign, continue to study and sample these rocks as they have for the past century.

It is now more than 70 years since Joseph Barrell gave us the first overview of what he called the Catskill Delta, and 20 years since the Pennsylvania Geological Survey convened "A Symposium of Middle and Upper Devonian Stratigraphy of Pennsylvania and Adjacent States." Both works stimulated research on these rocks and we hope that the papers collected here do likewise.

This volume is an outgrowth of a symposium on The Catskill Delta held at the 18th Annual Meeting of the Northeastern Section of the Geological Society of America in Kiamesha Lake, New York on March 24, 1983. Most of the papers presented at that Symposium are included here, in addition to others. Our arrangement has two sections. The first provides an overview of major features of the sequence while papers in the second section examine more specific aspects. Our scope is broad and the writers present either new ideas involving plate tectonics, paleogeography, and depositional models or sedimentologic and paleobiologic concepts not emphasized until recent years. However, by no means do we cover all of the ground. On the contrary, the obvious omissions give some indication of the need for additional research.

An explanation of the title of the volume is in order. Considering the varied terminology used to encompass the Central Appalachian, Middle and Late Devonian Catskill facies and its marine equivalents, one finds a substantial "glossary" (Table 1).

TABLE 1. GEOGRAPHIC AND SEDIMENTOLOGIC TERMS USED AS TITLES FOR THE MIDDLE- AND UPPER-DEVONIAN CLASTIC FACIES IN THE CENTRAL APPALACHIANS

Alluvial Coast	Coastal Plain
Facies Complex	Delta
Alluvial Plain	Delta Complex
Clastic Wedge	Delta Plain
Clinothem, Undathem, Fondothem	Muddy Coast
Coastal Alluvial Plain	Tectonic Delta-Complex
	Tide-Dominated Coast

Most writers make reference to local geographic or bathymetric features thought to have formed by deposition of the sediments involved, while others have used comprehensive terms to include all of the sediments included within the facies. Barrell used the term "delta" to convey that unity but as our understanding of these rocks has deepened through the efforts of several generations of geologists, it has become clear that Barrell's delta is both more and less than his term implied. For example, we do not find evidence for the one or two trunk streams he expected, instead the record appears to be one of many small streams. Also, the shore is represented by a muddy coast in some localities while at others it is either tide-dominated and sandy or deltaic. Further, we have found that the alluvial, shore and marine facies extend along the ancient coast for hundreds of kilometers, which is, as suggested by one of our reviewers, a pattern more reminiscent of the Texas Gulf Coast than it is of a delta.

In this situation, it is clear that our terminology is inadequate. Perhaps we are ignorant of some fundamental property of such sedimentary systems. Perhaps we are not asking the right questions. Whatever, it is clear that we are dealing with an ancient context of sedimentation and life which, with its processes, is not completely available for study in the modern world. The Devonian land surfaces, for example, were not covered by grass with all that that implies about sediment origin, erosion, and transport. We find too that one modern analogue of the Devonian geography and bathymetry, the southern coast of New Guinea and the Arafura Sea to the south, is virtually undocumented in the modern sedimentologic or biologic literature. Even if it were, its utility as an analogue of the Devonian world in anything like useful detail is not clear.

The choice of title for the *entire* sedimentary complex left to us then is *The Catskill Delta.* Wedded through those words are to one locality, and in spite of their carrying incomplete or even inaccurate implications, the title brought recognition of time, place, rock, fauna, and flora to the minds of our reviewers both here and abroad, no matter what their relationship (or lack of it) to the studies summarized in this volume. In that sense, we have the correct title. This may indeed be the last time we will see

Barrell's term as part of a paper- or book-title, but until something better comes along, perhaps it would be best to hark back and be flexible in our application of his thinking.

We dedicate this volume to Joseph Barrell, the last person to have what might be called an overview of the Catskill Delta and the first person to attempt a synthesis. Any serious students of these rocks or of similar ones in the Phanerozoic must read his papers: it will be an enlightening experience.

Our readers will soon notice that the editor's task did not include forcing authors into a single style of presentation and/or terminology. We tried instead, to present their work clearly and whole cloth for full expression of ideas. We ask, therefore, that the reader exercise perspicacity in the face of variability. Geologists should excel in that regard.

As editors, we realize full well that we have had much help in putting this volume together. Then too, our contributors were unusually cooperative and most were prompt. We gratefully acknowledge the assistance of Hobart and William Smith Colleges and their personnel, especially Ann Warner and Ann Rago. We further acknowledge the professionalism of our contributors.

Below are listed the names of reviewers. Their assistance is appreciated.

John R. L. Allen
Richard Bambach
Thomas Berg
Carleton Brett
Alex Brown
Robert Coe
Edward Cotter
John Dennison
Wallace DeWitt
David Dilcher
Alan Donaldson
Avery Drake
Roland Goldring
Michael Leeder
Thomas Lewis
Brad Macurda
Cathryn Newton
William Oliver
Douglas Patchen
Victor Rahmanian
Lawrence Rickard
John Rodgers
Bill Rogers
Bruce Selleck
John Sepkoski
James Skehan
Donn Gorsline
Francis Hueber
Kenneth Johnson
Rob Van der Voo
Gene Williams

Geological Society of America
Special Paper 201
1985

Middle and Upper Devonian stratigraphy within the Appalachian basin

W. D. Sevon
Pennsylvania Geological Survey
P.O. Box 2357
Harrisburg, Pennsylvania 17120

D. L. Woodrow
Department of Geosciences
Hobart and William Smith Colleges
Geneva, New York 14456

ABSTRACT

The problems of Middle and Upper Devonian stratigraphy within the Appalachian basin are briefly discussed with particular reference to biostratigraphic and lithostratigraphic approaches. Six stratigraphic correlation diagrams are presented.

INTRODUCTION

A complete appraisal of the complexities of Middle and Upper Devonian stratigraphic nomenclature within the Appalachian basin is beyond both our expertise and the intent of this paper. To attempt such an appraisal would be to enter a veritable thicket that has yet to be untangled. However, progress is being made toward that untangling and our purpose here is to present a brief stratigraphic comment and a general summary of the Middle and Upper Devonian stratigraphic nomenclature currently used in various states.

STRATIGRAPHIC COMMENT

Middle and Upper Devonian stratigraphy in the Appalachian basin has been approached from two sometimes conflicting viewpoints: biostratigraphy and lithostratigraphy. The differences in approach are exemplified by the stratigraphic nomenclatures used by the State Geological Surveys of New York and Pennsylvania. New York has emphasized paleontology and its assumed utility for time-equivalency correlation. The contemporaneous work of the Pennsylvania Geological Survey has emphasized correlation by lithology. Both approaches have merit, but when combined with the use of local geographic names, give rise to considerably different nomenclatures. In addition, both approaches are ultimately controlled by sedimentology. As facies change, correlation and its resultant nomenclature changes. The problem is least complicated in the areas farthest removed from the original source area because there was greater homogeneity in the depositional system. Greatest complexity occurs along the eastern outcrop margin where the depositional system was more diverse.

Even if this were the best of worlds and all of us could agree on a single approach to the problems of defining, tracing, and correlating geologic units, the matter would still be complicated in the Catskill Delta sequences. These rocks not only undergo facies changes, they are repetitive. The stratigraphic result is a large number of names, most of which are useful only over a limited area. However, this may reflect more accurately than we like to admit the true nature of the Delta's stratigraphic units.

STRATIGRAPHY

Stratigraphic correlations within individual states containing Appalachian basin Middle and Upper Devonian rocks have existed in some cases for nearly 150 years. However, the first useful regional correlation of the many stratigraphic names was that of Cooper and others (1942). Basin-wide stratigraphy was reviewed by Oliver and others (1969) and recent stratigraphic correlation charts have been issued by several state surveys (Rickard 1975; Berg and others 1983). At present the American Association of Petroleum Geologists, through its Committee on Stratigraphic Units of North America (COSUNA), is currently tabulating the continent's lithostratigraphy. We are grateful to Douglas Patchen for providing the COSUNA information (Childs in preparation) included here. We recognize that all aspects of stratigraphy are in a continuing state of refinement and that not all recently suggested changes (e.g., Ziegler and Klapper 1982) are included in the information presented here.

A series of correlation diagrams are presented in figures which follow. Figure 1 is an index map showing the areas for which specific stratigraphic columns are applicable and the lines of correlation. Figures 2 through 6 show stratigraphic columns arranged in lines normal to depositional strike and Figure 7 shows columns arranged parallel to depositional strike. Figure 2 is derived from Rickard (1975) and emphasizes the biostratigraphic approach to stratigraphy. Figure 3 is from Berg and others (1983) and emphasizes the lithostratigraphic approach. The columns presented in the figures were assembled by Sevon from information provided by the previously mentioned sources and any mistakes are entirely his fault. Additionally, the columns are simplified and are not intended as absolute guides to local stratigraphy: for that the reader should consult the sources cited.

We hope that these stratigraphic columns will: 1. provide the reader with some knowledge of the stratigraphic complexity of the Middle and Upper Devonian Catskill Delta works, and 2. allow the reader to better understand the position of the many stratigraphic names mentioned in this volume.

REFERENCES CITED

Berg, T. M., McInerny, M. K., Way, J. H., and MacLachlan, D. B., 1983, Stratigraphic correlation chart of Pennsylvania: Pennsylvania Geological Survey, 4th ser., General Geology Report 75.

Childs, O. E. (project director), in preparation, Correlation of stratigraphic units of North America: American Association of Petroleum Geologists.

Cooper, G. A. and others, 1942, Correlation of the Devonian sedimentary formations of North America: Geological Society of America Bulletin, v. 53, p. 1729–1794.

Oliver, W. A., Jr., de Witt, W., Jr., Dennison, J. M., Hoskins, D. M., and Huddle, J. W., 1969, Correlation of Devonian rocks units in the Appalachian basin: U.S. Geological Survey Oil and Gas Investigations Chart OC-64.

Rickard, L. V., 1975, Correlation of the Devonian rocks in New York State: New York Museum and Science Service, Map and Chart Series, No. 24.

Ziegler, W. and Klapper, G., 1982, Devonian series boundaries: decisions of the IUGS Subcommission: Episodes, v. 1982, p. 18–21.

MANUSCRIPT ACCEPTED BY THE SOCIETY JULY 20, 1984

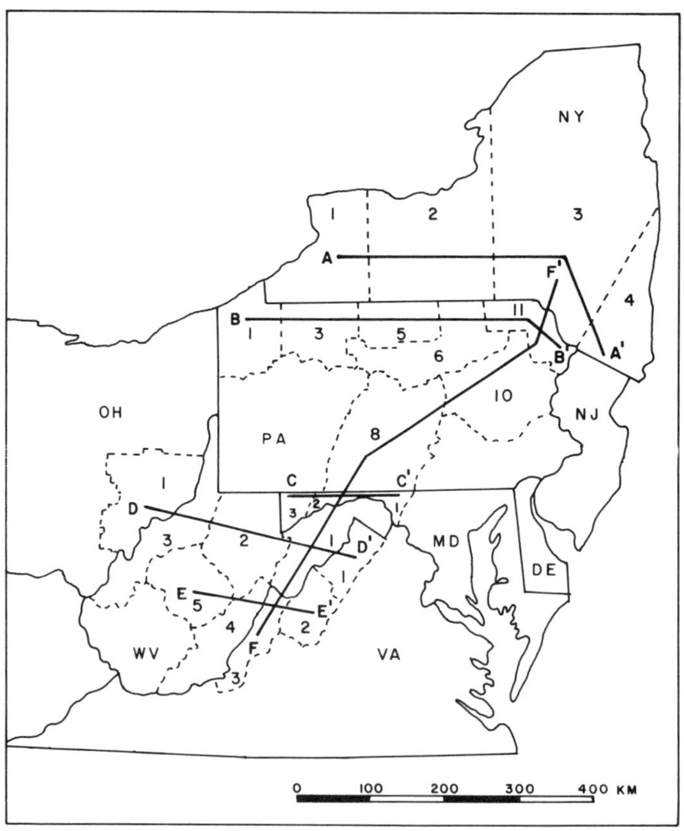

Figure 1. Location map for lines of stratigraphic correlation charts. Dashed lines and numbers represent arbitrary areas within each state for which specific stratigraphic terminology applies.

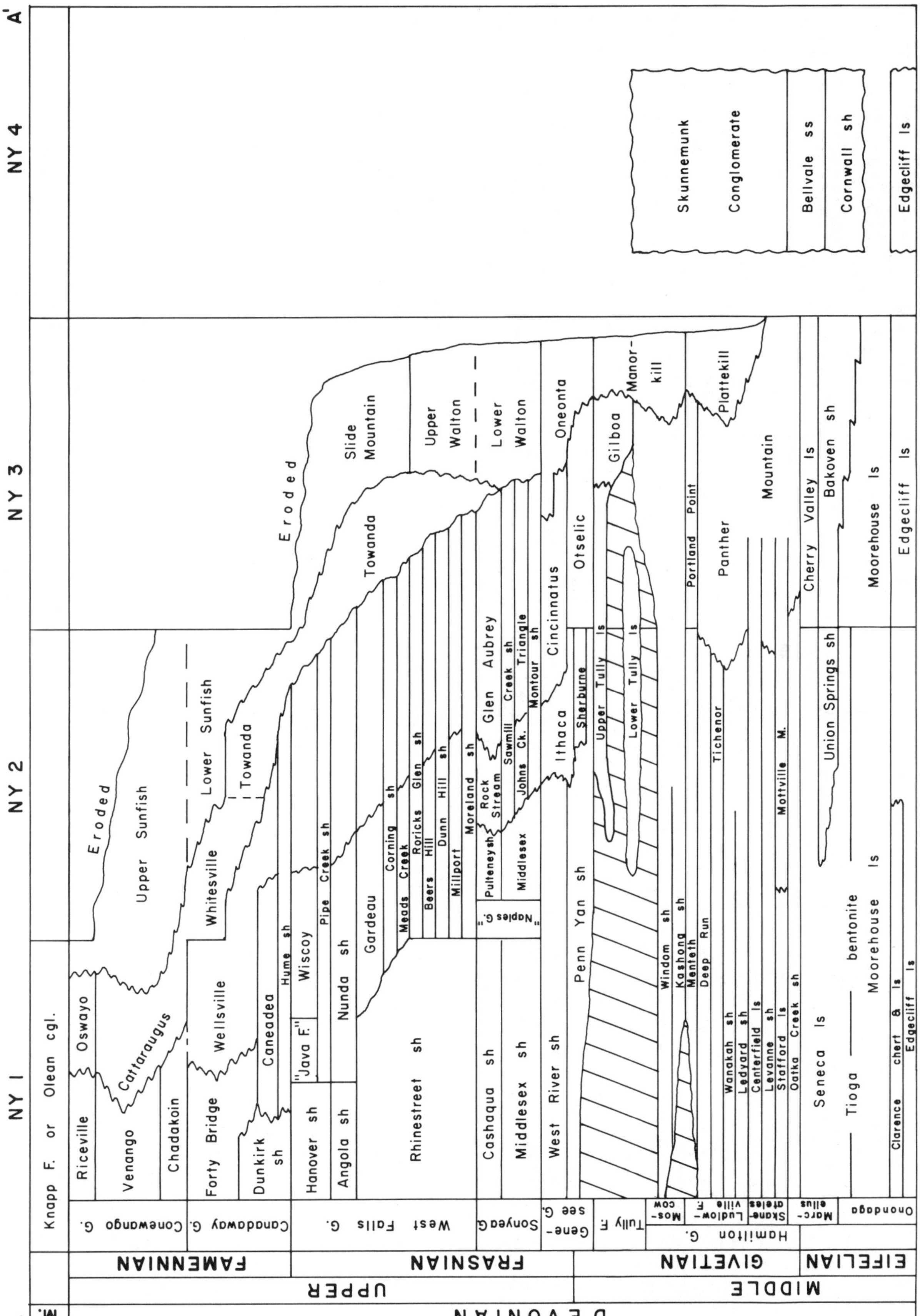

Figure 2. Stratigraphic correlation chart for the Middle and Upper Devonian in New York along line A–A' (Figure 1). Diagram simplified from Rickard (1975). Explanation: G–Group; F–Formation; M–Member; sh–shale; ls–limestone; ss–sandstone; cgl–conglomerate. Lithologies have not been capitalized for simplicity. Some units have only geographic names because the unit has mixed lithologies and stratigraphic rank was not given in sources used for compilation.

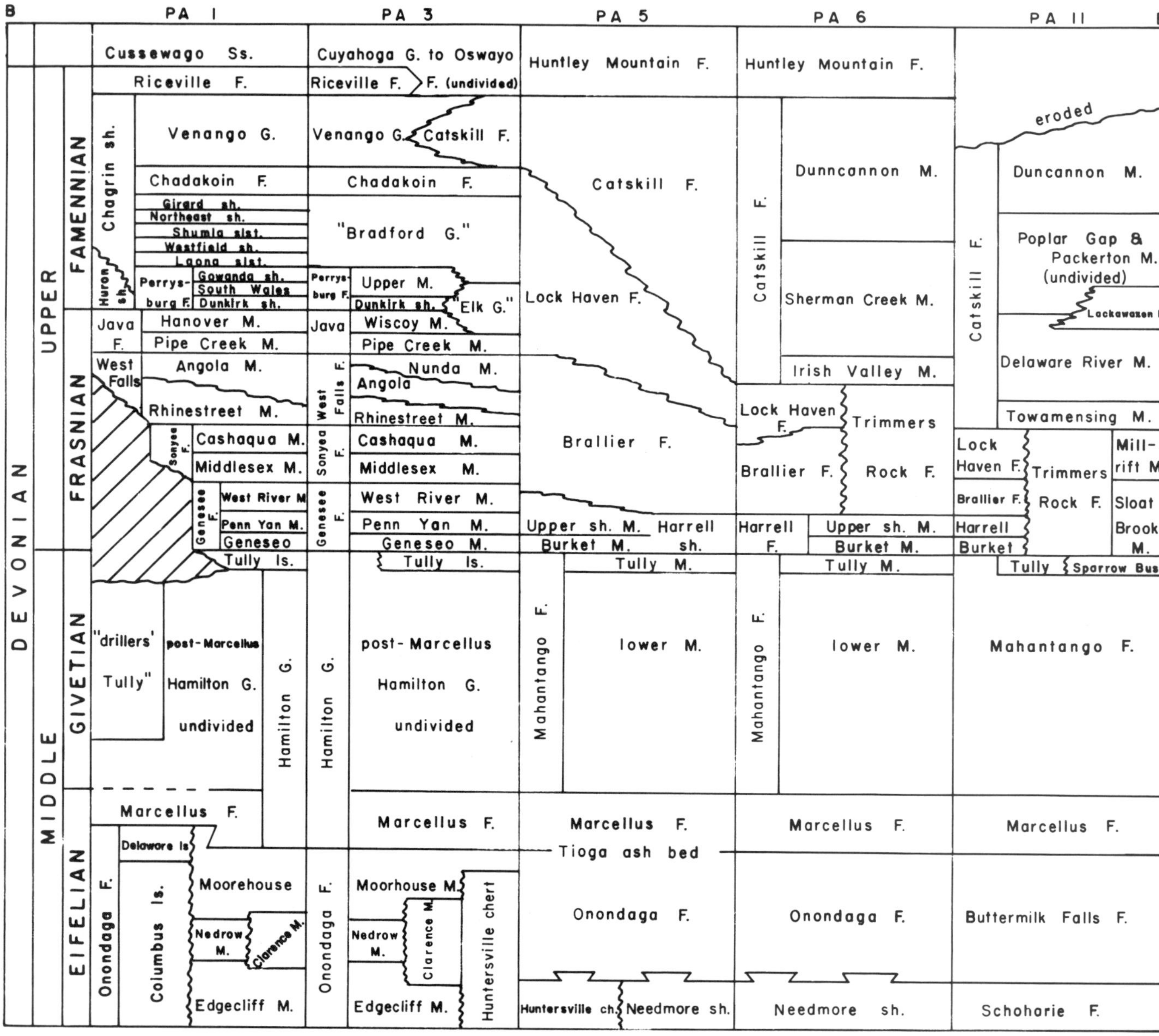

Figure 3. Stratigraphic correlation chart for the Middle and Upper Devonian in Pennsylvania along line B–B[1] (Figure 1). Diagram simplified from Berg and others (1983). Explanation: ch–chert.

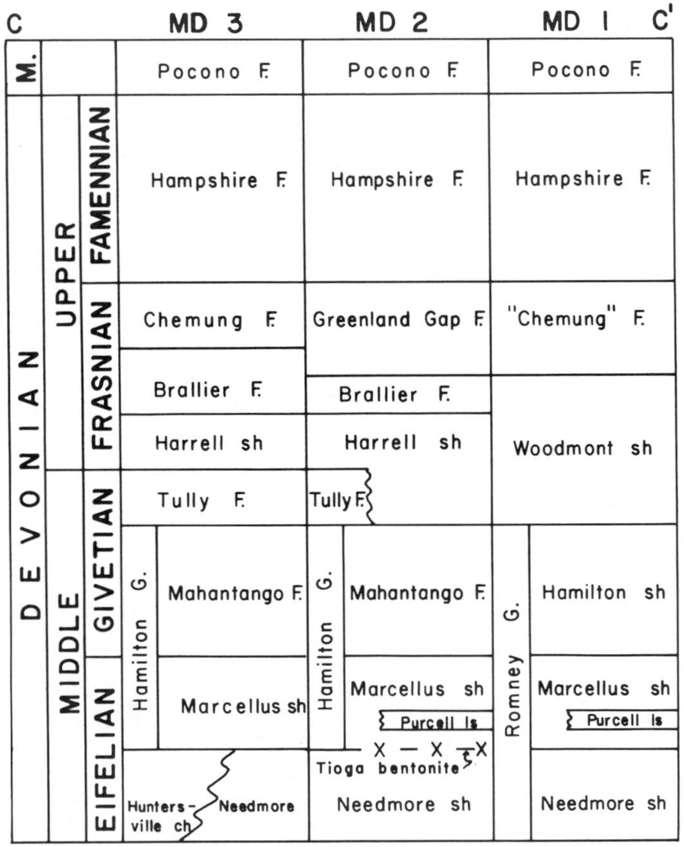

Figure 4. Stratigraphic correlation chart for the Middle and Upper Devonian of Maryland along line C–C[1] (Figure 1). Data from Childs (in preparation).

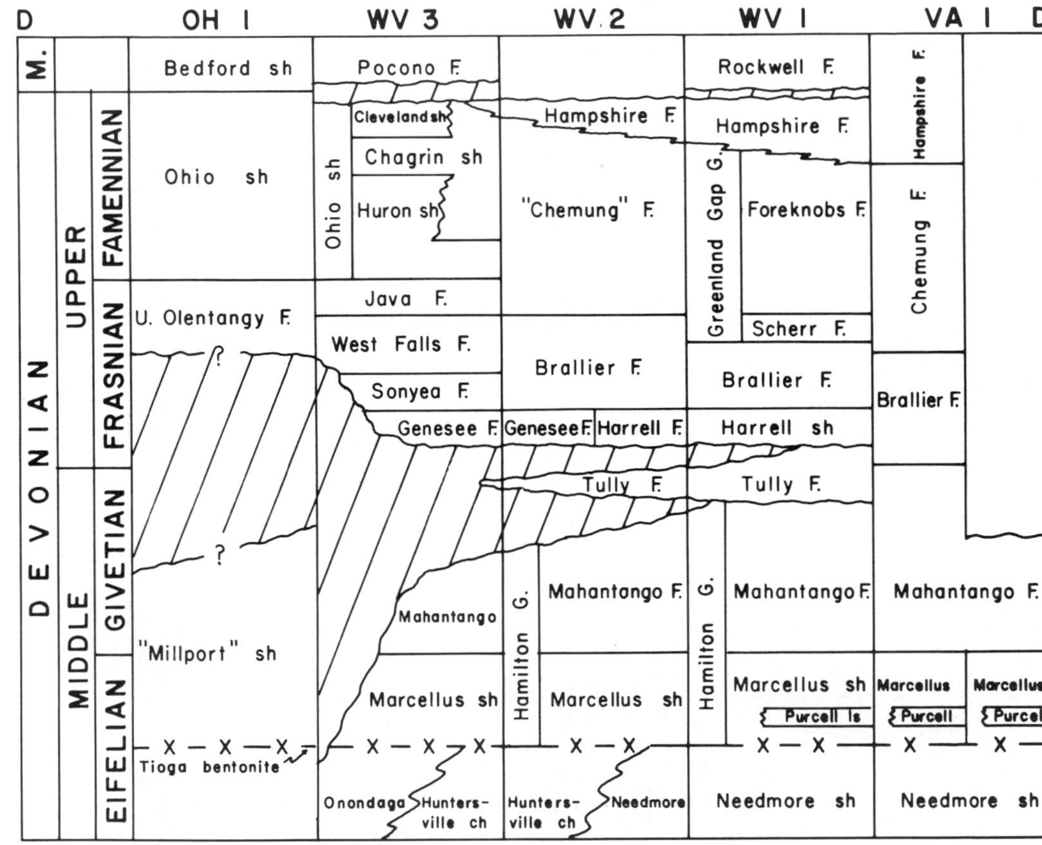

Figure 5. Stratigraphic correlation chart for the Middle and Upper Devonian in Ohio, West Virginia, and Virginia along line D–D[1] (Figure 1). Data from Childs (in preparation).

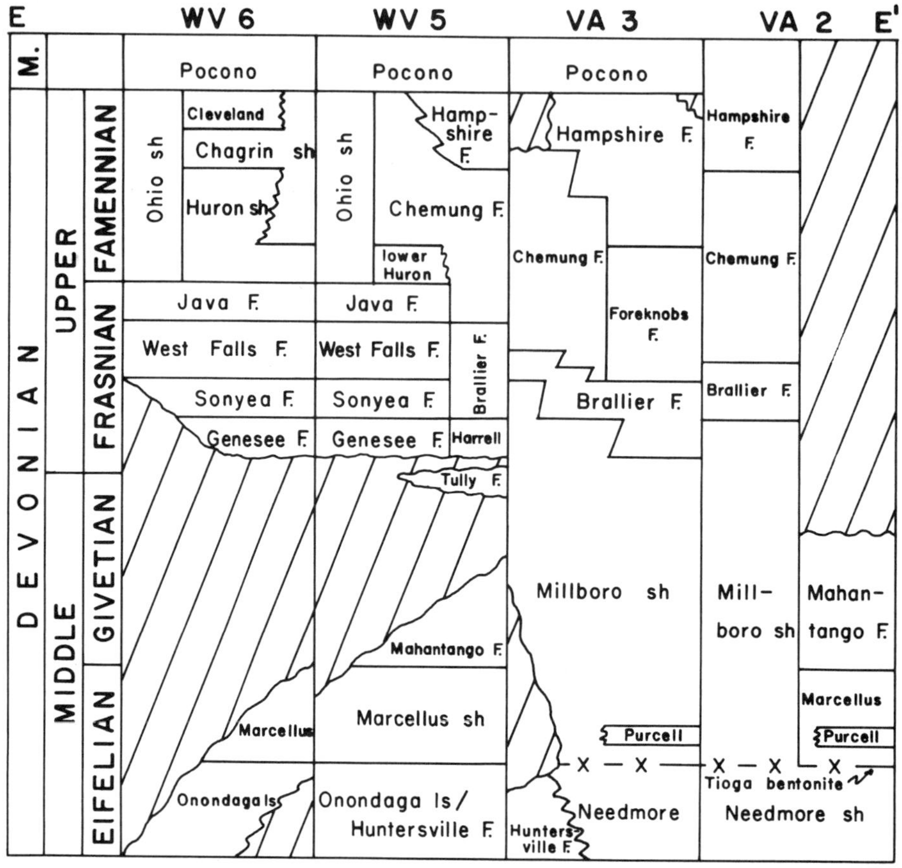

Figure 6. Stratigraphic correlation chart for the Middle and Upper Devonian in West Virginia and Virginia along line E–E[1] (Figure 1). Data from Childs (in preparation).

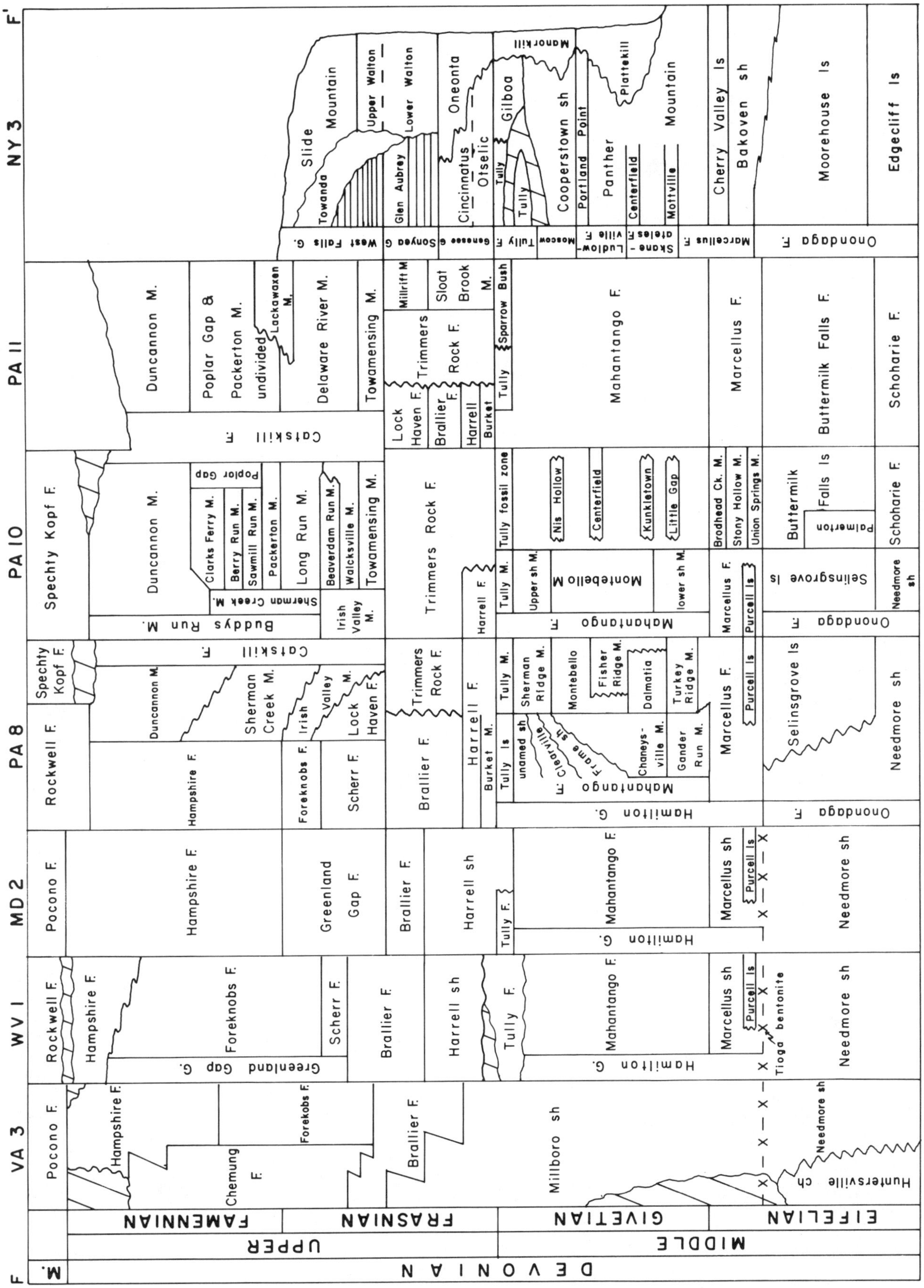

Figure 7. Stratigraphic correlation chart for Virginia, West Virginia, Maryland, Pennsylvania, and New York along line F–F[1] (Figure 1). Data from Figures 2, 3, 4, 5, 6, and Berg and others (1983).

Printed in U.S.A.

Geological Society of America
Special Paper 201
1985

Paleocontinental setting for the Catskill Delta

Dennis V. Kent
Lamont-Doherty Geological Observatory and
Department of Geological Sciences
Columbia University
Palisades, New York 10964

ABSTRACT

Paleomagnetic data confirm geological evidence for tropical paleolatitudes for deposition of the Catskill Delta and related Old Red deposits in Europe. A tectonic model constrained by paleomagnetic data suggests that the Catskill deposits are the product of the Acadian orogeny of the Northern Appalachians as a complex continent-continent collision between Armorica (Hercynian Europe), Laurentia (cratonic North America) and possibly Gondwana, with the Traveler terrane (central New England and New Brunswick) rotated and compressed in between.

INTRODUCTION

A low latitude setting for deposition of the mid- to late-Devonian Catskill Delta has been inferred from geologic evidence. The evidence indicates a depositional environment with warm to hot temperatures, geographically variable rainfall patterns, relatively high evaporation rates, and generally easterly winds (Woodrow and others 1973). Paleomagnetic data confirm low paleolatitudes for eastern North America in the Devonian and place additional constraints on the paleocontinental configuration during this time.

Figure 1 shows a paleocontinental reconstruction of the Atlantic-bordering continents for the mid- to late-Devonian based on an assessment of available paleomagnetic data from North America, Europe, and the southern continents (chiefly Africa and Australia). The various landmasses are positioned with respect to lines of paleolatitude according to averages of the individual paleomagnetic determinations. Paleolongitude is not known. Consequently, the relative positions of the landmasses are based on a variety of indirect criteria including minimum relative motion required to achieve the pre-Atlantic, Pangea configuration (e.g., Bullard and others 1965), geological evidence for the presence or absence of a closed ocean, and avoidance of overlap of older continental basements.

Cratonic North America straddled the equator at that time; indeed, paleomagnetic data obtained directly from Catskill rocks indicate southern hemisphere paleolatitudes of only 1° for exposures in New York (Kent and Opdyke 1978) and 4° in Pennsylvania (Van der Voo and others 1979). However, now adjacent areas of coastal New England and the Canadian Maritimes were approximately 20° farther south at this time. For example, metavolcanics in the Boston basin have magnetizations that were reset during the Devonian and indicate a southern hemisphere paleolatitude of about 23° (Schutts and others 1976). This difference in paleomagnetic directions led to the concept of the Acadia displaced-terrane (Kent and Opdyke 1978; 1979), one of three middle to late Paleozoic terranes now recognized in the Northern Appalachians on the basis of paleomagnetic data.

The Acadia terrane is considered to generally coincide with areas underlain by basement rocks of Avalonian affinity in the Northern Appalachians, including the Avalon Platform of Newfoundland (Kent 1982). The terrane appears to be bounded on the west by large fault zones, such as the Clinton-Newbury, Norumbega-Fredericton, and, with much less certainty, the Cabot Fault in western Newfoundland. Paleomagnetic evidence is compatible with the Acadia terrane reachings its present position with respect to North America by the Carboniferous; the approximately 2000 km of left-lateral motion required may have been taken up in these fault zones.

Inboard of the Acadia terrane, paleomagnetic data obtained from central Maine and southern Quebec have been interpreted as evidence for a rotated block, referred to as the Traveler terrane (Spariosu and Kent 1983). According to the geographic distribution of paleomagnetic sampling localities and through extrapolation of tectonostratigraphy, the Traveler terrane is thought to encompass a region bounded on the southeast by the Acadian terrane (along the aforementioned faults) and inboard by the Connecticut Valley-Gaspé synclinorium. The proposed Traveler

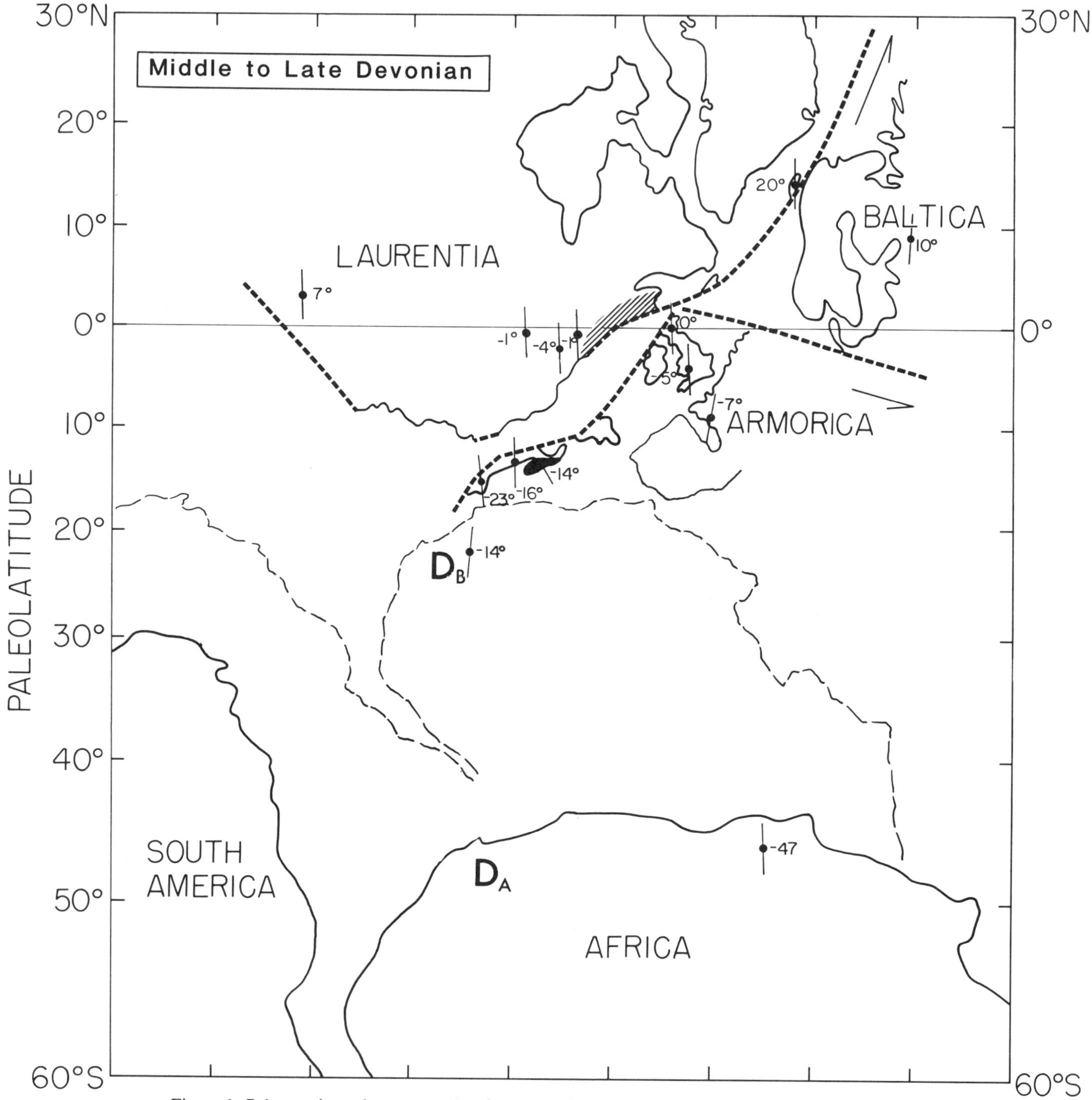

Figure 1. Paleocontinental reconstruction for the middle to late Devonian. Paleomagnetic sampling localities are indicated by filled circles with the intersecting lines showing orientation of paleomeridian and the adjacent numbers referring to calculated paleolatitudes (positive and negative for northern and southern hemisphere, respectively). These data are from the tabulation in Van der Voo and Scotese (1981) with the addition of data from the Gneiguira Formation of Mauritania (Kent and others, 1984), and secondary magnetization (Devonian?) from the Whiterock Formation of southern Nova Scotia (Spariosu and others 1984). Traveler terrane in northern Appalachians indicated by cross-hatchure and the Meguma terrane by dark shading. Position Da for Africa-South America based on paleomagnetic directions from Msissi norite of Morocco (47°S paleolatitude; Hailwood 1974); alternate position Db based on mean of paleopoles from Housetop Granite and Mulga Downs Group of Australia, and the Gneiguira Formation of Mauritania (14°S paleolatitude) (see Kent and others 1984, for discussion of southern continent paleomagnetic data).

block thus resembles the assemblage of Osberg's (1978) basements B and C, distinct from Laurentian Grenville basement to the west and Avalon basement to the east. The available paleomagnetic data suggest that the Traveler terrane rotated about 20° clockwise wile converging with cratonic North America during the Devonian, about the same time as Acadian deformation and metamorphism in the New England Appalachians (Spariosu and Kent 1983) and deposition of the Catskill Delta.

Outboard of both the Traveler and Acadia terranes is the third terrane that can be distinguished paleomagnetically, the Meguma of southern Nova Scotia. Although lithostratigraphic evidence strongly suggests that the Meguma was exotic to North America in the early Paleozoic (Schenk 1971; 1978), paleomagnetic data indicate that the Meguma experienced motion relative to North America even in the later Paleozoic. Recently obtained paleomagnetic results (Spariosu and others 1984) indicate that the Meguma was situated at about the same paleolatitude as Acadia, but also rotated approximately 20° counterclockwise in the Carboniferous to achieve its present position with respect to the inboard areas of North America.

Two of the three terranes identified as displaced on the basis of paleomagnetic data are composites, that is, assemblages of earlier terranes on the basis of tectonostratigraphic similarities and distinctions (Williams and Hatcher 1982; Keppie in press) According to Keppie (in press) the Acadia terrane is a composite of various Late Precambrian and early Paleozoic terranes that were successively added to the core of the Avalon. The Traveler terrane is similarly regarded as a composite of older terranes (e.g., Chain Lakes, Bronson Hill) that may themselves be composites. The amalgamations of the Acadia and Traveler terranes occurred, for the most part, in the early Paleozoic and tend to be associated with major deformational events recognized in the region, such as the Late Precambrian Cadomian, Late Cambrian–Early Ordovician Penobscot, and the Middle to Late Ordovician Taconic orogenic events. Unfortunately, there are few paleomagnetic data available to document the relative motions of terranes involved in most of these earlier amalgamation events. Instead, the presently available paleomagnetic data are primarily sufficient only to constrain relative motions of the composite terranes for the Devonian and younger time intervals.

In summary, the sequence of paleomagnetically-defined displacement in the Northern Appalachians is a clockwise rotation of the most inboard terrane (Traveler) in the Devonian, followed by a northward translation of both the Acadia and Meguma terranes and a counterclockwise rotation of the Meguma. The earlier rotation of the more inboard Traveler terrane can be associated spatially and temporally with the Acadian orogeny; the later northward translation of Acadia and the counterclockwise rotation of Meguma may be associated with an early phase of the Alleghanian orogeny.

PALEOCONTINENTAL RECONSTRUCTION

The significance of these postulated middle to late Paleozoic tectonic displacements in the northern Appalachians can be considered in the context of their circum-Atlantic paleocontinental setting. In the middle to late Devonian, paleomagnetic pole positions from Armorica (southern Britain, France, and Spain; Van der Voo 1982) and Baltica (Scandinavia and the Russian Platform) agree with those from Acadia (taking into account the opening of the Atlantic), but together are systematically displaced with respect to coeval paleopoles from cratonic North America. Thus Acadia apparently was not a separate plate or microcontinent by the Devonian; rather, it was part of eastern Euramerica, all of which was some 2000 km farther south with respect to western Euramerica (Laurentia) (Figure 1). The fault systems separating eastern and western Euramerica mostly lie in the Appalachians of New England and Maritime Canada but may include the Great Glen of northern Scotland (Van der Voo and Scotese 1981).

An interesting result of this offset between eastern and western Euramerica is that it brings many of the various Old Red deposition sites on the European side into lower, more equatorial paleolatitudes, such as those of the Catskill, than would be the case in a conventional Pangea juxtaposition (e.g., Bullard and others 1965).

Much uncertainty exists in the position of the facing margins of Gondwana (northern South America and Africa). A paleopole position from the Msissi norite of Morocco, cited as of Late Devonian (Famennian) age by Hailwood (1974), suggests that these facing margins were at high paleolatitudes (Da in Figure 1). Consequently, a wide ocean between the northern (Euramerican) and southern (Gondwanan) continental blocks is implied in the Devonian. Alternatively, recent paleomagnetic results from the Devonian Gneiguira sandstones of Mauritania suggest lower paleolatitudes for the facing margins of South America-Africa at this time (Kent and others 1984). Acceptance of these latter data leads to the conclusion that the ocean between the northern and southern continents was very small or effectively closed by some time in the Devonian (Db in Figure 1). Due to ambiguities in determination of the age of magnetization of virtually all mid-Paleozoic paleomagnetic data from Gondwana (discussed in Kent and others 1984) it is not yet possible to resolve which of the two Devonian positions for Gondwana illustrated in Figure 1 is more nearly correct.

PALEOMAGNETIC TECTONIC MODELS

The paleocontinental reconstruction shown in Figure 1 is applicable to the period after the Acadian (late Caledonian) orogeny and prior to the climatic Alleghanian (Hercynian) orogeny. On the basis of available paleomagnetic data for the preceding and succeeding time periods, it is possible to suggest how the Devonian configuration shown in Figure 1 might have been achieved and how it evolved into a Pangean assembly by the end of the Paleozoic. The paleocontinental frameworks for the Acadian and Alleghenian orogenies may thus be outlined.

After the Cambrian but prior to the Devonian, paleomagnetic evidence suggests that Armorica was a separate plate (Van der Voo 1979). Late Precambrian and early Paleozoic data match between Gondwana and Armorica and are thus compatible with them being adjacent and moving in unison (Hagstrum and others 1980). Results from Armorican parts of France and Spain indicate that this area was close to or connected with Baltica by the late Devonian but, according to position Da (Figure 1), was separated by a wide ocean from Gondwana (Jones and others 1979; Van der Voo 1982).

Originally, Van der Voo (1979) argued that Armorica collided with Laurentia during the Ordovician and gave rise to the Taconic orogeny. Alternatively, the available paleomagnetic data are also consistent with a collision between Armorica and Laurentia in the Devonian to account for the Acadian orogeny (Kent 1980). New paleomagnetic data for the Late Ordovician and Devonian of Armorica now show that Armorica did not collide with the northern continents at the time of the Taconic orogeny but instead did so at the time of the Caledonian or Acadian orogeny (Perroud and others 1983).

In the reconstruction shown in Figure 1, the docking position of Armorica is adjacent to the Traveler terrane in the northern Appalachians (with the outboard areas of eastern Massachusetts, Maine, New Brunswick, Nova Scotia and Newfoundland far to the south at this time). It is therefore possible to conceive of the Acadian orogeny as reflecting a complex continent-cntinent collision between Armorica and Laurentia with the Traveler block compressed and rotated in between. The voluminous Catskill deposits in North America and related Old Red deposits on the European side are likely the products of the erosion and uplift that resulted from this collision.

Euramerica was essentially assembled in the aftermath of the Acadian or Caledonian orogeny. However, apparently as the result of the progressive impingement of the African craton (Lefort and Van der Voo 1981), it was to undergo an internal rearrangement, principally along a sinistral megashear that appears to follow Acadian-Caledonian trends from New England through Newfoundland, Scotland and into Spitzbergen (Harland 1969; Watts 1983). Whether or not the Meguma, the most outboard terrane in the Appalachians, was once part of Africa remains to be verified by paleomagnetic data but its inferred counterclockwise rotation (Spariosu and others 1984) is consistent with its involvement within the hypothesized left-lateral megashear zone. The culmination of the collision of west Africa with North America seems to account for Alleghanian thrusting and faulting in the southern Appalachians (e.g., Cook and others 1981), after the Acadia promontory translated northward out of the direct path of collision.

DISCUSSION

It is important to point out that not all paleomagnetic data fit neatly into the interpretation of paleocontinental positions presented here. For example, controversy exists regarding the significance of numerous but generally aberrant paleopoles obtained primarily from igneous intrusions of mid-Paleozoic age from the northern Appalachians (see Discussion and Reply in Roy 1982; Kent and Opdyke 1982). Since the establishment of terrane displacements depends on the reliability of paleomagnetic data from the suspected terrane as well as reference paleopoles from the cratonic area, it follows that tectonic interpretations will need to be modified if either set of paleomagnetic data is shown to be erroneous. In this regard, Roy and Morris (1983) have recently suggested that the inferred difference in paleolatitude between cratonic North America and the Acadia terrane is due to an artifact of selective remagnetization, not the result of tectonic displacement. New paleomagnetic data from Carboniferous rocks from cratonic North America (e.g. Irving and Strong 1984; Kent and Opdyke, in press) now indicate that any significant tectonic offset of Acadia (as well as eastern Euramerica) with respect to Laurentia or western Euramerica was resolved by the early Carboniferous. Whether any offset existed in the late Devonian, as shown in Figure 1, is critically dependent on the reliability of the paleomagnetic results from the Catskill Formation, which include dual polarities and a positive fold test as evidence for an original magnetization (Van der Voo and others 1979). In any case, the postulated rotation of the Traveler terrane or of the Meguma terrane is still supported although it is conceivable that the northward translation of Acadia may turn out to have occurred in the Devonian and was perhaps associated with the Acadian orogeny.

Clearly, additional paleomagnetic data are needed to verify the paleocontinental positions and tectonic models outlined here. In addition to the problems mentioned above, adequate paleomagnetic data are still lacking for certain other areas and time intervals so that the paleopositions of some of the major continental blocks remain to be verified in critical intervals, for example, Gondwana from the Ordovician to the Early Carboniferous. There is a general consensus that Euramerica occupied low paleolatitudes in the mid-Paleozoic but exactly how Pangea was assembled and exactly how this might be related to major orogenic activity in plate tectonic terms remains a matter for future research. Acquisition of paleomagnetic data to refine the paleocontinental tectonic framework for the Ordovician Taconic orogeny, which might have been the time of assemblage of smaller, perhaps far-travelled terranes along an active continental margin of eastern North America, is also likely to provide important constraints for models of succeeding tectonic events in the circum-Atlantic region.

ACKNOWLEDGMENTS

Support for this work provided by the National Science Foundation, grant EAR82-12549.

REFERENCES CITED

Bullard, E. C., Everett, J. E., and Smith, A. G., 1965, A symposium on Continental Drift. IV. The fit of the continents around the Atlantic: Philosophical Transactions of the Royal Society of London, v. A258, p. 41–51.

Cook, F. A., Brown, L. D., Kaufman, S., Oliver, J. E., and Petersen, T. A., 1981, COCORP seismic profiling of the Appalachian orogen beneath the coastal plain of Georgia: Geological Society of America Bulletin, v. 92, p. 738–748.

Hagstrum, J. T., Van der Voo, R., Auvray, B., and Bonhommet, N., 1980, Eocambrian-Cambrian palaeomagnetism of the Armorican Massif, France: Geophysical Journal of the Royal Astronomical Society, v. 61, p. 489–517.

Hailwood, E. A., 1974, Palaeomagnetism of the Msissi Norite (Morocco) and the Palaeozoic Reconstruction of Gondwanaland: Earth and Planetary Science Letters, v. 23, p. 376–386.

Harland, W. B., 1969, Contribution of Spitsbergen to understanding of tectonic evolution of North Atlantic Region, *in* M. Kay, ed., North Atlantic: geology and continental drift: American Association of Petroleum Geologists Memoir 12, p. 817–851.

Irving, E. and Strong, D. F., 1984, Paleomagnetism of the Early Carboniferous Deer Lake Group, western Newfoundland: no evidence for mid-Carboniferous displacement of "Acadia": Earth and Planetary Science Letters, v. 69, p. 379–390.

Jones, M., Van der Voo, R., and Bonhommet, N., 1979, Late Devonian to Early Carboniferous Paleomagnetic Poles from the Armorican Massif, France: Geophysical Journal of the Royal Astronomical Society, v. 58, p. 287–308.

Kent, D. V., 1980, Formation of Euramerica in the Paleozoic (abs.): EOS, v 61, p. 220.

——1982, Paleomagnetic evidence for post-Devonian displacement of the Avalon Platform (Newfoundland): Journal of Geophysical Research, v. 87, p. 8709–8716.

Kent, D. O., and Opdyke, N. D., 1978, Paleomagnetism of the Devonian Catskill Redbeds: Evidence for motion of coastal New England Canadian Maritime region relative to Cratonic North America: Journal of Geophysical Research, v. 83, p. 4441–4450.

——1979, The Early Carboniferous paleomagnetic field of North America and its bearing on tectonics of the northern Appalachians: Earth and Planetary Science Letters, v. 44, p. 365–372.

——1982, Paleomagnetism of Siluro-Devonian rocks from eastern Maine: Reply: Canadian Journal of Earth Sciences, v. 19, p. 232–237.

——Multicomponent magnetizations from the Mississippian Mauch Chunk Formation of the Central Appalachians and their tectonic implications: Journal of Geophysical Research, in press.

Kent, D. V., Dia, O., and Sougy, J.M.A. 1984, Paleomagnetism of Lower-Middle Devonian and Upper Proterozoic-Cambrian(?) Rocks from Mejeria (Mauritania, West Africa: American Geophysical Union Geodynamics Series, v. 12, p. 99–115.

——in press, Multicomponent paleomagnetic directions from the Mississippian Mauch Chunk Formation of the central Appalachians and their tectonic implications: Journal of Geophysical Research.

Keppie, J. D., in press, The Appalachian Collage, *in* D. G. Gee and B. Sturt, eds., The Caledonide Orogen, Scandinavia and Related Areas: New York: John Wiley and Sons.

LeFort, J-P. and Van der Voo, R., 1981, A Kinematic model for the collision and complete suturing between Gondwanaland and Laurussia in the Carboniferous: Journal of Geology, v. 89, p. 537–550.

Osberg, Ph.H., 1978, Synthesis of the geology of the north eastern Appalachians, U.S.A., *in* IGCP Project 27, Caledonian-Appalachian Orogen of the North Atlantic Region: Geological Survey of Canada, Paper 78-13, p. 3–11.

Perroud, H., Van der Voo, R., and Bonhommet, N., 1983, Timing of the Paleozoic drift of the Armorica plate on the basis of new palaeomagnetic results: EOS American Geophysical Union Transactions, v. 64, p. 218.

Roy, J. L., 1982, Paleomagnetism of Siluro-Devonian rocks from eastern Maine: Discussion, Canadian Journal of Earth Sciences, v. 19, p. 225–232.

Roy, J. L., and Morris, W. A., 1983, A review of paleomagnetic results from the Carboniferous of North America; the concept of Carboniferous geomagnetic field horizon-markers: Earth and Planetary Science Letters, v. 65, p. 167–181.

Schenk, P. E., 1971, Southeastern Atlantic Canada, Northwestern Africa, and continental drift: Canadian Journal of Earth Sciences, v. 8, p. 1218–1251.

——1978, Synthesis of the Canadian Appalachians: *in* IGCP Project 27, Caledonian-Appalachian Orogen of the North Atlantic Region: Geological Survey of Canada paper 78-13, p. 111–136.

Schutts, L. D., Brecher, A., Hurley, P. M., Montgomery, C. W., and Krueger, H. W., 1976, A case study of the time and nature of paleomagnetic resetting in a mafic complex in New England: Canadian Journal of Earth Sciences, v. 13, p. 898–907.

Spariosu, D. J., and Kent, D. V., 1983, Paleomagnetism of the Lower Devonian Traveler Felsite and the Acadian orogeny in the New England Appalachians: Geological Society of America Bulletin, v. 94, p. 1319–1328.

Spariosu, D. J., Kent, D. V., and Keppie, J. D., 1984, Late Paleozoic Motions of the Meguma Zone, Nova Scotia: New Paleomagnetic Evidence: American Geophysical Union Geodynamics Series, v. 12, p. 82–98.

Van der Voo, R., 1979, Paleozoic assembly of Pangea: a new plate tectonic model for the Taconic, Caledonian, and Hercynian orogenies: EOS, American Geophysical Union Transactions, v. 60, p. 241.

——1982, Pre-Mesozoic Paleomagnetism and Plate Tectonics: Annual Review of Earth and Planetary Science, v. 10, p. 191–220.

Van der Voo, R. and Scotese, C., 1981, Paleomagnetic evidence for a large (2,000 km) sinistral offset along the Great Glen fault during Carboniferous time: Geology, v. 9, p. 583–589.

Van der Voo, R., French, A. N., and French, R. B., 1979, A paleomagnetic pole position from the folded Upper Devonian Catskill redbeds, and its tectonic implications: Geology, v. 7, p. 345–348.

Watts, D. R., 1983, Further Lower Carboniferous Paleomagnetic Results from the Billefjorden Region of Spitzbergen: EOS, v. 64, p. 216.

Williams, H. and Hatcher, R. D. Jr., 1982, Suspect terranes and accretionary history of the Appalachian orogen: Geology, v. 10, p. 530–536.

Woodrow, D. L., Fletcher, F. W., and Ahrnsbrak, W. F., 1973, Paleogeography and paleoclimate at the deposition sites of the Devonian Catskill and Old Red Sandstone Facies: Geological Society of America Bulletin, v. 84, p. 3051–3064.

Manuscript Submitted June 3, 1983
Revised Manscript Submitted February 10, 1984
Manuscript Accepted by the Society July 20, 1984

Printed in U.S.A.

Geological Society of America
Special Paper 201
1985

The Acadian orogeny and the Catskill Delta

Rodger T. Faill
Pennsylvania Geological Survey
P.O. Box 2357
Harrisburg, Pennsylvania 17120

ABSTRACT

The Middle Devonian Acadian orogeny affected the entire Appalachian orogen from Newfoundland to Alabama with varying intensities of deformation and metamorphism. Part of the erosional debris from this uplifted area of tectonism was shed westward into the adjacent Appalachian basin. The basin subsidence and filling varied along its length, but not in concert with the adjacent tectonism. In the Early Devonian, the basin was stable with very little subsidence. Beginning in the Middle Devonian and continuing throughout the Late Devonian, the eastern part of the basin (now Pennsylvania and central New York) underwent rapid subsidence, where the largest volume and coarsest sediments were deposited as the Catskill Delta. To the southwest, grain size and thickness progressively decrease to the thin sequence of black shales in eastern Tennessee and Kentucky. To the north (northern New York, eastern Ontario and southern Quebec), no real evidence exists that a Devonian basin developed—the sediment from the orogenic zone probably passed over the Laurentian shield to the Michigan basin and perhaps elsewhere.

The Acadian orogeny did not impinge upon the basin and its contained sediments, in distinct contrast with the Taconian and Alleghanian orogenies, both of which directly affected the basin, albeit in different ways. Reported Devonian structures within the basin are insignificant (growth folds), questionable (radiometric dating of faults), or probably incorrect (folding and angular unconformity). Acadian structures that occur between the basin and the main tectonic belt include open, upright folds and steep faults; cooling of Taconian metamorphic terrane during the Devonian and Early Carboniferous is also indicated for these rocks.

INTRODUCTION

The Middle and Late Devonian Catskill Delta occupies only a small part of the Appalachian basin, but it is the largest and coarsest extant accumulation of the erosional debris in North America shed from the mountains produced by the Acadian orogeny. The subaqueous deltaic sediments extend from east-central New York through Pennsylvania into Maryland, West Virginia, and western Virginia, westward into Ohio, and southward to eastern Tennessee; the subaerial portion of the delta is restricted to the central part of the basin (New York, Pennsylvania, Maryland, West Virginia and northwestern Virginia).

The Appalachian basin is part of an elongated depositional basin on the western flank of the Appalachian orogen, a complex mountain system on the eastern margin of the North American continent reaching from Newfoundland to Alabama (Figure 1). Various parts of this basin actively subsided from the Late Precambrian to the end of the Paleozoic, a span of more than 350 million years. In its earliest stages (Late Precambrian through Early Ordovician) it received clastic sediments primarily from the cratonic lands to the west; subsequently, its sediment came from the tectonically active Appalachian orogen on the eastern margin of the North American (Laurentian) craton.

The Appalachian mountain system was convulsed by three principal orogenies during the Paleozoic: the Middle to Late Ordovician Taconian, the Early to Middle Devonian Acadian, and the Late Carboniferous (?) to Permian Alleghanian. As a consequence of the first two orogenies, enormous volumes of terrigenous sediments were swept into the basin, creating respectively the Taconic clastic wedge ("Queenston delta") and the Catskill

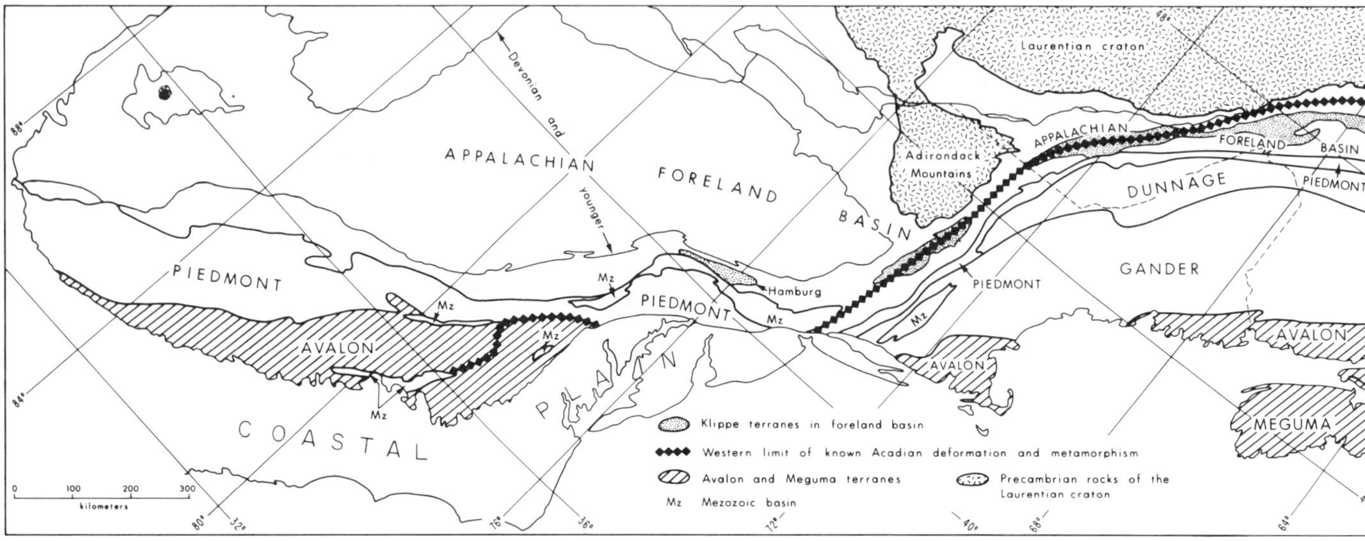

Figure 1. Generalized map of the Appalachian orogen, showing the various terranes (Piedmont, Dunnage, etc.), the Appalachian foreland basin, and the Laurentian craton. Note the contrast in areal extent between the foreland basin southwest of the Adirondacks, and the area to the northeast. In the southern Appalachians, the western limit of Acadian orogenesis is not shown (because it is not well known), but it may extend to near the western edge of the Piedmont terrane. Modified from Williams and Hatcher (1982) and Oliver and others (1967).

Delta. Prior to both of these orogenic events, sediment input was minimal and carbonate shelf or paralic deposition predominated over most of the basin. In contrast, prior to the Alleghanian, nonmarine deposition continued in the eastern part of the Appalachian basin because of continued tectonism in the orogen following the Acadian orogeny and the associated Catskill deposition. During the Carbonaceous marine sedimentation was intermittently widespread only in the western portion of the basin, the youngest of which (Ames Limestone) is Missourian (Late Carboniferous) in age.

Each of the three orogenies involved most of the length of the Appalachian orogen, but they differed from one another in their effects on the basin. Taconian tectonism reached out into the depositional basin in the form of klippen and nappes that were emplaced during deposition. Alleghanian tectonism also extended along the entire Appalachian orogen—its effect on the basin was in the form of decollement tectonics (but much more pervasively in the central and southern basin than in the northern). In contrast, the Acadian orogeny appears not to have affected the depositional basin at all. Superficially, it seems that the Acadian orogeny and the Catskill Delta were coincident in time but separate in space. However, they were intimately related as cause and effect.

THE ACADIAN OROGENY

The Acadian orogeny was first described for an area of considerable deformation and metamorphism around the Bay of Fundy, but it is recognized as well throughout the Gander and Dunnage terranes in central New England and the Canadian Maritime provinces (Figure 1). The axis of this orogeny lies in the Merrimack synclinorium (Rodgers 1970) that extends from eastern Connecticut through central Massachusetts, New Hampshire, central Maine, and thence into western New Brunswick (Figure 2). In New Hampshire, this synclinorium contains rocks primarily of Devonian age that have been raised to the sillimanite metamorphic grade of the amphibolite facies. To the northeast, progressively older rocks are exposed in the synclinorium and the metamorphic grade decreases to as low as the chlorite zone of the greenschist facies.

East of the Merrimack trough is a complex of metamorphosed Precambrian and Lower Paleozoic rocks belonging to the Avalon terrane. The fossils in this terrane have an Old World province affinity, indicating that during the Early Paleozoic the Avalon terrane was part of what is now Europe. Furthermore, paleomagnetic data indicate that during the Devonian, these rocks were some 2000 km southeast of their present position, relative to the North American craton, and that they moved to their present location during the Carboniferous (Kent 1982), well after the Acadian orogeny. Hence, these rocks apparently had no role in the development of the Catskill Delta.

Similarly, metamorphic and igneous rocks in northern Delaware exhibit Late Ordovician paleomagnetic poles that do not coincide with Late Ordovician poles of cratonic North America (Rao and Van der Voo 1980; Brown and Van der Voo 1983). This suggests that some terranes of the crystalline central and southern Appalachians were not initially part of Laurentia, but were originally island arcs, microcontinents, or even a fragment from a different plate (e.g., African). These were subsequently sutured to the North American plate during a later continental

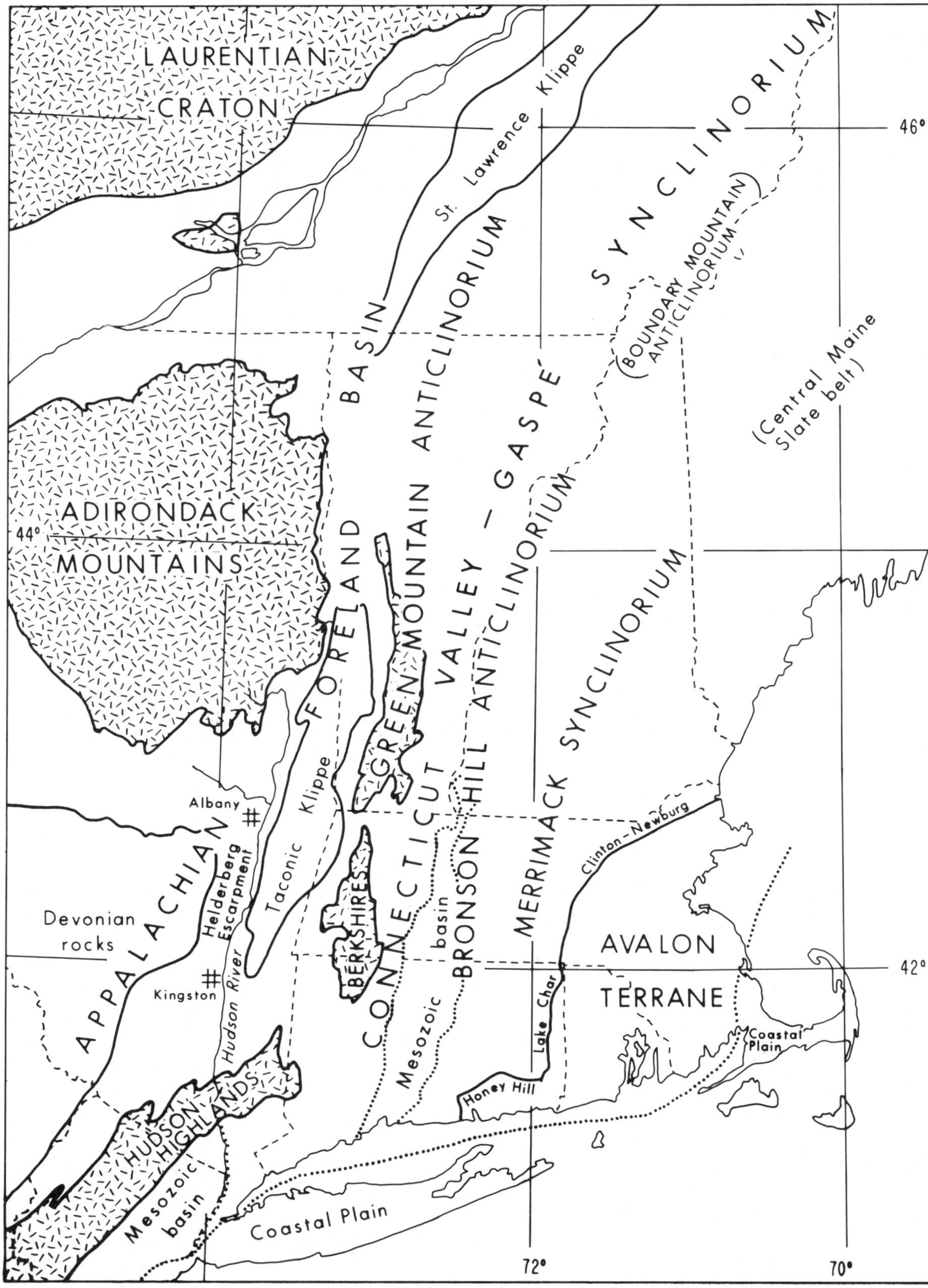

Figure 2. Generalized map of the southern portion of the northern Appalachians, delineating the principal tectonic structures. Simplified from Rodgers (1970).

collision and orogeny (probably Acadian, although geothermometer and geobarometer data from the Wilmington complex in northern Delaware and southeastern Pennsylvania have been used to argue for a Taconian, rather than Acadian, suturing [Srogi and others 1983]). It is not known whether or not this docking occurred early enough for these rocks to be a source for the Catskill Delta. It is also not known if these southern sutured rocks belong to the Avalon terrane or are part of a different microplate.

The possible absence of the Avalon terrane from eastern New England in the Devonian (Kent and Opdyke 1978) raises the possibility that, during the Devonian, the Gander and Dunnage terranes of central and western New England were elsewhere, and that the terrane that supplied the sediment for the Catskill Delta subsequently may have moved northeastward to New Brunswick, Newfoundland, or even Ireland and southern Scotland (e.g., Morris 1976). However, the continuity of Silurian and Lower Devonian stratigraphic units from the eastern flanks of the Green Mountain anticlinorium to the Merrimack trough suggests that these two terranes (Gander and Dunnage) were already in their present location by the beginning of the Silurian. Furthermore, paleomagnetic data from the Merrimack trough indicate that this terrane was accreted to the Laurentian shield during the Devonian, and not after. It was in fact caught between Laurentia and Armorica (western Europe) in a continent-continent collision (Kent 1985) that expressed itself as the Acadian orogeny. Therefore, some of the source of Catskill delta sediment was probably the tectonized rocks of central and western New England.

Five major structures exist in New England between the Precambrian Adirondacks (Laurentian shield) in the west and the bounding Carboniferous faults (e.g. the Lake Chargoggagoggmanchaugagoggchaubunagungamaugg [Dixon and Lundgren 1968] and the Clinton-Newbury faults) on the east (Rodgers 1970; Bradley 1983; see also Figure 2): the foreland basin of the Hudson Valley and Lake Champlain areas (the northern extension of the Appalachian basin); the Green Mountain-Berkshire anticlinorium; and the Merrimack synclinorium, the Bronson Hill anticlinorium (Piscataquis volcanic arc), and the Connecticut Valley-Gaspé synclinorium, which together comprise the Gander and Dunnage terranes (Williams and Hatcher 1982). The three troughs are similar in that they all extend, in one form or another, from southern New England northward into southeastern Quebec/New Brunswick, and that all three were repositories of sediments ranging in age from Cambrian to Lower Devonian. (In northeastern Maine and New Brunswick, the Merrimack trough bifurcates into the Fredericton and Aroostook-Matapedia troughs, separated by the Miramichi massif.) In the westernmost trough (the foreland basin), the pre-Middle Ordovician rocks consist predominantly of carbonate rocks in the west and a clastic sequence (Taconic) in the east. The other two troughs are almost completely clastic-filled with the middle one (Connecticut Valley-Gaspé) containing, in addition, thick sequences of volcanic rocks. One of the more significant aspects of the western two synclinoria is the presence of an unconformity between Middle Ordovician and younger rocks, reflecting the Taconian orogeny. The Merrimack trough may contain a similar unconformity, but one has not been identified because of the high metamorphic grade of the rocks there. Thus, in Cambrian to Middle Ordovician time western New England consisted of a carbonate bank building eastward from Laurentia—the present anticlinoria and synclinoria structures did not then exist. A volcanic arc, located in what is now the Bronson Hill anticlinorium, fed the back-arc basin of the Connecticut Valley synclinorium with volcanic-clastic sediments. Then, during the Middle Ordovician, the Taconian orogeny affected these terranes, producing large nappes and mantled gneiss domes cored with Oliverian granitoid rocks. It was at this time that the slope-rise sediments (Taconic sequence) west of the volcanic arc were incorporated into the Taconic klippen and other slices and were thrust westward over the carbonate platform (Stanley and Ratcliffe 1984). Concurrently, Barrovian metamorphism spread from the Taconian axis along the Bronson Hill area westward to the western flanks of the Green Mountain anticlinorium and the eastern part of the foreland basin (e.g., the Vermont marble belt) (Sutter and others 1984).

Deposition on what probably was a regional unconformity returned to western and central New England following erosion during the Late Ordovician, and continued into the Silurian, spreading into western New England. The Connecticut Valley, Bronson Hill, and Merrimack structures were still not formed. The Precambrian-cored Green Mountain anticlinorium was apparently a relative high area throughout the Silurian because uppermost Silurian rocks lie unconformably over deformed Ordovician rocks on its western flank. However, by the Middle Devonian, this highland was sufficiently eroded to allow terrigenous material to pass westward into the Appalachian basin.

The earliest sign of the Acadian orogeny developed in the Gaspé synclinorium in eastern Quebec, where clastic deposits of late Early Devonian age occur above a predominantly carbonate sequence (Rodgers 1970). By the Middle Devonian, deformation and metamorphism were widespread throughout central New England and the Maritimes, involving rocks as young as Eifelian (lower Middle Devonian) (Boucot 1968). Deformation was polyphase. Early nappe development and/or recumbent folding was followed by upwelling of gneiss-cored domes (some of the gneiss is probably of Precambrian age) centered in two belts, one in the Bronson Hill anticlinorium, the other in the Connecticut Valley trough. In contrast, the folds further north in Maine and southern New Brunswick are upright, and form open to isoclinal folds (Thompson and others 1968; Osberg and others 1968; and Ruitenberg and McCutcheon 1982). The syntectonic intrusion of the New Hampshire plutonic series rocks in the form of concordant sills and more irregular stocks was concentrated in the Merrimack trough, indicating that this area was the axis of the Acadian orogeny (Rodgers 1970). In New England, metamorphism was widespread, reaching westward to the east side of the Green Mountain anticlinorium, where it appears as a low-grade retrogression of the earlier Taconian metamorphism. The metamorphic grade increases eastward across the Connecticut Valley trough to the kyanite zone of the amphibolite facies around the gneiss domes and then drops abruptly farther east (on the east side of the trough) to greenschist facies. These rocks are separated from the amphibolite grade rocks further east by the post-Acadian Ammonoosuc fault. The sillimanite isograd of the amphibolite facies occurs on the crest of the Bronson Hill anticlinorium, and the amphibolite faces persists eastward across the Merrimack trough (Dixon and Lundgren 1968). This regional Barrovian metamorphism was largely coeval with the emplacement of the gneiss domes and the intrusion of rocks of the New Hampshire plutonic series. A late, post-metamorphic phase of deformation resulted in thrust faults and small, upright, steeply to gently plunging folds.

The five major structures in central and western New England continue more or less northward into Canada without profound change, with the aforementioned exception of the Merrimack trough. The apparent intensity of the Acadian orogeny decreased northward in western Maine and southeastern Quebec, but this may be an artifact of deeper erosion in the south. The Acadian was intense in eastern Maine and New Brunswick, where it was the major orogeny, easily surpassing the Taconian in its effects. A major aspect of the Acadian in the Canadian Maritimes is the enormous volume of predominantly granitic intrusions.

Tectonism in New England and the Canadian Maritimes did not cease in the Late Devonian, but continued with some interruptions into the Carboniferous. However, the tectonic style changed completely. Isolated basins developed in which thick sequences of Late Devonian to Pennsylvanian, predominantly non-marine sediments were deposited (Belt 1968). The basins are mostly rifts—strike-slip and normal faults are common. Metamorphism, where present, is very low grade. These features present an entirely different tectonic style during the Carboniferous and cannot be considered a continuation of the Acadian orogeny. Thus it can be said that the Acadian tectonism ended by the Late Devonian.

Southward from New England, the Acadian orogenic structures disappear under Long Island Sound and the Coastal Plain deposits. Clearly identifiable Acadian structures first reappear from under the younger cover in central Virginia (Drake 1980). The evidence for Acadian events in the southern Appalachians is largely isotopic. Sedimentary sequences of Devonian age containing angular unconformities are not present. Throughout much of the crystalline southern Appalachians, the major metamorphic event (as high as upper amphibolite grade) was during the Ordovician, accompanied by one or two phases of folding and faulting (Hatcher and others 1980). This was followed by a greenschist metamorphism of Devonian age (Hatcher 1978). On the other hand, the Piedmont in Alabama exhibits only one metamorphic event, in which Lower Devonian rocks and a granite (366 Ma) were metamorphosed to the sillimanite grade of the amphibolite facies (Thomas and others 1980) before 350 Ma (Tull, 1980). Similarly, the Talladega belt in Alabama contains rocks as young as Middle Devonian that exhibit a single, greenschist facies metamorphism—whole rock K-Ar ages range between 417 and 329 Ma suggest that an Early to Middle Devonian (Acadian?) event affected this belt as well (Tull 1982). Devonian ages have been obtained for rocks from the Inner Piedmont, the Charlotte belt and the Carolina Slate belt (Tull 1980) and these probably are cooling dates. In the Virginia Piedmont, a single metamorphism persisted throughout two periods of deformation, spanning from Late Ordovician to about 330 Ma (Bobyarchick and Glover 1979). Faults such as the Linville Falls, Fries, and Brevard faults were active during this Acadian episode (Hatcher 1978). In addition, the overriding of the North American craton by the Inner Piedmont-Blue Ridge thrust sheet, begun in the Taconian orogeny, was renewed in the Acadian (Hatcher and Zietz 1980).

In summary, the Acadian orogeny was active during the Middle and early Late Devonian along the entire Appalachian orogen, from Newfoundland to Alabama, manifesting itself by sedimentation, deformation, metamorphism, and igneous intrusion. The sole gap in this belt is opposite the central Appalachians—the rocks of the Acadian orogen are undoubtedly present, but are covered by the younger Coastal Plain deposits of New Jersey, Delaware, and Maryland.

THE DEPOSITIONAL BASIN

The Appalachian foreland basin persisted as a depositional trough on the west flank of the Appalachian orogen throughout most of the Paleozoic. Deposition began in the southern Appalachians during the Late Precambrian, and elsewhere during the Late Precambrian or Lower Cambrian, and persisted along its entire length into the Middle Ordovician. Following the Taconian orogeny, a marked contrast developed between the northern basin (that part north of the Adirondacks) and the central and southern portions. Despite the 1000 m of Middle and Upper Ordovician clastics in the Saint Lawrence Valley near and south of Montreal, overlaps and pinch-outs of Lower and Middle Paleozoic formations along the margins of the Laurentian shield and the Adirondacks indicate that they were relatively (and perhaps only intermittently) positive areas. They probably were not buried by thick sequences of sediments that have since been eroded (the Devonian fragments within the Monteregian intrusives near Montreal notwithstanding). This suggests that, from the Silurian onward, much of the northern basin (west of the central Gaspé peninsula, where 3 to 6 km of Silurian and Devonian sediments are preserved) remained rather narrow and underwent little or no subsidence. As a consequence, the westward-travelling sediment from the post-Taconian highlands appears to have been transported westward across the Laurentian craton and deposited in the Michigan basin. On the other hand, the Upper Ordovician, Siurian, and Lower Devonian sediment derived from these highlands (principally the Green Mountain anticlinorium) that travelled eastward was deposited in the large troughs that were to become the Gander and Dunnage terranes during the subsequent Acadian orogeny. In other words, whereas great thicknesses of

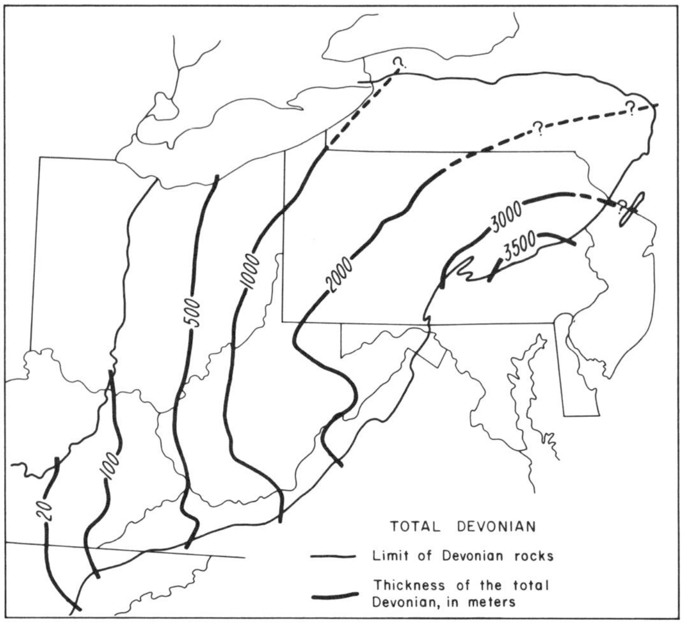

Figure 3. Isopach map of the total Devonian in the foreland basin, central Appalachians. Isopleths are dashed where only part of the Devonian remains. Green Pond synclinorium is the outlier straddling the New York–New Jersey boundary. Simplified from Oliver and others (1967).

sediment accumulated to the east in New England and New Brunswick, relatively thinner sequences appear to have accumulated in the Appalachian basin to the west in the Champlain/St. Lawrence Valley and northwestward on the Laurentian shield.

The foreland basin to the south is quite different. Following the Taconian orogeny, the basin in the central and southern Appalachians remained quite broad, and underwent a large amount of subsidence, particularly along its eastern portion of New York, Pennsylvania, Maryland, and Virginia. As a consequence, great thicknesses of Silurian and particularly Devonian sediments accumulated above largely undeformed (except on the southeast) Cambrian and Ordovician rocks. The southern part of the basin was continuous with the central part, except that the amount of subsidence seems to have been much less. In addition, the sediments accumulating there were finer-grained and much less in quantity.

Prior to the development of the Catskill Delta, the entire central and southern basin was a relatively quiet area, tectonically and depositionally. Following the extensive flysch and molasse deposits that resulted from the Taconian orogeny, sediment input fell off markedly in the late Early Silurian and paralic conditions dominated the basin. This situation persisted throughout Late Silurian and Early Devonian time. Late in the Early Devonian, terrigenous sedimentation, in the form of fine clays, spread along the eastern part of the basin from northeastern Pennsylvania to northwestern Virginia, but the return of basin-wide carbonate deposition in the late Onesquethawan indicates that this emergence of a source area southeast of the basin was a false start.

The beginning of the Acadian orogeny in the central and southern Appalachians was signaled in the earliest Middle Devonian by the Tioga ash fall. This major volcanic event (in the Talladega belt?) spread ash across the entire basin, with a depocenter in Virginia (Dennison and Textoris 1970, 1971). Immediately following this event and continuing throughout the Middle Devonian, terrigenous deposition, largely in the form of black and gray shales, pervaded the basin, reaching westward into Ohio and southwestward to Tennessee (Oliver and others 1967). During this time, sandy subaqueous delta deposits were formed along the eastern basin margin in central Pennsylvania (Faill and others 1978). Subaerial deposition first appeared in eastern New York during the middle Cazenovian (middle Middle Devonian) (Rickard 1975), but the subaerial part of the delta did not expand very far during the rest of the Middle Devonian.

After a brief pause in detrital input at the close of the Middle Devonian (Tully Limestone), terrigenous deposition again reached across the basin. Throughout the Late Devonian, the volume of material entering the basin was much greater than earlier, and the Catskill delta spread progressively westward throughout the central and southern parts of the Appalachian basin. The sedimentologic history of this basin is discussed in other papers in this volume. What is of interest is the overall subsidence history of this part of the basin as a reflection of basin tectonism.

SUBSIDENCE OF THE BASIN

The marine and the overlying subaerial deposits of the Catskill Delta grew progressively southwestward through the Middle and Late Devonian, but the amount of subsidence and sediment accumulation at any given time varied across the basin. For the most part subsidence and sediment accumulation were largest in the northeastern part of the basin and along the southeastern edge (eastern Pennsylvania to western Virginia, Figure 3). The rate of subsidence and basin filling also varied through time. Comparison of the Middle and Late Devonian basin subsidence with that of earlier time periods reveals striking differences in basin dynamics and correlation with Acadian tectonics to the east.

The distribution of thicknesses of Upper Silurian stratigraphic units across the Valley and Ridge and the Plateau provinces indicates that subsidence varied throughout the basin, being greatest in central Pennsylvania and diminishing to the northeast, northwest, and southwest. How far the basin extended southeast of the present outcrop limit is not known; between the Susquehanna and Schuylkill Rivers in Pennsylvania, the total Silurian and Lower Devonian section is quite thin (Auburn Promontory, Swartz 1939), suggesting either a long-term proximity of the basin margin or a persistent stability in this area. During the thirteen million year length of the Late Silurian (Harland and others 1982) paralic depositional conditions dominated the basin. Terrigenous input from the Taconian orogen was greatly reduced and restricted to the easternmost part of the basin. As a consequence the Upper Silurian rocks are primarily calcareous shale and limestone with extensive evaporite development in the cen-

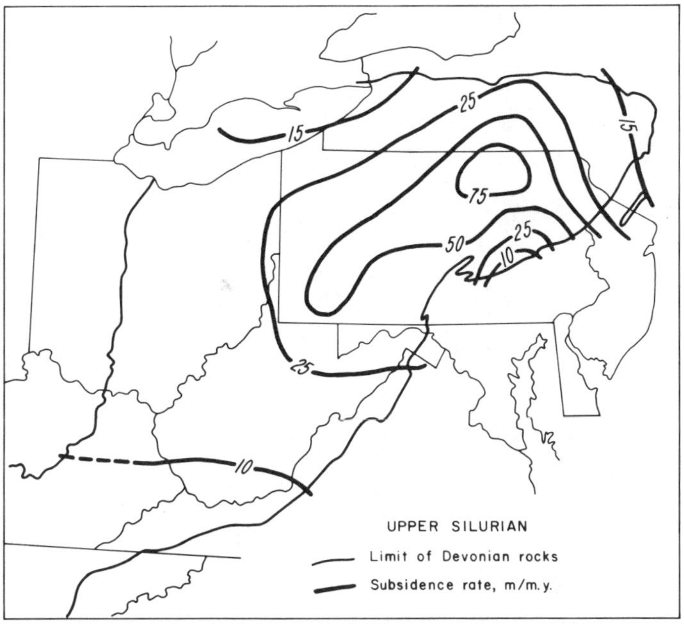

Figure 4. Average subsidence rate in the Appalachian foreland basin during the Late Silurian. The minimal values along the southeast margin in Pennsylvania represents the influence of the Auburn Promontory (Swartz 1939). Rates measured in meters/million years, calculated from thickness data and duration of Late Silurian (13 m.y., Harland and others 1982). New York thickness data from Rickard (1975); West Virginia data from Smosna and Patchen (1978).

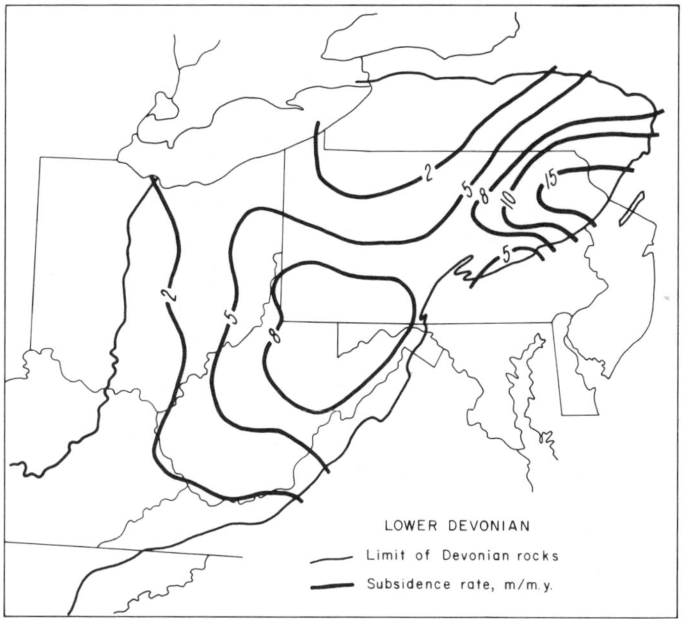

Figure 5. Average subsidence rate in the Appalachian foreland basin during the Early Devonian. The minimal values along the southeast margin in Pennsylvania represent the influence of the Auburn Promontory (Swartz, 1939). Rates measured in meters/million years, calculated from thickness data (Oliver and others 1967) and duration of Early Devonian (21 m.y., Harland and others 1982).

tral part of the basin from north-central Pennsylvania to northern West Virginia (clastics make up a greater proportion of the section further east). With maximum Upper Silurian thicknesses on the order of 1000 m, the subsidence rate ranged from less than 15 m/m.y. along much of the present outcrop margin to more than 75 m/m.y. in the basin center (Figure 4).

This subsidence pattern changed somewhat at the onset of the Devonian Period. Two areas of maximum subsidence developed. One consisted of a broad area centered in southwestern Pennsylvania and northeastern West Virginia, with a maximum sediment thickness of about 200 m; the other was developed in eastern Pennsylvania and adjacent New York, accumulating 400 m of sediments. The Early Devonian spanned a period of about 21 million years (Harland and others 1982), indicating that the average subsidence rate dropped to 8 m/m.y. or less over much of the basin (Figure 5), a dramatic decrease from that of the Late Silurian. A less pronounced decrease in subsidence rate occurred in the east, where a rate of 15 m/m.y. was sustained locally. This change is reflected in the character of the Lower Devonian rocks in the basin. Along the Helderberg escarpment west of the Hudson River, subsidence continued at a moderate pace (Figure 5), but so little terrigenous material entered the basin there that the section is predominantly carbonate (Oliver and others 1967; Rickard 1975). In southeastern New York and southwestward into Pennsylvania, the sandstones and shales in the section indicate some terrigenous input to the basin at this time (Epstein and others 1974), but this material did not spread very far into the basin. In central Pennsylvania and farther southwestward this stratigraphic interval consists of limestone, chert, and clean sandstone, suggestive of shallow water conditions with minimal sediment input, extensive sediment reworking, and very slow subsidence. The existence of a widespread pre-Onesquethaw unconformity (particularly in the western part of the basin) and the presence of isolated patches of Ridgeley (Oriskany) Sandstone beyond the main sand sheet (Oliver and others 1967) further emphasize the slowness, or even the occasional local absence, of subsidence. In the latter part of the Early Devonian, the first glimmering of tectonic activity was the appearance of a gray shale (Needmore) along the southeast margin from southern Pennsylvania to southwestern Virginia. However, carbonate deposition soon returned to most of the basin in the form of the Onondaga Limestone. As during the Late Silurian, the subsidence rate varied across the basin, being highest in the east and in the west central part of the basin, but even these were remarkably low.

The Middle Devonian was not much shorter than the Early Devonian—thirteen million years in duration (Harland and others 1982). Yet during the Middle Devonian, accumulation of sediment was much greater, and the basin center was no longer a site of maximum deposition—the greatest thicknesses occurred in the northeastern part of the basin and along the southeast margin. The thicknesses of these sediments, 500 to 800 m, are considerably greater than those of the underlying Lower Devonian units. The subsidence rate during the Middle Devonian had increased

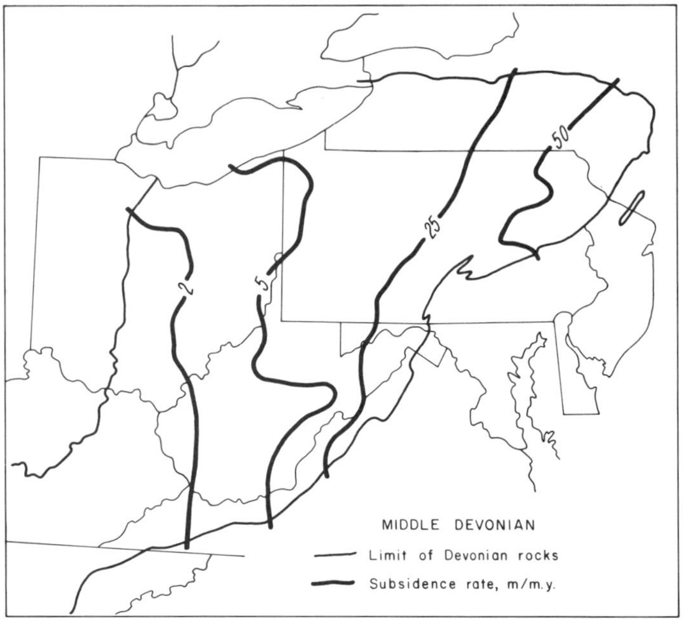

Figure 6. Average subsidence rate in the Appalachian foreland basin during the Middle Devonian. Rates measured in meters/million years, calculated from thickness data (Oliver and others 1967) and duration of Middle Devonian (13 m.y., Harland and others 1982).

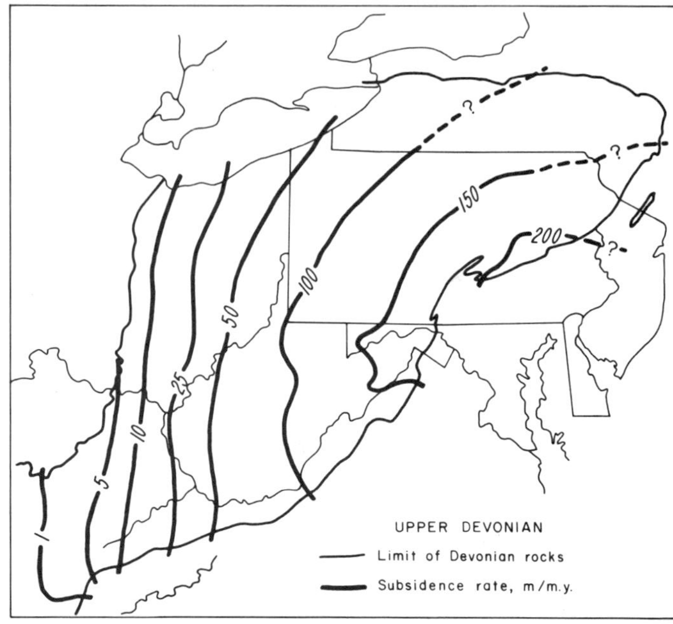

Figure 7. Average subsidence rate in the Appalachian foreland basin during the Late Devonian. Isopleths are dashed where Upper Devonian rocks are partly to completely absent. Rates measured in meters/ million years, calculated from thickness data (Oliver and others 1967) and duration of Late Devonian (14 m.y.) (Harland and others 1982).

to more than 50 m/m.y. (Figure 6). On the west side of the basin the subsidence rate hardly changed, and the pattern of sediment accumulation across the basin was a progressive (geometric) eastward increase. This change in subsidence pattern was paralleled by a concurrent change in the nature of the sediments. Termination of Onesquethawan carbonate deposition and the subsequent spread of black shales over much of the basin, punctuated by the Tioga ash fall, probably reflects the beginning of Acadian tectonism to the east in the Appalachian orogen. Dark shale deposition persisted throughout the Middle Devonian. At two locations along the southeast margin, sandstones and local conglomerates were introduced to the basin. In eastern New York, red beds of the Catskill facies, constituting the earliest subaerial sediments, were deposited continuously from the middle of the Middle Devonian (Rickard 1975). In addition, coarse detritus (Montebello Sandstone) was also introduced to the basin in central Pennsylvania by a delta system (Faill and others 1978). In New York, pebble lithologies and petrographic analyses show a substantial component of a sedimentary/low-grade metamorphic source, and a lesser component of higher-grade metamorphic source for the Catskill Formation (Allen and Friend 1968; see also Sevon 1985 for additional references). Hence, the source of the Catskill Delta sediments in New York was probably the low- to moderately high-grade metamorphic terrane in southern New England uplifted by the Acadian orogen; the closest source of the Montebello may have been in southern New Jersey, Delaware, and eastern Maryland, although strike-parallel transport from New England is also a possibility. The absence of moderately coarse detritus farther south along the eastern basin margin, and the absence of similar coarse clastics within the southern part of the basin (south of Virginia) suggest that the Acadian orogeny may not have extended farther south along the Appalachian orogen than Virginia during the Middle Devonian.

The Late Devonian period lasted about as long as the Middle Devonian—fourteen million years (Harland and others 1982)—but the amount of sediment accumulation was much larger: as much as 2000 to 3000 m. The rapid subsidence in the eastern part of the basin, begun in the Middle Devonian, spread throughout the rest of the basin during the Upper Devonian, yielding subsidence rates as much as an order of magnitude greater than during the Early or Middle Devonian (Figure 7). The pattern of the progressive eastward increase in sediment accumulation begun in the Middle Devonian persisted throughout the Upper Devonian. As large as the subsidence rate was during the Late Devonian, the rate of sediment input was substantially larger, resulting in the progressive westward building of both the subaerial and subaqueous portions of the Catskill Delta to its farthest western extent (Ohio, West Virginia, and western Virginia) by the end of the Devonian period.

Clearly, subsidence and sediment-input rates increased throughout the Appalachian basin during the Middle and Late Devonian (and during the Early Devonian in the east). The trends of eastward increasing regional subsidence (Figures 6 and 7) and general coarsening of the Catskill Delta sediments (and lateral equivalents) suggest that perhaps even greater subsidence and deposition of coarser sediments occurred to the east of the present outcrop margin. However, apart from the Green Pond outlier, no evidence remains to substantiate such speculation.

Indeed, it is uncertain whether the coarse-grained, Upper Devonian sediments of New York and Pennsylvania reached their present westerly locations because the hypothetical basin to the east became filled or stopped subsiding (and thus allowed the coarser sediment to travel further west) and/or because the mountains in the orogen increased in elevation and areal extent. Certainly the persistent rapid and voluminous sediment input into the Appalachian basin indicates that tectonic activity to the east (limited, perhaps, to only uplift and erosion) continued throughout the Late Devonian. Indeed, tectonism (of a different sort) and resultant coarse sedimentation in the Appalachian basin continued into the Carboniferous with the arrival and docking of the Avalon terrane east of the Merrimack synclinorium. At this same time (Early Carboniferous), coarse clastics appeared in the southern Appalachian basin for the first time since the Taconian orogeny, suggesting some orogenic activity east of the basin that resulted in uplift of that southern terrane.

One of the more remarkable aspects of the relation between the Acadian orogen and the accumulation of sediments in the Appalachian basin is the apparent lack of correspondence between tectonic intensity and the volume of adjacent sediments. The Acadian orogeny seems to have been most active in New England, yet from Albany, New York northward to the Gaspé peninsula, the narrow Appalachian basin contains little Middle and Upper Devonian sediment. In contrast, evidence of an Acadian tectonism in the central Appalachians is largely lacking (but possibly covered by the Coastal Plain deposits), yet an enormous volume of sediment accumulated in the adjacent foreland basin. To what degree Acadian orogenesis occurred in the southern Appalachians is still an open question, but the volume and coarseness of the basin sediments in the adjacent basin is much less (or even absent) than to the north.

The dearth of accumulated Middle and Upper Devonian rocks in the northern Appalachian basin was apparently caused by the stability (or later uplift) of the Laurentian (including the Adirondacks) terrane. The apparent lack of subsidence of these areas, despite the adjacent orogenic activity, precluded the development of a large basin. As a consequence material shed from the orogen may have passed over these terranes and accumulated in the Michigan basin to the west.

Southward from Albany, the Appalachian foreland basin subsided and provided a large area to accumulate the material shed from the orogen. Whether or not the greater subsidence in the central basin (relative to the southern) reflects greater (but now covered) tectonism adjacent to the basin or closer proximity to the intense orogenic zone in New England is not known. Whether or not the greater volume and coarseness of sediment in the central part similarly reflects greater tectonism or proximity to tectonism than farther south is also not known. Certainly, at the present level of erosion, the coarse sediment in the lowest Carboniferous (Pocono and its lateral equivalents) spread much farther south than did the Catskill Delta. This suggests that Acadian tectonism may have progressed southward through time (Rodgers 1967; Ettensohn 1984), but this may be an illusory conclusion—Alleghanian tectonism and subsequent erosion may have removed an earlier (Devonian) "Catskill" depositional basin east of the present Devonian outcrop.

Acadian Deformation at or Near the Basin Margin

It is evident that the Middle and Upper Devonian rocks of the central Appalachian basin in Pennsylvania and New York and the southern Appalachian basin from West Virginia to Alabama have undergone a deformation that in general decreases in severity in jumps at each structural front (Rodgers 1970; p. 64) from southeast to northwest. This deformation resulted in large folds and subordinate faults in the central Appalachians, and large faults with subordinate folds in the southern Appalachians (Rich 1934; Rodgers 1949; Nickelsen 1963; Gwinn 1964; Harris 1970; Rodgers 1971; Faill and Nickelsen 1973; Milici 1975; and Perry 1978). The deformation affected Carboniferous (and Permian) rocks as well as the Devonian and older rocks of the basin. These structures resulted from the Late Paleozoic Alleghanian orogeny that has recently been dated to be at least as old as Early Permian, but is probably Late Carboniferous (Van der Voo 1979). This apparent discrepancy between paleontologic and paleomagnetic ages may be due to experimental errors or it may indicate that the Alleghanian deformation was polyphase.

Evidence for a Middle to Late Devonian Acadian orogeny within the Appalachian basin or nearby is much less compelling. In the rocks of southern New York and the Piedmont of southeastern Pennsylvania, radiometric cooling dates (largely Devonian, but spanning from Silurian into the Carboniferous) and upright folds, steep faults, and widely spaced cleavages of probable post-Taconic age (Freedman and others 1964; Amenta 1974) suggest that the Acadian orogeny affected those rocks, albeit more weakly than did the Taconian. Closer to the basin, however, the effects of that orogeny are subtle, ambiguous, or absent. The folds and faults along the Helderberg escarpment in eastern New York and the deformation of the Green Pond synclinorium in New Jersey may or may not be Acadian in age. Farther to the southwest, firm evidence of Acadian tectonism is lacking in the belt of Precambrian, Cambrian and Ordovician rocks (the Reading Prong and Blue Ridge) between the Mesozoic basins and the Valley and Ridge province. In fact, it appears that a nearly complete geographic separation existed between the Acadian orogenic terrane and the basin that accumulated the erosional detritus produced by that orogeny. Closer examination of the evidence of this separation follows.

The Helderberg escarpment lies on the west side of the Hudson River in east-central New York, extending from Kingston northward to near Albany (Figure 2). It consists largely of Upper Silurian and Lower Devonian rocks, trending 010° azimuth, that have been extensively faulted and folded. A short distance (5 km) to the west, Middle and Upper Devonian rocks show no evidence of this deformation. The deformation is thus younger than Early Devonian, but how much younger is not known. In other words, this "Helderberg" belt of faults and folds could have been formed during either the Acadian or Alleghanian orogeny. The westernmost definitely Acadian deformation (ascertained by radiometric dating) occurs on the east flank of the Green Mountain anticlinorium in Vermont and northwestern Massachusetts, overprinting the earlier Taconian structures. The intensity of this deformation decreases to the west. There is some question about how much farther west Acadian structures extend. The post-Taconian Chatham fault just east of the Hudson River may be Acadian in age or younger (Ratcliffe and Bahrami 1976). Furthermore, the Acadian structures from Vermont southward into Massachusetts and Connecticut trend nearly north-south, a trend similar to that of the folds and faults along the Helderberg escarpment. It is this parallelism, as well as the geographic proximity, that underlies the long-held assumption that the Helderberg escarpment structures are of Acadian age. However, in the Carboniferous rocks of the Lackawanna basin to the southwest in Pennsylvania, two divergent fold trends are present—an early "Lackawanna" trend of 060 to 070° azimuth, which is overprinted by a later "Main" trend of 080 to 090° (Geiser and Engelder 1983). Northeastward, in the Devonian rocks in southeastern New York, cleavage of both of these trends are present. From there, the cleavage of the 060 to 070 trend can be traced farther northeastward, where it appears to gradually shift to a 010° trend at Kingston and thence northward to Albany along the Helderberg escarpment. Because the Lackawanna basin contains Late Carboniferous rocks, Geiser and Engelder (1983) argue that the early "Lackawanna" and the "Main" phases must both be Alleghanian in age. This, and the change in cleavage trends, suggests that the Helderberg escarpment structures may also be Alleghanian.

Southward in the Housatonic Highlands, the Berkshire allochthon, and the eastern Hudson Highlands (east of the Canopus fault), the Acadian event produced open folds, and perhaps amphiboltie-grade metamorphism (Hall 1980). As in northwestern New England, the Acadian orogeny increased in intensity to the east across the Fordham terrane and the southward extension of the Connecticut Valley-Gaspé synclinorium. West of the Canopus fault, in the western Hudson Highlands and the easternmost part of the Appalachian basin, the only certain Acadian effects were movement on high angle faults.

The Precambrian rocks of the western Hudson Highlands extend southwest across northwestern New Jersey in the New Jersey Highlands. The Green Pond synclinorium that lies within this Proterozoic terrane is an elongate northeast-trending outlier containing Paleozoic rocks from Cambrian to Middle Devonian in age. All the rocks in this outlier are folded and faulted, and it has long been argued whether the deformation was Acadian or Alleghanian in age (Offield 1967). The Middle Devonian age of the Acadian in New England, the lithic affinity of the Middle Devonian outlier rocks with those west of the Helderberg escarpment to the north, indicating they were part of the same basin (Kirby 1981), and the absence of any direct evidence of Acadian deformation in this belt suggest that the folding and faulting of the Green Pond synclinorium was Alleghanian rather than Acadian (with perhaps some Jurassic high-angle faulting).

Farther southwest, the Proterozoic terrane of the Reading Prong (here called the Durham and Reading Hills) crosses from northwestern New Jersey into southeastern Pennsylvania, bounded on the southeast by the Mesozoic Newark basin (as are the New Jersey Highlands). The Durham and Reading Hills and the Great Valley adjacent to the northwest, consist of a crystalline-cored nappe megasystem developed during the Taconian orogeny (Drake 1978). The isotopic ages of the Grenville crystalline rocks cluster around 900 to 800 Ma. No intense thermal disturbance has affected these rocks since then (Grauch and Ludwig 1979) and they exhibit a Paleozoic reheating to a maximum of only 200 to 250°C (Sutter and Dallmeyer 1982). Yet, this terrane has suffered post-Taconian deformation that is considered to be of Alleghanian age because these later folds parallel those of the Valley and Ridge province to the north (MacLachlan and others 1975; Drake 1980). Those folds to the north encompass the Carboniferous rocks of the Anthracite region and thus are Alleghanian. The lack of recognition of intervening tectonism suggests that the Acadian orogeny did not extend northwest of (what is now) the Mesozoic basin in New Jersey or Pennsylvania.

Structures in the northern Blue Ridge (South Mountain anticlinorium) are entirely congruent geometrically and kinematically with those in the Great Valley and Valley and Ridge provinces on its western flank, which are Alleghanian (Root 1970). Cleavage in the Blue Ridge province from York, Pennsylvania, southward to Charlottsville, Virginia, indicates a single post-Middle Devonian deformation as it pervades Middle Devonian rocks in the Massanutten syncline (Mitra and Elliott 1980). There is no apparent evidence of earlier Paleozoic tectonism, either Taconian or Acadian, in this terrane (Drake 1980).

INTRABASINAL ACADIAN TECTONICS

The paucity of evidence for Acadian deformation at or near the Appalachian basin margin would certainly suggest that little or no deformation affected the basin itself. It would seem that the basin had simply filled, its depositional history being determined solely by the interplay of subsidence and the nature and amount of sediment entering it. But Devonian deformation in the form of syndepositional folding, folding and erosion, faulting, and growth folds has been claimed for several areas within the basin. Each of these is examined in turn.

In northeastern Pennsylvania, divergent fold trends, channels on the flanks of small (40 m wide) anticlines, and soft-sediment structures are cited as evidence for Late Devonian syndepositional deformation (Woodrow 1968; Fletcher and Woodrow 1970). Soft-sediment structures are not necessarily of tectonic origin. Divergence of fold trends suggest more than one phase of folding, but do not time the deformation (e.g., the divergent trends in the Lackawanna basin). The appearance of channels on small anticlinal flanks may result more from the buttressing effect of channels on the development of later, flexural slip, Alleghanian folds than vice versa. The rarity of these fold-related channels (relative to the prevalence of non-fold-related channels), their small size, and the lack of reports of similar occurrences elsewhere in the basin underscore their equivocality.

The Anthracite region in eastern Pennsylvania encompasses rocks ranging in age from Devonian to Middle Carboniferous. Throughout central and eastern Pennsyvania, the Devonian-Mississippian Spechty Kopf Formation lies above the Late Devonian Catskill Formation. The Spechty Kopf is lithically transitional between the Catskill and the overlying Mississippian Pocono Formation. Along the Lehigh River, on the eastern edge of the Anthracite basin, there is evidence of erosion and profound changes in depositional conditions at the base of the Spechty Kopf (Sevon 1973), but the Spechty Kopf-Pocono contact is conformable (Epstein and others 1974). Along the Susquehanna River, on the western edge of the Anthracite basin, both the lower and upper contacts of the Spechty Kopf are conformable (Faill and Wells 1974) and sedimentation seems to have continued without interruption from the Catskill into the Pocono. In contrast, within the Anthracite basin itself, mapping by Trexler and Wood (1968) suggests that a marked angular unconformity (up to 75 degrees) exists between the Spechty Kopf and Pocono Formations (Trexler and others 1961; Wood and others 1969). If so, then it would indicate that the Spechty Kopf and older rocks were folded prior to the deposition of the overlying Pocono Formation, with erosion in places of more than 700 m of Spechty Kopf during the hiatus (Wood and others 1969). On the other hand the interpretation of an unconformity has been questioned by Hoskins (1979), who attributes the observed angular discordances to the later Alleghanian deformation. In addition, nowhere else within the Valley and Ridge or Plateau provinces has such an angular unconformity between the Spechty Kopf and Pocono Formations been proposed or suggested. Indeed, this contact is generally considered to be conformable, with little or no break in sedimentation (Edmunds and others 1979). Folding so severe as to result in an unconformity with a 75 degree angular divergence and 700 m of erosion would not be a local event, and would not be a subtle deformation that could be easily overlooked in other areas. For these reasons, the author has concluded that this proposed unconformity does not exist, and that there was no Late Devonian folding in the Anthracite portion of the Appalachian basin.

At this point it may be worth discussing Middle and Late Paleozoic unconformities in the central and southern parts of the Appalachian basin as clues to possible direct effects on the basin by the adjacent orogeny. The Devonian sequence is most complete in the central basin, and becomes less complete southwestward from Virginia to Alabama, where only scattered remnants are preserved (Oliver and others 1967). Subsequent to the Silurian-Devonian unconformity in northwestern Pennsylvania and western New York, no unconformities are recognized in the Devonian rocks except for an extended hiatus that spanned the Middle-Late boundary in the western part of the central Appalachian basin. Deposition seems to have been continuous, with no more than shifting environments in response to variations in subsidence and detrital input. The two most prominent shifts occurred at the end of the Middle Devonian (Tully limestone) and near the end of the Late Devonian (Riceville shale). These and lesser ones were only marine transgressions, and were probably responses to sudden increases in basin subsidence. Basin subsidence is, in a sense, a "tectonic event," but not one sufficiently strong to produce regional deformation or other evidence of orogenic activity. In the central basin, deposition was actually suspended only three times: immediately after the Lower Devonian Oriskany (Dennison and Head 1975); during the late Middle and early Late Devonian in western New York to Ohio (Oliver and others 1967); and locally in east-central Pennsylvania at the Catskill-Spechty Kopf boundary (Sevon 1973). No angular discordances are associated with any of these interruptions.

A much more significant hiatus developed in the Lackawanna syncline and to the west in north-central Pennsylvania during the late Early Carboniferous. This pre-Pottsville unconformity progressively cuts down-section toward the north through the Mauch Chunk and Pocono Formations, and the upper part of the underlying Catskill Formation (Edmunds and others 1979). This unconformity was produced by an apparent Late Mississippian uplift of the northern part of the basin to the north (centered in western New York?) with a consequent northward beveling of progressively older formations prior to deposition of the Early Pennsylvanian Pottsville Group. As with the older Devonian events no local folds or faults are associated with this unconformity.

In the southern Appalachian basin the role of unconformities is quite different. First, the Silurian-Devonian unconformity is profound, as Devonian rocks overlie Early Silurian and even older rocks (down to Lower Ordovician) from Tennessee southward. Second, erosion may have removed some of the previously deposited Devonian sediments. For example, isolated patches of sandstone of the Deerpark Stage (middle Lower Devonian) suggest that the Oriskany Formation (which was probably derived from the craton) was originally more widespread than as presently preserved (Oliver and others 1967). Third, Devonian sediments may not have been deposited. For much of the Lower and Middle Devonian sequence, the rocks thin greatly from West Virginia to Tennessee, and are essentially absent farther south except for local sandstone patches in northern Alabama (e.g., Frog Mountain Formation). During the Late Devonian, the Chattanooga black shales reached progressively farther southward, but the accumulations were quite small. One can speculate that the presently preserved southern basin may be the western edge of a much more extensive Devonian sequence, both areally and stratigraphically. In this scenario, sediment from a presumed Acadian orogen in the Carolinas and Georgia may have been trapped in an eastern, now eroded, part of the foreland basin and reached Tennessee and Alabama only in the latter part of the Devonian Period or the Carboniferous.

Apart from these unconformities, the Appalachian basin was a wide area of subsidence, fairly continuous in space as well as time, with no marked irregularities or prominent structures within it. Certainly the basin was not "layer-cake"—amounts and rates of subsidence varied across the basin, but they varied smoothly: during the Late Silurian, a salt depocenter developed in the basin center from central New York to southwestern Pennsylvania, but those deposits grade laterally into less saline deposits (Rickard 1975; Smosna and Patchen 1978); in the Early Devonian, a "no-sand" area developed in northwestern Pennsylvania, surrounded by and grading into the sandy Ridgeley (Oriskany) Formation (Abel and Heyman 1981); and numerous other lateral facies changes occur throughout the basin, but they too are gradational (see Berg and others 1983, and references cited therein). In short, it has yet to be demonstrated that sharp lateral lithologic boundaries exist that reflect intrabasinal tectonics such as faulting or folding within the Middle Paleozoic stratigraphy. Yet, there is a question that has been discussed informally over the years: are the folds of the Valley and Ridge and Plateau provinces strictly Alleghanian, or did they begin to develop earlier as growth folds, perhaps as far back as the Devonian Period? If these "Alleghanian" folds did begin as Devonian growth folds, then either the Alleghanian orogeny extended much farther back in time than is generally assumed, or the folds could be considered a marginal effect of thc Acadian orogeny.

First, it should be emphasized that it has not been demonstrated that the thickness of any Middle or Upper Paleozoic formation varies as a function of its position on major folds. For the most part stratigraphic thickness measurements are not precise and can easily be in error by 10 percent and more, but even so, available measurements do rule out growth folds of large amplitude. Low-amplitude growth folds are more difficult to demonstrate or refute—to demonstrate their presence requires examination of sedimentary aspects that are sensitive to low topographic relief.

Evidence for growth folding during the Devonian Period is not substantial for two basic reasons. First, the amplitude of the folds must not have been very large, even into the Late Carboniferous; otherwise the resulting topographic barriers would have had a profound effect on the depositional patterns (and thus the stratigraphy), an effect that has not been reported. Second, no basin-wide studies have been undertaken to evaluate the possible presence of growth folds. Several isolated, local structures have been cited as supporting evidence. The post-Spechty Kopf folding and subsequent erosion proposed by Wood and others (1969) and the evidence of Fletcher and Woodrow (1970) were discussed above and found unconvincing to this author.

In north-central and western Pennsylvania, the Middle Devonian black shales exhibit a variable gamma ray response, in which a higher radioactivity reflects a greater organic content in the shale. The total thickness of radioactive shale in each stratigraphic unit is generally irregular in areal distribution, except for the Middle Devonian shales in the southwestern part of the state. There, the greatest accumulations appear to be on anticlinal crests, suggesting that these crests may have been bathymetric highs that diverted currents away from these shallower, euxinic areas (Harper and Piotrowski 1978). Yet elsewhere in the state, no correspondence exists between radioactivity and structure. Heckel (1973) described an anticline at Smyrna (central New York, southeast of Syracuse) in the Tully Limestone and underlying Hamilton beds that was growing during deposition of the Lower Tully. As measured on the underlying *Vitulina* zone (Heckel 1973; Fig. 19), the fold grew about 20 m (in amplitude) during early Tully deposition (with a wavelength of 50 km, yielding an average dip of .02 degrees), was eroded and breached during a mid-Tully unconformity, and then shrank, subsiding about 10 m and accumulating on its crest the thickest upper Tully of the central New York outcrop belt. Clearly, the distribution and thicknesses of the Tully lithologies reflect shifting topographic controls as well as variation of sediment input, but this "anticline" appears to be more a consequence of minor differential vertical movements in the basin than a growth fold, particularly in that it does not correspond to any subsequent Alleghanian structure.

The Lackawanna syncline in northeastern Pennsylvania has long attracted the interest of geologists, in part because its trend is convex to the southeast, in striking contrast to the northwestward convexity of the other Valley and Ridge folds in Pennsylvania. Also, it appears (superficially) to be an infold of Carboniferous rocks, somehow "caught" within an Upper Devonian terrane as a consequence of some Late Paleozoic (post-Alleghanian?) basement tectonics. But seismic data clearly shows that the syncline is not a basement structure reflected in the overlying Paleozoic cover. Perpendicular to the syncline's trend, the top of basement is smooth, rising gradually both to the northwest and southeast, with the deepest point being approximately under the syncline. The apparent singular occurrence of the Lackawanna syncline in what appears to be Allegheny Plateau terrane is deceptive. Actually, the syncline is bounded on both sides by large Valley and Ridge anticlines within the Catskill rocks on either side with the Milton anticline on the northwest and the Berwick anticline on the southeast (Faill 1985). These anticlines do not appear on most geologic maps (e.g., Berg and others 1980) because the Catskill formation is 2000 m thick in this area, considerably greater than the structural relief between the anticlines and the Lackawanna syncline. Were the Catskill divisible into units 100 to 200 m thick, then the anticlines would appear on the map as boldly as do the Valley and Ridge folds to the west. In fact, the Lackawanna syncline and adjacent anticlines are the eastern terminations of major, first-order Valley and Ridge folds.

The apparent singularity of the Lackawanna syncline (i.e., the absence of Carboniferous rocks in the synclines on the other side of the anticlines) is a consequence of both the depth to basement and the thickness of the overlying Paleozoic cover (see cross-section in Faill 1985). To the northwest, the Paleozoic (pre-Carboniferous) section thins at nearly, but not quite the same rate as the basement rises. Hence, the Bernice syncline (just northwest of the Milton anticline) just misses containing a sizeable Carboniferous rock section. (Actually, a few isolated patches of the Early Carboniferous Mauch Chunk Formation occur at the tops of the highest mountains, and a more complete Carboniferous section is present farther west in the Bernice and other synclines because the basement is deeper there.) Southeast of the Lackawanna syncline the basement rises gradually but the Devonian section thickens. Hence, the youngest rocks preserved in the unnamed syncline southeast of the Berwick anticline belong to the lowermost part of the Duncannon Member (Catskill Formation). To the southwest, the basement deepens and the Carboniferous Anthracite fields appear at the surface. In fact, the eastern termination of the Anthracite basin is more an artifact of rising basement depth, and less a consequence of Alleghanian décollement tectonics.

The southeastward convexity of trend is not as well understood—it is paralleled on its northwest side by the southeastward convexity of the Milton anticline (Faill 1985). On the southeast side, the structure is more complex, with the "Main" 080 to 090° trend being predominant. This curvature of trend may be related in some manner to the northward termination of the wide Valley and Ridge, and Plateau structures, and to the change in structural trend near Kingston, in southeastern New York.

All this notwithstanding, the Lackawanna syncline remains a target for speculation. Woodrow and Fletcher (1967) made the hypothesis that the Lackawanna syncline influenced Upper Devonian depositional patterns. Glaeser (1974) pointed out that deep wells in (#1 Hudson Realty) or near (#1 Richards) the Lackawanna syncline have much greater thicknesses of Upper Devonian delta-front sandstones than do wells on either side, suggesting that the Catskill shoreline was somehow constrained to the Lackawanna syncline for an unusually long time. It is true that the #1 Richards well contains more delta-front sandstone than the #1 Goodwin well to the southwest, but both of these wells were drilled on the crest of the adjacent (to the northwest) Milton anticline. In other words, greater delta-front thicknesses occur in both anticlinal and synclinal hinges, and thus do not correlate with the later structures. The apparent persistence of the shoreline in this area must have had some other than tectonic cause, because the Lackawanna syncline probably did not exist during the Late Devonian (being an Alleghanian structure), and it could not have had any effect on Upper Devonian sedimentation.

Better evidence for growth folds during basin filling exists in the bituminous coal region of western Pennsylvania. Peat swamps are particularly sensitive to topography, and some correlation of coals, clay, and sandstone with Alleghanian folds is suggested by some recent work (Sholes and others 1979; Kent and Gomez 1971; and Williams and Bragonier 1974). It appears that folds were developing during deposition in the Appalachian basin before the main phase of Alleghanian deformation, but apparently their amplitudes were never more than a few meters. Furthermore, these effects are seen in coals of Upper Carboniferous age (Westphalian and Stephanian) and not of Devonian or even Lower Carboniferous age rocks. Thus, the growing folds within the basin must be considered as precursors to the Alleghenian orogeny, and not a result of, or an aftermath of the Acadian.

The evidence in favor of any intrabasinal Acadian tectonism is weak and isolated. Considerable evidence exists that argues strongly against such tectonics, at least against tectonics of more than a subtle scale. As with the coal measures, thin, widespread, shallow-water sandstone units are sensitive indicators of topographic relief in a basin. The Silurian Keefer sandstone extends across the central Appalachian basin with a thickness ranging from 2 to 10 m (Luttrell 1968). The isopach map shows no correlation of thickness with structure. Offshore sand bars are indicative of topographic high areas, but these bars occur on the flanks of the major anticlines, not on their crests. Similarly the Lower Devonian Ridgeley (Oriskany) Sandstone shows no thickness patterns that correlate with Alleghanian, or possibly earlier folds (Abel and Heyman 1981).

The Tully limestone makes an interesting break in this pattern, or lack of pattern. For the most part, the Tully thicknesses show no correspondence with the folds (Harper and Piotrowski 1978). For example, a linear belt of maximum thickness extends east-northeastward across north-central Pennsylvania oblique to the folds. On the other hand, another belt of maximum thickness follows the Chestnut Ridge anticline. Interestingly, just east of this belt is the westernmost extent of the Mahantango siltstones, implying that there may have been a topographic high along the axis of Chestnut Ridge, at least during the later part of the Middle Devonian. Yet the lateral persistence of the Tully across the basin indicates that whatever the relief was, it was not sufficiently large to interfere seriously with basinwide deposition patterns.

Paleocurrent direction measurements are useful indicators because currents are also sensitive to topographic relief, whether marine or subaerial. McIver (1970) measured flute casts, grooves and striations in the Upper Devonian marine "Portage" (Trimmers Rock and Brallier Formations) in New York, Pennsylvania, Maryland, Virginia, and West Virginia. These measurements exhibit a remarkably uniform westward direction across the entire basin, and in fact do not reflect the arcuation of the folds. Pelletier (1958) measured crossbed orientation in the Lower Carboniferous Pocono Formation, and Meckel (1967) in the basal Upper Carboniferous Pottsville Formation. Crossbeds in fluvial sandstones are notorious for exhibiting much greater variance in orientation than subaqueous current indicators in marine clastics, indicating that this type of sediment is much less sensitive as an indicator of local topography. Admittedly, their data has apparently not been statistically tested for possible local control of directions. But by the same token, neither has this data been used to demonstrate local control. At this time, one can only conclude that the overwhelming north-westward persistence of directions across the basin for both formations suggests that topographic highs, particularly ones produced by growing folds, were either nonexistent or were so low as to produce no significant shifts in the paleocurrent directions.

A structural candidate for use as evidence for Acadian tectonism in the basin is a thrust fault in the vicinity of McConnellsburg in south-central Pennsylvania (Pierce and Armstrong 1966). The town of McConnellsburg sits on the crest of a large first-order anticline cored by Ordovician carbonates. Around this carbonate valley lies a belt of Upper Ordovician shales. A shear zone exists at the base of the shale sequence all around this anticline. Samples taken from the mylonite in the shear zone and from undeformed rock above the shear zone yielded conventional K-Ar (whole rock) ages of 372 and 444 Ma, respectively. This would suggest that movement on this fault, which produced the mylonite, occurred in the middle of the Late Devonian. However, there may have been some argon loss from the sample that would bring the validity of this date into question. Secondly, the geometry of this fault and its direction of movement are congruent with other Alleghanian structures in the vicinity and thus the movement on this fault could have been of Alleghanian age rather than Acadian. The possibility of Acadian movement on this fault is currently unresolved. The present state of evidence indicates that, except for its subsidence and the sediment it received, the Appalachian basin was largely unaffected by the Acadian orogeny to the east.

Acadian Deformation in the Piedmont

Given the apparent absence of Acadian deformation in the Great Valley, Reading Prong, and Blue Ridge terranes in Pennsylvania, and given the northernmost definitely Acadian event of the southern Appalachians, consisting of folded, 385-415 Ma old plutonic rocks in central Virginia (Pavlides and others 1982), the question arises as to the presence of suspected Acadian deformation in the intervening Piedmont terrane in northern Virginia, Maryland, Delaware, and Pennsylvania that lies southeast of the Culpepper, Gettysburg, and Newark Mesozoic basins. This terrane, containing rocks of both Precambrian and Lower Paleozoic age, has undergone during the Paleozoic a complex history of metamorphism, folding, and faulting. Early recumbent folding and metamorphism were followed by later open folding, faulting, additional metamorphism, and basement uplift (Crawford and Crawford 1980).

The earliest structures in the Piedmont along the Susquehanna River consist of isoclinal folds with extensive flowage and schistosity that range from steeply southeast dipping in the southeast to subhorizontal in the northwest (Freedman and others 1964). The folds are associated with the nappe structures in the Great Valley to the north and are of Taconian age. These isoclinal folds and their axial surfaces were subsequently deformed into broad folds with sub-vertical axial surfaces and an associated foliation. The uplifts resulting from this deformation created the Tucquan anticline and the basement-cored Mine Ridge anticline. A third deformation of brittle fracture cleavage and thrusting is the expression of the Alleghanian orogeny in this area. Whereas the second deformation of broad uplift and vertical foliation could be late Taconian, or even early Alleghanian, the contrast in structural style and the numerous Devonian cooling ages (see below) suggest that this second deformation may be Acadian.

East of the Susquehanna, the Piedmont is more complex—two distinct Precambrian terranes (Honey Brook Upland and the West Chester–Avalon massif) are present, both surrounded by (probably) Lower Paleozoic metamorphosed sediments. The Precambrian rocks have undergone Grenville metamorphism and deformation. South of the West Chester-Avalon massif, a second, separate, Late Precambrian metamorphism affected the gneisses and anorthosite within and surrounding the Wilmington complex. All these terranes were overprinted by a third, Early Paleozoic (Taconian) metamorphic episode, consisting of a retrograde greenschist facies metamorphism in the Honey Brook Upland and adjacent Lower Paleozoic rocks, and a more intense amphibolite facies metamorphism to the south (Crawford and Crawford 1980).

Farther east, in the Philadelphia area, a complex history of five deformations is reported for the rocks southeast of the West Chester-Avondale massif by Amenta (1974). The first two consist primarily of isoclinal, recumbent folds and are associated with a regional amphibolite facies metamorphism, probably Taconian. These events were overprinted by a second amphibolite grade metamorphism (kyanite and sillimanite zones), folding, and basement uplift, possibly of Acadian age. The last two deformations occurred during the waning stages of this second metamorphism. The correlation of these events with those along the Susquehanna River is tentative.

More than 50 radiometric age determinations have been made of rocks from the Piedmont province (Lapham and Root 1971; Faill 1985). Fifteen of these (mostly U-Pb on zircon, also K-Ar on muscovite and/or biotite) fall within the Devonian Period, from 408 to 360 Ma (Harland and others 1982), and another five (K-Ar, whole rock, muscovite, or biotite) are Early Carboniferous, from 360 to 320 Ma (Harland and others 1982). A few ages (6) are even younger, but they, being mostly lead-alpha determinations, are probably not valid.

The remaining cluster within two general groups, 1060 to 820 Ma (15) and 550 to 460 Ma (9). A fair number of these ages are of questionable quality (lead-alpha on zircon or monazite, Rb-Sr on biotite, and K-Ar on biotite), but the clustering could reflect the significant tectonic events of the Piedmont terrane in Pennsylvania.

The oldest isotopic ages represent the Grenville Orogeny and the following cooling/uplift as preserved in the Precambrian Baltimore Gneiss and gneisses of the Honey Brook Upland, and in the zircons from the schists of the (probably) Lower Paleozoic Wissahickon Group (Crawford and Crawford 1980). The Early Paleozoic ages represent the dynamothermal metamorphism of the Taconian orogeny. Apparently the heating from this orogeny persisted in southeastern Pennsylvania throughout the Silurian and Devonian, because there is no isotopic evidence for a subsequent deformation (the Alleghanian was apparently a relatively cold deformation in Pennsylvania). The muscovite ages are younger than the hornblende ages from the same samples (Sutter and Dallmeyer 1982), reflecting the lower blocking temperatures of muscovite (~300°C) versus that of hornblende (~500°C).

This cooling is probably a consequence of uplift of the Piedmont terrane and erosion of the overlying cover resulting in a

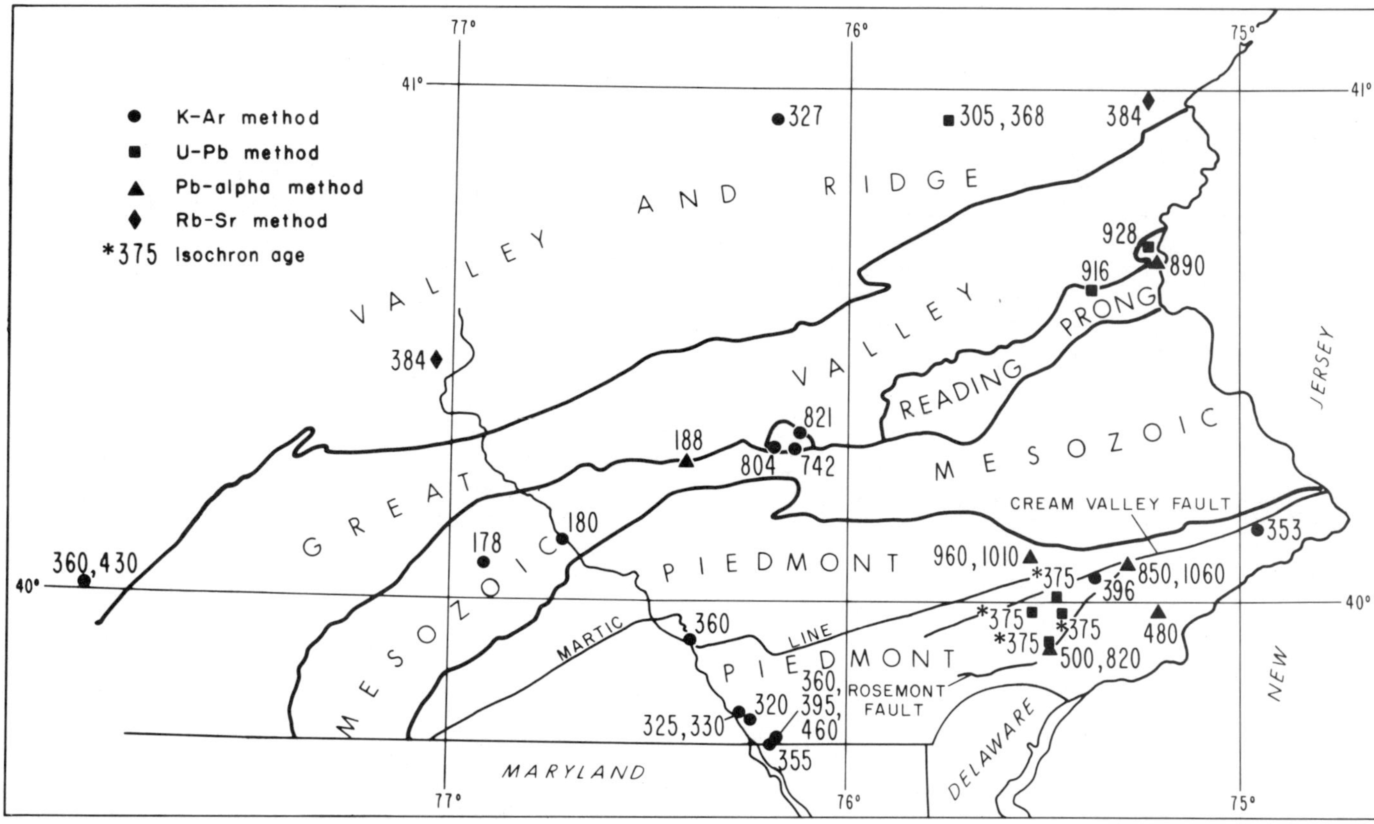

Figure 8. Radiometric age determinations of rocks from several geologic provinces in southeastern Pennsylvania. In the Piedmont province, the cluster of dates within the Devonian Period probably reflects cooling (because of uplift?) of that terrane from the Taconian metamorphism which widely affected this province. See Faill (1985) for specific details and references.

general region-wide cooling of the rocks. Additional evidence of uplift of the Piedmont can be seen in the rock fragments and minerals such as schist, serpentinite, hornblende, and epidote reported from the Pottsville Formation (Montgomery 1954), indicating that the high grade metamorphic rocks of the Piedmont region were exposed to erosion by the beginning of Late Carboniferous time, and perhaps as early as Early Carboniferous (kyanite and staurolite in the Mauch Chunk Formation, Avery A. Drake 1984, personal communication).

The evidence suggests that the Acadian orogeny manifested itself in the Piedmont of Pennsylvania and Maryland in a rather limited way. The igneous activity (primarily plutonic) and widespread metamorphism, characteristic of the Acadian orogeny in New England and Virginia was simply not present in Pennsylvania and Maryland. Rather, the uplift, the moderately brittle folding, and the evidence of cooling during the Devonian suggest that this terrane was on the margin of the Acadian event that showed itself primarily by vertical tectonics, somewhat similar to that in the western Hudson Highlands.

Some of the faults and folds in the Piedmont are quite clearly post-Taconian, but their exact age is ambiguous—Acadian, Alleghanian, or between the two. The Cream Valley fault on the north side of the West Chester—Avalon massif (Figure 8) is one of the more prominent structures of this category. It is a vertical fault with a very straight trace, trending east-northeast. The northeast and southwest terminations of the Cream Valley fault are currently unknown. It is overlapped by the Cretaceous Coastal Plain deposits in New Jersey to the northeast. To the southwest, it disappears within the metamorphic Wissahickon terrane north of the Woodville dome (west end of the West Chester-Avalon massif). Smear lineations near the fault indicate that at least some of its movement was strike-slip (Anderson and others 1964). Movement on the fault is later than Taconian because it offsets rocks of the greenschist facies metamorphism produced during the Taconian orogeny (Crawford and Crawford 1980). On the other hand, its orientation and direction of movement is not concordant with the later Alleghanian movements in the area. Thus bracketed, it may be an Acadian or post-Acadian structure. Current thinking within the plate tectonics paradigm indicates that the Avalon terrane of northern Appalachians was moved to its present position during the Early Carboniferous in what was largely strike-slip tectonics. Whereas none of the presently exposed Piedmont is thought to be part of the Avalon terrane, the tectonics of that docking could well have extended cratonward, possibly exhibiting itself as movement on the Cream Valley fault.

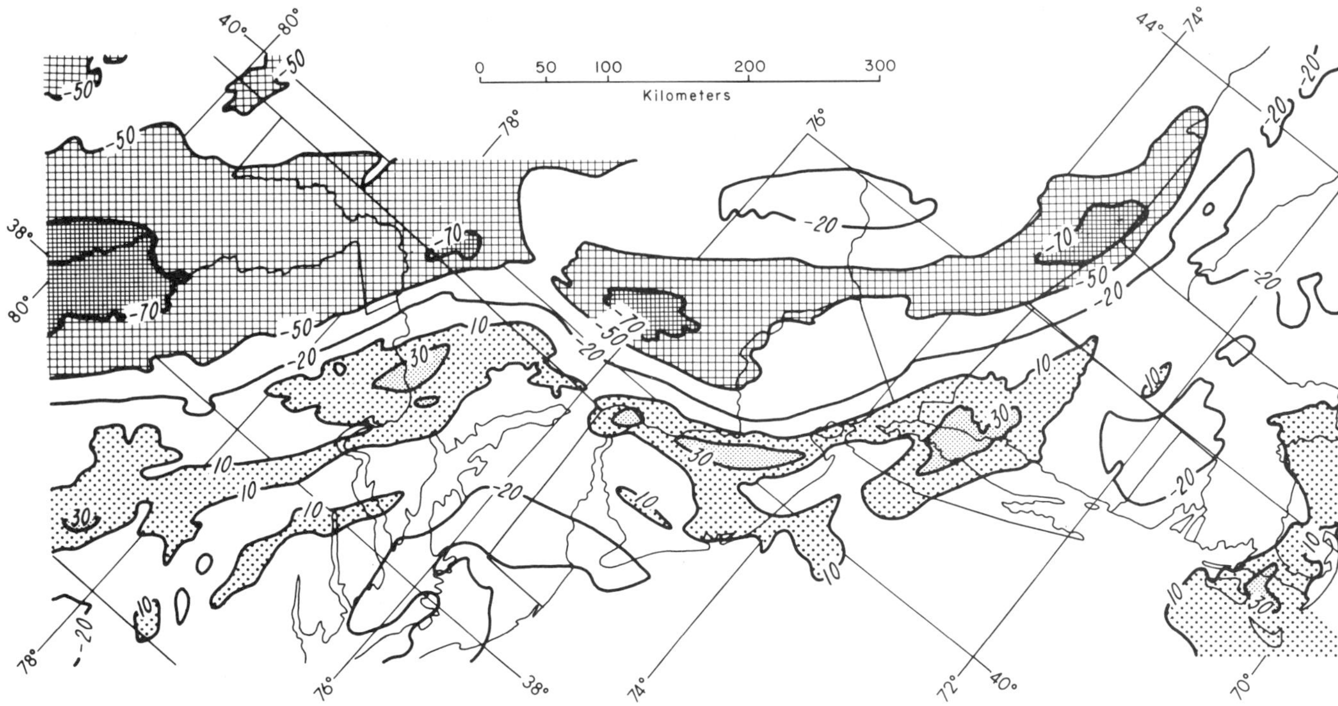

Figure 9. Bouguer anomaly gravity map, central Appalachians, in milligals. The prominent "ridge" of gravity highs (stipple patterns) is paralleled by a "trough" of gravity lows (hatchure patterns). Both appear to be offset along a northwest trending zone that passes through southern New Jersey–northern Delaware–southeastern Pennsylvania. Simplified from Haworth and others, (1980).

However, if the Cream Valley fault were utilized during the Avalon docking, its present east-northeast trend would require docking movements with a large east-west component, in distinct contrast to the more northerly trend of inferred movements in southern New England (e.g., along the Lake Char and Clinton-Newbury faults). On the other hand, it must be remembered that the Piedmont has been affected by the subsequent Alleghanian deformation, and that the present positions and orientations of Piedmont structures may not be as they were during pre-Alleghanian time. It would be worthwhile, then, to reconstruct this area palinspasticly to the Early Carboniferous.

Alleghanian Transport of Acadian Structures

The Alleghanian orogeny in the central and southern Appalachians resulted in décollement tectonism, in which the Paleozoic cover moved northwestward on a subhorizontal fault(s) (décollement(s)) above the basement rocks. Fault splays from this basal décollement created the folds, stepfaults, and other thrusts that are characteristic of the Valley and Ridge, and Plateau provinces. Recent seismic data indicate that in the southern Appalachians perhaps as far eastward as the Kings Mountain belt (Iverson and Smithson 1982) or the Carolina Slate belt the décollement extends underneath the Blue Ridge and Inner Piedmont provinces (Cook and others 1979). Considering the amount of northwestward transport of the Great Valley, Valley and Ridge, and Plateau province rocks in Pennsylvania and the presence of southeast-dipping thrust faults of Alleghanian age in the Great Valley and the Piedmont, it is probable that a décollement extends beneath the Blue Ridge and at least part of the Piedmont in Pennsylvania as well. The rocks west of South Mountain (Blue Ridge) have been calculated to have been transported about 80 km (Gwinn 1970). Subhorizontal seismic reflectors beneath South Mountain and the Gettysburg basin to the east indicate that the Blue Ridge itself is not rooted, but has moved northwestward an additional minimum 40 km. Adding the movements on the several Alleghanian thrust faults in the Piedmont (20 km?/50 km?) leads to the conclusion that the Piedmont rocks of southeasternmost Pennsylvania moved northwestward between 150 to 200 km during the Alleghanian. This transport is comparable to that estimated for the southern Appalachians (225 km, Hatcher and Zietz 1980). Northeastward across Pennsylvania, the decrease in folding (as measured by the width of Valley and Ridge province) and faulting suggests that this northwestward Alleghanian transport decreased to the northeast. That is, the Baltimore domes may have been transported 150 to 200 km, whereas northern New Jersey moved much less.

This Alleghanian northwestward movement may be reflected in the patterns of the gravity data along the east coast of the North American continent. The most prominent features on this map is the long, continuous gravity anomaly high with a deep anomaly low on its western side (Figure 9; see also Haworth and others 1980). This gravity trough (low) concides with the Appalachian miogeocline in the Valley and Ridge province, the Great

Valley, and the Precambrian terranes from the Berkshires to the Blue Ridge in Virginia. It is thought to reflect a thickened granitic crust. The adjacent anomaly high is interpreted to be an expression of underlying mafic crust. Thus the intervening steep gradient probably represents the eastern edge of the North American craton (Hatcher and Zietz 1980).

One of the more striking aspects of this gravity anomaly high is the double bend in southern New Jersey, southeastern Pennsylvania, northern Delaware, and northeastern Maryland. North and south of this bend the anomaly high trends 010° to 030°; within the bend the trend veers to the east as much as 080° (Figure 9). This bend may have been an original feature, perhaps as old as Precambrian, or it may have been acquired at a later time, beginning as a linear feature that was subsequently offset along a northwest trending zone. The latter possibility (subsequent offset) would in effect represent a right-lateral offset between the southern and northern Appalachians. This could be accomplished either by simple offset (strike-slip fault) of the crust, or by greater crustal shortening south of the offset zone ("transform fault") relative to the north. Simple offset is unlikely because the offset zone of the gravity anomalies can be traced northwestward only as far as Lake Erie—the displacement would have to be absorbed by some intracratonic shortening, for which no evidence seems to exist. The alternative, greater crustal shortening in the southern Appalachians, does not appear to be Taconian or Acadian, because these orogenies were most intense in New England and northward, and seem not to have been strong enough in the south to have produced the bend. On the other hand, the westward displacement during the Alleghanian was much greater in the southern Appalachians than in the northern, which leads to the supposition that the offset of the initially linear gravity anomaly high is an Alleghanian artifact. This supposition is reinforced by the coincidence of the bend and the northern termination of the southern and central Appalachian structures formed during the Alleghanian. It is as if the craton edge from Delaware southward had been impinged upon during continental collision, and driven northwestward relative to the northern Appalachians. If so, the resultant crustal thickening would have elevated overlying cover rocks, and thereby provided the gravitational energy needed to drive the Alleghanian décollement tectonism. Another coincidence is that the offset of the anomaly peak is approximately 180 km, commensurate with the estimated crustal shortening for the cover rocks.

Whether or not the craton boundary was shifted northwestward during the Alleghanian is moot. If one accepts this supposition, then one needs to examine the consequences of this Alleghanian displacement on the Devonian paleogeography. Palinspastic restoration by realigning the offset gravity anomaly crest to its presumed initial linearity would produce an eastward shift of preexisting central and southern Appalachian features. In the restoration shown in Figure 10, the offsetting zone was assumed to trend 300 degrees azimuth for several reasons: 300 degrees is approximately perpendicular to the trend of the anomaly crest; the average transport direction of the cover rocks was about 300 degrees; and a line of anomaly terminations crosses Pennsylvania, from the offset northwestward (at approximately 300 degrees azimuth) to Lake Erie (Hildenbrand and others 1982). Surface expression of such an offset is not obvious. Some dextral faulting along the lat 40°N zone has been noted (Root and Hoskins 1977), but the amounts of displacement are not nearly commensurate with the presumed crustal offset. Perhaps the strongest surface expression is the arcuation of the Alleghanian structures, with the concomittant decrease in shortening from 175(?) km in south-central Pennsylvania (above the southern boundary of the offsetting zone) to a mere 10 or so km in northern New Jersey and northeastern Pennsylvania (above the northern boundary of the offsetting zone).

Be that as it may, given the palinspasticly restored Devonian paleogeography (at least of the cover rocks), all the geologic provinces of southeastern and southcentral Pennsylvania, Maryland and Delaware were located farther to the southeast (Figure 11) during the Devonian. The southeastern edge of the Appalachian basin is displaced the least, the Blue Ridge was just west of Chesapeake Bay, and the Baltimore gneiss domes were located along the present coast in southeastern Delaware. Of particular note is the Cream Valley fault—presently trending 070, the differential movements within the offsetting zone may have rotated it from an original 040 degrees, a trend more commensurate with the Carboniferous (Avalon) faults in New England.

DISCUSSION

It is quite obvious that the Acadian orogeny and the Catskill Delta are intimately related, not only with respect to time, but also with respect to geography. The earliest known evidence for the Acadian is in the Gaspé Peninsula of Quebec, in rocks deposited during the latter part of the Early Devonian. The tectonism spread southward throughout New England and perhaps into the central and southern Appalachians by the end of the Eifelian (early Middle Devonian). In the Appalachian basin, some terrigenous sedimentation was occurring during the Onesquethawan (late Early Devonian), particularly along the southeastern margin but it was at the beginning of the Middle Devonian that clastic deposition began in earnest and spread across the entire basin. The termination of the Acadian orogeny is more difficult to ascertain. In New England, granites that cross-cut early nappes and high-grade metamorphic rocks yield ages that cluster around 380 Ma (early Upper Devonian). These rocks themselves were deformed and metamorphosed by the late dome stage (Hall and Robinson 1982). The clustering of ages around 360-370 Ma (spanning the late Devonian-Early Carboniferous boundary) (Lyons and Faul 1968) probably represent cooling (Naylor 1975). That an orogenic area southeast of the basin remained topographically positive throughout the remainder of the Devonian is obvious from the nature of the sediments filling the basin. Just before the end of the Devonian, Catskill deposition terminated, and was replaced by the coarse clastics of the Spechty Kopf and younger formations that continued to enter the basin

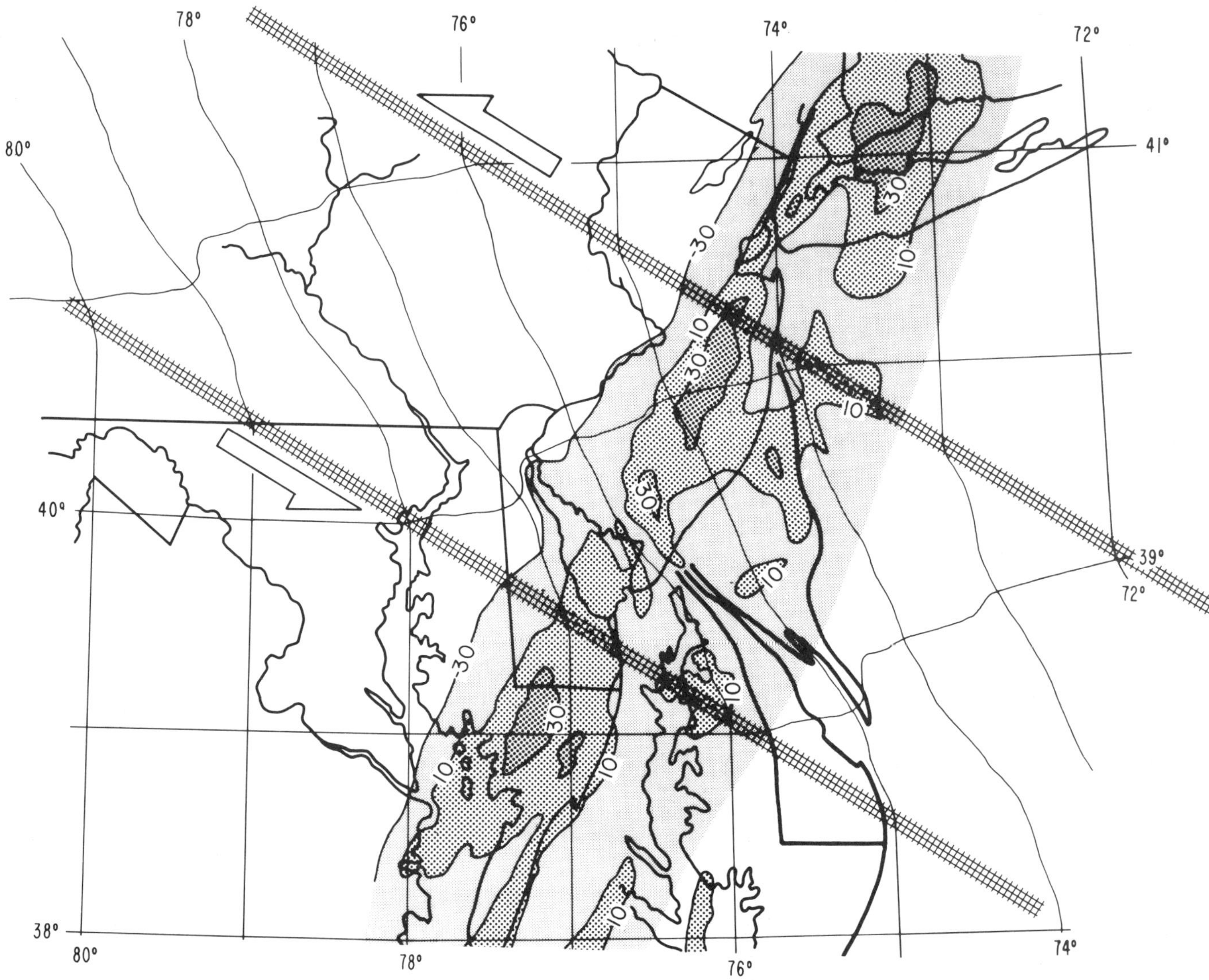

Figure 10. Palinspastic restoration of Bouguer anomaly gravity high to a presumed initially linear "ridge". The offsetting zone is nearly 200 km wide, and its trend is 300° azimuth. Offset is not constant across the zone, as can be judged by the irregular traces of the latitude and longitude lines. The heavy line in the lower right represents the palinspasticly restored location of the New Jersey–Delaware–Maryland coast and Delaware Bay.

from the southeast well into the Carboniferous. Tectonism occurred in New England at this time but it was of a quite different nature—one dominated by strike-slip faulting and the formation of local basins of sediment accumulation. This Maritime disturbance is quite separate from the Acadian orogeny and reflects a change from a continental collisional tectonism to that of transcurrent tectonism which moved the Avalon terrane into its present position.

The Catskill Delta and the Acadian orogeny were curiously related by geography as well. They are separate in space (i.e., they are now). The Appalachian foreland basin occupies the western flank of the Appalachian orogen, extending its entire length from Newfoundland to Alabama. The northern part of the basin, however, is much narrower, and subsided much less during the Devonian than did that part of the basin south of central New York. As a consequence, little sediment accumulated adjacent to the orogenic zone, but instead crossed the Laurentian shield and accumulated in sites such as the Michigan basin. In contrast, southward from central New York, great thicknesses of sediment accumulated in the basin. The thickness, coarseness and character, however, was not constant along the length of the basin (Figure 3; Oliver and others 1967) at least not in the preserved basin. Southwestward from New York and Pennsylvania to eastern Tennessee, not only does the accumulated Devonian thickness decrease dramatically, but the sediments change from coarser marine and continental clastics to finer gray and black marine shales. But is this an accurate view of the southern basin during the Devonian, or is this a distortion because of subsequent tecto-

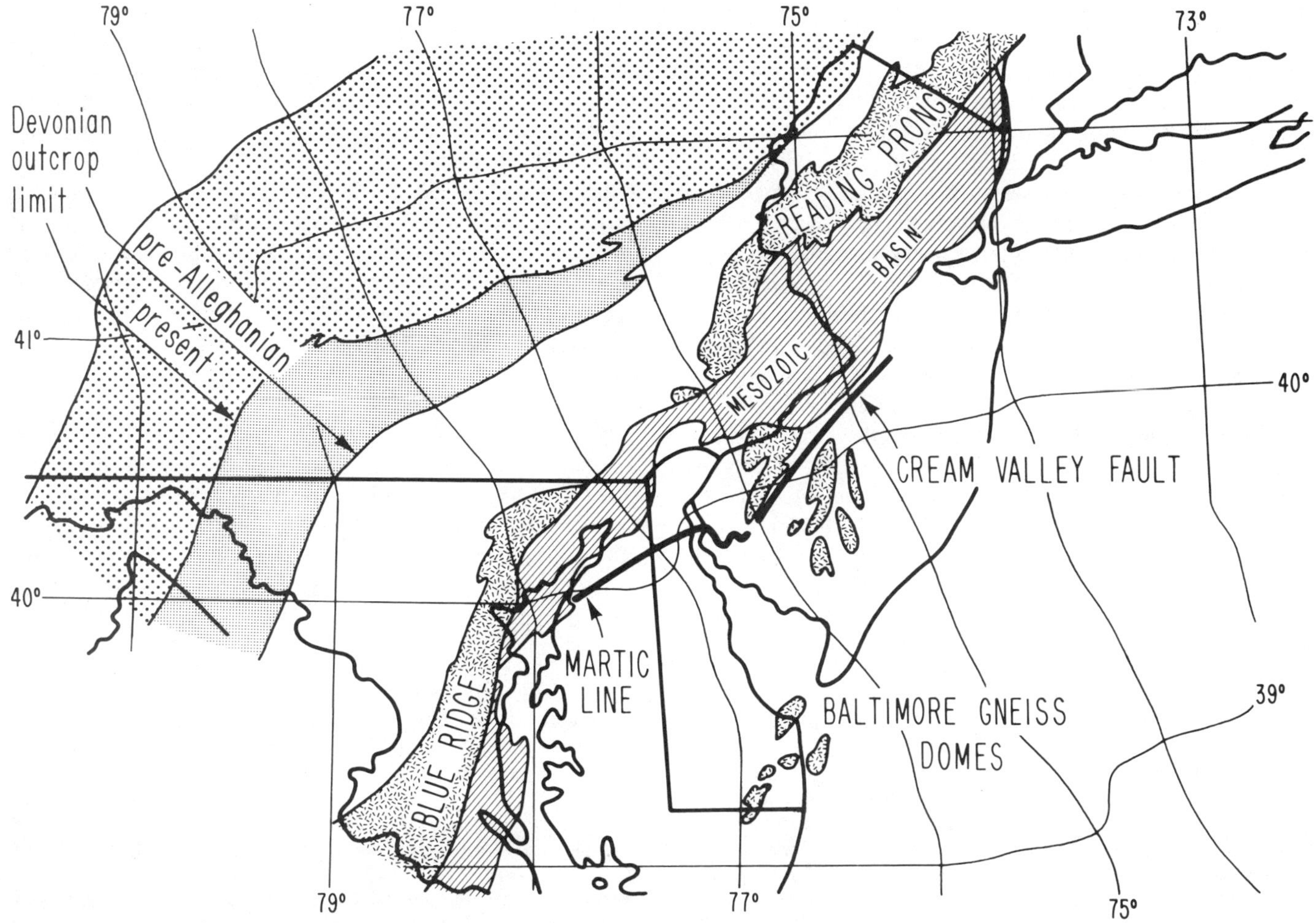

Figure 11. Present geologic provinces and features in their presumed Late Devonian positions, using the palinspastic restoration of Figure 10. The Baltimore gneiss domes have the largest southeastward displacement, the Blue Ridge less so, and the Valley and Ridge the least, reflecting the progressive decrease in Alleghanian transport of the domes, Blue Ridge, and Valley and Ridge, respectively. 'Devonian outcrop limit' refers to the present southeast margin, and not to the southeast margin of the Appalachian foreland basin in the Late Devonian.

nism? The Acadian orogeny may well have extended along the entire length of the Appalachian orogen, although in the south it apparently was not quite the dynamometamorphic event as it was in southern New England. Yet an orogeny strong enough to produce a greenschist facies metamorphism (and locally sillimanite grade–amphibolite facies) should have more consequence than a few tens of meters of dark shale some distance from the orogen. The detritus from the orogen may have passed through to a depositional site farther west, as it did in the northern basin, but the fine-grained character of the preserved Devonian rocks suggests that this did not happen. Additionally, no Devonian counterpart to the Michigan basin exists in southcentral U.S. The other possibility is that a deeper basin did originally exist east of the preserved Devonian rocks, but that it has since been removed by erosion. Examination of the isopleth lines on the total Devonian isopach map (Figure 3) or on the Middle and Upper Devonian isopach maps of Oliver and others (1967; Figures 8 and 9) shows a distinct divergence between the nearly north-south trending isopleths and the northeast-southwest trend of the preserved basin. This suggests that the southern Appalachian basin may have extended considerably farther to the southeast than at present, that its eastern portion may have subsided much more than the western (as in Pennsylvania), that detritus coarser than shale (and in much greater amounts) was derived from the southern Acadian orogen, and even that a Catskill Delta with subaerial deposition may have developed there. The subsequent complete erosion of this hypothetical southern Catskill Delta, and the preservation of the Pennsylvania-New York Catskill Delta, may simply be a consequence of the later Alleghanian tectonism. The large amount of crustal shortening achieved by movement on deep décollements under the Inner Piedmont, Blue Ridge, and Valley and Ridge terranes in the southern Appalachians during the Alleghanian probably generated a large uplift, not only of the orogenic zone but also of the adjacent Appalachian basin. Uplift may have been sufficient, especially in the eastern part, to completely erode the upper portion, which may have included younger (post-Silurian?) rocks. To the north, in Pennsylvania, the amount of crustal shortening was significantly less, with consequently much less uplift. Thus much more of the upper portion of the basin, including the Catskill Delta rocks, is still preserved.

At the beginning of this section it was stated that the Catskill Delta and the Acadian orogeny were intimately related, but not in the way the Taconian and Alleghanian orogenies were intimately related to the Appalachian basin. The Taconian consisted essentially of two parts: a zone of metamorphism and deformation within the Appalachian Piedmont, and an adjacent zone of nappe development and klippe emplacement. One distinguishing feature of the latter is that the nappes and klippen were carried out from the main orogenic zone into the adjacent foredeep basin. Thus the nappes and klippen, structures born of the orogeny, became incorporated with the other finer-grained detritus that was filling the Appalachian foredeep. The Alleghanian orogeny did not produce nappes and klippen that could slide into the basin, partly because the basin was subaerial and in a completely different depositional framework. But the Alleghanian did affect the depositional basin structurally through the décollement process. The far reaching faults and associated folds extended throughout perhaps more than 90 percent (volumetrically) of the Paleozoic cover, involving rocks from Cambrian to Pennsylvanian in age. In contrast, the Acadian orogeny accomplished neither of those. Even though it was a strong orogeny, producing a tectonism (at least in the north) comparable to, or even exceeding the Taconian, no deformation or deformational products (e.g., klippen) reached into the Appalachian basin. It seems almost as if a tectonic barrier existed between the Acadian orogenic zone and the adjacent depositional basin.

ACKNOWLEDGMENTS

The author is particularly indebted to John Rodgers and Avery A. Drake, Jr. for critically reading the manuscript, and for their extensive comments and probing questions. Appreciation is also expressed for the helpful comments by Donald L. Woodrow, William D. Sevon, Arthur A. Socolow, and Thomas M. Berg.

REFERENCES CITED

Abel, K. K. and Heyman, L., 1981, The Oriskany Sandstone in the subsurface of Pennsylvania: Pennsylvania Geological Survey, 4th ser., Mineral Resource Report 81, 9 p.

Allen, J.R.L. and Friend, P. F., 1968, Deposition of the Catskill facies, Appalachian region: with notes on some other Old Red Sandstone basins: Geological Society of America Special Paper 106, p. 21–74.

Amenta, R. V., 1974, Multiple deformation and metamorphism from structural analysis in the eastern Pennsylvania Piedmont: Geological Society of America Bulletin, v. 85, p. 1647–1660.

Anderson, T. H., Drake, D. E., and Wise, D. U., 1964, Megapetrofabric of the Coatesville-Doe Run area, Pennsylvania: Pennsylvania Academy of Science Proceedings, v. 38, p. 174–182.

Belt, E. S., 1968, Post-Acadian rifts and related facies, eastern Canada, *in* Zen, E-an, White, W. S., Hadley, J. B., and Thompson, J. R., Jr., eds., Studies in Appalachian geology: northern and Maritime: New York, Interscience Publishers, p. 95–113.

Berg, T. M., Edmunds, W. E., Geyer, A. R., Glover, A. D., Hoskins, D. M., MacLachlan, D. B., Root, S. I., Sevon, W. D., and Socolow, A. A., 1980, Geologic map of Pennsylvania: Pennsylvania Geological Survey, 4th ser., Map 1.

Berg, T. M., McInerney, M. K., Way, J. H., and MacLachlan, D. B., 1983, Stratigraphic correlation chart of Pennsylvania: Pennsylvania Geological Survey, 4th ser., General Geology Report 75.

Bobyarchick, A. R. and Glover, L., III, 1979, Deformation and metamorphism in the Hylas zone and adjacent parts of the eastern Piedmont in Virginia: Geological Society of America Bulletin, Part I, v. 90, p. 739–752.

Boucot, A. J., 1968, Silurian and Devonian of the northern Appalachians, *in* Zen, E-an, White, W. S., Hadley, J. B., and Thompson, J. R., Jr., eds., Studies of Appalachian geology: northern and Maritime: New York, Interscience Publishers, p. 83–94.

Bradley, D. C., 1983, Tectonics of the Acadian Orogency in New England and adjacent Canada: Journal of Geology, v. 91, p. 381–400.

Brown, P. M. and Van der Voo, R., 1983, A paleomagnetic study of Piedmont metamorphic rocks in northern Delaware: Geological Society of America Bulletin, v. 94, p. 815–822.

Cook, F. A., Albaugh, D. S., Brown, L. D., Kaufman, S., Oliver, J. E., Hatcher, R. D., Jr., 1979, Thin-skinned tectonics in the crystalline southern Appalachians; COCORP seismic-reflection profiling of the Blue Ridge and Piedmont: Geology, v. 7, no. 12, p. 563–567.

Crawford, M. L. and Crawford, W. A., 1980, Metamorphic and tectonic history of the Pennsylvania Piedmont: Journal of the Geological Society of London, v. 137, p. 311–320.

Dennison, J. M. and Head, J. W., 1975, Sea level variations interpreted from the Appalachian basin Silurian and Devonian: American Journal of Science, v. 275, p. 1089–1120.

Dennison, J. M. and Textoris, D. A., 1970, Devonian Tioga tuff in northeastern United States: Bulletin Volcanologique, v. 34, p. 289–294.

Dennison, J. M. and Textoris, D. A., 1971, Devonian Tioga Tuff, *in* Johnson, R. W., Jr., Milton, C., and Dennison, J. M., Field trip to the igneous rocks of Augusta, Rockingham, Highland, and Bath Counties, Virginia: Virginia Division of Mineral Resources Information Circular 16, p. 64–68.

Dixon, H. R. and Lundgren, L. W., Jr., 1968, Structure of eastern Connecticut, *in* Zen, E-an, White, W. S., Hadley, J. B., and Thompson, S. R., Jr., eds., Studies of Appalachian geology: northern and Maritime: New York, Interscience Publishers, p. 219–229.

Drake, A. A., Jr., 1978, The Lyon Station-Paulins Kill Nappe—The frontal structure of the Musconetcong Nappe system in eastern Pennsylvania and New Jersey: U.S. Geological Survey Professional Paper 1203, 20 p.

—— 1980, The Taconides, Acadides, and Alleghenides in the central Appalachians, *in* Wones, D. R., ed., The Caledonides in the U.S.A.: Blacksburg, Virginia Polytechnic Institute and State University, Memoir no. 2, p. 179–187.

Edmunds, W. E., Berg, T. M., Sevon, W. D., Piotrowski, R. C., Heyman, L., and Rickard, L. V., 1979, The Mississippian and Pennsylvanian (Carboniferous) Systems in the United States—Pennsylvania (Carboniferous) Systems in the United States—Pennsylvania and New York: U.S. Geological Survey Professional Paper 1110-B, 33 p.

Epstein, J. B., Sevon, W. D., and Glaeser, J. D., 1974, Geology and mineral resources of the Lehighton and Palmerton quadrangles, Carbon and Northampton Counties, Pennsylvania: Pennsylvania Geologic Survey, 4th ser., Geologic Atlas 195 cd, 460 p.

Ettensohn, F. R., 1985, The Catskill Delta complex and the Acadian Orogeny: A model, *in* Woodrow, D. L. and Sevon, W. D., eds., The Catskill Delta: Geological Society of America Special Paper 201, (this volume).

Faill, R. T., 1985, Tectonic map of Pennsylvania: Pennsylvania Geological Survey, 4th ser., Map, in press.

Faill, R. T., Hoskins, D. M. and Wells, R. B., 1978, Middle Devonian stratigraphy in central Pennsylvania—a revision: Pennsylvania Geological Survey, 4th ser., General Geology Report 70, 28 p.

Faill, R. T., and Nickelsen, R. P., 1973, Structural geology, *in* Faill, R. T. and others, eds., Structure and Silurian and Devonian stratigraphy of the Valley and Ridge province in central Pennsylvania: Guidebook, 38th Annual Field Conference of Pennsylvania Geologists, Pennsylvania Geological Survey, p. 9–38.

Faill, R. T. and Wells, R. B., 1974, Geology and mineral resources of the Millerstown quadrangle, Perry, Juniata, and Snyder Counties, Pennsylvania: Pennsylvania Geological Survey, 4th ser., Atlas 136, 276 p.

Fletcher, F. W. and Woodrow, D. L., 1970, Geology and economic resources of the Pennsylvania portion of the Milford and Port Jervis 15-minute quadrangles: Pennsylvania Geological Survey, 4th ser., Atlas 223, 64 p.

Freedman, J., Wise, D. U. and Bentley, R. D., 1964, Pattern of folded folds in the Appalachian Piedmont along Susquehanna River: Geological Society of America Bulletin, v. 75, p. 621–638.

Geiser, P. and Engelder, T., 1983, The distribution of layer parallel shortening fabrics in the Appalachian foreland of New York and Pennsylvania: Evidence for two non-coaxial phases of the Alleghanian orogeny: Geological Society of America Memoir 158, p. 161–175.

Glaeser, J. D., 1974, Upper Devonian stratigraphy and sedimentary environments in northeastern Pennsylvania: Pennsylvania Geological Survey, 4th ser., General Geology Report 63, 89 p.

Grauch, R. I. and Ludwig, K. R., 1979, Uranium mineralization at Easton, Pennsylvania: Pennsylvania Geology, v. 10, no. 5, p. 14–16.

Gwinn, V. E., 1964, Thin-skinned tectonics in the Plateau and northwestern Valley and Ridge provinces of the central Appalachians: Geological Society of America Bulletin, v. 75, p. 863–900.

——1970, Kinematic patterns and estimates of lateral shortening, Valley and Ridge and Great Valley provinces, central Appalachians, south-central Pennsylvania *in* Fisher, G. W., Pettijohn, F. J., Reed, J. C., Jr., and Weaver, K. N., eds., Studies of Appalachian Geology: central and southern: New York, Interscience Publishers, p. 127–146.

Hall, L. M., 1980, Basement-cover relations in western Connecticut and southeastern New York, *in* Wones, D. R., ed., The Caledonites in the U.S.A.: Blacksburg, Virginia Polytechnic Institute and State University, Memoir no. 2, p. 299–306.

Hall, L. M. and Robinson, P., 1982, Stratigraphic-tectonic subdivisions of southern New England, *in* St. Julien, P. and Beland, Jr., eds., Major structural zones and faults of the northern Appalachians: Geological Association of Canada Special Paper 24, p. 15–41.

Harland, W. B., Cox, A. V., Llewellyn, P. G., Picton, C.A.G., Smith, A. G., and Walters, R., 1982, A geological time scale: Cambridge, Cambridge University Press, 128 p.

Harper, J. A. and Piotrowski, R. G., 1978, Stratigraphy, extent, gas production, and future gas potential of the Devonian organic-rich shales in Pennsylvania: U.S. Department of Energy, Morgantown Energy Technology Center, DOE-METC/SP-78/6 Vol. 1, p. 310–329.

Harris, L. D., 1970, Details of thin-skinned tectonics in parts of Valley and Ridge and Cumberland Plateau provinces of the southern Appalachians, *in* Fisher, G. W., Pettijohn, F. J., Reed, J. C., Jr., and Weaver, K. N., eds., Studies of Appalachian Geology: central and southern: New York, Interscience Publishers, p. 161–173.

Hatcher, R. D., Jr., 1978, Tectonics of the western Piedmont and Blue Ridge, southern Appalachians: review and speculation: American Journal of Science, v. 278, p. 276–304.

Hatcher, R. D., Jr., Butler, J. R., Fullager, P. D., Secor, D. T., and Snoke, A. W., 1980, Geologic synthesis of the Tennessee-Carolinas-northeast Georgia southern Appalachians, *in* Wones, D. R., ed., The Caledonides in the U.S.A.: Blacksburg, Virginia Polytechnic Institute and State University, Memoir no. 2, p. 83–90.

Hatcher, R. D., Jr. and Zietz, I., 1980, Tectonic implications of regional aeromagnetic and gravity data from the southern Appalachians, *in* Wones, D. R., ed., The Caledonides of the U.S.A.: Blacksburg, Virginia Polytechnic Institute and State University, Memoir no. 2, p. 235–244.

Haworth, R. T., Daniels, D. L., Williams, H., and Zietz, I., 1980, Bouguer gravity anomaly map of the Appalachian orogen: St. Johns, Newfoundland Memorial University of Newfoundland, Map no. 3.

Heckel, P. H., 1973, Nature, origin, and significance of the Tully Limestone: Geological Society of America Special Paper 138, 244 p.

Hildenbrand, T. G., Simpson, R. W., Godson, R. H., and Kane, M. F., 1982, Digital colored residual and regional Bouguer gravity maps of the conterminous United States with cut-off wavelengths of 250 km and 1000 km: U.S. Geological Survey Geophysical Investigations Map GP-953 A.

Hoskins, D. M., 1979, Alternative interpretation of the Catskill-Pocono contact in the Antracite area of Pennsylvania (abs.): Geological Society of America, Abstracts with Programs, v. 2, no. 1, p. 25.

Iverson, W. P. and Smithson, S. B., 1982, Master decollement root zone beneath the southern Appalachians and crustal balance: Geology, v. 10, no. 5, p. 241–245.

Kent, B. H. and Gomez, M., 1971, Paleotopographic-structural controls on thickness and composition of Pittsburgh coal in southwestern Pennsylvania (abs.): Geological Society of America, Abstracts with Programs, v. 3, no. 7, p. 622.

Kent, D. V., 1982, Paleomagnetic evidence for post-Devonian displacement of the Avalon Platform (Newfoundland): Journal of Geophysical Research, v. 87, p. 8709–8716.

——1985, Paleogeographic setting of the Catskill Delta, *in* Woodrow, D. L. and Sevon, W. D., eds., The Catskill Delta: Geological Society of America Special Paper 201, (this volume).

Kent, D. V. and Opdyke, N. D., 1978, Paleomagnetism of the Devonian Catskill red beds: evidence for motion of the coastal New England-Canadian Maritime region relative to cratonic North America: Journal of Geophysical Research, v. 83, p. 4441–4450.

Kirby, M. W., 1981, Sedimentology of the Middle Devonian Bellvale and Skunnemunk Formations in the Green Pond outlier in northern New Jersey and southeastern New York [M.S. thesis]: New Brunswick, New Jersey, Rutgers University, 109 p.

Lapham, D. M. and Root, S. I., 1971, Summary of isotopic age determinations in Pennsylvania: Pennsylvania Geological Survey, 4th ser., Information Circular 70, 29 p.

LeFort, J. P. and Van der Voo, R., 1981, A kinematic model for the collision and complete suturing between Gondwanaland and Laurussia in the Carboniferous: Journal of Geology, v. 89, p. 537–550.

Luttrell, E. M., 1968, An analysis of the Silurian Keefer sandstone of Pennsylvnia [Ph.D. thesis]: Princeton, New Jersey, Princeton University, 111 p.

Lyons, J. B. and Faul, H., 1968, Isotope geochronology of the northern Appalachians, *in* Zen, E-an, White, W. S., Hadley, J. B., and Thompson, J. B., Jr., eds., Studies of Appalachian Geology: northern and Maritime: New York, Interscience Publishers, p. 305–318.

MacLachlan, D. B., Buckwalter, T. V. and McLaughlin, D. B., 1975, Geology and mineral resources of the Sinking Spring quadrangle, Berks and Lancaster Counties, Pennsylvania: Pennsylvania Geological Survey, 4th ser., Geological Atlas 117d, 228 p.

McIver, N. L., 1970, Appalachian turbidities, *in* Fisher, G. W., Pettijohn, F. J., Reed, J. C., Jr., and Weaver, K. N., eds., Studies of Appalachian Geology: central and southern: New York, Interscience Publishers, p. 69–81.

Meckel, L. D., 1967, Origin of Pottsville conglomerates (Pennsylvanian) in the central Appalachians: Geological Society of America Bulletin, v. 78, p. 223–258.

Milici, R. C., 1975, Structural patterns in the southern Appalachians: Evidence of a gravity slide mechanism for Alleghanian deformation: Geological Society of America Bulletin, v. 86, p. 1316–1320.

Mitra, G. and Elliott, D., 1980, Deformation of basement in the Blue Ridge and the development of the South Mountain cleavage, *in* Wones, D. R., ed., The Caledonides in the U.S.A.: Blacksburg, Virginia Polytechnic Institute and

State University, Memoir no. 2, p. 307–311.

Montgomery, A., 1954, Uranium minerals of the Mauch Chunk area, Pennsylvania: Pennsylvania Academy of Sciences Proceedings, v. 28, p. 102–110.

Morris, W. A., 1976, Transcurrent motion determined paleomagnetically in northern Appalachians and Caledonides and the Acadian orogeny: Canadian Journal of Earth Sciences, v. 13, p. 1236–1243.

Naylor, R. S., 1975, Age provinces of the northern Appalachians: Annual Review of Earth and Planetary Sciences, v. 3, p. 387–400.

Nickelsen, R. P., 1963, Fold patterns and continuous deformation mechanisms of the central Pennsylvania folded Appalachians, *in* Tectonics and Cambrian-Ordovician stratigraphy in the central Appalachians of Pennsylvania: Guidebook, Pittsburgh Geological Society and Appalachian Geological Society, Field Conference, p. 13–29.

Offield, T. W., 1967, Bedrock geology of the Goshen-Greenwood Lake area, New York: New York State Museum, Map & Chart Series, no. 9, 77 p.

Oliver, W. A., Jr., DeWitt, W., Jr., Dennison, J. M., Hoskins, D. M., and Huddle, J. H., 1967, Devonian of the Appalachian basin, United States, *in* Oswald, D. H., ed., International Symposium on the Devonian System, vol. 1: Calgary, Alberta Society of Petroleum Geologists, p. 1001–1040.

—— 1971, Isopach and lithofacies maps of the Devonian in the Appalachian basin: Pennsylvania Geological Survey, 4th sers., Progress Report 182.

Osberg, P. H., Moench, R. H., and Warner, J., 1968, Stratigraphy of the Merrimack synclinorium in west-central Maine, *in* Zen, E-an, White, W. S., Hadley, J. B., and Thompson, J. R., Jr., eds., Studies of Appalachian geology: northern and Maritime: New York, Interscience Publishers, p. 241–253.

Pavlides, L., Stern, T. W., Arth, J. G., Muth, K. G., and Newell, M. F., 1982, Middle and Upper Paleozoic granitic rocks in the Piedmont near Fredericksburg, Virginia: Geochronology: U.S. Geological Survey Professional Paper 1231-B, 9 p.

Pelletier, B. R., 1958, Pocono paleocurrents in Pennsylvania and Maryland: Geological Society of America Bulletin, v. 79, p. 1033–1064.

Perry, W. J., Jr., 1978, Sequential deformation in the central Appalachians: American Journal of Sciences, v. 278, p. 518–542.

Pierce, K. L. and Armstrong, R. L., 1966, Tuscarora fault, an Acadian (?) bedding plane fault in central Appalachian Valley and Ridge province: American Association of Petroleum Geologists Bulletin, v. 50, p. 385–390.

Rao, K. V. and Van der Roo, R., 1980, Paleomagnetism of a Paleozoic anorthosite from the Appalachian Piedmont, northern Delaware: possible tectonic implications: Earth and Planetary Science Letters, v. 47, p. 113–120.

Ratcliffe, N. M. and Bahrami, B., 1976, Chatham fault: A reinterpretation of the contact relationships between the Giddings Brook and Chatham slices of the Taconic allochthon in New York: Geology, v. 4, p. 5–60.

Rich, J. L., 1934, Mechanics of low angle overthrust faulting as illustrated by Cumberland thrust block, Virginia, Kentucky, and Tennessee: American Association of Petroleum Geologists Bulletin, v. 18, p. 1584–1596.

Rickard, L. V., 1975, Correlation of the Silurian and Devonian rocks in New York State: New York State Museum and Science Service, Map and Chart Series no. 24, 16 p.

Rodgers, J., 1949, Evolution of thought on structure of middle and southern Appalachians: American Association of Petroleum Geologists Bulletin, v. 33, p. 1643–1654.

—— 1967, Chronology of tectonic movements in the Appalachian region of eastern North America: American Journal of Science, v. 265, p. 408–427.

—— 1970, The Tectonics of the Appalachians: New York, Wiley-Interscience, 271 p.

—— 1971, Evolution of thought on structure of middle and southern Appalachians: Second paper: West Virginia University and West Virginia Geological Survey: A seminar on Appalachian structure, p. 1–15.

Root, S. I., 1970, Structure of the northern terminus of the Blue Ridge in Pennsylvania: Geological Society of America Bulletin, v. 81, p. 815–830.

Root, S. I. and Hoskins, D. M., 1977, Latitutde 40°N fault zone, Pennsylvania: A new interpretation: Geology, v. 5, p. 719–723.

Ruitenberg, A. A. and McCutcheon, S. R., 1982, Acadian and Hercynian structural evolution of the southern New Brunswick, *in* St. Julien, P. and Beland, J., eds., Major structural zones and faults of the northern Appalachians: Geological Association of Canada Special Paper 24, p. 131–148.

Sevon, W. D., 1973, Glaciation and sedimentation in the Late Devonian and Early Mississippian of Pennsylvania (abs.): Geological Society of America, Abstracts with Programs, v. 5, no. 2, p. 218–219.

—— 1985, Nonmarine facies of the Middle and Late Devonian Catskill coastal alluvial plain, *in* Woodrow, D. L. and Sevon, W. D., eds., The Catskill Delta, Geological Society of America Special Paper 201, (this volume.

Sholes, M. A., Edmunds, W. E. and Skema, V. W., 1979, The economic geology of the Upper Freeport Coal in the New Stanton area of Westmoreland County, Pennsylvania: A model for coal exploration: Pennsylvania Geological Survey, 4th ser., Mineral Resource Report 75, 51 p.

Smosna, R. A. and Patchen, D. G., 1978, Silurian evolution of central Appalachian basin: American Association of Petroleum Geologists Bulletin, v. 62, p. 2308–2328.

Srogi, L., Wagner, M. E., Lutz, T. M., and Hamre, J., 1983, Metamorphic and tectonic history of a Paleozoic granulite facies terrane, Delaware-Pennsylvania Piedmont (abs): Geological Society of America, Abstracts with Programs, v. 15, no. 6, p. 694.

Stanley, R. S. and Ratcliffe, N. M., 1985, Tectonic synthesis of the Taconian orogeny in western New England: Geological Society of America Bulletin, v. 96, in press.

Sutter, J. F. and Dallmeyer, R. D., 1982, Interpretation of $^{40}Ar/^{39}Ar$ ages from the Appalachian Grenville terrane (abs.): Geological Society of America, Abstracts with Programs, v. 14, no. 1 and 2, p. 87.

Sutter, J. F., Ratcliffe, N. M., and Mukasa, S. B., 1984, $^{40}Ar/^{39}Ar$ and K-Ar data bearing on the metamorphic and tectonic history of western New England: Geological Society of America Bulletin, v. 95, in press.

Swartz, F. M., 1939, Keyser limestone and the Helderberg group, *in* Willard, B., The Devonian of Pennsylvania: Pennsylvania Geologic Survey, 4th ser., General Geology Report 19, p. 29–91.

Thomas, W. A., Tull, J. F., Bearce, D. N., Russell, G., and Odom, A. L., 1980, Geologic synthesis of the southernmost Appalachians, Alabama and Georgia, *in* Wones, D. R., ed., The Caledonides of the U.S.A.: Blacksburg, Virginia Polytechnic Institute and State University Memoir no. 2, p. 91–97.

Thompson, J. B., Jr., Robinson, P., Clifford, T. N., and Trask, N. J., Jr., 1968, Nappes and gneiss domes in west-central New England, *in* Zen, E-an, White, W. S., Hadley, J. B., and Thompson, J. R., Jr., eds., Studies of Appalachian geology: northern and Maritime: New York, Interscience Publishers, p. 203–218.

Trexler, J. P. and Wood, G. H., Jr., 1968, Geologic map of Lykens quadrangle, Dauphin, Lebanon, and Schuylkill Counties, Pennsylvania: U.S. Geological Survey Geologic Quadrangle Map CQ-701.

Trexler, J. P., Wood, G. H., Jr., and Arndt, H. H., 1961, Angular unconformity separates Catskill and Pocono Formations in western part of Anthracite region, Pennsylvania: U.S. Geological Survey Professional Paper 424B, p. B94–B88.

Tull, J. F., 1980, Overview of the sequence and timing of deformational events in the southern Appalachians: Evidence from the crystalline rocks North Carolina to Alabama, *in* Wones, D. R., ed., The Caledonides in the U.S.A.: Blacksburg, Virginia Polytechnic Institute and State University, Memoir no. 2, p. 167–177.

—— 1982, Stratigraphic framework of the Talladega slate belt, Alabama Appalachians: Geological Society of America Special Paper 191, p. 3–18.

Van der Voo, R., 1979, Age of the Alleghenian folding in the central Appalachians: Geology, v. 7, p. 297–298.

Williams, E. G. and Bragonier, W. A., 1974, Controls of early Pennsylvanian sedimentation in western Pennsylvania, *in* Briggs, G., ed., Carboniferous of southeastern United States: Geological Society of America Special Paper 148, p. 135–152.

Williams, H. and Hatcher, R. D., Jr., 1982, Suspect terranes and accretionary history of the Appalachian orogen: Geology, v. 10, no. 10, p. 530–536.

Wood, G. H., Jr., Trexler, J. P. and Kehn, T. M., 1969, Geology of the west-central part of the southern Anthracite field and adjoining areas, Pennsylvania: U.S. Geological Survey Professional Paper 602, 150 p.

Woodrow, D. L., 1968, Stratigraphy, structure, and sedimentary patterns in the Upper Devonian of Bradford County, Pennsylvania: Pennsylvania Geological Survey, 4th ser., General Geology Report 54, 78 p.

Woodrow, D. L. and Fletcher, F. W., 1967, Late Devonian paleogeography in southeastern New York and northeastern Pennsylvania, *in* Oswald, D. H., ed., International Symposium on the Devonian System, vol. 2: Calgary, Alberta Society of Petroleum Geologists, p. 1327–1334.

Manuscript Accepted by the Society July 20, 1984

Printed in U.S.A.

Geological Society of America
Special Paper 201
1985

The Catskill Delta complex and the Acadian Orogeny: A model

Frank R. Ettensohn
Department of Geology
University of Kentucky
Lexington, Kentucky 40506

ABSTRACT

The Catskill Delta complex is interpreted to be the aggregate of delta-alluvial wedges and associated facies that developed in the central Appalachians and on adjacent parts of the stable craton from the Early-Middle Devonian transition to the Middle Mississippian during the Acadian orogeny.

Recent interpretations of the Acadian orogeny suggest that it probably was related to oblique convergence and transcurrent movement along a major strike-slip fault zone separating the eastern margin of the North American landmass from a linear continental fragment called the Avalon terrane. Distribution of clastic wedges and basinal deposits resulting from this orogeny support a general southwestward progression of orogeny and indicate that the major clastic wedges emanated from areas near promontories on the continental margin during successive phases of Acadian deformation. Three and possibly four such tectophases have been noted. Each tectophase appears to represent increased convergence or possible collision between a specific continental promontory and the Avalon terrane, but some delta development occurred continually along many parts of the orogen in response to each tectophase.

The four tectophases are: (1) Collision near the St. Lawrence promontory during the Early-Middle Devonian transition with initiation of the Catskill Delta complex represented by the Needmore and Esopus shales and associated clastics near promontories. (2) Southward migration of deformation and collision near the New York promontory during the Middle Devonian with the development of a large peripheral basin having an east-dipping, western paleoslope. This basin was filled with cyclic delta clastics and carbonates of the Hamilton Group and Tully Limestone. (3) Southward migration of deformation and collision near the Virginia promontory during the Late Devonian to earliest Mississippian accompanied by intense clastic influx of the Genesee-through-Canadaway groups. As a result, the basin was progressively filled from the east so that basinal environments migrated westward out of the peripheral basin and onto adjacent parts of the stable craton. Eventually the basin was filled and a regional west-dipping paleoslope was established. (4) Migration of deformation southward from the Virginia promontory during the Early to Middle Mississippian as basinal environments in cratonic seas were destroyed and Pocono and equivalent clastic wedges essentially filled the epicontinental sea. Middle Mississippian carbonates mark the end of the Acadian orogeny and Catskill Delta complex.

INTRODUCTION

The close relationships between the Catskill delta complex and the Acadian orogeny are so well known that they are almost taken for granted, but the exact nature of these relationships is largely unknown. However, information derived from a recent study of the black shales of the Catskill Delta complex, when combined with new knowledge about the Appalachian orogen, has the potential to elucidate some of these relationships. This possibility became apparent when I realized that successive basinal, black-shale units not only transgressed westward in time, but also transgressed southward at about the same time and rate as the Acadian orogeny was taking place (Rodgers 1967). Moreover, prominent disconformities in the delta-complex sequence seem to corresopnd with major phases of Acadian tectonism. These and other observations form the basis for a series of interpretations that are presented as "first approximations" towards understanding the complex interrelationships between the Catskill delta complex and the Acadian orogeny.

The Catskill delta complex was completely coupled to the Acadian orogeny, both in terms of provenance and depositional setting. Based on the models of Beaumont (1981) and Jordan (1981), topographic relief resulting from Acadian deformation was the fundamental source of the delta sediments, whereas deformational loading in the orogen was largely responsible for the subsiding peripheral basin in which these sediments were deposited. As a result, periods of uplift in the Acadian Mountains were accompanied by conjoined episodes of subsidence in the peripheral basin. Because of interrelationships between paleoclimate, paleogeography, and tectonics during the Middle and Late Devonian of the central Appalachians (Ettensohn and Barron 1981), periods of maximum subsidence in the peripheral basin are reflected by black, fissile, delta-basin shales, which are easily recognized in surface and subsurface geologic records. Hence, because these basinal, black shales reflect subsidence, which in turn reflects topograhic relief and deformation in the adjacent Acadian orogen, the spatial and temporal distribution of these shales and their relationships to other lithologic units provide information about the nature, extent, and timing of Acadian orogenic events. Using the distribution and stratigraphic relationships of successive black-shale units in relation to the tectonic framework, a history of the Catskill delta, the peripheral basin, and Acadian orogen in the central Appalachians is developed in four phases.

TECTONIC FRAMEWORK

Williams and Hatcher (1982) suggest that the Acadian orogeny was the product of oblique convergence or major transcurrent movement along a sinistral strike-slip fault zone separating the North American craton from a microcontinent called the Avalon terrane. The most recent work (Van der Voo 1983; Perroud and others 1984) indicates that the Avalon terrane probably was part of a larger plate called Armorica and that Avalonian parts of Armorica collided with North America in the Devonian. The occurrence of major transcurrent movement at this time is supported by paleomagnetic differences between the two terranes (Kent and Opdyke 1978, 1980; Irving 1979; Van der Voo and others 1979; Van der Voo and Scotese 1981). The Avalon terrane and other parts of Armorica are thought to have originated on the western margin of Africa (Gondwana) (Skehan 1973; Skehan and Murray 1980; Rast and Skehan 1983; Perroud and others 1984) and to have moved northeasterly along the southeast margin of the North American craton on a major strike-slip fault zone (Fig. 1) until the Late Carboniferous. Although movement along strike-slip faults is dominantly horizontal, a vertical or dip-slip component is often present. The resulting transpression or convergent strike-slip can lead to folding, thrusting and vertical uplift of mountainous zones (Reading 1980). The fact that Acadian deformation migrated southwestward in time along the orogen (Rodgers 1967) favors the predominance of oblique convergence. By the Early Devonian, the onset of Acadian deformation had reached the Appalachian orogen. Although the onset may have occurred as early as the Gedinnian, Donohoe and Pajari (1973) suggested that a late Siegenian to Emsian age (Sawkill Stage) is more likely. Nevertheless, by the Early Mississippian, the effects of the Acadian orogeny were apparent along the entire Appalachian orogen (Rodgers 1967; Johnson 1971).

The Acadian orogeny was originally defined as Middle Devonian in age, but the duration of the orogeny has commonly been extended to include supposedly related events in the Late Devonian and Early Mississippian of the central Appalachians. Now that a model has been developed that clearly relates Middle Devonian, Late Devonian, and Early Mississippian events, I consider all events associated with oblique convergence along the fault zone to be Acadian.

Although the orogeny was time transgressive, three and possibly four major phases of intense deformation have been recognized from various locations along the orogen (Boucot and others 1964; Johnson 1971). Although deformation apparently occurred along the entire orogen to some extent during each phase, major deformation and topographic relief were concentrated in different geographic areas during each successive phase as indicated by intensity of deformation and the location of clastic wedges. Moreover, these sites of intense deformation migrated southwestward with time. The first tectophase occurred during the Early-Middle Devonian transition and was concentrated in Maine and the Canadian Maritime provinces; the second, during the Middle Devonian in central-eastern New York and eastern Pennsylvania; the third, during the Late Devonian and earliest Mississippian in south-central Pennsylvania and the Virginias; and a possible fourth, during the Early to the early Middle Mississippian (Osagean) in the Virginias and southward. Distinct clastic wedges associated with each tectophase emanated from each of three cratonic promontories (Thomas 1977) thus suggesting that the promontories formed the loci of successive tectophases.

The promontories were originally protuberances formed by rifting when the continental margin was established in the Late Precambrian (Dewey and Burke 1974; Rankin 1976; Thomas

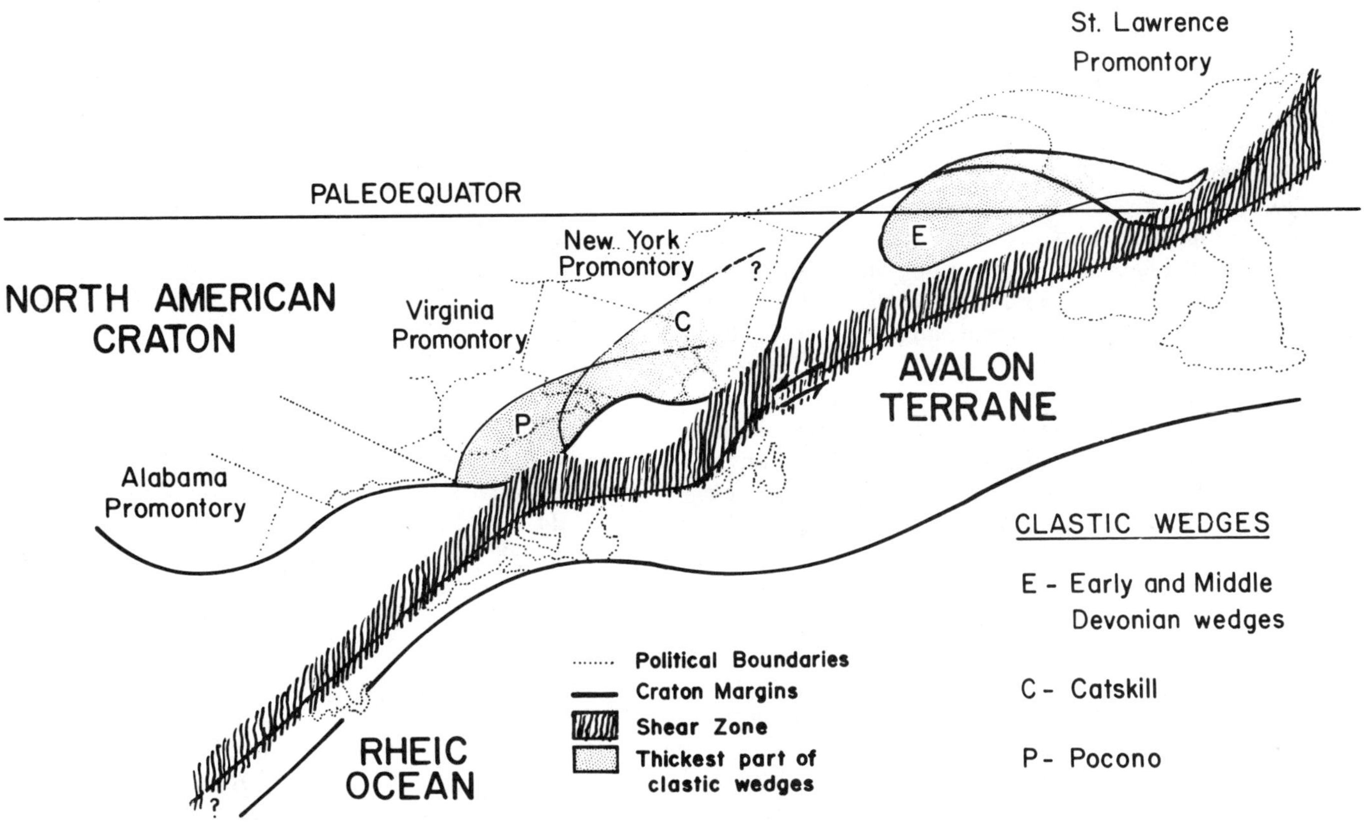

Figure 1. Schematic diagram showing the interpreted position of Avalonian terranes in Armorica relative to the North American craton throughout much of the Devonian. In the position shown, the Avalon terrane has essentially collided with the St. Lawrence promontory, although increased convergence at the other promontories also would result in deformation at those points. Because of oblique convergence, the locus of collision and intense deformation on the North American side of the collision boundary would have migrated southwestward in time to successive promontories. Major clastic wedges appear to emanate successively from each promontory and to fill parts of adjacent reentrants (parts of diagram adapted from Colton 1970; Poole 1976; Thomas 1977; Kent and Opdyke 1978).

1977, 1983). The promontories would have collided first during any type of convergence event and would have been subject to greater compression, whose intensity would have created areas of greater deformation and resulting relief (Dewey and Burke 1974; Dewey and Kidd 1974). Deformation would have been less intense in adjacent reentrants and subsidence may have predominated (Dewey and Kidd 1974; Thomas 1977). Hence, the promontories probably formed the major source areas, and parts of the peripheral basin coincident with the reentrants were the areas of greatest sediment accumulation. I suggest that the various tectophases of the Acadian orogeny, the formation and distribution of the Catskill and related clastic wedges, and the formation and distribution of basinal, black-shale deposits are all related to increased convergence or possible collision between the Avalon terrane and successive promontories. Because of the strike-slip movement, collisions were of the "hit and run" nature so that no long-term docking occurred. Collision with the St. Lawrence promontory would have been followed successively by collision with the New York and Virginia promontories because of oblique convergence. Convergence with the Alabama promontory may have occurred during the Early and Middle Mississippian, but I will discuss only the first three collisions. Presumably, deformation resulting from collision with the St. Lawrence promontory would have been the most intense, because it was the most pronounced promontory (Thomas 1977: Fig. 1). The nature of the clastic wedges and peripheral basins suggests that successive collisions apparently were less intense. The progressive decrease in intensity of each successive collision may be related to a progressive decrease in size of each successive promontory (Thomas 1977: Fig. 1).

In addition to the intense deformation near the promontory involved in collision, some transpression and resulting deformation were probably generated along other parts of the fault during each phase of movement. Temporally related but unequally developed clastic wedges, peripheral-basin deposits, and unconformities occur at all promontories and are thought to be evidence for deformation and uplift, at locations other than at the promontory undergoing collision. For example, the Esopus and Needmore basinal deposits and small, associated clastic wedges developed at the New York and Virginia promontories respectively at the same time as docking occurred at the St. Lawrence promontory. The low intensity of the deformation at the two southern promontories suggests that major collision, and perhaps docking, occurred only at the St. Lawrence promontory.

Each tectophase appears to be represented by a pattern of four stages that can be discerned in the sedimentary record of the Catskill Delta complex: *Stage 1.* Inception of tectonism and formation of a peripheral basin through rapid subsidence: transgression dominates throughout the basin and is reflected by basinal black-shales. *Stage 2.* Impending collision or the southward migration of deformation from one promontory to the next due to oblique convergence: regression dominates, particularly near the east side of basins, as subsidence progressively decreases and clastic influx increases. Because movement on the fault and the resulting southward migration of deformation were episodic, this stage may be represented by minor, transgressive-regressive cycles that mirror the larger four-stage tectophase model. *Stage 3.* Collision accompanied by a period of widespread regional uplift and development of a regional disconformity, even in parts of the peripheral basin (Dickinson 1974: 22). The disconformity is best developed in parts of the peripheral basin nearest the involved promontory. *Stage 4.* Return to Stage 1 or a period of tectonic quiescence represented by widespread carbonate deposition in slowly transgressing seas or stillstand conditions.

The remainder of the paper treats the relationships between the Catskill Delta complex, its basinal deposits, and the Acadian orogeny in terms of this four-stage tectophase model.

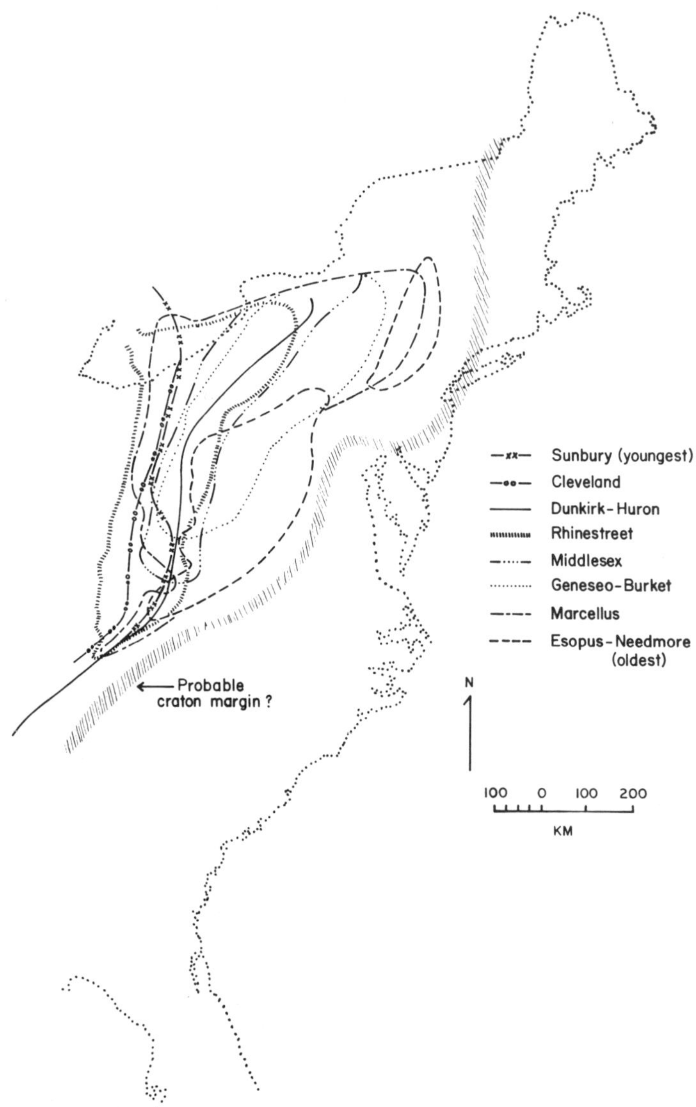

Figure 2. Approximate aerial distribution of successive black-shale units in parts of the Appalachian Basin relative to the probable, former cratonic margin. Note the successive westward and southward migration of Late Devonian (Geneseo through Cleveland) black-shale units. Only the eastern boundaries of the Dunkirk, Huron, Cleveland, and Sunbury shales are shown because the greater portion of these shales occur on adjacent parts of the craton to the west.

FIRST TECTOPHASE

The first tectophase began in the northern Appalachians when collision occurred between the Avalon terrane and the St. Lawrence promontory. Subsidiary tectonism and resulting sedimentation occurred to the southwest at the positions of the other promontories, but were less intense. This collision occurred either during the late Emsian (Boucot and others 1964) or during the late Siegenian to early Emsian (Donohoe and Pajari 1973). The timing of deposition of the dark Esopus and Needmore shales (Figure 2) in the central Appalachians supports the latter interpretation. Intense deformation occurred in northern New England and the Maritime provinces and represents transpressional deformation during which a peripheral basin was formed in the Gaspé Peninsula, northern New Brunswick and northern New England. This basin was subsequently filled with clastic sediments (Cumming 1967; Poole 1976; Bradley 1982). Initial sedimentation on the Catskill Delta complex commenced in the central Appalachians.

Stage 1

Oblique convergence at the New York and Virginia promontories was sufficient to produce two small peripheral basins through deformation and deformation loading. These basins were coeval, but not connected. The Needmore Shale was deposited in the southern of the two basins and the Esopus in the northern, and both shales pinch out into the Schoharie clastics in the Pennsylvania reentrant (Epstein and Epstein 1967: Fig. 5).

Both the Needmore and the Esopus shales represent the first basinal and prodeltaic deposits of the Catskill delta complex (Dennison 1971; Fisher 1979). Thin, black-shale, basinal facies (anaerobic) at the base of the Needmore are overlain by gray, calcareous shales and argillaceous carbonates deposited in dysaerobic to aerobic, open-shelf to prodelta and shallow open-marine environments (Newton 1979; Inners 1979). Esopus sediments are similar although more sandy throughout. The Esopus grades upwards into the increasingly calcareous and fossiliferous Carlisle Center and Schoharie formations and culminates in limestones of the Onondaga Formation (Willard 1939; Epstein and Epstein 1967; Rickard 1975; Fisher 1979). The facies indicate that dysaerobic conditions dominated the Esopus and Carlisle Center formations, and aerobic to marginally aerobic conditions dominated the Schoharie and Onondaga formations.

Stage 2

Stage two is poorly developed in the Needmore and Esopus basins. In neither basin was regressive, delta progradation as complete as it was in later tectophases, for delta-front and delta-platform deposits appear to be absent. Instead, deeper-water carbonates and calcareous shales (Schoharie and, locally, lower Onondaga equivalents) followed by shallow-water carbonates directly overlie prodeltaic shales (upper parts of Needmore and the Carlisle Center Formation). Uplift adjacent to the basins apparently was insufficient to produce major prograding clastic wedges, and carbonate deposition returned to the basins.

Stage 3

Collision during this phase was apparently restricted to the area of the St. Lawrence promontory and is represented by intense late Emsian deformation (Boucot and others 1964). This phase of collision is represented in the central Appalachians, by a late Emsian disconformity at the base of the Onondaga and its equivalents. Although this disconformity is not uniformly developed throughout the central Appalachians, it is best developed on the basin margins in parts of New York and Pennsylvania near the New York promontory and in the Virginias near the Virginia promontory (Dennison and Head 1975: Fig. 7; Rickard 1975: Plate 3).

Stage 4

The shallow-water Onondaga limestone and its equivalents represent tectonic quiescence and a slow transgression over the unconformity; this marks the end of the first Acadian tectophase in the central Appalachians.

SECOND TECTOPHASE

The second Acadian tectophase is represented by the rocks of the Hamilton Group and Tully Limestone and is approximately late Eifelian through Givetian (Casenovia-through-Taghanic stages) in age. The tectophase culminated in the collision of the Avalon terrane with parts of the New York promontory during the Middle–Late Devonian transition, and the resulting Catskill Delta complex is best developed in south-central New York and adjacent parts of Pennsylvania. Subsidiary development of the Catskill Delta complex related to the Virginia promontory is represented by the Mahantango Formation of south-central Pennsylvania.

During this period, the basin was greatly enlarged through combination and expansion of parts of the Needmore and Esopus basins formed during the first tectophase. The Marcellus basin migrated westward relative to the northern basin (Esopus) and has a greater westward extent in the north (Figure 2), thus reflecting greater westward migration of Acadian deformation and basin infilling in the area of the New York promontory.

Stage 1

The beginning of the second tectophase is indicated by the black Marcellus Shale deposited first in the northeastern parts of the peripheral basin during the latest Southwood or Onesquethaw Stage. Intertonguing of the Marcellus and underlying Onondaga indicates slow but steady subsidence as the dark shales progressively displaced the carbonate facies westward in an overlapping relationship (Schwietering 1978). Aerobic, dysaerobic, and anaerobic conditions existed during this facies progradation and resulted in considerable variation within the Marcellus as a whole (Cooper 1930, 1933; Butts 1940; Dennison 1963; Cate 1963; Butts and Edmundson 1966). However, the net result was the replacement of aerobic, non-clastic sedimentation with dysaerobic to anaerobic clastic sedimentation in a westward progression.

Stage 2

During the late Eifelian and Givetian stages, collision with the New York promontory intensified and subsidence in the peripheral basin progressively declined. Sedimentation during this stage of tectonism is represented by calcareous silty shales and carbonates of the Hamilton, Big Bend or Moscow facies. These facies reflect delta-front and delta-platform environments that exhibit four small-scale, cyclic alternations. The subcycles reflect a progressive decline in water depth (ineffectiveness of subsidence), and an increase in oxygenation (Figure 4) and consist of alternations of major siltstone and shale lithofacies that represent subsidence and transgression. In each subcycle the rocks change from dark gray shale with *Leiorhynchus* fauna, upward through blue-gray and gray-green shales to limestone, and represent transitions from marginally anaerobic to dysaerobic conditions in prodelta environments to aerobic conditions in delta-front and delta-platform environments. The limestones apparently reflect shallow-water delta destruction characterized by slight transgression and reduced clastic input (McCave 1973; Baird 1979).

Three such subcycles occur in the Hamilton Group (Cleland 1903; Cooper 1930, 1933; Figure 4, this paper). An incomplete subcycle may occur in the Moscow Formation (Cooper 1930; Patchen and Dugolinsky 1979; Baird and Brett 1981) and even smaller subcycles may be indicated in the Moscow by the presence of an unconformity between its Windom and Kashong shale members (Baird and Brett 1981). Similar subcycles in coeval units farther south are indicated by the work of Dennison and Head (1975), and Hasson and Dennison (1977). Marked decline in subsidence and sea level occurred during the Late Givetian (Tioughnioga Stage) and signaled the end of Stage 2.

Stage 3

Stage 3 is represented by a prominent disconformity at the Middle–Late Devonian transition that is best developed in parts of New York nearest the New York promontory (Dennison and Head 1975). No disconformity exists in the central Appalachian Basin, but Dennison and Head (1975) present stratigraphic evidence for uplift and sea level drop in this area.

Stage 4

Stage four of the second tectophase, tectonic quiescence, is represented by the late Givetian (Taghanic Stage) Tully Limestone, which is present in some form throughout large parts of the central Appalachian peripheral basin. The Tully Limestone appears to represent a period of slow transgression in clastic-deficient waters on the recently uplifted delta platforms.

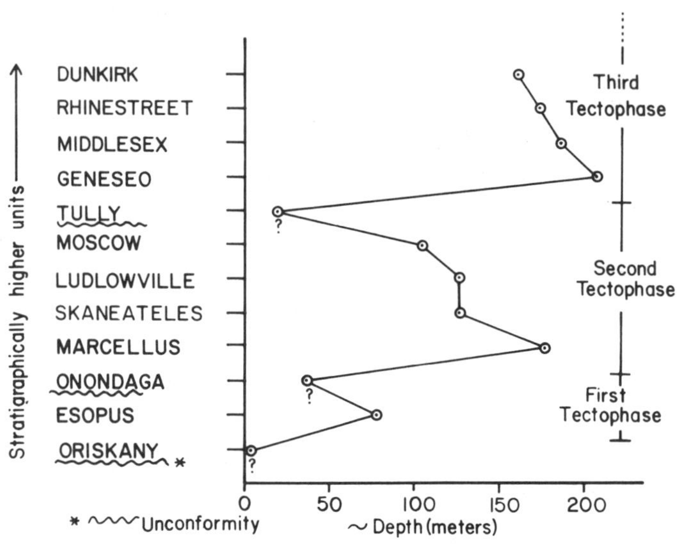

Figure 3. Approximate depositional depths for major black-shale and Hamilton clastic units calculated using the model of Klein (1974). The model does not work for carbonates (Oriskany, Onondaga, Tully), so these depths are crude estimates. Note that the tectophases are bracketed by disconformities and exhibit a depth cycle of increase and decrease. Note also that the maximum depths for each successive tectophase increase, a response interpreted to be related to the increasing proximity and intensity of deformation and tectonism.

THIRD TECTOPHASE

The third tectophase is represented by the rocks of the Genesee-through-Canadaway groups and their equivalents, and is approximately Frasnian-through-earliest Tournaisian (Senecan-through-Kinderhook series) in age. The tectophase represents the southward migration of deformation and collision with the Virginia promontory during the Devonian-Mississippian transition. The progressive southward migration of successive peripheral basins (Figure 2) indicates that major deformation and deformational loading also migrated southward from the New York promontory. The relatively widespread distribution of anaerobic peripheral-basin facies between the New York and Virginia promontories (Figure 2) may indicate more or less continuous collision along the entire length of the area separating the two promontories. This may be related to the small size of the reentrant separating the two promontories (Figure 1) or the possibility of small microcontinental fragments near the reentrant (Thomas 1977: Fig. 3).

Stage 1

The beginning of the third tectophase is represented by the black Geneseo and Burket shales that appear to represent the maximum in Acadian subsidence. This is indicated by depth estimations of over 200m (Figure 3) based on the model of Klein (1974). This period of maximum Acadian subsidence and transgression was related to a period of major global transgression and cratonic submergence (Sloss and Speed 1974; Dennison and Head 1975) called the Taghanic Onlap by Johnson (1970). The onlap apparently was related to the Acadian and other approximately coeval tectonic events (Antler and Ellesmerian in North America) that reflected increased plate convergence and spreading rates (Sloss and Speed 1974). Onlap began as early as the late Middle Devonian west of the Virginia promontory (Johnson 1970; Dennison and Head 1975), but was not apparent in the New York area until the early Late Devonian (Frasnian, Finger Lakes Stage) when the Tully carbonate platform subsided abruptly below the pycnocline, setting the stage for deposition of black, basinal Geneseo and Burket shales.

Stage 2

After the initial stage of major subsidence (Geneseo-Burket), the second stage was characterized by decreasing rates of subsidence and increasing rates of clastic influx caused by the southward migration of the locus of collision and major deformation. This stage, however, is cyclic, characterized by alternating basinal black shales and coarser, more proximal deltaic clastics. The cyclicity apparently reflects episodic movement on the fault zone, as well as episodes of deformation in the orogen. Beginning with the Geneseo Shale, four major cycles are present in the Appalachian Basin (Pepper and deWitt 1950, 1951; Pepper and others 1956; Colton and deWitt 1958; deWitt and Colton 1959, 1978; Rickard 1975; Figure 5, this paper). The Cleveland Shale begins the fifth and final cycle of this tectophase, but it occurs only in extreme western parts of the Appalachian Basin (Figure 2).

The base of each cycle is defined by a black-shale unit. These black shales, except for the Dunkirk (Patchen and Dugo-

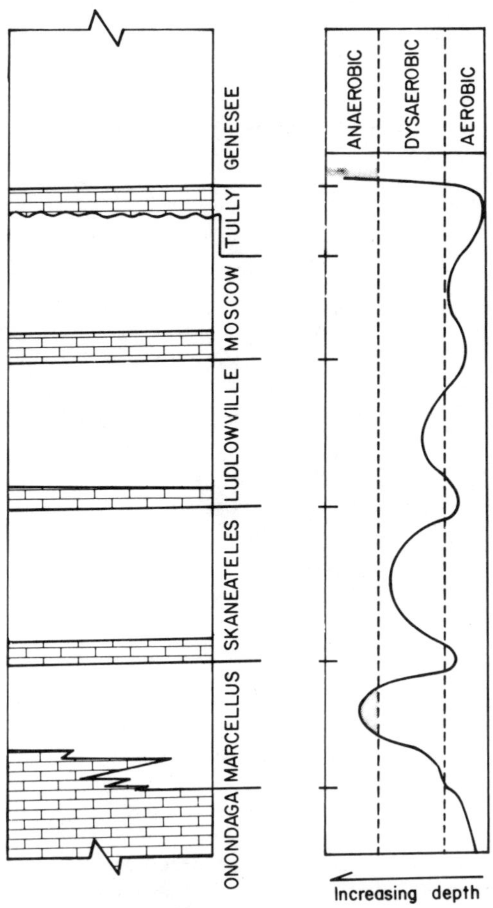

Figure 4. Interpretation of water stratification during the time of deposition of various Hamilton clastic units. Compare to interpreted depths during second tectophase on Figure 3. The overlying Genesee Group begins with the black Geneseo Shale.

linsky 1979: 157), have sharp basal contacts, indicating abrupt subsidence, rapid transgression, and the cessation of abundant clastic influx accompanying initiation of tectonism (see Ettensohn this volume). In places, the basal contacts may be marked by pyritic lag horizons reflecting erosion and reworking of underlying units by the rapidly transgressing seas. Black-shale units grade upward into a prograding wedge of deltaic clastics. The wedges form the upper half of each cycle (Figure 5).

After Geneseo time, however, the wedges of deltaic clastics in each successive cycle became progressively more dominant as subsidence decreased and clastic influx increased. Decreased subsidence and increased clastic influx are also supported by: (1) Depth approximations of basinal black-shale units show a steady decrease from the Geneseo to the Dunkirk (Figure 3). (2) The number of gray-shale interbeds, representing brief returns to dysaerobic, open-shelf and prodelta environments, increases in the thick central portions of each successively younger black-shale unit (see Pepper and deWitt 1951; Pepper and others 1956; Colton and deWitt 1958; deWitt and Colton 1978). (3) The basal contact of the Dunkirk, the last major black-shale unit in the Appalachian Basin, is not sharp, but gradational with underlying clastics (Figures 5, 6). (4) Each successive clastic wedge thickens, and the pronounced transgressive-regressive-transgressive symmetry of environments and lithologies in the clastic wedges of the lower two cycles (Genesee and Sonyea groups, Figure 5; see Pepper and deWitt 1950; Colton and deWitt 1958; Rickard 1964, 1975; Thayer 1974; Fig. 2; deWitt and Colton 1978) is poorly developed or absent in clastic wedges of the last two cycles (West Falls and Canadaway groups, Figure 5; see Pepper and others 1956, and Rickard 1964, 1975).

As deformation and deformational loading migrated southward, so did successive peripheral basins, as indicated by the distribution of the Late Devonian black-shale units (Figure 2). Despite the continued southward movement of deformation, subsidence in southern parts of the Acadian peripheral basin was never as great as it was to the north in the area of the New York promontory. Decreased subsidence is indicated by the thinner nature of the shales to the south, their reduced aerial extents in the south (Figure 2), their marginally anaerobic nature, and the early (post-Burket) dominance of clastics in southeastern and south-central parts of the basin. Because peripheral-basin subsidence is controlled by loading in the adjacent orogen (Beaumont 1981; Jordan 1981), decreasing subsidence must indicate that transpression and resulting deformation were substantially reduced in the area of the Virginia promontory. I suggest that this reduction was related to the less pronounced, more elongate nature of the promontory (Figure 1).

Each of the enclosed peripheral basins (Figure 2) caused a restriction in circulation during black-shale deposition and acted as a confining sink for coarser clastics debouched into them. However, by the latest Frasnian, subsidence in the basin had declined to the point that open-shelf and prodeltaic Angola and Hanover shales prograded beyond the peripheral basin onto adjacent parts of the stable craton in eastern Ohio and eastern Kentucky, where they are called the Upper Olentangy Shale (Wallace and others 1977, 1978, Kepferle and others 1978; Roen and others 1978a, b; Figure 5, this paper). As the basin filled from the east with clastics, the locus of deposition for each successive basinal black-shale unit migrated westward with time (Figures 2, 5). By the end of Dunkirk deposition, subsidence could no longer keep pace with clastic influx and basinal black-shale deposition was essentially "pushed out" of the peripheral basin by clastic influx onto adjacent parts of the stable craton (Figure 5). At this point, the basin was dominated along its entire length by westwardly prograding clastic wedges, represented by the Brallier, Chagrin, Catskill, Hampshire, and equivalent formations.

Thus, the basin essentially filled to overflowing, forcing subsequent clastics to prograde beyond its western confines. Once this happened, the east-dipping, western paleoslope of the basin was destroyed, and a regional west-dipping paleoslope was established (Figure 5). The Dunkirk Shale represents the last major basinal black-shale unit in the Appalachian peripheral basin. Subsequent Catskill delta basin-facies largely developed on adjacent parts of the stable craton in deepening eipcontinental seas and are

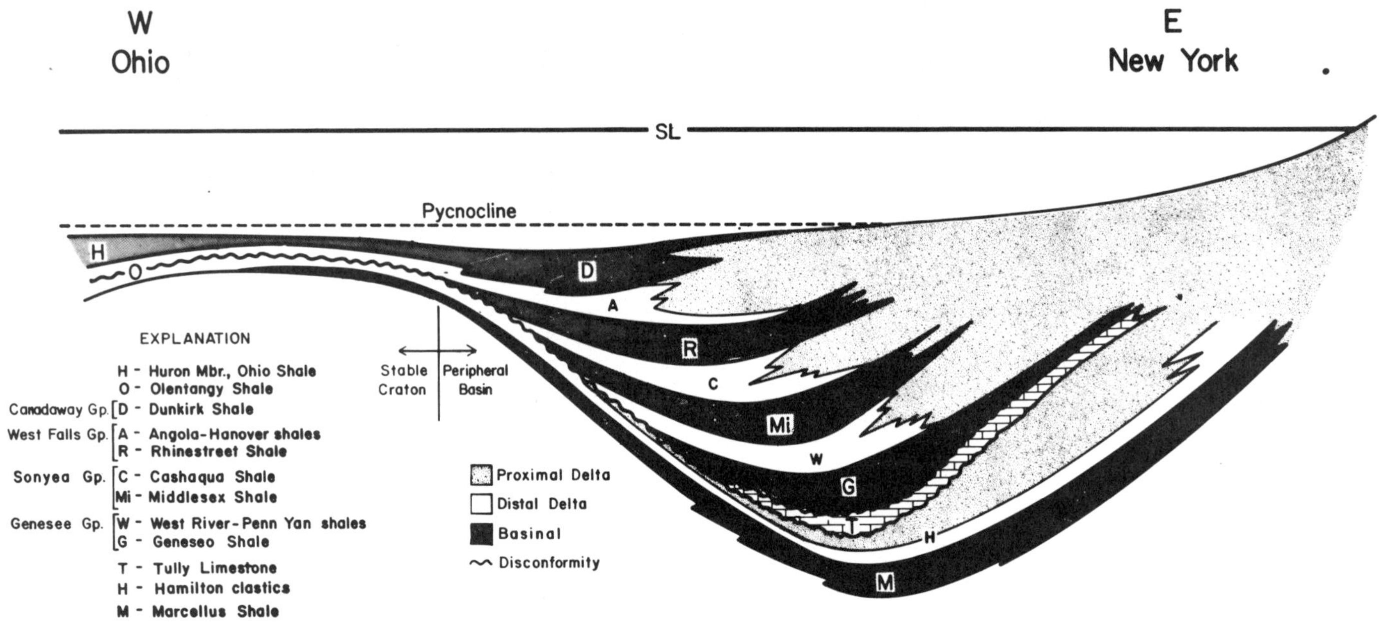

Figure 5. Schematic diagram showing the general relationships between proximal-delta-distal-delta, and basinal facies for the Middle and Upper Devonian Catskill Delta complex on a cross-section from New York to Ohio. Aside from the Hamilton-age delta (rocks below disconformity), all Late Devonian delta facies were confined to the peripheral basin until the post-Rhinestreet (Angola-Hanover shales) clastic influx. As subsidence in the peripheral basin declined and depth in the epicontinental sea increased after Rhinestreet deposition, both black shales and intervening coarser clastic wedges migrated westward onto the craton (Huron Shale). During Dunkirk-Huron deposition, the pycnocline, or zone separating anaerobic waters below from aerobic waters above, was no longer restricted to the peripheral basin as during earlier black-shale deposition. The Olentangy Shale of Ohio is a Middle-Upper Devonian green-shale unit formed from westward equivalents of the Hamilton Group (Middle Devonian) and westward equivalents of the West Falls Group (Upper Devonian). A subtle disconformity separates Middle and Upper Devonian parts of the unit. Not drawn to scale.

represented by Ohio, New Albany, and Chattanooga shales (Figure 5).

Stages 3 and 4

Sometime near the Devonian-Mississippian transition, collision in the area of Virginia promontory occurred, indicated by the presence of a disconformity, locally separating the Catskill and Pocono clastic wedges. The resulting uplift apparently was mild, for the disconformity is largely absent throughout most of the Appalachian Basin. However, a disconformity is present near the New York promontory in Pennsylvania (Sevon 1969, 1979, 1981) and is relatively widespread in areas adjacent to the Virginia promontory (Colton 1970: Fig. 8B) where deformation was most intense. The presence of clastics on either side of the disconformity and the continuity of clastic deposition during this time elsewhere in the basin, indicate the apparent absence of Stage 4.

FOURTH TECTOPHASE?

Most workers have cited evidence for only three Acadian tectophases, the last one of which ended in the earliest Mississippian. I believe that evidence exists for another, wholly Mississippian tectophase of the Acadian orogeny, reflecting continued southward migration of convergence and deformation along the fault zone. The stages are not everywhere well developed, and in places the resulting Mississippian clastic wedge is difficult to differentiate from the older Catskill wedge.

Stage 1

Stage 1 is represented by the black, basinal Sunbury Shale in extreme western parts of the peripheral basin. The Sunbury is the last basinal black-shale unit and, unlike earlier basinal shale units, its eastern boundary migrated eastward with time, particularly in the area of central West Virginia to northeastern Tennessee (Figure 2). This area appears as a prominent eastward bulge that corresponds to the southern half of the Virginia promontory (Figure 2) and probably reflects an area of renewed deformation on southern parts of the Virginia promontory. In this bulge, basinal black-shale deposition apparently continued well beyond the duration of the Sunbury elsewhere and is included in parts of the Big Stone Gap Member of the Chattanooga Shale (Elam 1981; Kepferle and others 1981). A band of fossiliferous shallow open-marine and marginal marine sediments surrounding the bulge is included in the Price Formation (Kreisa and Bambach 1973).

Stage 2

Stage 2 is represented by the lower Tournaisian (Kinderhook and Osage) clastics of the Pocono, Price, Grainger, and Borden formations. The Pocono wedge forms the center of this complex and corresponds to the length of the Virginia promontory from south-central Pennsylvania to southwestern Virginia (Figure 1). Pocono equivalents like the Borden prograded from the northeast and north onto central parts of the craton as far west as Illinois.

Stages 3 and 4

Whether collision ever occurred any farther south than the Virginia promontory (e.g., Alabama promontory) is uncertain. A disconformity probably separates the Chattanooga Shale from the overlying Maury and Fort Payne formations in north-central Alabama and central Tennessee (Weller and others 1948; Hass 1956). This disconformity, however, seems to occur in the wrong position, unless the area was subject to starved-basin conditions and/or erosion during Stage 2. Farther north, early Middle Mississippian (Osagean) clastics and locally Ft. Payne carbonates appear to overlie Sunbury equivalents in uppermost parts of the Chattanooga Shale conformably. A disconformity, however, does separate the Pocono from overlying Middle Mississippian carbonates in the central Appalachians (Weller and others 1948; Adams 1970). Nonetheless, the Middle Mississippian carbonates that overlie the basinal or proximal delta clastics of the Fourth Tectophase mark the end of the Acadian orogeny and the beginning of a period of tectonic quiescence that endured throughout much of the Middle and Late Mississippian.

CONCLUSIONS

The Catskill delta complex is interpreted to have a much longer duration and more widespread distribution than is generally thought. The delta complex began near the Early-Middle Devonian transition with the Esopus and Needmore shales in the north-central and central Appalachians and ended in the Middle Mississippian with the Borden, Price, and Pocono formations of the east- and north-central Midcontinent. The delta complex was deposited in four episodes that correspond with four suggested tectophases in the Acadian orogeny. Each tectophase exhibits four stages that are reflected in the sedimentary record of each delta episode. The occurrence and distribution of delta-basin deposits (black shales), clastic wedges, major carbonate units, and regional disconformities, as well as the estimated depth of basinal deposits, are used to define the tectophases and their deltaic responses. The distribution of these deposits indicates that each tectophase was the result of collision of Avalonian terranes in Armorica with successive promontories on the southeast margin of the North American landmass. Because of oblique convergence, the Avalon terrane collided successively with the promontories from northeast to southwest, so that Acadian tectonism and major delta episodes migrated southwestward with time. Each of the deltaic episodes can be associated with collision at a promontory.

ACKNOWLEDGMENTS

I would like to thank N. Rast for valuable discussion and references relative to portions of the tectonic framework. The editors, John Rodgers, and J. W. Skehan, S. J., reviewed the manuscript and made many valuable suggestions. Margie A. Palmer and Beth Ettensohn helped with manuscript preparation.

Parts of this research were supported by a contract (DE-AC21-76ET-12040) from the Department of Energy, Morgantown Energy Technology Center and by a grant from the Graduate School, University of Kentucky.

REFERENCES CITED

Adams, R. W., 1970, Loyalhanna Limestone—cross-bedding and provenance, *in* Fisher, G. W., and others, eds., Studies of Appalachian geology: central and southern: New York, Interscience Publishers, p. 83–100.

Baird, G. C., 1979, Sedimentary relationships of Portland Point and associated Middle Devonian rocks in central and western New York: New York State Museum Bulletin No. 433, 24 p.

Baird, G. C., and Brett, C. E., 1981, Submarine discontinuities and sedimentary condensation in the upper Hamilton Group (Middle Devonian): examination of marine shelf and paleoslope deposits in the Cayuga Valley: New York Stage Geological Association, 53rd Annual Meeting Guidebook, p. 115–145.

Beaumont, C., 1981, Foreland basins: Geophysical Journal of the Royal Astronomical Society, v. 65, p. 291–329.

Boucot, A. J., Field, M. T., Fletcher, R., Forbes, W. H., Naylor, R. S., and Pavlides, L., 1964, Reconnaissance bedrock geology of the Presque Isle Quadrangle, Maine: Maine Geological Survey, Quadrangle Mapping Series No. 2, 123 p.

Bradley, D. C., 1982, Subsidence in Late Paleozoic basins in the northern Appalachians: Tectonics, v. 2, p. 107–123.

Butts, C., 1940, Geology of the Appalachian Valley in Virginia: Virginia Geological Survey Bulletin 52, pt. I, 568 p.

Butts, C., and Edmundson, R. S., 1966, Geology and mineral resources of Frederick County: Virginia Division of Mineral Resources, Bulletin 80, 142 p.

Cate, A. S., 1963, Lithostratigraphy of some Middle and Upper Devonian rocks in the subsurface of southwestern Pennsylvania: Pennsylvania Topographic and Geologic Survey, ser. 4, General Geology Report G39, p. 229–240.

Cleland, H. F., 1903, Fauna of the Hamilton Formation of the Cayuga Lake section in central New York: U.S. Geological Survey Bulletin 206, 112 p.

Colton, G. W., 1970, The Appalachian Basin its depositional sequences and their geologic relationships, *in* Fisher, G. W., and others, eds., Studies of Appalachian geology: central and southern: New York, Interscience Publishers, p. 5–47.

Colton, G. W., and deWitt, W., Jr., 1958, Stratigraphy of the Sonyea Formation of Late Devonian age in western and central New York: U.S. Geological Survey Chart OC-54.

Cooper, G. A., 1930, Stratigraphy of the Hamilton Group of New York: American Journal of Science, ser. 5, v. 19, p. 116–134, 214–236.

——1933, Stratigraphy of the Hamilton Group in eastern New York: American

Journal of Science, ser. 5, v. 26, p. 537–551.

Cumming, L. M., 1967, Devonian of Canadian Appalachians and New England States, *in* Oswald, D. H, ed., International symposium on the Devonian System, Calgary, 1967: Calgary, Alberta Society of Petroleum Geologists, v. 1, p. 1041–1055.

Dennison, J. M., 1963, Stratigraphy between Tioga Bentonite and Pocono Formation along Allegheny Front in Maryland and West Virginia: Pennsylvania Topographic and Geologic Survey, ser. 4, General Geology Report G39, p. 213–327.

—— 1971, Petroleum related to Middle and Upper Devonian deltaic facies in central Appalachians: American Association of Petroleum Geologists Bulletin, v. 55, p. 1179–1193.

Dennison, J. M., and Head, J. W., 1975, Sea-level variations interpreted from the Appalachian Basin Silurian and Devonian: American Journal of Science, v. 275, p. 1089–1120.

Dewey, J. F., and Burke, K.C.A., 1974, Hot spots and continental breakup: implications for collisional orogeny: Geology, v. 2, p. 57–60.

Dewey, J. F., and Kidd, W.S.F., 1974, Continental collisions in the Appalachian-Caledonian belt: variations related to complete and incomplete suturing: Geology, v. 2, p. 543–546.

deWitt, W., Jr., and Colton, G. W., 1959, Revised correlations of lower Upper Devonian rocks in western and central New York: American Association of Petroleum Geologists, v. 43, p. 2810–2828.

—— 1978, Physical stratigraphy of the Genesee Formation (Devonian) in western and central New York: U.S. Geological Survey Professional Paper 1032-A, 22 p.

Dickinson, W. R., 1974, Plate tectonics and sedimentation; *in* Dickinson, W. R., ed., Tectonics and sedimentation: Society of Economic Paleontologists and Mineralogists Special Publication No. 22, p. 1–27.

Donohoe, H. V., and Pajari, G., 1973, The age of the Acadian deformation in Maine-New Brunswick: Maritime Sediments, v. 9, p. 78–82.

Elam, T. D., 1981, Stratigraphy and paleoenvironmental aspects of the Bedford-Berea sequence and the Sunbury Shale in eastern and south-central Kentucky [M.S. thesis]: Lexington, University of Kentucky, 155 p.

Epstein, J. B., and Epstein, A. G., 1967, Geology in the region of the Delaware to Lehigh Water Gaps: 32nd Field Conference of Pennsylvanian Geologists, Guidebook, p. 1–89.

Ettensohn, R. F., and Barron, L. S., 1981, Depositional model for the Devonian-Mississippian black shales of North America: a paleoclimatic-paleogeographic approach, *in* Roberts, T. G., ed., G.S.A. Cincinnati '81 field trip guidebooks, volume II: economic geology, structure: American Geological Institute, p. 344–361.

Fisher, D. W., 1979, Devonian stratigraphy and paleoecology in the Cherry Valley, New York region: Joint annual meeting, New York State Geological Association and New England Intercollegiate Geological Conference, Guidebook, p. 20–46.

Hass, W. H., 1956, Age and correlation of the Chattanooga Shale and Maury Formation: U.S. Geological Survey Professional Paper 286, 47 p.

Hasson, K. O., and Dennison, J. M., 1977, Devonian Harrell and Millboro shales in parts of Pennsylvania, Maryland, West Virginia, and Virginia, *in* Schott, G. L., and others, eds., Proceedings, first eastern gas shales symposium: U.S. Department of Energy, Morgantown Energy Research Center, MERC/SP-77/5, p. 634–636.

Inners, J. D., 1979, The Onesquethaw Stage in south-central Pennsylvania and nearby areas: 44th Annual Field Conference of Pennsylvania Geologists, Guidebook, p. 38–55.

Irving, E., 1979, Paleopoles and paleolatitudes of North America and speculations about displaced terranes: Canadian Journal of Earth Sciences, v. 16, p. 669–694.

Johnson, J. G., 1970, Taghanic onlap and the end of North American Devonian provinciality: Geological Society of America Bulletin, v. 81, p. 2077–2106.

—— 1971, Timing and coordination of orogenic, epeirorogenic, and eustatic events: Geological Society of America Bulletin, v. 82, p. 3263–3298.

Jordan, T. E., 1981, Thrust loads and foreland basin evolution, Cretaceous, western United States: American Association of Petroleum Geologists Bulletin, v. 65, p. 2506–2520.

Kent, D. V., and Opdyke, N. D., 1978, Paleomagnetism of the Devonian Catskill redbeds: evidence for motion of the coastal New England-Canadian Maritime region relative to cratonic North America: Journal of Geophysical Research, v. 83, p. 4441–4450.

—— 1980, Paleomagnetism of Siluro-Devonian rocks from eastern Maine: Canadian Journal of Earth Sciences, v. 17, p. 1653–1665.

Kepferle, R. C., Potter, P. E., and Pryor, W. A., 1981, Stratigraphy of the Chattanooga Shale (Upper Devonian and Lower Mississippian) in vicinity of Big Stone Gap, Wise County, Virginia: U.S. Geological Survey Bulletin 1499, 20 p.

Kepferle, R. C., Wilson, E. N., and Ettensohn, F. R., 1978, Preliminary stratigraphic cross section showing radioactive zones in the Devonian black shales in the southern part of the Appalachian Basin: U.S. Geological Survey Chart OC-85.

Klein, G. deV., 1974, Estimating water depths from analysis of barrier island and deltaic sedimentary sequences: Geology, v. 2, p. 409–412.

Kreisa, R. D., and Bambach, R. K., 1973, Paleogeographic development of a portion of southwestern Virginia during the lower Mississippian [abs.], *in* Southeastern Section, 22nd Annual Meeting, Geological Society of America Abstracts with Programs, v. 5, no. 5, p. 410–411.

McCave, I. N., 1973, The sedimentology of a transgression: Portland Point and Cooksburg members (Middle Devonian, New York State): Journal of Sedimentary Petrology, v. 43, p. 484–504.

Newton, C. R., 1979, Aerobic, dysaerobic, and anaerobic facies within the Needmore Shale (Lower to Middle Devonian): 44th Annual Field Conference of Pennsylvania Geologists, Guidebook, p. 56–60.

Patchen, D. G., and Dugolinsky, B. K., 1979, Guidebook, Middle and Upper Devonian clastics, central and western New York State: U.S. Department of Energy, Eastern Gas Shales Project, 170 p.

Pepper, J. F., and deWitt, W., Jr., 1950, Stratigraphy of the Upper Devonian Wiscoy Sandstone and equivalent Hanover Shale in western and central New York: U.S. Geological Survey Chart OC-37.

—— 1951, Stratigraphy of the Late Devonian Perrysburg Formation in western and west-central New York: U.S. Geological Survey Chart OC-45.

Pepper, J. F., deWitt, Jr., and Colton, G. W., 1956, Stratigraphy of the West Falls Formation of Late Devonian age in western and west-central New York: U.S. Geological Survey Chart OC-55.

Perroud, H., Van der Voo, P., and Bonhommet, N., 1984, Paleozoic evolution of the Armorica plate on the basis of paleomagnetic data: Geology, v. 579–582.

Poole, W. H., 1976, Plate tectonic evolution of the Canadian Appalachian region: Geological Survey of Canada Paper 76-1B, p. 113–126.

Rankin, D. W., 1976, Appalachian salients and recesses: Late Precambrian continental breakup and the opening of the Iapetus Ocean: Journal of Geophysical Research, v. 81, p. 5605–5619.

Rast, N., Skehan, S. J., J. W., 1983, Avalon Plate in North America, Europe and Africa: The Caledonide Orogen-International Geological Correlation Program Project Number 27, Morocco and Paleozoic orogenesis, Résumé des Communications, p. 29.

Reading, H. G., 1980, Characteristics and recognition of strike-slip fault systems, *in* Ballance, P. F., and Reading, H. G., eds., Sedimentation in oblique-slip mobile zones: International Association of Sedimentologists Special Publication No. 4, p. 7–26.

Rickard, L. V., 1964, Correlation of the Devonian rocks in New York State: New York State Museum and Science Service, Geological Survey Map and Chart Series No. 4.

—— 1975, Correlation of the Silurian and Devonian rocks in New York State: New York State Museum and Science Service, Geological Survey Map and Chart Series 24.

Rodgers, J., 1967, Chronology of tectonic movements in the Appalachian region of eastern North America: American Journal of Science, v. 265, p. 408–427.

Roen, J. B., Wallace, L. G., and deWitt, W., Jr., 1978a, Preliminary stratigraphic cross section showing radioactive zones of the Devonian black shales in eastern Ohio and west-central Pennsylvania: U.S. Geological Survey Chart OC-82.

—— 1978b, Preliminary stratigraphic cross section showing radioactive zones in the Devonian black shales in the central part of the Appalachian Basin: U.S. Geological Survey Chart OC-87.

Schwietering, J. F., 1978, Preliminary model of Catskill Delta in West Virginia, *in* Schott, G. L., and others, eds., Proceedings, first eastern gas shales symposium: U.S. Department of Energy, Morgantown Energy Research Center, MERC/SP-77/5, p. 195–205.

Sevon, W. D., 1969, Sedimentology of some Mississippian and Pleistocene deposits of northeastern Pennsylvania, *in* Subitzky, S., ed., Geology of selected areas in New Jersey and eastern Pennsylvania: Rutgers University Press, New Brunswick, p. 214–234.

—— 1979, Polymictic diamictites in the Spechty Kopf and Rockwell Formations, *in* Dennison, J. M., Hasson, K. O., Hoskins, D. M., Jolly, R. M., and Sevon, W. D., Devonian shales in south-central Pennsylvania and Maryland: Guidebook, 44th Annual Field Conference of Pennsylvania Geologists, Pennsylvania Geological Survey, Harrisburg, 116 p.

—— 1981, The Middle and Upper Devonian clastic wedge in northeastern Pennsylvania: Guidebook, 52nd Annual Meeting of the New York State Geological Association, Binghamton, p. 31–63.

Skehan, S. J., J. W., 1973, Subduction zone between the Paleo-American and Paleo-African plates in New England: Geofiscia Internacional, v. 13, p. 291–308.

Skehan, S. J., J. W., and Murray, D. P., 1980, Geologic profile across southeastern New England: Tectonophysics, v. 69, p. 285–319.

Sloss, L. L., and Speed, R. C., 1974, Relationships of cratonic and continental-margin tectonic episodes, *in* Dickinson, W. R., ed., Tectonics and sedimentation: Society of Economic Paleontologists and Mineralogists Special Publication No. 22, p. 98–119.

Thayer, C. W., 1974, Marine paleoecology in the Upper Devonian of New York: Lethaia, v. 7, p. 121–155.

Thomas, W. A., 1977, Evolution of Appalachian-Ouachita salients and recesses from reentrants and promontories in the continental margin: American Journal of Science, v. 277, p. 1233–1278.

—— 1983, Continental margins, orogenic belts and intracratonic structures: Geology, v. 11, p. 270–272.

Van der Voo, R., 1983, Paleomagnetic constraints on the assembly of the Old Red Continent: Tectonophysics, v. 91, p. 271–283.

Van der Voo, R., and Scotese, C., 1981, Paleomagnetic evidence for a large (~2000 km) sinistral offset along the Great Glen fault during Carboniferous time: Geology, v. 9, p. 583–589.

Van der Voo, R., French, A. N., and French, R. B., 1979, A paleomagnetic pole position from the unfolded Upper Devonian Catskill red beds and its tectonic implications: Geology, v. 7, p. 345–348.

Wallace, L. G., Roen, J. B., and deWitt, W. Jr., 1977, Preliminary stratigraphic cross section showing radioactive zones in the Devonian black shales in the western part of the Appalachian Basin: U.S. Geological Survey Chart OC-80.

—— 1978, Preliminary stratigraphic cross section showing radioactive zones in the Devonian black shales in southeastern Ohio and west-central West Virginia: U.S. Geological Survey Chart OC-83.

Weller, J. M., and others, 1948, Correlation of the Mississippian formations of North America: Geological Society of America Bulletin, v. 59, p. 91–196.

Willard, B., 1939, The Devonian of Pennsylvania: Pennsylvania Geological Survey, ser. 4, Bulletin G. 19, 481 p.

Williams, H., and Hatcher, R. D., Jr., 1982, Suspect terranes and accretionary history of the Appalachian orogen: Geology, v. 10, p. 530–536.

Manuscript Accepted by the Society July 20, 1984

Printed in U.S.A.

Geological Society of America
Special Paper 201
1985

Paleogeography, paleoclimate, and sedimentary processes of the Late Devonian Catskill Delta

Donald L. Woodrow
Department of Geoscience
Hobart and William Smith Colleges
Geneva, New York 14456

ABSTRACT

The Late Devonian Catskill Delta is made up of marine and non-marine facies built up on the flank of the tectonic Appalachian Peninsula during assembly of the Old Red (Laurasian) Continent. Much of the continent was under the influence of tropical climates showing a wide range of rainfall. Over the delta, the climate was either tropical wet and dry or desert, due in part, to a rainshadow effect caused by the mountains to the east. Streams showed great variations in discharge and an extended period of drought was an annual event over the region.

Alluvial processes were dominant on land. Braided streams deposited the coarsest sediments on alluvial fans and sinuous, channelized streams deposited sand and mud on the alluvial plains. Interfluves on the alluvial plains were sufficiently long-lived to permit the formation of carbonate soils. Plants were most common near stream courses. Fine sand and mud were carried across the shoreline in distributaries to the floor of the adjacent Catskill Sea. Deltaic processes, wave-related processes, and tides shaped the shore. Wave-related processes and bioturbation modified and reworked the shallow marine sediments while turbidity currents and slow deposition from suspension were most effective over the rest of the basin.

INTRODUCTION

The purpose of this paper is to provide a synthesis of interpretations dealing with the paleoclimatology, paleogeography, and sedimentary processes typical of the subaerial and subaqueous parts of the Catskill delta. Only the Late Devonian is considered. Thorough review of the rock and fossil record on which the interpretations are based is given in other papers in this volume and in the papers cited. No attempt has been made to offer complete citation lists; instead I have cited recent papers that provide both a sample of the data and entrée to the massive Catskill literature.

Paleoclimate is discussed at a level that integrates data from the entire Late Devonian with emphasis on the better-known parts of the rock record in New York and Pennsylvania. Paleogeography is considered in a transect from the source area across the subaerial part of the delta to the various parts of the marine environment. For each subdivision in that transect, the main features of the rocks are tabulated, the geographic elements are described, and the sedimentary processes are discussed.

PALEOGEOGRAPHY OF THE OLD RED CONTINENT

As Kent has pointed out elsewhere in this volume, the Old Red (Laurasian) Continent was assembled by the end of the Devonian. This assembly would last, with some modifications, until its breakup in the early Jurassic, a span of nearly 180,000,000 years. The Old Red Continent was made up of Baltica, Armorica and Laurentia (North America). The configuration and placement of the Continent are shown in Figure 1. External basins of clastic deposition existed from the Soviet Union west to Ireland, in Maritime Canada, in the eastern United States, and in Arctic Canada. Interior basins of clastic deposition developed in Scandinavia, Scotland, eastern Greenland, Spitsbergen, and, perhaps, the Barents Sea (Allen 1979; Dineley 1975, 1979). Thick marine sequences lapped onto the continent on the south, southwest, and west (House 1975).

The late Devonian geography of the Old Red Continent is shown in Figure 2. Highlands existed along the north and eastern coasts in a belt of varying width extending from the southwestern

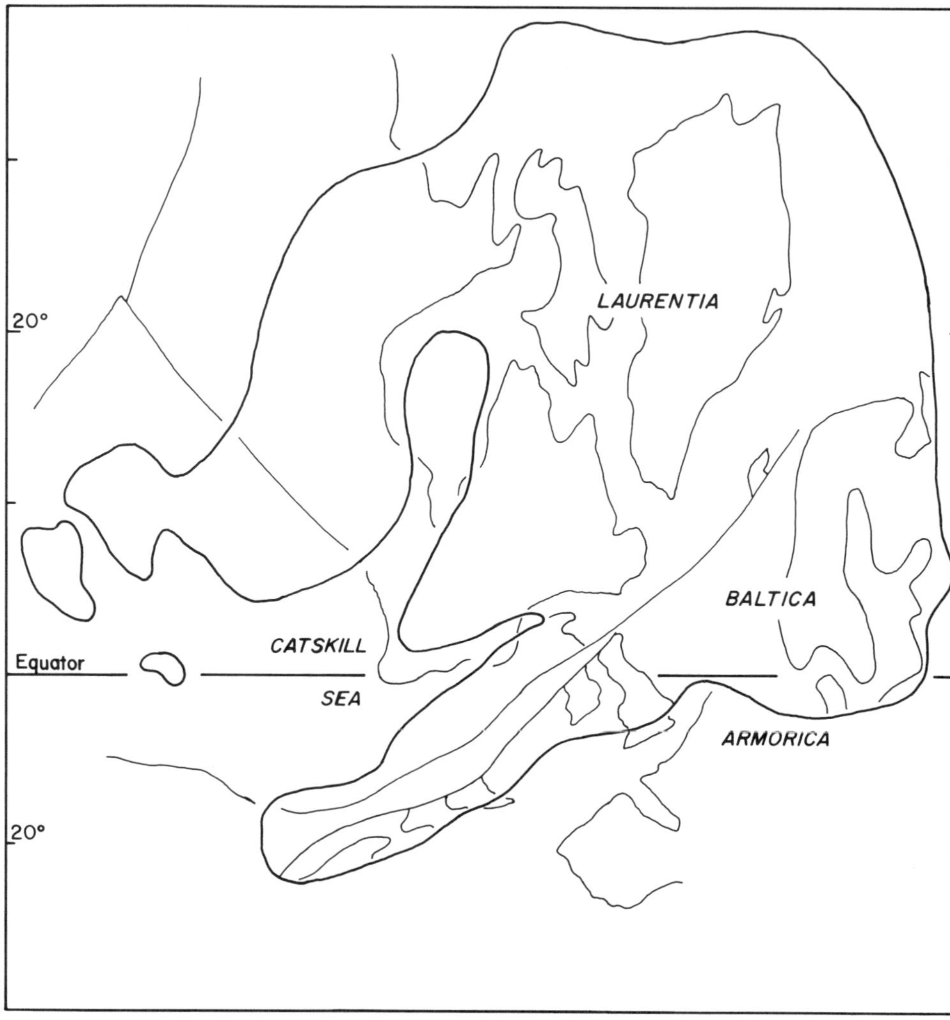

Figure 1. Configuration and placement of the Old Red Continent in the Late Devonian. Modern continental boundaries: light lines (Kent, this volume). Old Red Continent boundary: heavy lines.

tip of the Appalachian Peninsula northeast to a broad upland in Scandinavia. Highlands also existed in the north but whether the two were joined in northern Greenland is unknown. The rest of the continent comprised an interior surface of low relief and low altitude that extended across central and eastern Canada and western Greenland. Extensive alluvial plains built up on both sides of the highlands. The coastline is shown as a simple feature but there is good evidence to suggest that its configuration was complicated locally. For example, the Middle and Upper Devonian record in the North Sea, elsewhere in the British Isles and in northern Europe suggests that at least some of the interior basins opened to the sea.

Shallow, epicontinental seas encroached on the west and south. One of these, the Catskill Sea (Woodrow and Isley 1983), was separated from the world ocean by the Appalachian peninsula, a southwestern extension of the Caledonides formed by plate interaction in Maritime Canada (Kent this volume; Faill this volume). To the southwest, the Catskill Sea joined the world ocean and to the west but exchange with the world ocean was limited by an island chain. To the north, the rock record is incomplete so the actual extent of any epicontinental sea is speculative. However, it is clear that evaporites and carbonates of Devonian age underlying what is now Hudson's Bay were deposited in a highly saline bay. That bay probably was connected to the world ocean through the Catskill Sea to the south.

PALEOCLIMATE

Two methods of paleoclimatic interpretation are available. In one approach, the land/sea distribution based on paleomagnetic data is modified to account for tectonics and then a paleoclimatic interpretation is derived either by analogy with modern climatic patterns or by use of synthetic climatic models built from modern data. This is the approach used by Scotese and others (1979) to arrive at paleogeographic and paleoclimatic interpretations for the Phanerozoic world. A second approach, based on fossil and lithologic data, deduces both the paleogeography and paleoclimate from the record provided by presumed climatic in-

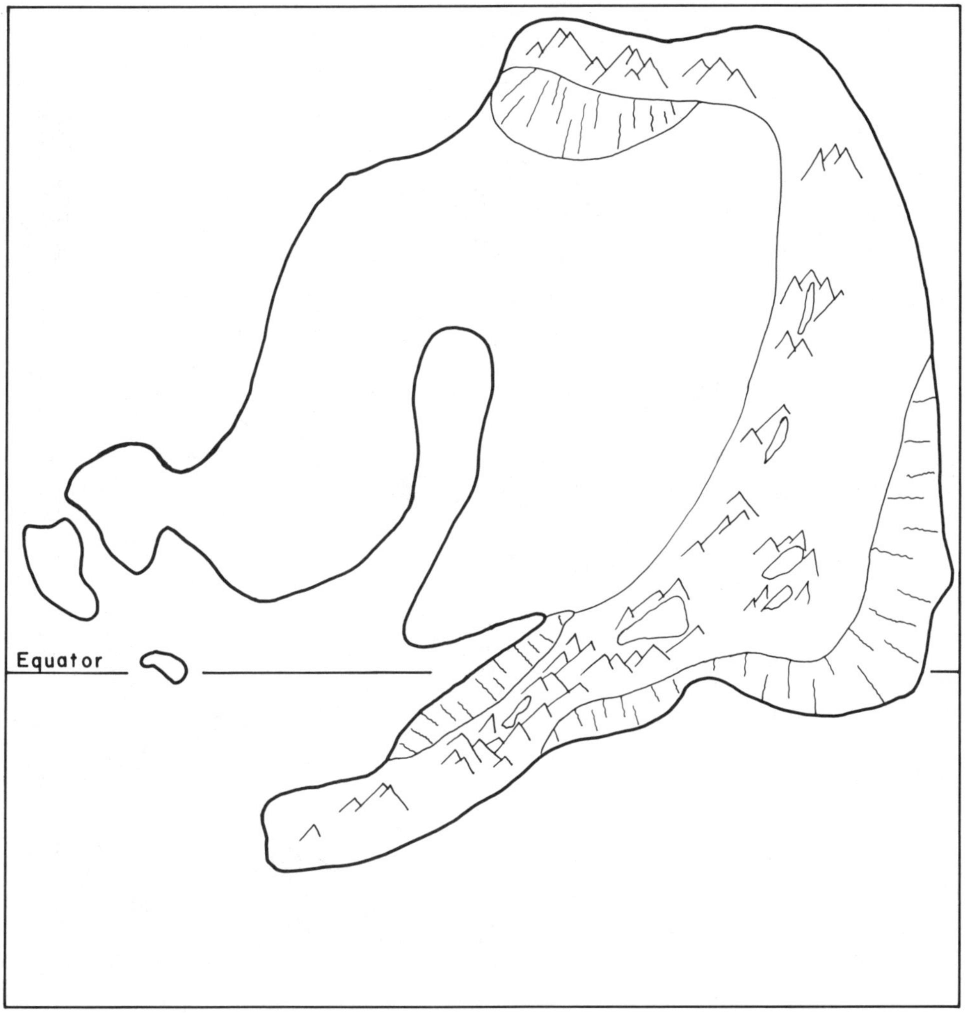

Figure 2. Physical geography of the Old Red Continent. Shoreline placement is speculative on the northeast coast, the south-eastern side of the Appalachian Peninsula and west of the highlands in eastern Canada and Greenland. Interior basins suggestive only. Volcanoes near interior basins. Lined areas - alluvial plains.

dicator fossils and rocks. This approach is essentially that used by Heckel and Witzke (1979) for the Devonian world. A combination of both approaches was used by Woodrow and others (1973) who worked with a paleomagnetically derived fit of North Atlantic continents and placed on it the highlands required by tectonics for the Late Devonian. Paleoclimate was deduced for this configuration by reasoning from the modern world's climatic patterns. That picture was then "calibrated" with the fossil and lithologic data. Important in this record are the presence, absence, and distribution of the paleoclimatic indicators given in Table 1.

The Late Devonian paleoclimate interpretation for the Appalachian region is shown in Figure 3. The major features of the climatic patterns of interest here are the climatic zones defined by the rainfall, temperature, and pressure distributions. Many of these effects are the result of the rainshadow developed in the lee of the Appalachian highlands (Ettensohn this volume; Ettensohn and Barron 1981).

TABLE 1. PALEOCLIMATIC INDICATORS OF THE TERRESTRIAL ROCK RECORD

Mudcracks	Faunal diversity
Rainprints	Dipnoan (lung-fish) trace fossils
Paleosols (calcretes)	
Oxidized sediments	Paleobotanic data

It is clear from this interpretation that the Catskill Delta, the local landmass of which it was a part, and the adjacent Catskill Sea were under the influence of the southeast trade winds in a tropical-dry or savannah-like climatic zone. Rainfall was seasonal and periods of extensive drying occurred each year.

PHYSICAL GEOGRAPHY AND SEDIMENTARY PROCESSES OF THE CATSKILL DELTA AND ITS ENVIRONS

Figure 4 displays an interpretation of the physical geography

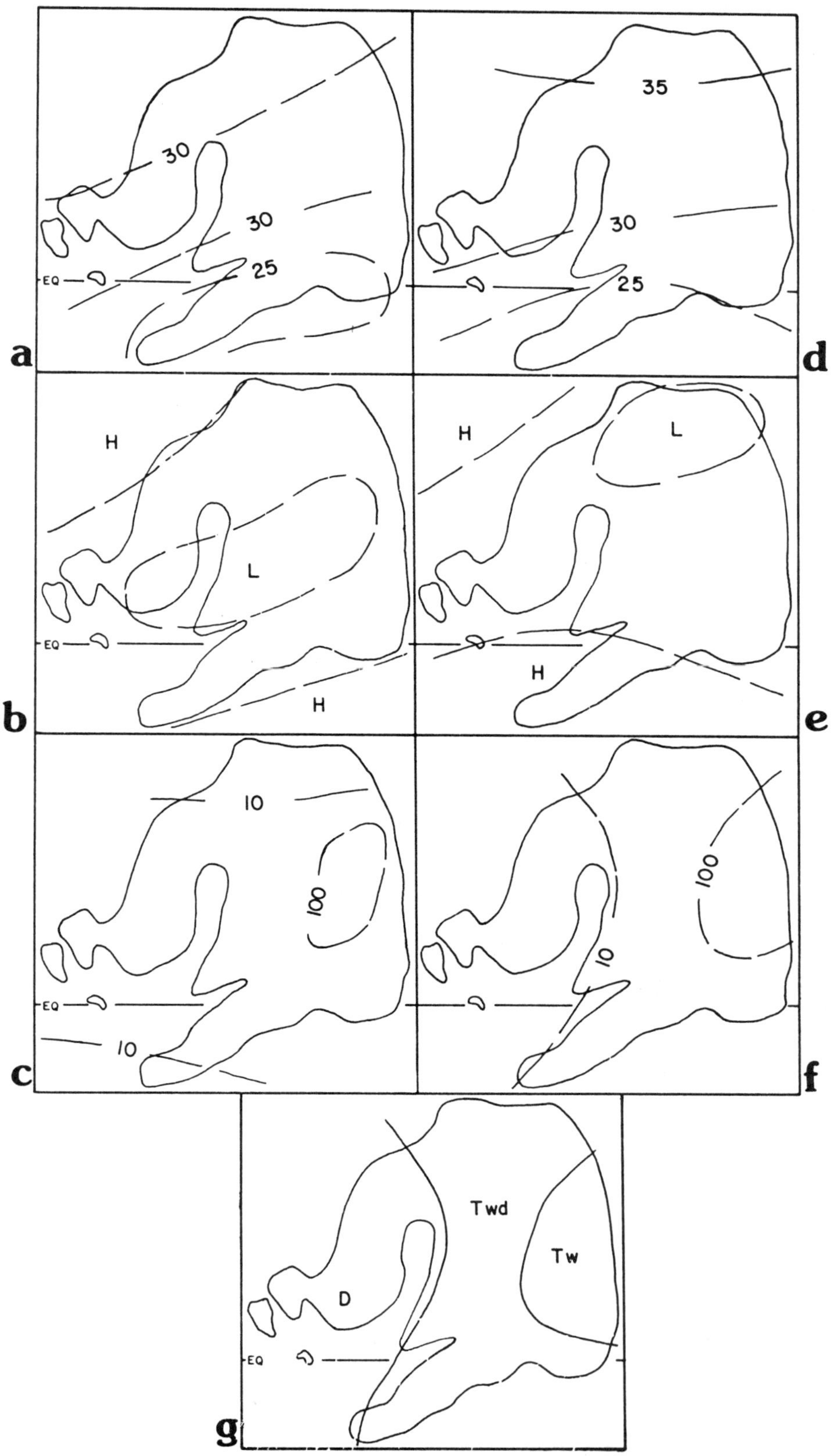

Figure 3. Paleoclimate on the Old Red Continent during the Late Devonian. (a) January temperatures (°C), (b) January barometric pressure, (c) January rainfall (mm), (d) July temperature, (e) July barometric pressure, (f) July rainfall, (g) climatic zones: Twd: tropical wet and dry, Tw: tropical wet, D: dry. Values are suggestive only. Adapted from Woodrow and others, 1973 with the assistance of W. F. Ahrnsbrak.

Figure 4. Panorama of the Catskill Delta and environs as seen from a vantage point above south-central Pennsylvania. Shoreline trends northeast. Length of shore: approximately 300 km. Basin margin and upper clinoform stippled.

of the Catskill alluvial plain, its adjacent uplands, and the configuration of the floor of the adjoining Catskill Sea. This is a view of the western side of the Appalachian Peninsula (Figure 1).

Highlands

The existence of the highland source area has long been established (e.g., Barrell 1913, 1914; Schuchert 1923). The evidence for the highland includes the Catskill Delta itself and the paleocurrent indicators contained in it (Meckel 1970; Potter and others 1980). Petrologic and paleomagnetic data are available to constrain interpretation with respect to the character of the rocks exposed in the highlands and with respect to the extent of the highlands themselves (Burtner 1963; Buttner 1968; Sevon 1979; Kent this volume). One possible interpretation of the highland's geography is summarized as follows:

1. The highlands were made up of uplifted, early Paleozoic sedimentary rocks, some metamorphosed, and their immediately subjacent crustal rocks. Sedimentary unroofing did not provide detritus from deeper crustal rocks. Uplift was apparently the expression of plate interaction.

2. The width of the uplift was over 100 km but less than 1000 km given the geometry of the interacting plates (Kent this volume). The highland width would be comparable to the width of the present-day Alps or Andes, but that comparison should not be taken to imply tectonic similarity.

3. The elevation of the mountains was temporally variable along the length of the uplift, with the greatest elevations in the regions of maximum deformation. Elevations at any specific place varied with time as a result of the varying intensity of plate interactions.

4. The aerial view of the highland may have been like that on the south side of the Himalayas or the north side of the Swiss Alps where major rivers flow parallel to fold axes and faults and break across structure in only a few places (Figure 4). Smaller, more ephemeral streams probably headed in the front ranges between the major points of debouchment.

5. Flow in the streams draining the highlands fluctuated greatly during the year. Some streams may not have been perennial because of the highland rainshadow effect. Similarly, the rainshadow effect severely limited plant cover on these slopes (See Banks and others, this volume). Sediments were introduced to the streams as sheet wash from slopes on which soils were thin and vulnerable to erosion.

6. The location of at least the major streams was fixed, as evidenced by the relative permanency of the major depocenters (Dennison this volume; Sevon 1979, 1985; Sevon and Thompson 1982; Rahmanian 1979).

7. The existence of fan-like deposits in the coarsest nonmarine facies of the delta suggests that the alluvial plain/highland boundary was a prominent, structurally controlled topographic feature.

Alluvial Fans

Thick masses of very coarse-grained sediments arrayed in fan-like symmetry abutted the highlands and extended to the northwest at various locations along the highland front. The coarsest sedimentary rocks now exposed at or near the erosional edge on the southeast form the basis for this interpretation. Their main features are given in Table 2.

That the alluvial fans had variable geometries and surface extent is indicated by the discontinuous nature of the rock bodies thought to have formed on them. Data from some of the resulting rock bodies can be used to reconstruct the areal dimensions of fans from which fan slope and the drainage area of the stream(s) building the fans can be roughly calculated (Table 3).

None of the rock bodies under consideration can be traced for more than a few tens of kilometers basinward from the southeastern edge of the exposure suggesting that the fans occupied a narrow zone of varying width parallel to the highland/plain boundary. That zone was probably never more than 20 km wide in the early Late Devonian, but it appears to have expanded with time. A modern analogue may exist on the south side of the

TABLE 2. MAIN FEATURES OF THE VERY COARSE-GRAINED ROCKS DEPOSITED WITH FAN-LIKE SYMMETRY AT THE SOUTHEASTERN EDGE OF THE CATSKILL DELTA

Unit thicknesses of several meters or more
Unit impersistence along strike over distances of 10-20 km
Granule - cobble conglomerates and very coarse sandstones
Durable clasts: quartz, quartzite, fine-grained sedimentaries, fine-grained metamorphics
Sorting: good; Roundness: moderate - good; Imbrication: rare
Individual strata discontinuous over distances of a few meters
Thick cross-sets, shallow channels
Shale and mudstone make up less that 20% of rock units

Examples:

Slide Mountain, Walton (NY)
Lackawaxen, Packerton (PA)
Hampshire (MD, WV, VA)

TABLE 3. THE GEOMETRY OF THREE ALLUVIAL FANS IN THE CATSKILL DELTA

	Twilight Park Conglomerate[1] (Oneonta Fm.,NY	Delaware River Formation[2] (northeast PA)	Lackawaxen Conglomerate[3] (northeast PA)
Fan radius	18 km	19 km	26 km
Fan area[4]	500 km^2	560 km^2	1061 km^2
Slope of Fan[5]	8 m/km	7 m/km	6 m/km
Drainage area	460 km^2	500 km^2	775 km^2

Notes: [1]Conglomerate/very coarse sandstone (F. W. Fletcher, personal communication).

[2]Coarse sandstone, siltstone. Gray-green unit with red bounding facies indicating a permanently high water table (Fletcher and Woodrow, 1967; Sevon and others, in prep.).

[3]Conglomerate, coarse sandstone (Fletcher and Woodrow, 1967; Sevon and others, in prep.).

[4]Assuming semi-circle.

[5]From: Bull, 1964, Fig. 54; mudstone/shale-rich.

TABLE 4. MAIN FEATURES OF ROCKS THOUGHT TO HAVE BEEN DEPOSITED ON THE CATSKILL ALLUVIAL PLAIN

Medium- to coarse-grained sandstones; lithic wackes
Discontinuous sand-bodies many meters thick
Upward-fining cycles: conglomerate-sandstone-mudstone; greenish-gray to red, bottom to up
Cross-strata on many scales, parting lineation
Body fossils rare
Trace fossils: large and small burrows (Beaconites?); root casts
Vertebrate fossils; plant imprints and casts
Bioturbation in mudstones

Examples:

Walton, Oneonta (NY); Catskill, Shohola, Sunfish, Huntley Mountain (PA); Hampshire (WV, VA, MD).

Himalayas where piedmont fans are developed (Collinson, 1978).

The coarse-grained units making up the fan-bodies provide a record of traction processes. Bar forms, sand waves, and small channels, all indicative of channelized flow, have been noted. Evidence for mass-flows, sieve-net deposition, or mud-flows is not established. Evidence for overbank deposition is uncommon. Apparently, braided streams carried clastics across this zone of fan development with the coarsest fraction trapped on the fans and the finer materials carried on to the alluvial plain.

Alluvial Plain

The alluvial plain is the most widely developed part of the subaerial Catskill Delta (Figure 5) and most of the red-colored units were emplaced on it. The main features of the rocks deposited on this surface are given in Table 4.

The rocks and their sequences indicate deposition by streams flowing in sinuous channels separated by broad interfluves (Woodrow and others 1973; Bridge and Gordon this volume), although braided rivers were active across wide areas of the alluvial plain in central and northeastern Pennsylvania (Epstein and others 1974). The sinuous channels migrated laterally by erosion and avulsion but most interfluves existed for as long as 10,000 to 20,000 years, sufficient time to permit development of carbonate soils (Woodrow and others 1973). A plant cover existed, at least near the streams (Banks and others this volume). Small lakes (oxbows) and ponds made by floods dotted this surface. None were long-lived. Dipnoan and other fish lived in some of them, surviving drought by aestivation. Their plates and bones were left as part of the geologic record.

The slope of the alluvial plain surface was quite shallow. It must have been less than 0.2 m/km near the shoreline (Woodrow and others 1973). The surface had little relief except for levees built up on the flanks of the streams. At times of extensive braided river development, the alluvial fans apparently broadened at the expense of the alluvial plains. Incursions of coarse sediment may have resulted from invigoration of streams during times of heightened tectonic activity. During those times the line between fan and plain was indistinct and fans may have developed that were comparable in scale to those on the southern flanks of the Himalayas, where the fan radii are measured in many tens of kilometers (Collinson 1978).

Traction processes were restricted to stream channels, while much of the interfluve silts and muds were deposited from suspension. Point bars and lateral accretion sets developed while sands and gravels in channels were moved as sand waves or ripples. The coarsest material carried by these streams was made up of shale and siltstone clasts and carbonate nodules, all derived from within the alluvial plain by collapse of interfluve materials on cutbanks during channel migration. Longitudinal bars and large-scale bed forms developed in the braided channels. The coarse- and medium-grained sands carried by the streams were derived in the main from outside the alluvial plain.

Seasonal rainfall meant that there were wide fluctuations in the water table. Over much of the year the surficial materials were exposed to an oxidizing environment. As a result, iron-rich silicates were partially or completely oxidized and carbonate nodules formed in association with developing soils.

Shoreline

Of all the geographic boundaries on the Delta surface, the shore might well be expected to be the most obvious in the rock record. However, it is often the most difficult to identify. In some

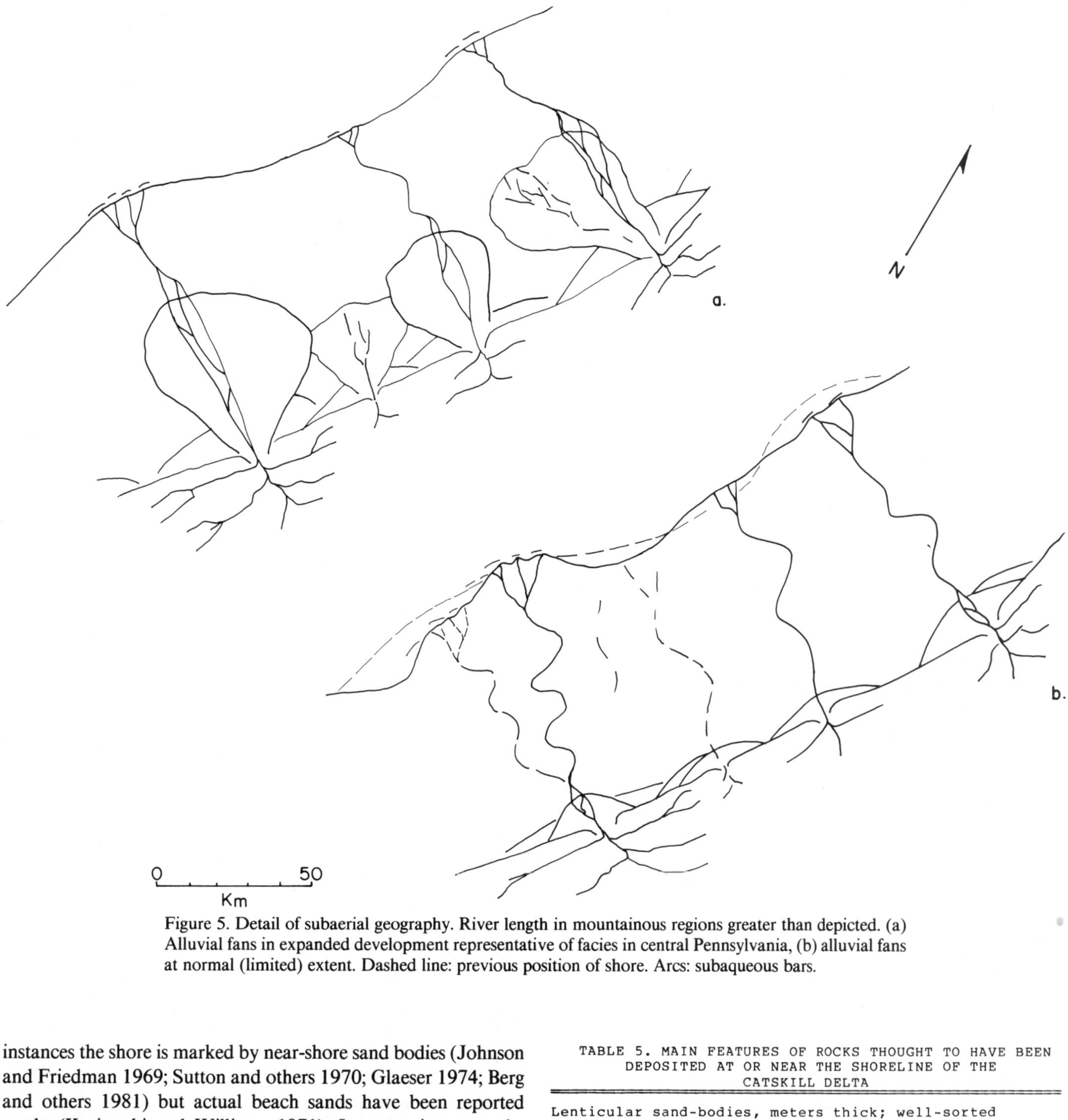

Figure 5. Detail of subaerial geography. River length in mountainous regions greater than depicted. (a) Alluvial fans in expanded development representative of facies in central Pennsylvania, (b) alluvial fans at normal (limited) extent. Dashed line: previous position of shore. Arcs: subaqueous bars.

instances the shore is marked by near-shore sand bodies (Johnson and Friedman 1969; Sutton and others 1970; Glaeser 1974; Berg and others 1981) but actual beach sands have been reported rarely (Krajewski and Williams 1971). In many instances the shore is marked only by subtle changes in rock character or color or by the presence/absence of body fossils in vertical sequence (Table 5). It is for this reason that Walker and Harms (1971) refer to the "muddy shoreline" of the Catskill Delta in central Pennsylvania.

The outline of the shore at any one time along its length must have been complicated because it resulted from the interaction between distributaries, waves, tides, and subsidence due to

TABLE 5. MAIN FEATURES OF ROCKS THOUGHT TO HAVE BEEN DEPOSITED AT OR NEAR THE SHORELINE OF THE CATSKILL DELTA

Lenticular sand-bodies, meters thick; well-sorted
Quartz arenites, lithic arenites
Shell-rich sands (coquinites); limited faunal diversity
Thin sequences of gray, fossiliferous sandstone interbedded with red mudstone containing root impressions and/or carbonate nodules
Low-angle cross-strata, herringbone cross-strata, flaser ripples, climbing ripples
Vertebrate debris; coalified plant debris
Skolithus-like burrows; *Diplocraterion*

Examples:

Thin sands or sandstone units within many of the basin-margin sequences.

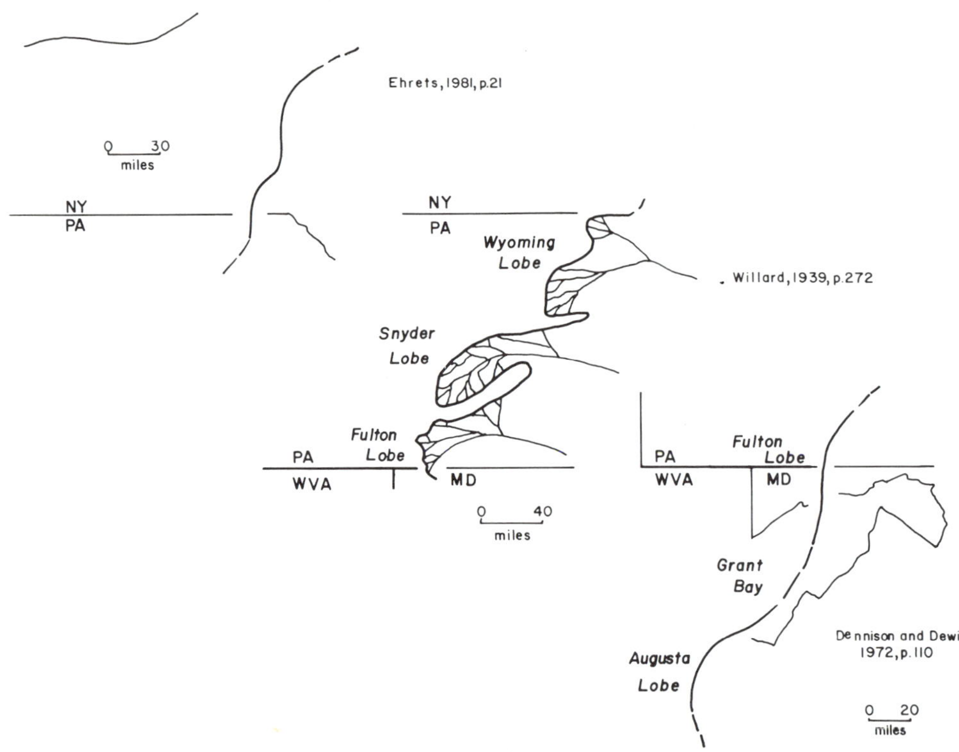

Figure 6. Catskill shorelines as envisaged by three writers.

compaction and tectonics. One need only look to the plan of the modern shoreline adjacent to the Mississippi Delta to get an idea of likely complexity. Yet, all portrayals of Catskill Delta shorelines are simple (Figure 6).

An example will serve to clarify the dilemma faced in trying to *map* a shoreline in the Catskill Delta. Willard (1934) and Willard and others (1939) produced a map, widely cited since publication, showing an "early Chemung" shore line based on the first appearance of the brachiopod *Spirifer disjunctus* (now referred to as *Cyrtospirifer chemungensis* and allied species). Willard's shoreline has not been confirmed in northeastern Pennsylvania, and in central Pennsylvania his approach yielded "delta lobes" (Figure 6), with shapes unlike any modern delta and trends that are very close to those of the local structure. The problem is that the first appearance of the brachiopod and nonmarine strata are time-transgressive across a great part of the Upper Devonian (Rickard 1975). As a result, tracing a shoreline based on such criteria for distances of tens or hundreds of kilometers is to follow facies through time. Accurate depictions of paleogeography do not lie in that direction. Perhaps the search for a shoreline in other than vertical sequences is not a reasonable problem to pursue because there may not be sufficient time-constrained data.

Waves, tides, and deltaic processes associated with distributaries shaped the shoreline although the published evidence for each of these processes is scattered. Waves as agents forming near-shore sand bars are well documented (Sutton and others 1970; Krajewski and Williams 1971; Glaeser 1974; Johnson and Friedman 1969; Berg and others 1981). The bars are relatively thin sand bodies, lens-shaped in outcrop, and of unknown lateral extent. Evidence for tidal processes is more widely available but the significance of tides as shapers of the shoreline has yet to be determined. Tidal flat sequences have been described by Woodrow and Fletcher (1967), Humphreys and Friedman (1975), Rahmanian (1979), and Berg and others (1981) in both New York and Pennsylvania. Tidal currents are thought to have been causative factors in the formation of shelly, near-shore bars produced during the Late Devonian in northcentral Pennsylvania (Woodrow and others 1981; Bridge and Droser this volume).

It is unlikely that tidal range along the Catskill Delta shoreline exceeded two meters (Woodrow and Isley 1983), so the dearth of thick bar sequences or of deep, sediment-filled channels is not unexpected. Given the location of the Catskill Sea under the influence of the southeast trades (offshore winds), waves striking the shoreline would normally have been small (two meters or less) (Woodrow and others 1973; Ettensohn and Barron 1981). Even under the highly unusual circumstances of a tropical cyclone or squall-line advancing to the northeast across the shallow Catskill Sea, wind-driven wave height would not be expected to exceed 4-5 meters (wind speed 40-70 kmph, fetch 1000 km, wind blowing for 2+ days; Bialek 1966; table 2).

The Floor of the Catskill Sea

The floor of the Catskill Sea will be discussed in two parts: the basin-margin and the clinoform/basin floor (Woodrow and

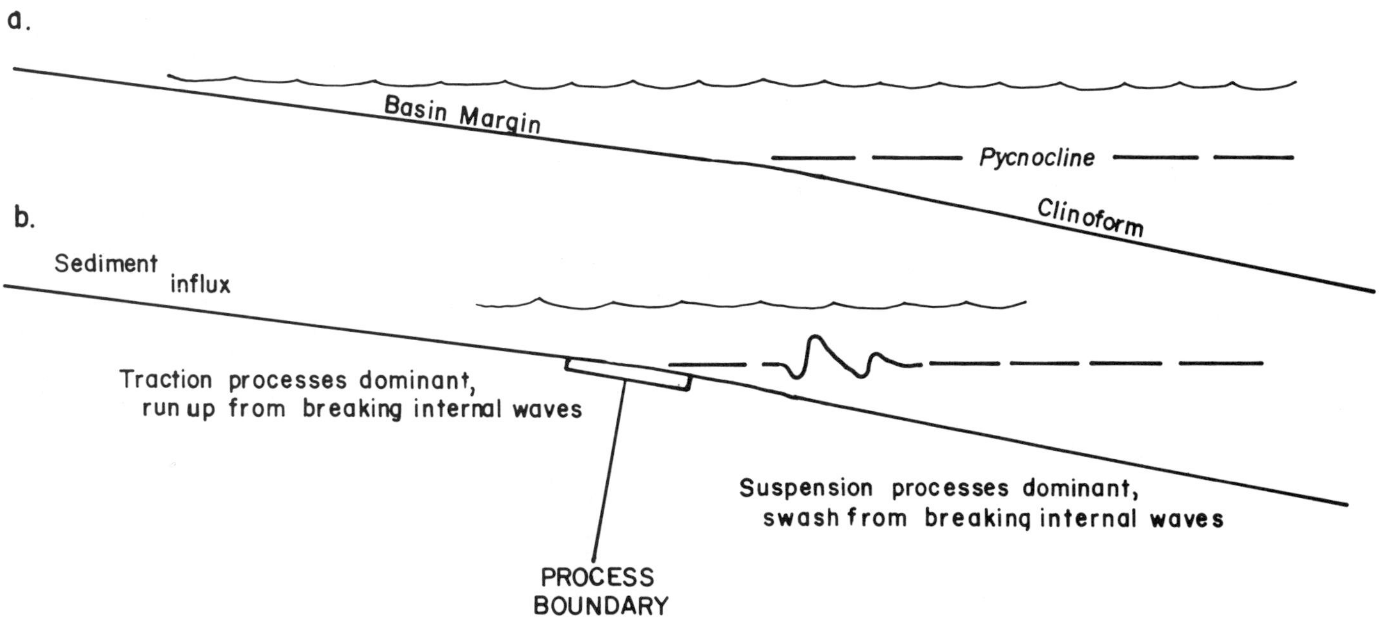

Figure 7. Process separation in the Catskill Sea. Wind- and tide-driven currents active above the pycnocline; turbidity currents active below. Internal waves operate as a sediment-modifying agent at the pycnocline/sea floor junction. (a) Configuration of sea floor and relation to pycnocline. (b) Pycnocline contact with sea floor as process boundary.

Isley 1983; Figure 7a). The *basin margin* is a surface of gentle gradient extending from the shoreline to a point where the gradient steepens. The *clinoform* extends from there to the *basin floor.* The development of these sea-floor features is not in fixed proportion either from place to place or through time (Figure 8). For example, the basin margin is well developed in New York, little developed over much of the central Pennsylvania, and moderately developed south of there. With the filling of the basin through the Late Devonian, it is likely that the basin margin and clinoform expanded at the expense of the deep basin (Sutton and McGhee this volume; Potter and others 1980).

Basin Margin. The configuration of this surface is the most complicated on the sea floor. It is the product of the deposition of sediment from distributaries, reworking by both physical and biological processes, and subsidence (Sutton and others 1970; McGhee and Sutton 1981). Marine transgression and regression forced the displacement of process regimes over considerable distances, introducing additional complexity. The basin margin configuration at any one time and place was the product of the most recent deposition event modified by compaction, reworking or sediment scour. Many environments of deposition have been named for this surface: delta platform, delta-front, prodelta, and interdeltaic (Sutton and others 1970; Glaeser 1974; Sevon 1981; Rahmanian 1979) and these are juxtaposed horizontally as a result of shoreline development and vertically as a result of marine transgression/regression. The result is a complex mosaic of facies. Table 6 summarizes the main features of these rocks.

Processes active on this surface were wave-driven currents, tidal effects, and bioturbation. Distributaries delivered mud, fine sand, and, occasionally, fine gravel to the basin margin. Much of the fine sand was arrayed in bar forms near distributaries and is characterized by upward-coarsening sequences. In central Pennsylvania, the sand was moved immediately down the clinoform as turbidity currents and deposited on the basin floor. In many instances, these sands were deposited on water-logged sediments and subsequently foundered to form ball-and-pillow structures, Mud, either derived directly from distributaries or reworked and entrained by wave-driven currents, was carried to the deeper parts of the basin or was moved by long-shore drift onto the shore between distributaries.

TABLE 6. MAIN FEATURES OF THE ROCKS THOUGHT TO HAVE BEEN DEPOSITED ON THE BASIN MARGIN OF THE CATSKILL SEA

Laminated, fine-grained sandstones and siltstones
Lithic arenites, lithic wackes
Medium- and dark-gray, silty shales and mudstones
Thin, discontinuous, quartz-rich, granule conglomerates
Thin coquinites and coquinite laminae in sandstones small-scale cross-strata, hummocky cross-strata, wave ripples, interference ripples, scour marks
Diverse and abundant invertebrate fauna; plant debris; trace fossils
Bioturbation
Examples:
Otselic, Glen Aubrey, Wellsville, Venango (NY); Towanda, Lock Haven, Trimmers Rock (in part) (PA); Fore Knobs, Scherr, "Chemung" (WV, VA, MD).

Wave-driven currents operated on the basin margin where they were capable of agitating and reworking bottom sediments. Their effects are seen in wave ripples, hummocky cross-strata, and in formation of thin- to medium-bedded, fine-grained, parallel-laminated sandstones with sharp bottom contacts as well

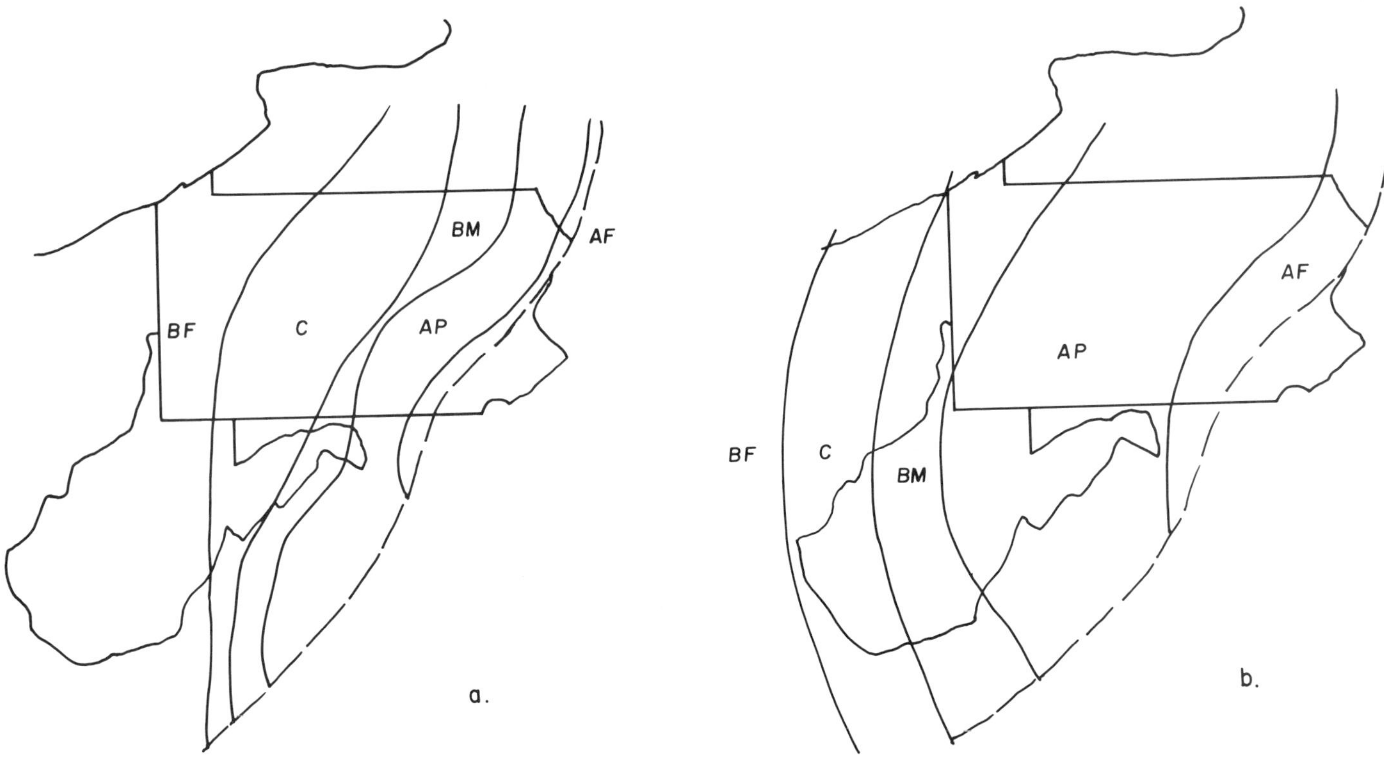

Figure 8. Evolution of geographic elements of the Catskill Delta. Af: alluvial fans, Ap: alluvial plains, Bm: basin margin, C: clinoform, Bf: basin floor. Dashed line is outcrop edge. (a) Early Late Devonian, (b) Latest Devonian.

as in the development of lenticular coquinites or coquinite laminae in sandstones. These observations are compatible with two interpretations: either wave effects destroyed most of the evidence of fair-weather strata deposition or fairweather effects on depositional processes were limited.

Tidal effects on sediments away from the shore must have been limited if the assumption of a tidal range of less than two meters is correct.

Organic structures are common in many of the rocks formed on this part of the ocean floor. Most mudstones and many siltstones lack stratification and show extreme bioturbation while burrows, tracks, and trails are common features of most sandstones. Given the range of trace fossils, the animals responsible must have been a diverse group (Miller 1979; Bridge and Droser this volume).

Clinoform and Deep Basin

When taken together these features made up most of the area of the Catskill Sea Floor. The clinoform is the surface basinward from the basin margin and the basin floor is the deepest part of the Catskill Sea. The surfaces were nearly featureless. No canyons or significant channels, levees, slump scars, or prominent fans have been documented from the clinoform or its base.

Sediments deposited on these surfaces are finer grained, thinner bedded, and more persistent as individual strata than those seen in the basin margin sequence (Table 7). This reflects a limited range of sedimentary processes active on these surfaces, probably as a result of the development of a pycnocline (density stratification) in the Catskill Sea (Ettensohn and Barron 1981,

TABLE 7. MAIN FEATURES OF THE ROCKS THOUGHT TO HAVE BEEN DEPOSITED ON THE CLINOFORM AND BASIN FLOOR OF THE CATSKILL SEA

Clinoform:

- Laminated, fine-grained sandstones (turbidites)
- Dark gray and black shales
- Green "slope" shales
- Flute casts, groove casts and other sole marks; Bouma cycle b - e subdivisions common; convolute bedding
- Shell debris in groove casts; plant debris
- Bioturbation rare

Examples:

Sherburne, Rock Stream, Penn Yan, Wiscoy, Forty Bridge (NY) Brallier, Trimmers Rock (in part) (PA, WV, VA, MD)

Basin Floor:

- Black shale; gray silty shale; siltstone
- Thin, persistent limestones
- Invertebrates rare; bioturbation very rare
- Plant debris
- Conodonts, cephalopods, pteropods, brachiopods

Examples:

Geneseo, Middlesex, Genundewa, Rhinestreet, Dunkirk (NY) Harrell, Rhinestreet, Dunkirk (PA, WV, VA, MD)

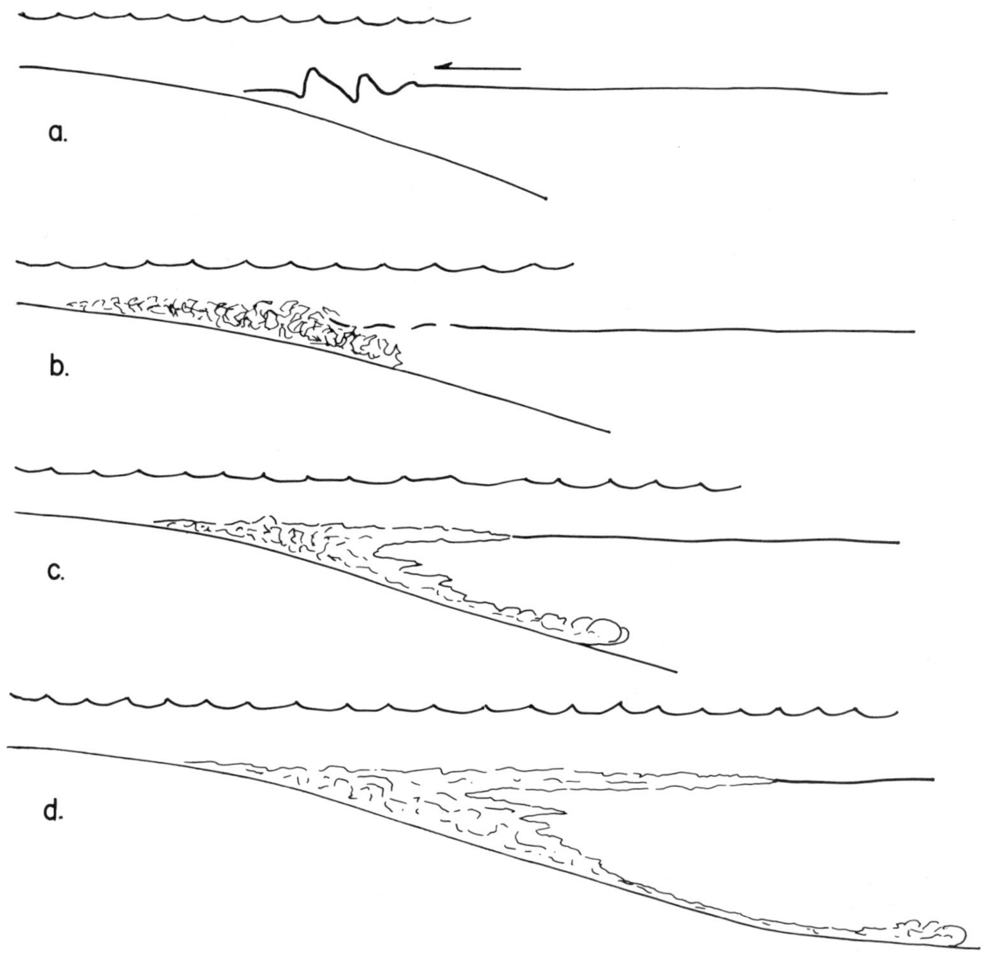

Figure 9. A mechanism likely to have generated density currents in the Catskill Sea. (a) internal wave near sea floor, (b) internal wave shoals and breaks; sediment agitated and entrained, (c) Density current loosed; finest fraction trapped on pycnocline, (d) sand sheet at basin margin; turbidites on clinoform and basin floor. Finest fraction setting.

Woodrow and Isley 1983) and the separation of process formed as a result (Fig. 7b; Woodrow and Isley 1983; Woodrow 1983). The pycnocline restricted the vertical transfer of mechanical energy and, as a result, only turbidity currents carried sandy sediments to depth. These currents apparently flowed unrestrained by channels (Woodrow and Isley 1983; Lundegard and others this volume) and appear to have been derived from a linear source either on the basin margin or at the basin margin/clinoform junction. Some may have been triggered by storm waves entraining sediment from the shore and basin margin and others may have been triggered by internal wave/sediment interactions. Whatever the cause, the effect was to loose a sediment mass down the clinoform to the basin floor (Figure 9). Finer sediment trapped at the pycnocline rained down long after turbidity current's demise. Evidence for reworking of the turbidites by other bottom currents (contour currents, for example) has not been documented. Thus, sediment was delivered to the clinoform mainly by traction processes and to the deep basin mainly by suspension processes. Deep basin sediments are, in part, hemipelagic. Since the benthos was highly restricted in those waters, bioturbation was severely limited.

SUMMARY

The Catskill Delta formed on the northwest side of a tectonic peninsula, part of the Old Red Continent, which was raised by plate interaction centered in Maritime Canada and New England. Nearly all of the Continent was equatorial. South of the Continent and separated from the world ocean by the Appalachian Peninsula was the Catskill Sea. To the west the Catskill Sea was shallow or broken by islands. Only to the southwest was there a clear connection to the world ocean.

Climate over the delta was strongly effected by the rainshadow caused by the mountains on the Peninsula. The mountains effectively blocked the southeasterly trade winds causing a tropical wet or dry climate or a desert climate. An extended discharge period of drought with resultant highly variable stream discharge was an annual feature.

The rock record indicates that the subaerial Catskill Delta consisted of alluvial fans, alluvial plains, and shoreline. Streams originated in the tectonic highlands adjacent to the Delta and poured out on to alluvial fans where braided streams deposited gravel and coarse sand on the fans. These streams were apparently fixed in location but variable in capacity. Alluvial plains with channelized, sinuous streams separated by well-established interfluves adjoined the fans and extended from them to a region of distributaries. The alluvial plains were the major features of the subaerial delta. Surface slopes were 6–8 m/km on the fans and less than 1–3 m/km on the alluvial plain.

On these surfaces, sand and gravel deposition occurred through traction processes and mainly in stream channels while deposition from suspension occurred on the interfluves. Carbonate soils formed on the alluvial plain sediments.

The marine rock record indicates deposition on the basin margin, the clinoform, and the basin floor. Fine sand and mud were carried to the basin margin by distributaries, then reworked by waves, tides and bioturbation. Transgression and regression displaced marine environments as did the effects of subsidence. The resulting facies patterns formed a complex facies mosaic.

On the basin margin the sands were emplaced by traction processes. Most muds were deposited from suspension and were subsequently reworked either by physical processes or the benthos. Wave-related effects are particularly common; hummocky cross-strata are found in many sections.

The clinoform, a surface of slight gradient but steeper than the basin margin, extended seaward to the basin floor. Both the clinoform and the basin floor appear to have been essentially featureless. Evidence for canyons, deep channels, levees, slumped masses of sediments, and subaqueous fans is absent.

The clinoform/basin margin boundary apparently coincided often with a surface of stratification developed in the Catskill Sea. That surface formed an effective process boundary. Above it sands were moved by the effects of wave-related traction processes while below it sands were moved mainly by turbidity currents. These currents flowed unrestrained by channels and appear to have been derived from a linear source. Some may have been triggered by internal waves striking the sea floor. The distal ends of turbidity currents moved into the deep basin and carried very fine sediments. The deep basin received not only terrigenous sediment but also quantities of basin-derived organics combined in hemipelagic muds.

REFERENCES CITED

Allen, J.R.L., 1979, Old Red Sandstone facies in external basins, with particular attention to southern Britain, *in* House, M. R., Scrutton, C. T., and Bassett, M. G ., eds., The Devonian System: The Paleontological Association, Special Papers in Paleontology, no. 23, p. 65–80.

Banks, H. P., Grierson, J. D., and Bonamo, P. M., 1985, The Flora of Catskill Clastic Wedge: Where we are with the evidence, *in* Woodrow, D. L. and Sevon, W. D., eds., The Catskill Delta: Geological Society of America, Special Paper 201.

Barrell, J., 1913, The Upper Devonian delta of the Appalachian Geosyncline, Part I: The Delta and its Relation to the Interior Seas: Geological Society of America Bulletin, v. 24, p. 429–472.

—— 1914, Part III: The relations of the delta to Appalachia: Geological Society of America Bulletin, v. 25, p. 225–253.

Berg, T. M., Crowl, G. H., Edmunds, W. E., Luce, P. B., Sevon, W. D., Wilshusen, J. P., and Woodrow, D. L., and Pohn, H., 1981, Geology of Tioga and Bradford Counties, Pennsylvania: Guidebook for the 46th Annual Field Conference of Pennsylvania Geologists, Pennsylvania Topographic and Geologic Survey, Harrisburg, Pennsylvania, 162 p.

Bialek, E. L., 1966, Handbook of Oceanographic Tables: U.S. Naval Oceanographic Office, Special Publication SP-68, 427 p.

Bridge, J. S. and Droser, M., 1985, Depositional environments of an unusual limestone within the chemung magnafacies: *in* Woodrow, D. L. and Sevon, W. D., eds., The Catskill Delta: Geological Society of America Special Paper 201.

Bridge, J. S. and Gordon, E. S., 1985, Quantitative reconstruction of ancient river systems in the Oneonta Formation, Catskill magnafacies: *in* Woodrow, D. L. and Sevon, W. D., eds., The Catskill Delta: Geological Society of America Special Paper 201.

Bull, W. B., 1964, Geomorphology of segmented alluvial fans in western Fresno County, California: United States Geological Survey Professional Paper 352E, 129 p.

Burtner, R. L., 1963, Paleocurrent and petrographic analysis of the Catskill facies of southeastern New York and northeastern Pennsylvania: Ph.D. Thesis, Harvard, University.

Buttner, P.J.R., 1968, Proximal continental rhythmic sequences in the Genessee Group (lower Upper Devonian) of southeastern New York: in, Klein, G. D., ed., Late Paleozoic and Mesozoic Sedimentation, northeastern North America, Geological Society of America Special Paper 106, p 109–126.

Collinson, J. D., 1978, Alluvial sediments: *in* Reading, H. E., ed., Sedimentary Environments and Facies; New York, Elsevier, p. 15–79.

Dennison, J. M., DeWitt, W., Jr., Hasson, K. O., Hoskins, D. M., and Head, J. W., 1972, Stratigraphy, sedimentology and structure of Silurian and Devonian rocks along the Allegheny Front in Bedford Co., PA., Allegheny Co., MD., and Grant Co., W. Va.: Guidebook for the 37th Annual Field Conference of Pennsylvania Geologists; Harrisburg, Pennsylvania, Pennsylvania Topographic and Geologic Survey, 114 p.

Dineley, D. L., 1975, North Atlantic Old Red Sandstone—some implications for Devonian paleogeography: *in* Yorath, C. J., Parker, E. R., and Glass, D. J., eds., Canada's Continental Margins and Offshore Petroleum Exploration: Canadian Society of Petroleum Geologists, Memoir No. 4, p. 773–790.

—— 1979, Tectonic setting of Devonian sedimentation: *in* House, M. R., Scrutton, C. T., and Bassett, M. G., eds., The Devonian system: The Paleontological Society, Special Papers in Paleontology No. 23, p. 49–63.

Ehrets, J. R., 1981, The West Falls Group (Upper Devonian) Catskill Delta Complex: stratigraphy, environments and sedimentation: *in* Enos, P., ed., Guidebook for field trips in southcentral New York, 53rd Annual Meeting, New York State Geological Association, p. 3–30.

Epstein, J., Sevon, W. B., and Glaeser, J. D., 1974, Geology and mineral resources of the Lehighton and Palmerton Quadrangles, Carbon and Northampton Counties, Pennsylvania: Pennsylvania Geological Survey, 4th ser., Atlas 195CD, 460 p.

Ettensohn, F. R., 1985, Framework for development of Catskill delta complex basin-facies, *in* Woodrow, D. L. and Sevon, W. D., eds., The Catskill Delta, Geological Society of America, Special Paper 201.

Ettensohn, F. R., and Barron, L. S., 1981, Depositional model for the Devonian-Mississippian black-shale sequence of North America: A Tectonic-climatic approach: Department of Energy, Morgantown (W. Va.) Energy Technology Center, Document No. 12-42-2, 80 p.

Faill, R. T., 1985, The Acadian Orogeny and the Catskill Delta: *in*, Woodrow, D. L., and Sevon, W. D., The Catskill Delta, Geological Society of America

Special Paper 201.

Glaeser, J. D., 1974, Upper Devonian stratigraphy and sedimentary environments in northeastern Pennsylvania: Pennsylvania Topographic and Geologic Survey, 4th ser., General Geology Report 63, 89 p.

Heckel, P. E., and Witzke, B. J., 1979, Devonian world paleogeography from distribution of paleoclimatic indicators: *in* House, M. R., Scrutton, C. T., and Bassett, M. G., eds., The Devonian system: The Paleontological Society, Special Papers in Paleontology No. 23, p. 99–123.

House, M. R., 1975, Facies and time in the Devonian tropical areas: Proceeding of the Yorkshire Geological Society; v. 40, p. 233–288.

Humphreys, M., and Friedman, G. M., 1975, Late Devonian Catskill Delta Complex in northcentral Pennsylvania: *in* Broussard, M. L., ed., Deltas, Models for Exploration; Houston Geological Society, p. 369–379.

Johnson, K. G., and Friedman, G. M., 1969, Tully clastic correlatives (Upper Devonian) of New York State: a model for recognition of alluvial, dunes (?), nearshore (bar and lagoon) and offshore sedimentary environments in a tectonic delta complex: Journal of Sedimentary Petrology, v. 39, p. 451–465.

Kent, D. V., 1985, Paleocontinental setting for the Catskill Delta: *in* Woodrow, D. L. and Sevon, W. B., eds., The Catskill Delta, Geological Society of America, Special Paper 201.

Krajewski, S. A., and Williams, E. G., 1971, Upper Devonian sedimentation in Susquehanna Co. (Pa.): Guidebook, 36th Annual Field Conference of Pennsylvania Geologists: Pennsylvania Geological Survey, Harrisburg, p. 1–76.

McGhee, G. R. and Sutton, R. G. 1981, Late Devonian marine ecology and zoogeography of the central Appalachians and New York: Lethaia, v. 14, p. 27–43.

Meckel, L. D., 1970, Paleozoic alluvial deposition in the central Appalachians: a summary: *in* Fisher, G. W ., Pettijohn, F. J., and Reed, J. C., eds., Studies of Appalachian Geology—Central to Southern: New York, Wiley Interscience, p. 49–67.

Miller, M. F., 1979, Paleoenvironmental Distribution of Trace Fossils in the Catskill Delta Complex, New York State: Paleogeography, Paleoclimatology, Paleoecology, v. 28, p. 117–141.

Lundegard, P. D., Samuels, N. D., and Pryor, W. D., 1985, Upper Devonian turbidite sequence, Central and Southern Appalachians - contrasts with submarine fans: *in* Woodrow, D. L. and Sevon, W. B., eds., The Catskill Delta, Geological Society of America, Special Paper 201.

Potter, P. E., Maynard, J. P., and Pryor, W. A., 1980, Final report of special geological, geochemical, and petrological studies of the Devonian shales in the Appalachian Basin: Department of Energy, Morgantown, W. Va. Energy Technology Center, 86 p.

Rahmanian, V. D., 1979, Stratigraphy and sedimentology of the Upper Devonian Catskill and Uppermost Trimmers Rock Formations in central Pennsylvania [Ph.D. thesis]: Pennsylvania State University, 396 p.

Rickard, L. V., 1975, Correlation of the Upper Devonian and Silurian rocks in New York State: New York State Museum and Science Service, Map and Chart Series, no. 24.

Schuchert, C., 1923, Sites and nature of the North American Geosyncline: Geological Society of America Bulletin, v. 34, p. 151–229.

Scotese, C., Bambach, R. K., Barton, C., Van der Voo, R., and R. Ziegler, 1979, Paleozoic base maps: Journal of Geology, v. 87, p. 217–278.

Sevon, W. D., 1985, Non-Marine Facies of the Middle and Upper Devonian Catskill Clastic Wedge: *in* Woodrow, D. L., and Sevon, W. D., eds., The Catskill Delta: Geological Society of America Special Paper 201.

——1981, The Middle and Upper Devonian clastic wedge in northeastern Pennsylvania: *in* Enos, ., ed., Guidebook for field trips in south-central New York; New York State Geological Association, p. 31–64.

——1979, Devonian sediment-dispersal systems in Pennsylvania (abs.), Geological Society of America Abstracts with Programs, v. 11, no. 1, p. 53.

Sutton, R. G., Bowen, Z. P., and McAlester, A. L., 1970, Marine shelf environments of the Upper Devonian Sonyea Group of New York: Geological Society of America Bulletin, v. 81, p. 2975–2992.

Sutton, R. G. and McGhee, G. R., Jr., 1985, The Evolution of Frasnian Marine "community-types" of south-central New York: *in* Woodrow, D. L. and Sevon, W. B., eds., The Catskill Delta, Geological Society of America, Special Paper 201.

Thompson, A. M., and Sevon, W. D. 1982, Comparative sedimentology of Paleozoic clastic wedges in the central Appalachians, USA, Guidebook for Excursion 19B, International Association of Sedimentologists, 11th International Sedimentological Congress, Hamilton, Ontario, 136 p.

Walker, R. G., and Harms, J. G., 1971, The Catskill Delta:—prograding muddy shoreline in central Pennsylvania: Journal of Geology, v. 79, p. 381–399.

Willard, B., 1934, Early Chemung shoreline in Pennsylvania: Geological Society of America Bulletin, v. 45, p. 897–908.

——Swartz, F. M., and Cleaves, A. B., 1939, The Devonian of Pennsylvania: Pennsylvania Geologic Survey, 4th ser., Bulletin G-19, 481 p.

Woodrow, D. L., 1983, Topography and sedimentary processes in an epicontinental Sea: *in* Stanley, D. J., Moore, G., eds., The Shelf Break: Critical Interface on Continental Margins: Society of Economic Paleontologist and Mineralogists Special Paper No. 33, p. 159–166.

Woodrow, D. L., Enos, P., Bottjer, R. J., and Minero, C., 1981. Bioclastic carbonates in the Catskill clastic wedge: *in* Enos, P., ed., Guidebook for field trips in south-central New York, New York State Geological Association, p. 201–229.

Woodrow, D. L., and Fletcher, F. W., 1967, Late Devonian paleogeography in southeastern New York and northeastern Pennsylvania: *in* Oswald, D. M., ed., International Symposium on the Devonian System: Alberta Society of Petroleum Geologists, v. 2, p. 1327–1334.

Woodrow, D. L., Fletcher, F. W., and Ahrnsbrak, W. F., 1973, Paleogeography and paleoclimate at the deposition sites of the Devonian Catskill and the Old Red Facies: Geological Society of America Bulletin, v. 84, p. 3051–3064.

Woodrow, D. L., and Isley, A. M., 1983, Facies, Topography and Sedimentary processes in the Catskill Sea (Devonian), New York and Pennsylvania: Geological Society of America Bulletin, v. 94, p. 459–470.

Manuscript Accepted by the Society July 20, 1984

Printed in U.S.A.

Geological Society of America
Special Paper 201
1985

Controls on development of Catskill Delta complex basin-facies

Frank R. Ettensohn
Department of Geology
University of Kentucky
Lexington, Kentucky 40506

ABSTRACT

Dark, commonly black, fissile shales represent the basinal facies of the Catskill Delta complex. This dark-shale facies originated in the Appalachian Basin during the Emsian (late Early Devonian) and subsequently spread beyond the Appalachian Basin thus suggesting the importance of regional controls such as paleogeography, paleoclimate, and tectonic regime.

The Catskill delta debouched into a subsiding peripheral basin on the eastern margin of a nearly enclosed, equatorial sea. Enclosure prevented deep-water incursions from other water bodies and enhanced the production and concentration of organic matter. Existence in a warm, rainy equatorial belt increased organic productivity and probably formed a nearly permanent thermohaline water stratification that prevented vertical circulation and resulted in anaerobic bottom conditions.

Episodes of uplift and deformational loading in the rising Acadian Mountains resulted in closely following or concomitant periods of subsidence through isostatic adjustment in the adjacent peripheral basin. Hence, episodic Acadian uplift not only formed the source of the Catskill clastic wedge, but was also responsible for forming the peripheral basin that received the wedge. Basin subsidence isolated deltaic sediments from other parts of the sea, enhanced water stratification, and augmented net transgression. Rapid aggradation of sediments in nearshore areas due to transgression and the formation of a rainshadow due to uplift in the mountains caused decreased sedimentation on the delta, while concomitant basin subsidence produced abrupt deepening, strong water stratification (bottom anoxia), and migration of clastic-deficient (dark-shale) basinal environments shoreward. During intervening periods of tectonic quiescence, subsidence eventually halted, and erosion outstripped the effects of uplift in the mountains, so that the rainshadow was destroyed and abundant clastic influx resumed. As a result, the delta prograded rapidly across former basinal environments. Decreasing subsidence, shallower water, and abundant clastic influx destroyed basinal, bottom anoxia, so that marginally- to highly-oxygenated bottom conditions resumed throughout all delta environments. Five such major cycles of dark, basinal shales alternating with coarser clastics occur in the Catskill Delta complex of the Appalachian Basin. The progressive westward migration of successive cycles reflects the westward migration of Acadian deformation and accompanying basin infilling.

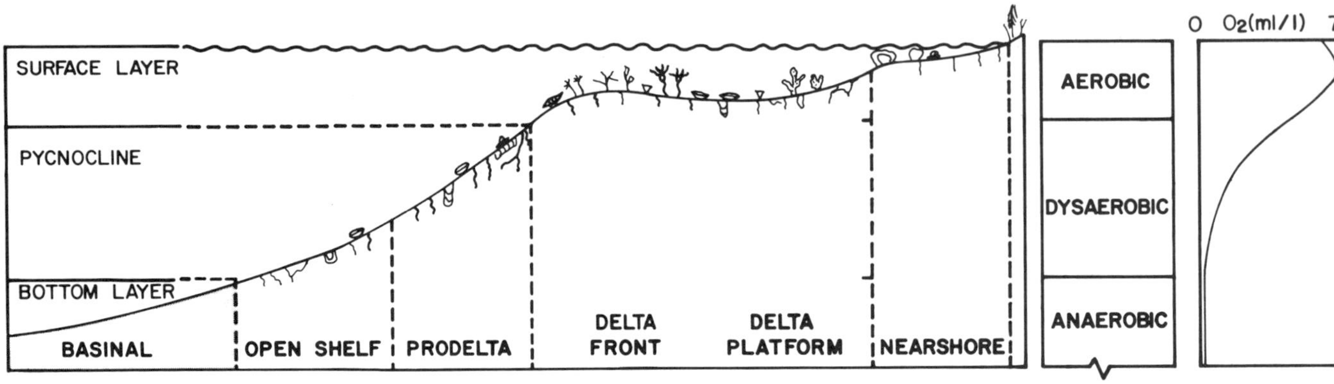

Figure 1. Schematic diagram showing the nature of a stratified water column and the general correspondence between layers in the water column and various deltaic subenvironments. Oxygen content is an estimate based on the Black Sea (adapted from Bowen and others 1974; Rhoads and Morse 1971; Byers 1977).

INTRODUCTION

Although deltas represent one of the best understood sedimentary settings, they are extremely complex because of the potential interplay of fluvial, wave, and tidal processes, as well as a number of less easily defined regional controls. Deltaic models predict a lateral distribution and vertical succession of deltaic subenvironments and facies that vary to some extent depending upon local and regional controls (e.g., Klein 1975: Fig. 78). The Catskill Delta complex represents a series of prograding clastic wedges that were among the earliest to be deciphered in terms of the lateral and vertical succession of deltaic subenvironments. Basinal black shales were traced eastward into slope-edge facies ("Chagrin" or "Portage" facies), which, in turn, grade eastward into coarser platform facies ("Hamilton" or "Big Bend" facies). Platform facies were then traced eastward into even coarser marginal marine ("Chemung" or "Smethport" facies) and terrestrial, shelf-edge ("Catskill" facies) facies (Caster 1934; Rickard 1964). This lateral sequence of transitions represents major time-transgressive facies changes in the Catskill Delta during progradation from the Middle Devonian to the Early Mississippian, and has been well documented in many earlier studies (Barrell 1913 1914; Broughton and others 1962; Caster 1934; Rickard 1975). The generalized facies scheme of Bowen and others (1974) and Thayer (1974) for the Catskill Delta will be used in this study (Figure 1).

In contrast, the influence of the larger, regional controls on the depositional history of the Catskill Delta has not been so well documented. I content that these controls can best be studied by paleogeographical reconstruction (Figure 2) and examination of basinal deposits. The basinal facies are critical to such construction. They represent the deepest part of the delta-facies continuum, deposited well below normal wave and storm wave bases, and thus their potential for recording subtle information about regional controls is greater than that of other deltaic subenvironments, even though some local events in the source area may be blurred due to distance. In addition, biotic perturbations at the sediment-water interface in these basinal deposits are nearly absent. Information is recorded not only in the sediments themselves, but also in their distribution, thickness, and lateral facies relationships.

The Catskill Delta basin-facies is composed almost wholly of black to dark gray sparsely fossiliferous fissile shales, variously called the "*Leiorhynchus* facies" (Cleland 1903; Cooper 1930), the "Marcellus facies" (Cooper 1933) or the "Cleveland facies" (Caster 1934; Rickard 1964). This contrasts markedly with the off-shore or basinal facies of most modern and many ancient deltas, which consist of gray, silty clays and/or turbiditic siltstones. Deposition of this unique black-shale basinal facies began in the Appalachian Basin near the end of Early Devonian time and, by the latest Devonian, the facies had migrated out of the Appalachian Basin across east-central United States into the Southwest and possibly into west-central Canada. This study, however, will be concerned only with those black shales in the Appalachian Basin.

PALEOGEOGRAPHY

Following the collision of the North America and northwestern European landmasses in the Silurian, most of North America was part of a large, equatorial continent recently called Laurussia (Ziegler and others 1979; Scotese and others 1979), although terrestrial portions of the continent have been called the Old Red Sandstone landmass (Figure 2). The Silurian collision of Europe and North America produced a nearly north-south range of mountains dividing Laurussia into two parts (Figure 2). The inception of the Acadian Orogeny by at least the Early Devonian (Siegenian) in the northern Appalachians (Donohoe and Pajari 1973) continued the southwestward extension of the mountain belt along the southeastern edge of the North American landmass (Figure 2). By the late Early Devonian (Emsian) uplift had begun in the central Appalachians, and by late Givetian to early Fras-

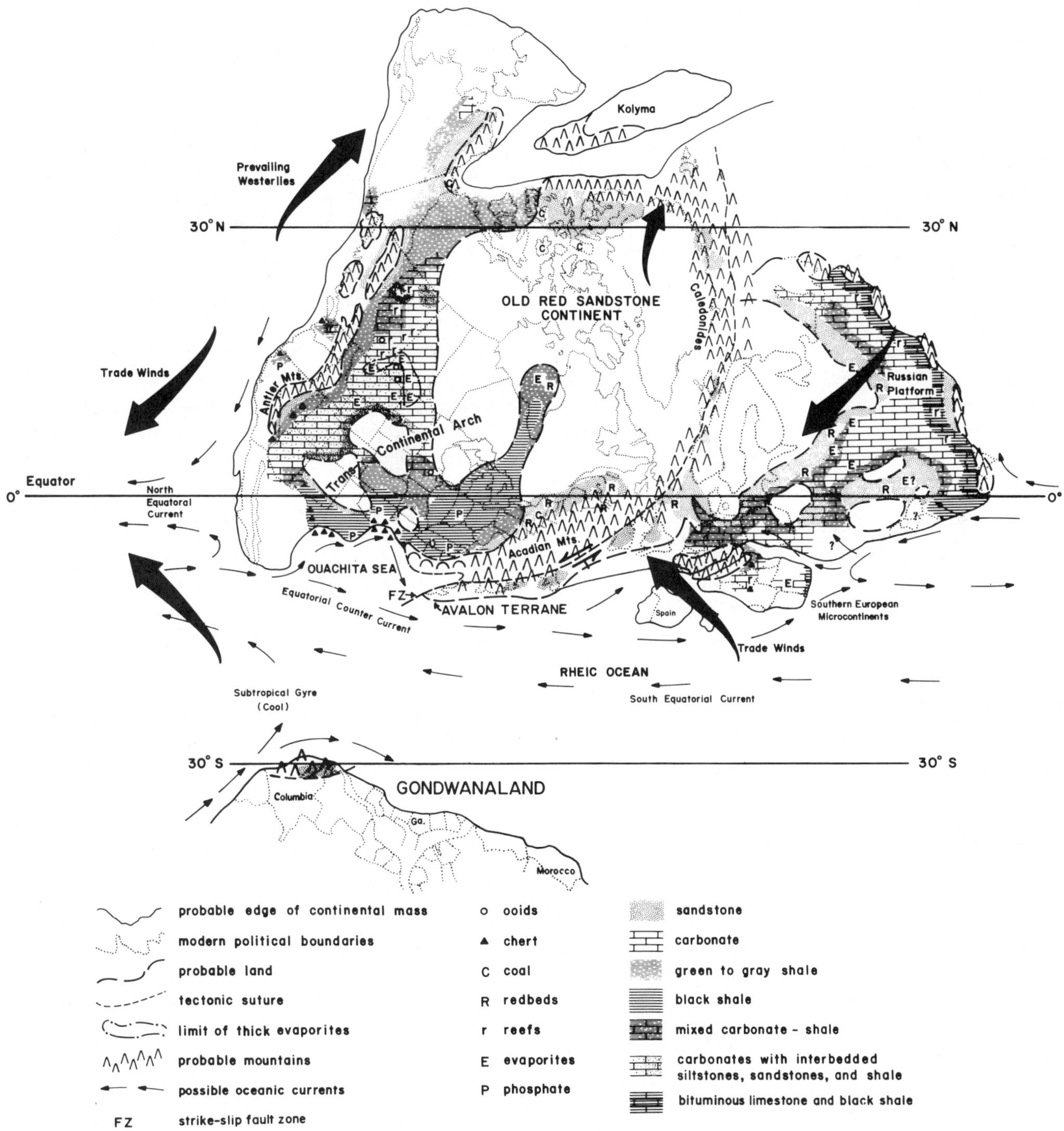

Figure 2. Generalized Late Devonian paleogeography and lithofacies for the Laurussia landmass. The Acadian Mountains are most likely related to transpression generated through oblique convergence and transcurrent movement on the strike-slip fault separating North America and the Avalon terrane (adapted from Ettensohn and Barron 1981).

nian, Acadian convergence was great enough to have created a moutain range estimated at least 4 km high (Bradley 1982). The Acadian Orogeny slowly proceeded southwestward with time (Rodgers 1967), so that by the latest Devonian and Early Mississippian, the Acadian Mountains, and perhaps a lower belt of highlands to the south, extended along the southeastern margin of the North American landmass; this completed the near enclosure of the epicontinental sea into which the Catskill Delta complex prograded (Figure 2). To the north the sea was bound by the Old Red Sandstone Continent, and to the west, by the Trans-Continental Arch. Even though shallow seas may have covered all or parts of the above two features, they still provided effective barriers to deep-water circulation and the exchange of coarse clastics. In the present-day Marathon-Ouachita area, the epicontinental sea opened to the ocean. The presence of a very cherty shale facies in the Marathon-Ouachita region (e.g., Lowe 1975) and its absence elsewhere in nearby parts of the epicontinental sea, suggest that subaequeous extensions of southern parts of the Trans-Continental Arch, Ozark Dome, and southern Acadian Mountains may have formed a low barrier separating a cherty facies on the cratonic margin from a black-shale facies in the epicontinental sea (Figure 2); however, such a barrier was not necessary for the separate deposition of either facies.

During the Late Devonian and Early Mississippian, orogenic events coeval with the last phase of the Acadian Orogeny occurred on the western and northern margins of the North American landmass. Although tectophases of the Ellesmerian orogeny to the north and of the Antler orogeny to the west appear to correspond approximately to those of the Acadian orogeny (Johnson 1971), the three orogens were separated by large expanses of gently uplifted, but low-lying craton that may have been covered by shallow seas.

Sediments derived from the Acadian orogen were deposited in the epicontinental sea east of the Trans-Continental Arch, but Catskill Delta complex deposits were restricted to the Appalachian peripheral basin and were largely unrelated to sedimentation elsewhere in the epicontinental sea until the Late Devonian (Famennian), when the complex prograded beyond the confines of the peripheral basin. Until this time, subsidence in the peripheral basin was apparently sufficient to restrict all deltaic subenvironments to the peripheral basin (Ettensohn and Barron 1981). While the Catskill Delta complex was confined to the peripheral basin, adjacent parts of the craton were characterized by shallow-water carbonate deposition and/or erosion, except on the southern cratonic margin where chert deposition dominated.

PALEOCLIMATE

The Catskill Delta complex developed during a global period of warmth and transgression (Sloss and Speed 1974; Vail and others 1977; Fisher 1981). Most workers agree that Laurussia was an equatorial continent during the Devonian, but differences exist regarding placement of the paleoequator (Heckel and Witzke 1979: 119–120). The position of the Late Devonian paleoequator in Figure 2 is based on the paleomagnetic data of Kent and Opdyke (1978) and Van der Voo and others (1979), as well as on the paleogeographic reconstructions of Ziegler and others (1979) and Scotese and others (1979). The distribution patterns of thick marine carbonates and clastics, reefs, evaporites, and silica-rich sediments relative to this equator are in good general agreement with climatically related patterns of sediment distribution described by Lisitzin (1972), Lees (1975), Heckel and Witzke (1979), and Habicht (1979). The Late Devonian distribution of these facies in Laurussia has been described in detail by Ettensohn and Barron (1981, 1983).

The most important aspect of the paleogeographic and paleoclimatic reconstructions is that the Catskill Delta and its accompanying basinal dark-shale deposits occur within five degrees of the paleoequator. The equatorial belt is a warm, humid belt characterized by heavy rainfall, more than 200 cm per year in the modern equatorial zone (Strahler 1960). This precipitation, largely of the convectional type (Figure 3), forms as moisture-laden trade winds undergo convectional rising due to equatorial heating, and condensation in the rising air causes extensive precipitation. The presence of mountains in the path of converging trade winds can add an orographic effect to this precipitation (Figure 3). If the trade winds encounter a high mountain range before converging on the equator, they are forced to rise on the windward side of the mountains where water vapor in the air will condense giving rise to orographic precipitation (Figure 3).

In the reconstruction shown in Figure 2, trade winds converging on the paleoequator over North America would have been forced to rise at the Acadian and Caledonian mountains and drop most of their moisture on the windward side of the mountains (orographic precipitation, Figure 3). This situation explains the thick sequences of non-marine Old Red Sandstone within and on the European side of the Acadian and Caledonide mountains, as well as clastics transported eastward from these rainy mountains into coastal-plain and shallow marine portions of the western Russian platform (Figure 2). Hence, eastern parts of North America, including most of the Appalachian peripheral basin, were located in the rainshadow of the Acadian and Caledonide mountains.

The only thick clastic sequence west of these mountains is the Catskill Delta complex. At first it may seem anomalous to have a major tropical delta complex in a rainshadow, but as previously indicated, this delta complex was located wholly in the rainy, equatorial belt within five degrees on either side of the paleoequator (Figure 2) where moisture-laden air normally rises anyway due to convection. Because the air would have risen convectionally in this belt, the Acadian Mountains would not have formed as effective an orographic barrier in the equatorial area as that to the north and south. Hence, the rainshadow effect apparently was greatly diminished in the equatorial Catskill area and a thick clastic sequence accumulated there. The mountains may have had some rainshadow effect periodically, for the Catskill facies contains indicators of periodic aridity and restricted rainfall such as dipnoan trace fossils, carbonate paleosols, and redbeds (Woodrow and others 1973), although some of these

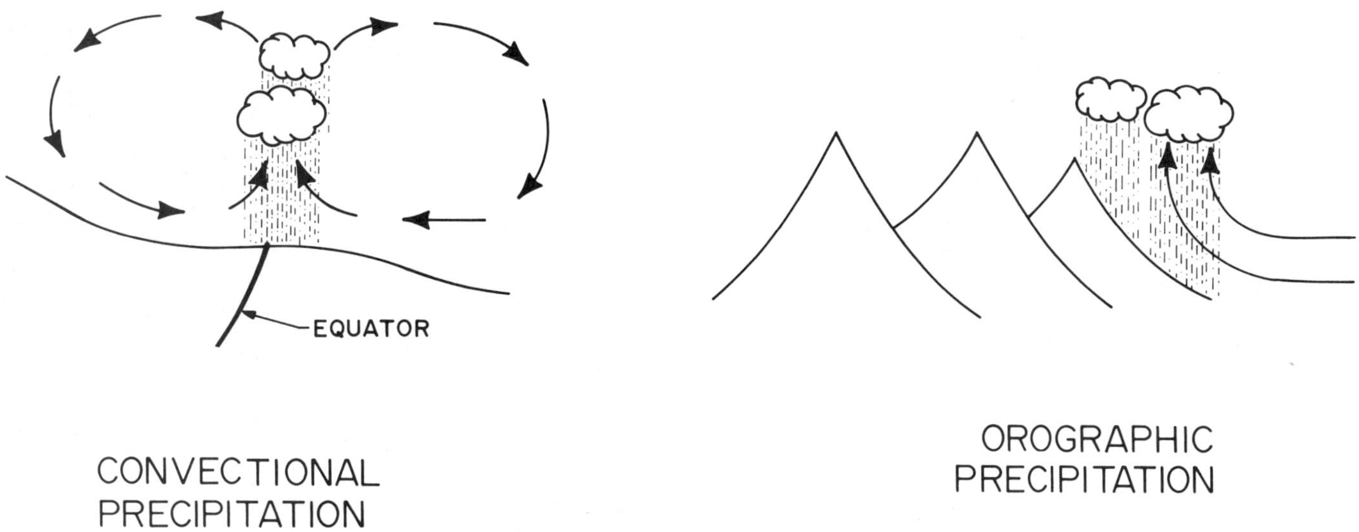

Figure 3. A comparison between the nature of convectional and orographic precipitation.

may have been seasonal indicators. Such periods of aridity with restricted rainfall and runoff may reflect episodes of tectonic uplift in the Acadian orogen, whereas periods of humidity with increased rainfall, runoff, and clastic sedimentation may reflect tectonically quiescent periods in the orogen (Ettensohn and Barron 1981). I suggest that alternation between these two states is probably responsible for the pronounced cyclicity characterizing marine deltaic clastics (Ettensohn and Barron 1981) and terrestrial clastics (e.g., Fletcher 1967) in the Catskill Delta complex.

TECTONIC REGIME

Deposition of the Catskill Delta complex and accompanying basinal sediments occurred during a period of worldwide transgression (Vail and others 1977; Fisher 1981) probably related to increased convergence at plate margins and accelerated spreading rates (Sloss and Speed 1974). Although the period was dominated by transgression, it was also a time of many short-term eustatic fluctuations. The deposition of the basinal dark-shale facies across large parts of North America is closely related to this period of dominantly transgressive seas and to the orogenies occurring at the margins of the continent. This facies began in the Appalachian area during the Emsian (Needmore and Esopus shales) at the inception of the Catskill Delta complex and the Acadian Orogeny.

Although the Acadian Orogeny commonly has been ascribed to a subduction mechanism, the recent work of Williams and Hatcher (1982) suggests that Acadian deformation is probably related to oblique convergence and major transcurrent movement involving the Avalon terrane. The Avalon terrane is now thought to have been the western part of the Armorica plate (Van der Voo 1983: Perroud and others 1984), which apparently moved northeastward (in terms of both present-day and Devonian continental positions) along the southeastern margin of North America on a major sinistral strike-slip fault zone (Figure 2) until the Late Carboniferous. Paleomagnetic differences between the North American craton and the Avalon terrane support such major transcurrent movement (Kent and Opdyke 1978, 1980; Irving 1979; Van der Voo and others 1979). Movement along these large fault zones can produce some of the same orogenic effects that characterize subduction mechanisms (Reading 1980).

By the Early Devonian, the Acadian orogeny had begun in the northern Appalachians (Donohoe and Pajari 1973), and I suggest that the three-part subdivision of the Acadian Orogeny noted by Johnson (1971) is related to the successive collision of the Avalon terrane with promontories (Dewey and Burke 1974; Dewey and Kidd 1974; Thomas 1977) on the southeastern margin of North America during oblique convergence. Because convergence was oblique, successive phases of collision migrated from the northeast to southwest with time. Although some deformation apparently accompanied all phases of oblique convergence along the fault zone, major tectophases or episodes of orogenic activity appear to represent times of actual continental collision. Hence, the Early Devonian tectophase represents collision of the Avalon terrane with the St. Lawrence promontory; the Middle-Late Devonian tectophase represents its collision with the New York promontory; and the Late Devonian-Early Mississippian tectophase represents its collision with the Virginia promontory (see Thomas 1977, for origin of promontories). Even though major collision occurred only at one promontory during each tectophase, every promontory apparently experienced lesser, but coeval, episodes of compression and deformation during the collisions. (A more detailed account of this sequence of events and its effects on sedimentation in the Catskill Delta is presented in another paper in this volume by the author.)

The most important aspect of such a tectonic regime is that the oblique convergence and periodic collision resulting from the strike-slip movement can produce deformation and deformational loading on nearby parts of the craton. The mass load of the resulting fold-thrust mountain belt produces a subsiding peripheral basin that develops on the cratonic side of the mountain range because of regional isostatic adjustment to the load by the lithosphere (Beaumont 1981; Jordan 1981). Hence, a close relationship exists between the mountain belt, which produces a clastic wedge, and its adjacent peripheral basin, which receives the wedge (Beaumont 1981). Moreover, as deformation continues, the fold-thrust belt can migrate towards the craton, causing the deformation of earlier peripheral-basin sediments and the subsequent, progressive migration of newer peripheral basins and their contained clastic wedges toward the craton.

The age and distribution of the presumed oldest basinal facies of the Catskill Delta complex (Needmore and Esopus shales: Dennison 1971; Fisher 1979) suggest that the oblique convergence that marks the initiation of the Acadian orogeny in the central Appalachian area began during the Emsian. Deformation and topographic relief accompanying oblique convergence were apparently greatest in the area of the promontories because: (1) these are the areas from which the major clastic wedges emanated, and (2) because the distribution and depth indicators of the sediments accumulated in the resulting peripheral basins confirm that these basins were deeper and larger in areas adjacent to the promontories.

Thus, the Catskill Delta complex largely represents a clastic wedge produced through collision of the Avalon terrane with the New York promontory. The wedge accumulated in parts of the peripheral basin associated with adjacent structural reentrants (Dewey and Kidd 1974; Thomas 1977). Although the Catskill delta complex actually began developing during the late Early Devonian in response to uplift and loading accompanying oblique convergence on the fault zone, the major phase of delta development occurred in the late Middle Devonian and early Late Devonian (Eifelian-Frasnian; Cazenovia-Cohocton stages), when actual collision with the New York promontory probably occurred. The actuality of collision is indicated by the magnitude of the deformation (Bradley 1982), the deep, peripheral basin that developed opposite the New York promontory in the early Late Devonian, the magnitude of the Late Devonian clastic wedge that filled the basin, and the presence of a prominent unconformity representing regional uplift at the Middle-Late Devonian transition (Ettensohn and Barron 1981). (These aspects are examined in detail in another paper in this volume.)

ORIGIN OF BLACK, BASINAL FACIES

Part of the uniqueness of the Catskill Delta complex basin-facies is the abundance of organic matter that gives these shales their dark color. Organic matter composes (by volume) approximately one-third of the blackest shales, which is consistent with the 15 to 20 percent organic matter indicated by chemical and petrographic analyses (Conant and Swanson 1961; Potter and others 1980). Accumulation of so much organic matter requires a source of organic matter, a means of preserving that organic matter, and a means of preventing excessive clastic dilution.

An abundant source of organic matter is readily available in the plankton that inhabit the surface waters of seas and oceans nearly everywhere. The equatorial setting suggested for the Catskill Delta complex (Figure 2) was conducive to greatly enhanced plankton productivity because of greater solar radiation and the possibility of equatorial upwelling. This productivity is nearly doubled in a near-shore area because of the influx of terrestrial nutrients (Ryther 1969). In addition, the influx of terrestrial nutrients is greatly increased in inland seas (Bordovskiy 1964).

Nearby terrestrial areas represent another major source of organic matter for adjacent seas. The humid, equatorial belt affecting the Catskill Delta complex would have enhanced terrestrial plant growth because of the effects of increased solar radiation and greater precipitation. The climate would also have increased the amount of runoff carrying terrestrial nutrients and organic matter into adjacent seas.

The sources of organic matter described above can be greatly enhanced during a period of transgression. Transgression increases the surface area of epicontinental seas. This has an ameliorating effect on climate and increases the area available for surface marine productivity. Transgression also increases the size of nearshore areas and the proximity of these areas to sources of terrestrial nutrients. Moreover, because transgressions commonly inundate low-lying coastal and alluvial plains there is a greater likelihood that more terrestrial organic matter will be transported directly into the open sea. The increased basin depth that generally accompanies transgression also increases the possibility of upwelling and the preservation potential of the organic matter.

Most of the conditions conducive to increased influx of organic matter were present in the paleogeographic and paleoclimatic setting of the epicontinental sea into which the Catskill Delta complex prograded. Evidence for both marine and terrestrial organic material is amply present in the black basinal shales, and a study of the carbon isotopes from some of these shales (Maynard 1981) indicates that the proportion of non-marine carbon in the organic matter increases in a northeasterly direction toward the source area.

High organic productivity and organic influx may increase the likelihood of forming organic-rich basinal sediments, but they are not crucial elements. What is critical, however, is a means of preserving the organic matter from the effects of physical oxidation, biologic consumption, and clastic dilution. When large amounts of organic debris are sedimented in a nearly enclosed basin with little or no replenishment of oxygen from outside sources, the oxygen demand may exceed that produced by the biological system, and free oxygen is depleted at depth. Moreover, in water stratified because of a density difference between surface and bottom waters (Figure 1), vertical replenishment of oxygen to the bottom does not occur and anaerobic, reducing conditions develop, resulting in the preservation of organic matter. The lack of free oxygen precludes most metazoan life, and only anaerobic and facultative anaerobic life can exist.

The Catskill Delta complex developed in a warm setting ideal for thermohaline density stratification and hindrance of oxygenation at depth by density currents. Such stratification was apparently widespread in the peripheral basin during various stages in the development of the Catskill Delta complex, as indicated by the abundance of preserved organic material and the near absence of benthic life in the basinal, black-shale facies (Byers 1977; Heckel and Witzke 1979). It is doubtful that surface waters ever cooled sufficiently to sink and displace bottom waters. The result was a stratification of warmer, lighter, oxygenated waters at the surface and colder, denser waters on the bottom. The thermocline, a transition zone between surface and bottom temperature layers, is likely to be permanent in equatorial seas (Degens and Stoffers 1976).

The rainy, humid climate that periodically affected the equatorial Catskill Delta complex may have aided in establishing or enhancing stratification. In enclosed seas with a large fresh-water influx, a positive water balance is established, and the excess outflow of lighter, fresh or brackish water floats above the denser, saltier bottom waters. This situation can impede vertical circulation and cause the loss of bottom oxygenation (Deuser 1974; Demaison and Moore 1980). Such conditions occur in the Baltic and Black seas where black, organic-rich muds are presently accumulating (Segerstrale 1957; Caspers 1957; Deuser 1974). The halocline is a zone of transition between the lighter, fresh to brackish surface waters and denser, more saline, bottom waters. In modern oceanic waters, thermoclines and haloclines approximately coincide forming a zone of changing salinity and density known as the pycnocline (Byers 1977; Fig. 1).

The pycnocline also corresponds to the zone of rapid downward decrease in oxygen concentration between surface and bottom layers (Fig. 1). The dysaerobic facies of Rhoads and Morse (1971) develops where this zone intersects the bottom.

The dysaerobic facies together with the overlying aerobic facies and underlying anaerobic facies form vertical and lateral sequences of sediments and biota that are easily identified in the rock record. The presence of such a sequence indicates a stratification of the water column and changing depths during deposition. Each deltaic progradation within the Catskill Delta complex commonly exhibits this vertical sequence. The sequence is generally so consistent that aerobic, dysaerobic and anaerobic zones can be related to specific deltaic subenvironments (Figure 1) based on faunal content and sediment type. For example, during the Late Devonian, most basinal environments are represented by the anaerobic (black-shale) facies. Some basinal environments in Middle Devonian phases of delta development were probably dysaerobic. The anaerobic facies is well-known and is characterized by black, organic-rich shales with no benthic fauna, although planktic, epiplanktic, and nektic forms may be preserved.

Dysaerobic facies commonly characterize the open-shelf and prodelta environments and may include black to dark grayshales or gray to grayish green, silty shales and siltstones. Although benthic organisms are present, the low oxygen concentrations make epifaunal life difficult and most of the benthic life is infaunal. The infaunal organisms derive some measure of protection from harsh surface conditions by establishing their own local habitats within burrows. Hence, bioturbation and trace fossils characterize this facies. Shelled epifauna are generally scarce, but a few eurytopic articulate and inarticulate brachiopods and pelecypods may occur. The *Leiorhynchus* fauna that is commonly ascribed to an epiplanktic life mode probably represents a eurytopic fauna adapted for epifaunal life in dysaerobic environments. Thayer (1981) has indicated that some articulate brachiopods may survive quite well in poorly oxygenated or temporarily anoxic environments, and I believe that a *Leiorhynchus* fauna occurring in dark shales probably represents a sparse epifauna living under such conditions in basinal or open-shelf environments.

The aerobic facies that include a diverse fauna of trace fossils and shelled invertebrates, characterizes the shales, siltstones, sandstones, and carbonates of the delta front, delta platform, and various nearshore environments.

The organic-rich nature of the basinal sediments suggests that they were protected from the effects of excess clastic dilution. Although increased sedimentation may enhance preservation of organic matter by burial sufficiently rapid to prevent consumption or oxidation, this process is effective only up to some poorly known level at which point dilution occurs. Evidence from modern and ancient examples suggests that extremely high rates of clastic sedimentation effectively dilute the amount of organic matter (Brumsack 1980), particularly in areas of high deltaic sedimentation (Demaison and Moore 1980) leading to barren gray- or greenish gray-shale facies that are also seen in Catskill basinal settings.

The fact that the organic-rich, basinal facies of the Catskill delta complex are so well developed indicates that, at least periodically, mechanisms developed to prevent excessive clastic dilution. The term "starved-basin" frequently has been applied to this organic-rich facies and is partially appropriate for the slow rate of clastic sedimentation reflected by the shales. The highly fossiliferous facies occurring higher on the delta (e.g., Harrington 1970; Bowen and others 1974; Thayer 1974) may also reflect times of reduced clastic influx. Bowen and others (1974) and Thayer (1974) have recognized that periods of delta progradation and increased clastic influx had a marked influence on the abundances and types of fuanas present.

Three mechanisms were most likely responsible for preventing excessive clastic dilution during the formation of basinal facies. Perhaps the most significant was the fact that the basinal facies was deposited in the rainshadow of the Acadian Mountains. Although the rainshadow probably was not as effective in the equatorial belt for reasons already discussed, periods of greater tectonic uplift and relief would have produced more effective rainshadows, thereby periodically inhibiting abundant precipitation and runoff on the western flank of the mountains where the delta complex formed. As will be discussed in the following section, reasonable evidence exists to suggest that the periods of highest mountain development coincided with periods of maximum dark-shale or basinal sedimentation and minimal delta development (Ettensohn and Barron 1981).

The distribution of these black shales also reflects major periods of transgression (Sutton and others 1970; Ettensohn and Barron 1981), another mechanism to prevent clastic dilution. During transgression, net sea-level rise would have deepened all former delta environments and inundated nearshore estuarine, fluvial, or flood-plain environments. This kind of transgression of nearshore marine and nonmarine environments by deeper, dark-mud deposition is common in the geologic record of the Catskill Delta complex (Sutton and others 1970). Sea-level rise and inundation of nearshore marine and nonmarine environments would not only cause an increased rate of floodplain aggradation, but would also disturb the marine profile of equilibrium in nearshore areas and cause a higher rate of clastic deposition close to source areas (McCave 1973). These two effects of transgression would have decreased clastic influx to more distal portions of the deltas, particularly in the offshore, basinal environments. Evidence suggests that transgression in the Catskill Delta complex was essentially concomitant with both uplift in the Acadian Mountains and basinal, black-shale deposition.

Finally, it is possible that folding on the eastern margins of the peripheral basin between the highest mountains and the beginning of the deltas, and perhaps even in nearshore areas, formed barriers to the debouchment of abundant clastics in nearshore areas. According to Beaumont (1981), the migration of deformation in an orogen will eventually encompass the proximal margin of the peripheral basin, causing the deformation of older peripheral-basin sediments and influencing the deposition of contemporaneous sediments. Syn-sedimentary folding of Catskill Delta complex sediments has been noted in the Middle and Upper Devonian rocks of Pennsylvania (Harper and Piotrowski 1978, 1979) and New York (Brett and Baird 1982), and Heckel (1973) postulated the presence of such a nearshore, structural barrier to explain the absence of clastics during Tully deposition. The so-called Sherburne Bar of Clark (1904: Plates A and B) and Grabau (1917, 1919) also may be the result of such folding and occupies the right position for the structure necessitated by Heckel's model. Although structures may have played some role in impeding sediment influx to nearshore areas, they were probably not as important as the two other mechanisms.

The coexistence of the three mechanisms for impeding clastic influx during times of maximal black-shale deposition is not accidental. Although coincidence of a clastic source area in the New York promontory with the equatorial climatic zone and the resulting Catskill Delta complex are fortuitous, the correspondence of major orogenic episodes with periods of major black-shale deposition in the Appalachian peripheral basin is not. These aspects are all integrally related parts of a tectonic-climatic-sedimentologic regime modeled in the following section

A TECTONIC-CLIMATIC MODEL

Origin of the organic-rich, basinal facies is closely related to the origin of the more clastic-rich, proximal delta facies. In fact, a cyclic alternation of the dark, basinal facies and lighter, proximal facies characterizes the Catskill Delta complex (Figure 4). The cycles are regional in nature and for the most part are mappable at the formation level, thus making them macro-scale variations in the terminology of Wetzel (1982). Five major cycles occur in the Catskill Delta complex starting with the Middle Devonian Hamilton Group and concluding with the Upper Devonian Canadaway Group. Smaller subcycles are present in most of the major cycles (Ettensohn and Barron 1981; Figure 4). Although younger cycles (Cleveland and Sunbury shales; Figure 4) occur on adjacent parts of the craton, the maximum development of all delta facies occurs in the five cycles of the Appalachian peripheral basin, a reflection of its proximity to and developmental relationships with the Acadian orogen. A model expressing these cyclic relationships is developed in three phases (Figure 5).

Phase 1

Phase 1 was initiated by a period of increased uplift and orogeny in the Acadian Mountains. Orogeny was apparently started by an episode of movement on the large strike-slip fault separating the Avalon terrane from North America (Figure 2) and the resulting oblique convergence. By the Middle-Late Devonian transition, collision of the New York promontory and the Avalon terrane was apparently in the progress (Ettensohn and Barron 1981), and the cyclicity apparent in the Hamilton-through-Canadaway groups (Figure 4) is largely a response to various phases of this collision (see paper elsewhere in this volume by the author).

Transpression related to oblique convergence or collision in the area of the promontory apparently caused episodic uplift in the Acadian Mountains through thrusting and folding. Uplift in the mountains created a more effective orographic barrier, thereby decreasing precipitation in the mountains and subsequent westwardly clastic influx to the Catskill Delta complex. At the same time, smaller folds on the distal, western flanks of the mountains, and perhaps in eastern portions of the peripheral basin itself, may have created additional barriers to the debouching of clastics onto the Catskill Delta complex.

Deformational loading, a product of folding and thrusting in the mountains, produced concomitant or closely following subsidence in the adjacent peripheral basin through isostatic adjustment. Beaumont (1981) indicates that deformational loading begins near the site of convergence and advances toward the craton in a series of pulses. During any one pulse, the lag time between the inception of deformation and its westward migration, and for that matter, the westward migration of any resulting clastics, must have been greater than any lag time between the inception of deformation and resulting subsidence in the peripheral basin, because clastic-deficient, basinal and other deep-water environments rapidly transgressed eastward over former proximal-delta and alluvial-plain environments during each tectonic pulse in the Catskill delta complex. For example, in Figure

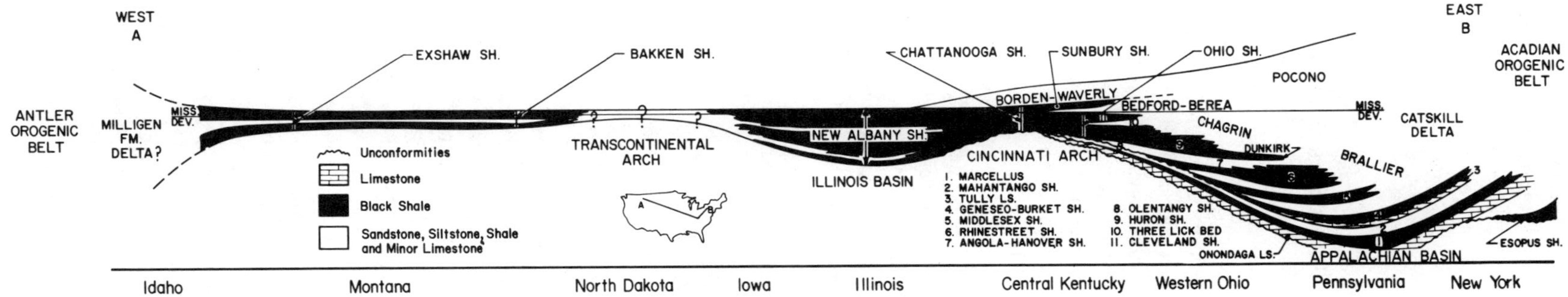

Figure 4. Schematic cross-section showing the distribution of black shales and related coarser clastics facies from New York to Idaho. The large body of coarser clastics cyclically intertonguing with the black shales in the east is the Catskill Delta Complex. Not drawn to scale (adapted from Ettensohn and Barron 1981).

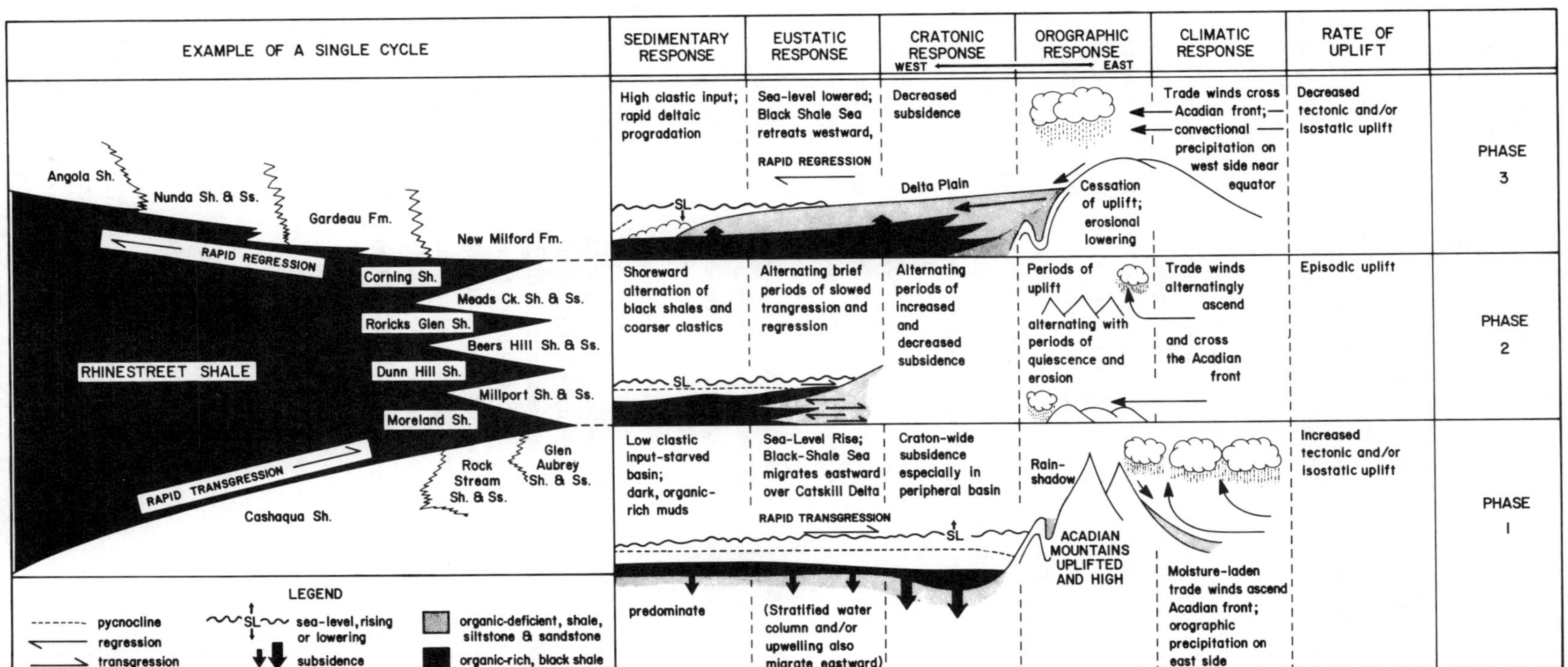

Figure 5. A three-phase model for the cyclicity of black shales and coarser clastics in the Catskill Delta complex illustrated to the left with a typical cycle from the Catskill Delta in New York. Each cycle is thought to begin with an episode of increased tectonism. The height of the newly uplifted mountains, smaller structures on the western flank of the mountains and concomitant transgression all form barriers to westward clastic influx during phase 1; by phase 3, all of these effects are destroyed (adapted from Ettensohn and Barron 1981).

5, the dark Moreland Shale, a proximal intertongue of the Rhinestreet, sharply overlies delta-platform and nearshore environments in the Rock Stream and Glen Aubrey formations (Rickard 1975). Similarly, Sutton and others (1970), Bowen and others (1974), Thayer (1974), and Rickard (1975) indicate that proximal, dark-shale intertongues of the Middlesex and Geneseo shales, although not entirely anaerobic, sharply overlie delta front-through-alluvial plain sediments. The net result was the expansion of marine environments at the expense of terrestrial equivalents and rapid aggradation in transitional and non-marine environments.

The subsidence and resulting transgressive episodes were apparently rapid, for the contacts of the black basinal shales with underlying clastics or carbonates are generally sharp, and any shallow-water deposits are generally condensed into basal-lag deposits. The initial subsidence may have been so rapid, and the halt to clastic influx so complete, that the only sediments to accumulate during these initial stages were composed wholly of material reworked from underlying units along with sparse organic debris sedimented from the water column. The Leicester Pyrite probably represents such a deposit (Brett and Baird 1982).

In addition to its transgressive effects, subsidence that outstrips sedimentation will at some point create a basin deep enough to restrict vertical circulation and initiate stratification, especially in the warm, humid, equatorial zone. A deepening basin in a warm climate not only restricts vertical circulation, but also restricts horizontal, deeper-water circulation, so that the basin geometry itself assumes the restricting function normally ascribed to bars, barriers, and threshholds. The result is a restricted, subsiding basin in an already nearly enclosed, equatorial sea (Figure 2) with little clastic input. In the absence of major clastic influx, sedimentation of the abundant organic matter produced in upper parts of the warm, equatorial seas would predominate. Subsidence and thermohaline stratification in the equatorial zone would result in the preservation of much of this organic matter. In such circumstances, it would have been unlikely for the basin to accumulate anything other than clastic-deficient, black shales, and this is also the origin of the widespread organic-rich, dark-shale facies of the Catskill Delta complex. During the initial phase of major subsidence and transgression, basinal and other deeper-water facies transgressed over former nearshore and non-marine strata and attained their most widespread distribution. For example, Sutton and others (1970) pointed out that deeper-water, dark-shale deposits of the Sonyea overlap underlying non-marine clastics by ten miles. However, this phase of maximum transgression was short-lived and represents maximum subsidence. As subsidence declined and regression occurred, the basinal facies retreated to deeper parts of the basin where maximum thicknesses of black shale occur.

During this discussion it has been assumed that transgression was related to subsidence with no mention of possible eustatic controls. In the peripheral basin, especially in areas adjacent to promontories, I believe that subsidence probably was the major factor. However, a number of the transgressive phases correspond to major regional transgressions that effected large parts of the North American craton, the most notable being the early Frasnian transgression or Taghanic onlap. Such transgressions must be related to larger-scale eustatic controls. Hence, some degree of eustatic control may have been inherent in every cycle.

Phase 2

Phase 2 is represented by the series of dark shales intertongued with open-shelf or prodelta shales that mark the eastern termination of most black-shale units (Figure 5). In most instances, Phase 2 probably encompassed the largest amount of time in each cycle based on the relative thickness of strata and on the presence of many laterally and vertically gradational contacts between dark shales and lighter, coarser clastics. The presence of coarser clastic intertongues indicates that subsidence and resulting transgression (black-shale deposition) no longer dominated the cycle. Instead, shorter periods of subsidence and transgression by sediment-starved seas alternated with short periods of tectonic quiescence and regression. By this phase, major tectonism and uplift began to decline, and thus resulting deformation and uplift slowed and became episodic (Figure 5). Some deformation also would have advanced farther westward, and major clastic influx would have begun to reach the peripheral basin. This combination of events would have reduced transgression and the rain-shadow effect while increasing clastic influx. The result was an episodic, but moderate, expansion of the deltas.

Phase 3

During this phase, deformation had probably attained its farthest westward extent, but movement on the fault zone had slowed to the point that resulting uplift in the mountains could no longer keep pace with weathering and erosion, and the Acadian mountains experienced a net lowering. At the same time, reduction of deformational loading in the mountains caused a concomitant reduction of subsidence in the adjacent peripheral basin. Beaumont (1981) even suggests that peripheral basins may experience uplift during erosion of the adjacent mountains. With subsidence and transgression effectively halted, and the mountains reduced to an even less effective orographic barrier through decreased uplift and increased erosion, all major barriers to clastic influx were destroyed, so that vast amounts of clastics prograded westwardly into the peripheral basin forming one of the large Catskill delta complexes. These effects were no doubt enhanced by the arrival of major clastic influx from the more eastwardly deformation of earlier phases and by the proximity of the most recent deformation. This combination of events would have caused major expansion of the delta and rapid regression during this phase. The regressions were probably not as rapid as the transgressions of Phase 1, because the upper parts of each black-shale unit are not sharply terminated, but intertongue with overlying deltaic facies.

As deltaic progradation and regression continued, the deep,

anaerobic basinal facies was eventually destroyed. However, in parts of the peripheral basin distant from the tectonically active promontory, black-shale deposition may have continued largely uninterrupted into the next cycle. The abundant clastic influx and accompanying pulses of dense, sediment-laden, oxidizing water effectively diluted or destroyed most organic matter. The cessation of subsidence and beginning of uplift in the basin, coupled with abundant sediment influx, means that subsidence was no longer sufficient to support a completely stratified water column with a deep, anaerobic layer. As regional uplift and sediment accumulation continued, environments changed progressively from anaerobic basinal to dysaerobic open-shelf and prodelta environments to aerobic delta-front, delta-platform, and even to terrestrial environments, especially in eastern parts of the basin where sediment influx was greatest. Progradation was not as active in the central and western parts of the basin and fine-grained clastics were the main deposit, with the result that the sea floor never built up much above the dysaerobic open-shelf facies. In fact, the dominantly dysaerobic, open-shelf facies contains many thin layers of black, fissile, organic-rich shale (e.g., Pepper and deWitt, 1950; Colton and deWitt 1958; deWitt and Colton 1978) that indicate that the boundary between the anaerobic and dysaerobic layers was always close to the sediment-water interface in central and western parts of the basin. However, the dominance of dark gray to gray, bioturbated shales throughout the open-shelf facies implies that most of the time this boundary was below the sediment-water interface.

The three-phase model just described was repeated five times in the Appalachian peripheral basin; each cycle begins with a major black-shale unit. The first occurrence is represented by the Hamilton Group. Although the Hamilton Group may be treated as a single cycle beginning with the Marcellus Shale (Figure 4), it is far more complex and consists of at least four partial cycles in which basinal environments were incompletely or only locally developed. Four Upper Devonian cycles were completely developed, and each lasted approximately two to three million years based on the chronology of Harland and others (1982). The bases of the four cycles are defined in ascending order by the Geneseo-Burket, Middlesex, Rhinestreet, and Dunkirk black shales (Figure 4). Each cycle represents the development of a new peripheral basin within the larger Appalachian basinal area. During each cycle, the locus of subsidence and sedimentation shifted westward to accommodate the westward migration of basin infilling and deformation in the Acadian Mountains. Hence, the eastern limit of each successive basinal, black-shale unit occurs farther westward than that of previous black-shale units (Figure 4). By the latest Devonian the basinal facies had been effectively "pushed" out of the peripheral basin into deepening cratonic seas as the peripheral basin filled to overflowing with more proximal deltaic clastics.

CONCLUSIONS

The Catskill Delta complex is composed of cyclically alternating dark shales and coarser clastics throughout much of the Appalachian Basin. Each cycle can be related to an episode of tectonism within larger Acadian tectophases, and five such major cycles occur in the Middle and Upper Devonian rocks of the Appalachian Basin. The clastic-deficient black shales represent a phase of rapid subsidence and transgression concomitant with or closely following the initiation of tectonism and uplift. The coarser clastics, on the other hand, represent a phase of rapid deltaic progradation and regression accompanying greatly decreased tectonism and subsidence near the end of a tectonic episode. The greatest part of each cycle, however, is probably represented by an intermediate or transitional phase recording: (1) a period of episodic tectonism and uplift and (2) a lag time between the inception of tectonism and the westward migration of deformation and the resulting clastic wedge. A three-phase model is constructed to show how interaction of Acadian tectonism with paleoclimate and paleogeography produced the cyclic sedimentary response in the Catskill Delta complex.

ACKNOWLEDGMENTS

This paper is based on parts of earlier research presented in the 1981 Geological Society of America Annual Meeting Guidebook Series. I wish to thank the editors, Carlton E. Brett and John M. Dennison for much constructive criticism. Danna L. Antoine, Beth M. Ettensohn and Margie A. Palmer helped with manuscript preparation.

This research was partially funded by a contract (DE-AC21-76ET-12040) from the Department of Energy, Morgantown Energy Technology Center and a grant from the Graduate School, University of Kentucky.

REFERENCES CITED

Barrell, J., 1913, The Upper Devonian delta of the Appalachian geosyncline: American Journal of Science, ser. 4, v. 36, p. 430- 472.

—— 1914, The Upper Devonian delta of the Appalachian geosyncline—part II: American Journal of Science, ser. 4, v. 37, p. 87–109.

Beaumont, C., 1981, Foreland basins: Geophysical Journal of the Royal Astronomical Society, v. 65, p. 291–329.

Bordovskiy, O. K., 1965, Accumulation and transformation of organic substance in marine sediments: Marine Geology, v. 3, p. 3–114.

Bowen, Z. P., Rhoads, D. C., and McAlester, A. L., 1974, Marine benthic communities in the Upper Devonian of New York: Lethaia, v. 7, p. 93–120.

Bradley, D. C., 1982, Subsidence in late Paleozoic basins in the northern Appalachians: Tectonics, v. 2, p. 107–123.

Brett, C. E., and Baird, G. C., 1982, Upper Moscow-Genesee stratigraphic relationships in western New York: New York State Geological Association, 54th Annual Meeting, Guidebook, p. 19–63.

Broughton, J. G., Fisher, D. W., Isachsen, Y. W., and Rickard, L. V., 1962, The geology of New York State: New York State Museum and Science Service, Geological Survey Map and Chart Series No. 5, 42 p.

Brumsack, H. J., 1980, Geochemistry of Cretaceous black shales from the Atlantic Ocean (DSDP Legs 11, 14, 36, and 41): Chemical Geology, v. 31, p. 1–25.

Byers, C. W., 1977, Biofacies patterns in euxinic basins: a general model: Society of Economic Paleontologists and Mineralogists, Special Publication No. 25, p. 5–17.

Caspers, H., 1967, Black Sea and Sea of Azov, *in* Hedgpeth, J. W., ed., Treatise on marine ecology and paleoecology, v. 1: Geological Society of America Memoir 67, p. 801–890.

Caster, K. E., 1934, The stratigraphy and paleontology of northwestern Pennsylvania, Bulletins of American Paleontology, v. 21, 185 p.

Clarke, J. M., 1904, Naples fauna in western New York, part 2: New York State Museum Memoir 6, 454 p.

Cleland, H. F., 1903, Fauna of the Hamilton Formation of the Cayuga Lake section in central New York: U.S. Geological Survey Bulletin 206, 112 p.

Colton, G. W., and deWitt, W., Jr., 1958, Stratigraphy of the Sonyea Formation of Late Devonian age in western and central New York: U.S. Geological Survey Chart OC-54.

Conant, L. C., and Swanson, V. E., 1961, Chattanooga Shale and related rocks of central Tennessee and nearby areas, U.S. Geological Survey Professional Paper 357, 91 p.

Cooper, G. A., 1930, Stratigraphy of the Hamilton Group of New York: American Journal of Science, ser. 5, v. 19, p. 116–134, 214–236.

—— 1933, Stratigraphy of the Hamilton Group of eastern New York: American Journal of Science, ser. 5, v. 26, p. 537–551.

Degens, E. T., and Stoffers, P., 1976, Stratified waters as a key to the past: Nature, v. 263, p. 22–27.

Demaison, G. J., and Moore, G. T., 1980, Anoxic environments and oil source bed genesis: American Association of Petroleum Geologists Bulletin,v. 64, p. 1179–1209.

Dennison, J. M., 1971, Petroleum related to Middle and Upper Devonian deltaic facies in central Appalachians: American Association of Petroleum Geologists Bulletin, v. 55, p. 1179–1193.

Deuser, W. G., 1974, Evolution of anoxic conditions in Black Sea during Holocene, *in* Degens, E. T. and Ross, D. A., eds., The Black Sea—geology, chemistry and biology: American Association of Petroleum Geologists Memoir 20, p. 133–136.

Dewey, J. F., and Burke, K.C.A., 1974, Hot spots and continental breakup: implications for collisional orogeny: Geology, v. 2, p. 57–60.

Dewey, J. F., and Kidd, W.S.F., 1974, Continental collisions in the Appalachian-Caledonian belt: variations related to complete and incomplete suturing: Geology, v. 2, p. 543–546.

deWitt, W., Jr., and Colton, G, W,m 1978, Physical stratigraphy of the Genesee Formation (Devonian) in western and central New York: U.S. Geological Survey Professional Paper 1032A, 22 p.

Donohoe, H. V., Jr., and Pajari, G., 1973, The age of the Acadian deformation in Maine-New Brunswick: Maritime Sediments, v. 9, p. 78–82.

Ettensohn, F. R., and Barron, L. S., 1981, Depositional model for the Devonian-Mississippian black shales of North America: a paleoclimatic-paleogeographic approach, *in* Roberts, T. G., ed., G.S.A. Cincinnati '81 field trip guidebooks, volume II: economic geology, structure: American Geological Institute, p. 344–361.

—— 1983, A tectonic-climatic approach to the deposition of the Devonian-Mississippian black-shale sequence of North America: Proceedings, 1982 Eastern Oil Shale Symposium, Kentucky Department of Energy and University of Kentucky Institute for Mining and Minerals Research, Lexington, p. 5–37.

Fisher, A. G., 1981, Climatic oscillations in the biosphere, *in* Nitecki, M. H., ed., Biotic Crises in ecological and evolutionary time: New York, Academic Press, p. 103–131.

Fisher, D. W., 1979, Devonian stratigraphy and paleoecology in the Cherry Valley, New York region: Joint annual meeting, New York State Geological Association and New England Intercollegiate Geological Conference, Guidebook, p. 20–46.

Fletcher, F. W., 1967, Middle and Upper Devonian clastics of the Catskill Front, New York, Guidebook, New York State Geological Association, 39th Annual Meeting, p. C1-C29.

Grabau, A. W., 1917, Stratigraphic relationships of the Tully Limestone and the Genesee Shale in eastern North America: Geological Society of America Bulletin, v. 28, p. 945–958.

—— 1919, Significance of the Sherburne Sandstone in Upper Devonian stratigraphy: Geological Society America Bulletin, v. 30, p. 423–470.

Habicht, J.K.A., 1979, Paleoclimate, paleomagnetism, and continental drift: American Association of Petroleum Geologists Studies in Geology No. 9, 31 p.

Harland, W. B., Cox, A. V., Llewellyn, P. G., Pickton, C.A.G., Smith, A.G., and Walters, R., 1982, A geologic time scale: Cambridge, Cambridge University Press, 131 p.

Harper, J. A., and Piotrowski, R. G., 1978, Stratigraphy, extent, gas production, and future gas potential of the Devonian organic-rich shales in Pennsylvania: Preprints, Second Eastern Gas Shales Symposium, Morgantown, West Virginia, METC/SP-78/6, v. 1, p. 310–329.

—— 1979, Stratigraphic correlation of surface and subsurface Middle and Upper Devonian, southwestern Pennsylvania: Guidebook, 44th Annual Field Conference of Pennsylvania geologists, Pennsylvania Geological Survey, p. 18–37.

Harrington, J. W., 1970, Benthic communities of the Genesee Group (Upper Devonian): New York State Geological Association, 42nd Annual Meeting, Guidebook, p. A1–A18.

Heckel, Ph.H., 1973, Nature, origin, and significance of the Tully Limestone: Geological Society America Special Paper 138, 244 p.

Heckel, P. H., and Witzke, B. J., 1979, Devonian world paleogeography determined from distribution of carbonate and related lithic, paleoclimatic indicators, *in* House, M. R. and others, eds., The Devonian System: Special Papers in Palaeontology No. 23, p. 99–123.

Irving, E., 1979, Paleopoles and paleolatitudes of North America and speculations about displaced terrains: Canadian Journal of Earth Sciences, v. 16, p. 669–694.

Johnson, J. G., 1971, Timing and coordination of orogenic, epeirogenic, and eustatic events: Geological Society of America Bulletin, v. 82, p. 3263–3298.

Jordan, T. E., 1981, Thrust loads and foreland basin evolution, Cretaceous, western United States: American Association of Petroleum Geologists Bulletin, v. 65, p. 2506–2520.

Kent, D. V., and Opdyke, N. D., 1978, Paleomagnetism of the Devonian Catskill redbeds: evidence for motion of the coastal New England-Canadian Maritime region relative to cratonic North America: Journal of Geophysical

Research, v. 83, p. 4441–4450.
—— 1980, Paleomagnetism of Siluro-Devonian rocks from eastern Maine: Canadian Journal of Earth Sciences, v. 17, p. 1653–1665.
Klein, G. deV., 1975, Sandstone depositional models for exploration for fossil fuels: Champaign, Illinois, Continuing Education Publishing Company, 109 p.
Lees, A., 1975, Possible influence of salinity and temperature on modern shelf carbonate sedimentation: Marine Geology, v. 19, p. 159–198.
Lisitzin, A. P., 1972, Sedimentation in the world ocean: Society of Economic Paleontologists and Mineralogists Special Publication No. 17, 218 p.
Lowe, D. R., 1975, Regional controls on silica sedimentation in the Ouachita System: Geological Society of America Bulletin, v. 86, p. 1123–1127.
McCave, I. N., 1973, The sedimentology of a transgression: Portland Point and Cooksburg members (Middle Devonian), New York State: Journal of Sedimentary Petrology, v. 43, p. 484–504.
Maynard, J. B., 1981. Carbon isotopes as indicators of dispersal patterns in Devonian-Mississippian shales of the Appalachian Basin: Geology, v. 9, p. 262–265.
Pepper, J. F., and deWitt, W., Jr., 1950, Stratigraphy of the Upper Devonian Wiscoy Sandstone and equivalent Hanover Shale in western and central New York: U.S. Geological Survey Chart OC-37.
Perroud, H., Van der Voo, R., and Bonhommet, N., 1984, Paleozoic evolution of the Armorica plate on the basis of paleomagnetic data: Geology, v. 12, p. 579–582.
Potter, P. E., Maynard, J. B., and Pryor, W. A., 1980, Final report of special geological, geochemical, and petrological studies of the Devonian shales in the Appalachian Basin: Cincinnati, University of Cincinnati, 86 p.
Reading, H. G., 1980, Characteristics and recognition of strike-slip fault systems, *in* Ballance, P. F., and Reading, H. G., eds., Sedimentation in oblique-slip mobile zones: International Association of Sedimentologists Special Publication No. 4, p. 7–26.
Rhoads, D. C., and Morse, J. W., 1971, Evolutionary and ecologic significance of oxygen-deficient marine basins: Lethaia, v. 4, p. 413–428.
Rickard, L. V., 1964, Correlation of the Devonian rocks in New York State: New York State Museum and Science Service, Geological Survey Map and Chart Series No. 4.
—— 1975, Correlation of the Silurian and Devonian rocks in New York State: New York State Museum and Science Service, Geological Survey Map and Chart Series No. 24.
Rodgers, J., 1967, Chronology of tectonic movements in the Appalachian region of eastern North America: American Journal of Science, v. 265, p. 408–427.
Ryther, J. H., 1969, Photosynthesis and fish production in the sea: Science, v. 166, p. 72–76.
Scotese, C. R., Bambach, R. K., Barton, C., and others, 1979. Paleozoic base maps: Journal of Geology, v. 87, p. 217–268.
Segerstrale, S. G., 1957, Baltic Sea, *in* Hedgpeth, J. W., ed., Treatise on marine ecology and paleoecology, v. 1: Geological Society of America Memoir 67, p. 751–800.
Sloss, L. L., and Speed, R. C., 1974, Relationships of cratonic and continental-margin tectonic episodes, *in* Dickinson, W. R., ed., Tectonics and sedimentation: Society of Economic Paleontologists and Mineralogists Special Publication No. 22, p. 98–119.
Strahler, A. N., 1960. Physical geography (second edition): New York, John Wiley & Sons, 534 p.
Sutton, R. G., Bowen, Z. P., and McAlester, A. L., 1970. Marine shelf environments of the Upper Devonian Sonyea Group of New York: Geological Society of America Bulletin, v. 81, p. 2975–2992.
Thayer, C. W., 1974. Marine paleocology in the Upper Devonian of New York: Lethaia, v. 7, p. 121–155.
—— 1981, Ecology of living brachiopods, *in* Broadhead, T. W., ed., Lophophorates, notes for a short course: University of Tennessee Department of Geological Sciences Studies in Geology 5, p. 110–126.
Thomas, W. A., 1977, Evolution of Appalachian-Ouachita salients and recesses from reentrants and promontories in the continental margin: American Journal of Science, v. 277, p. 1233–1278.
Vail, P. R., Mitchum, R. M., Jr., and Thompson, S., III, 1977, Stratigraphy and global changes of sea level, part 4: global cycles of relative changes of sea level, *in* Payton, C. E., ed., Seismic stratigraphy-applications to hydrocarbon exploration: American Association of Petroleum Geologists Memoir 26, p. 83–97.
Van der Voo, R., 1983, Paleomagnetic constraints on the assembly of the Old Red continent: Tectonophysics, v. 91, p. 271–283.
Van der Voo, R., French, A. N., and French, R. B., 1979, A paleomagnetic pole position from the unfolded Upper Devonian Catskill red beds and its tectonic implications: Geology, v. 7, p. 345–348.
Wetzel, A., 1982, Cyclic and dyscyclic black shale formation, *in* Einsele, G., and Seilacher, A., eds., Cyclic and event stratification: Berlin, Springer-Verlag, p. 431–455.
Williams, H., and Hatcher, R. D., Jr., 1982, Suspect terranes and accretionary history of the Appalachian orogen: Geology, v. 10, p. 530–536.
Woodrow, D. L., Fletcher, F. W., and Ahrnsbrak, W. F., 1973, Paleogeography and paleoclimate at the deposition sites of the Catskill Delta and Old Red facies: Geological Society of America Bulletin, v. 84, p. 3051–3064.
Ziegler, A. M., Scotese, C. R., McKerrow, W. S., Johnson, M. E., and Bambach, R. K., 1979, Paleozoic paleogeography: Annual Reviews of Earth and Planetary Sciences, v. 7, p. 473–502.

MANUSCRIPT ACCEPTED BY THE SOCIETY JULY 20, 1984

Printed in U.S.A.

Geological Society of America
Special Paper 201
1985

Nonmarine facies of the Middle and Late Devonian Catskill coastal alluvial plain

W. D. Sevon
Pennsylvania Geological Survey
P.O. Box 2357
Harrisburg, Pennsylvania 17120

ABSTRACT

Nonmarine Catskill facies were deposited by many rivers flowing northwestward from an eastern source area and comprise a suite of sedimentary rocks that (1) thin and decrease in average grain size from southeast to northwest, (2) increase in average grain size from base to top, (3) are dominantly red in color, (4) are variable both laterally and vertically, and (5) were deposited mainly by either meandering or braided streams. The marine to nonmarine transition is a complex interval composed of various facies that represent repeated alternations of marine and nonmarine depositional environments. It is frequently many tens of meters thick. Meandering-stream deposits comprise sandy to gravelly, crossbedded, channel facies and finer-grained, overbank facies arranged in fining-upward sequences. Delta-plain meandering-stream deposition was by streams of low gradient, high sinuosity, shallow water depth, and low erosional capability. The cycles generated are usually only a few meters thick and are dominated by finer-grained floodbasin sediments. Alluvial-plain meandering-stream deposition was by streams of moderate to high gradient, moderate sinuosity, moderate water depth, and moderate to large erosional capability. The generated cycles are frequently many meters thick, dominated by the coarse-grained, channel facies, and show evidence of considerable channel erosion. The braided-stream facies comprises trough-crossbedded sand deposited in shallow to moderate water depths on an alluvial plain. These deposits are almost exclusively gray in color and are sandwiched between the delta-plain and alluvial-plain meandering-stream sequences. The end of Catskill deposition was accompanied by local erosion of the alluvial plain, basin-wide transgression, and demise of the Acadian Mountains as a source area.

INTRODUCTION

Development and growth of the Catskill coastal alluvial plain in response to the Acadian orogeny during the Middle and Late Devonian resulted in deposition of a large accumulation of nonmarine sediments in the Appalachian basin. The purpose of this paper is to present an overview of the nonmarine facies. The information presented is derived from the literature and from my own work. The term *coastal alluvial plain* as used here refers to coalesced, multiple, contiguous sediment-dispersal systems operating in the same sedimentary basin at approximately the same time. The name *Catskill* is added because of its historic and continued use in relation to the whole Middle and Late Devonian depositional system in general and the nonmarine facies in particular.

SOURCE AREA

Paleocurrent, isopach, and lithofacies data indicate derivation of Catskill Delta clastics from an eastern source area, as discussed elsewhere in this volume. This source area, the Acadian Mountains, consisted of uplifted and metamorphosed Precambrian(?), Cambrian, and Early Ordovician rocks as well as volcanics and intrusives. This composition has been confirmed by

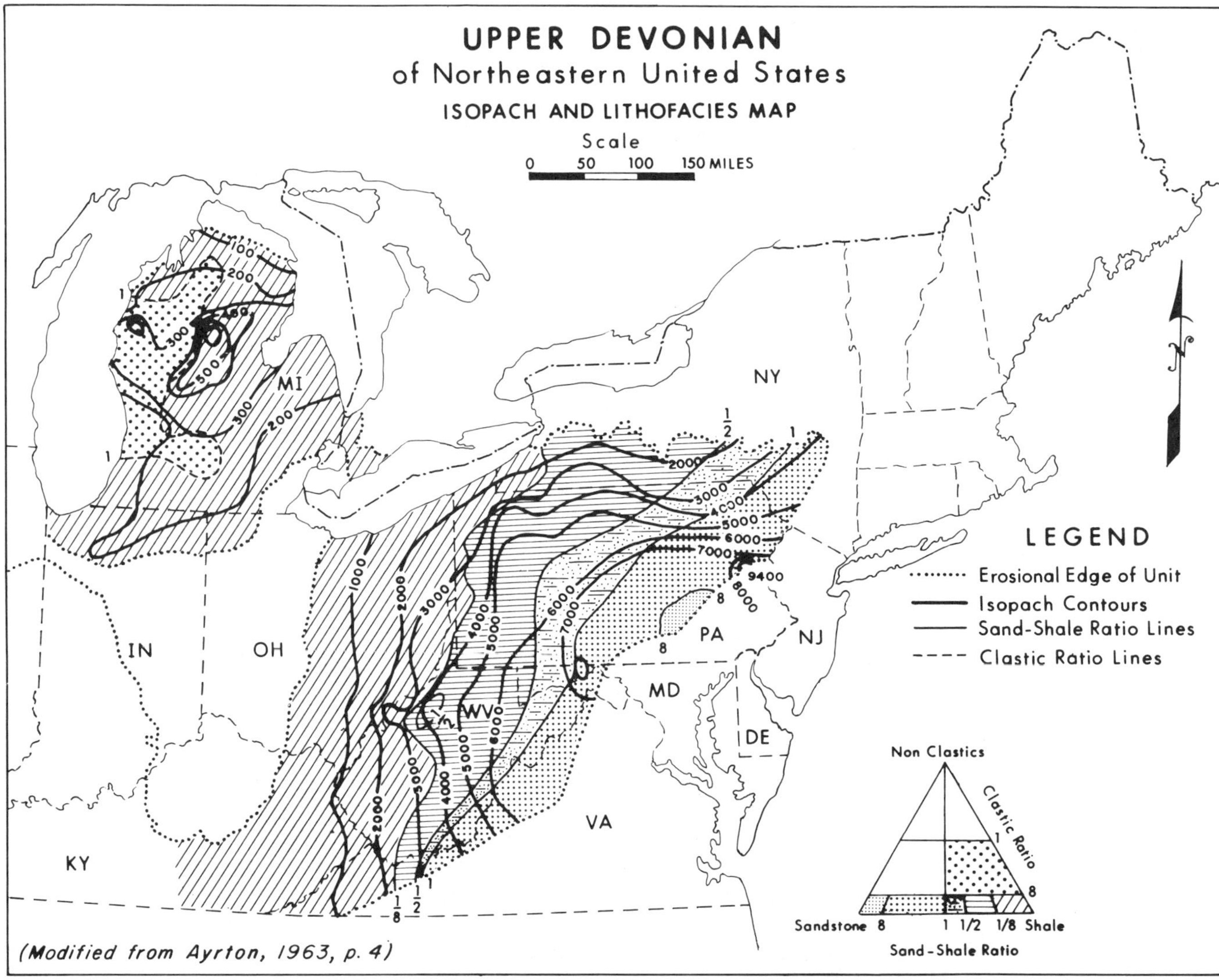

Figure 1. Isopach and lithofacies map for the Upper Devonian of northeastern United States.

identification of coarse clasts (Barrell 1914; Mencher 1939; Sevon 1969; Perry and deWitt 1977; Sevon and others 1978; Seaman 1979; and Kirby 1981) and thin-section petrography (Mencher 1939; Lucier 1966; Sulenski 1969; Kramers and Friedman 1971; Humphreys and Friedman 1975; and Sevon and others in press) of Catskill rocks. These studies indicate a mixed terrane of low-grade metamorphic and sedimentary rocks and a variable occurrence of feldspar. Metaquartzites are particularly common as coarse clasts and illite-chlorite-rich rock fragments are common in thin section.

Estimates of the position of the fall line for the Acadian Mountains during the Devonian range from 40 km (Lucier 1966) to 204 km (Pelletier 1958) east of the present outcrop in Pennsylvania and New York. Its actual position is uncertain, but estimates of 50 to 100 km seem reasonable.

THE CATSKILL COASTAL ALLUVIAL PLAIN

The large quantities of sediment (Figure 1) contributed by the Acadian Mountains were accompanied by marked subsidence in the eastern part of the basin and presumably even greater thicknesses of Middle and Late Devonian sediment now lost to erosion were deposited east of the present outcrop margin.

Sediments deposited in the Appalachian basin during the Middle and Late Devonian are often attributed to the Catskill Delta with the implication that a single delta of unspecified nature was responsible for all of the sediments. However, Barrell (1913) early recognized that many streams were involved in Late Devonian deposition. Willard (1934) was the first to define the position of specific delta lobes, but comparison of the shape and position of those lobes with the Devonian geology of Pennsylva-

Figure 2. Interpreted centers of sediment-dispersal systems entering the central Appalachian basin during the Devonian.

nia indicates that at least one of them, his Snyder, has no factual basis. I recently identified the position of the centers of seven Late Devonian sediment-dispersal systems in Pennsylvania along the eastern outcrop margin (Figure 2) (Sevon and others 1978; Sevon 1979a). Recognition of these sediment-dispersal systems is based upon interpretation of local and regional variations in thickness, facies, and sedimentary structures of many lithostratigraphic units.

Locations of the positions of sediment-dispersal systems in New York are derived from Burtner (1964) and McCave (1968), and those in Virginia–West Virginia from Dennison (1970) and Dennison and deWitt (1972).

The many Catskill sediment-dispersal systems are comparable in size, areal distribution, and depositional facies to those of the lower Wilcox Group (Eocene) in Texas (Fisher and McGowan 1969) as well as the modern depositional systems in the same area (Fisher and others 1969). This similarity supports the concept of a Catskill coastal alluvial plain and may provide guidance for unraveling of Catskill complexity.

The specific depositional history of each distributory system and the interrelationships between systems have yet to be worked out. Available information suggests that each system operated independently and that while some systems may have had true delta character (Mazzullo 1973; Rahmanian 1979), others did not (Allen and Friend 1968; Walker and Harms 1971). It also seems probable that individual sediment-dispersal systems changed character during the process of continued progradation for the following reasons: (1) The intensity and timing of orogeny was laterally variable resulting in differential uplift in the Acadian Mountains and differential sediment supply to the basin (Ettensohn 1985). (2) The length of each of the numerous rivers varied through time in relation to itself and the other rivers. This variation resulted in differences in gradient, grain size of the transported sediment, and, possibly, sediment quantity. These differences, in turn, caused variations in the character of the depositional environments associated with each stream. (3) The interaction between adjacent sediment-dispersal systems was variable.

Despite the considerable lithologic variations involved in

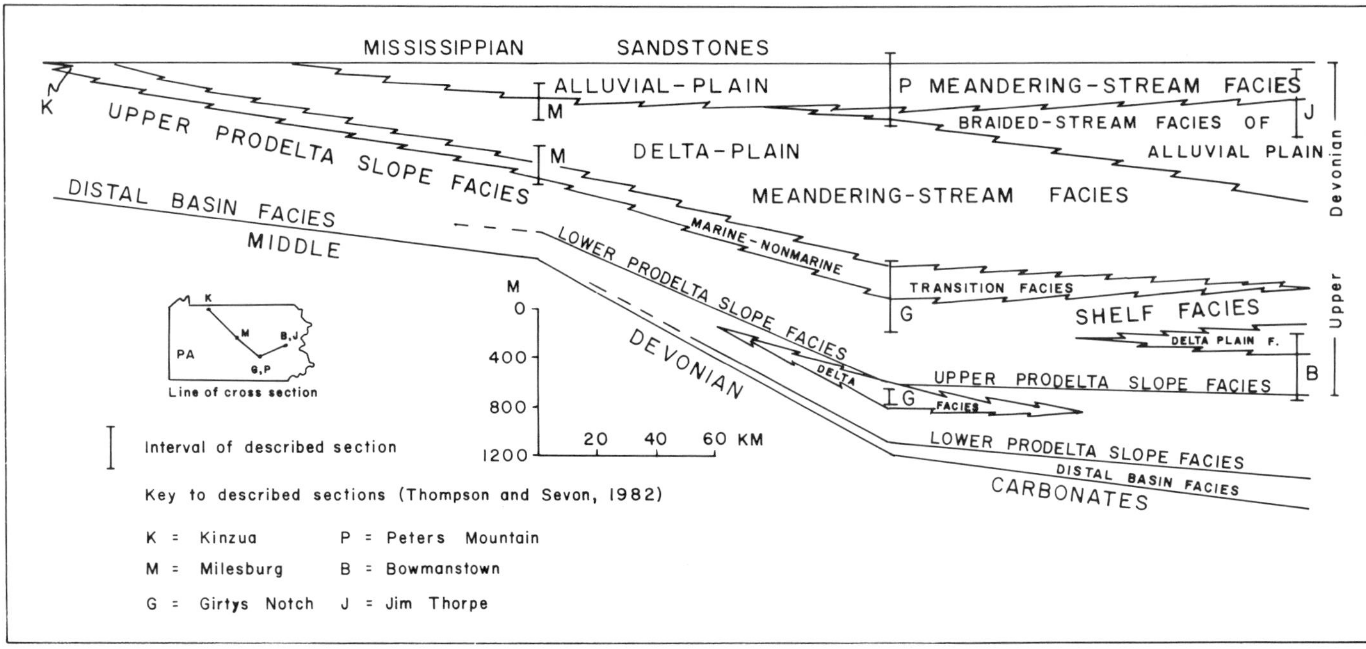

Figure 3. Cross section of Upper Devonian, coastal alluvial plain facies in Pennsylvania (modified from Thompson and Sevon 1982).

Catskill rocks, my work and the work of many others cited here indicate that there is an overall regional pattern to the sequence of sediments deposited in the coastal alluvial plain.

NONMARINE FACIES OF THE CATSKILL COASTAL ALLUVIAL PLAIN

In their simplest form, the nonmarine facies of the Catskill coastal alluvial plain in Pennsylvania represent an orderly progression of change in depositional environments from those of the marine-nonmarine transition upwards through a delta-plain meandering-stream facies to an alluvial-plain braided-stream facies and culminating in an alluvial-plain meandering-stream facies (Figure 3). The Catskill nonmarine sequence ends in braided-stream facies in the Catskill Mountains of New York. However, a large amount of rock has been eroded there (Friedman and Sanders 1982; Lakatos and Miller 1983) and if stratigraphic correlations with Pennsylvania (Fletcher and Woodrow 1970) are correct, then more than half of the Late Devonian nonmarine sequence originally deposited in the Catskill Mountains area has since been eroded. The alluvial-plain meandering-stream facies may be part of that eroded sequence. To the south of Pennsylvania only delta-plain meandering-stream facies were deposited.

There is considerable variation and complexity within the progression present in Pennsylvania and New York. In addition to the regional variations discussed in the previous section, the orderly progression of nonmarine facies is interrupted numerous times by return to a marine facies, particularly in areas far removed from the source area (Sutton 1963; Johnson and Friedman 1969; Sutton and others 1970; Berg 1975; Dennison and Head 1975; Berg and others 1977; Rahmanian 1979).

Facies of the marine-nonmarine transition

Rocks representing the marine-nonmarine transition zone of the Catskill coastal alluvial plain are the most environmentally complex and difficult to interpret of any in the sequence. At best, two or more sequences in the same region that are about the same distance from the original fall line may be comparable in general features (Rahmanian 1979). At worst, sequences covering a larger area and located at different distances from the original fall line may be strikingly different (Walker 1972). In addition to the reasons for variation mentioned earlier, variation also resulted from the balance between sediment supply and basin subsidence. In areas such as the Catskill Mountains and Port Jervis, New York and Carbon County, Pennsylvania sediment supply exceeded subsidence and the shoreline regressed rapidly with the preservation of few, often difficult to interpret, environments of deposition. In areas such as east-central Pennsylvania, where subsidence generally exceeded fluctuating sediment supply for a long period of time, the shoreline transgressed and regressed many times. In this area, interpreted as a muddy shoreline by Walker (1971), the marine and nonmarine environments of deposition are usually readily identifiable, but the position of the shoreline in the sequence is usually obscure. In a few areas (e.g., northeastern Pennsylvania) sediment supply and subsidence achieved an extended period of balance and moderately-thick, easily-interpreted facies were deposited (Krajewski and Williams 1971). Table 1 summarizes the various interpretations of the marine-nonmarine

TABLE 1. ENVIRONMENTS OF DEPOSITION IDENTIFIED IN ROCKS OF THE CATSKILL COASTAL ALLUVIAL PLAIN MARINE-NONMARINE TRANSITION ZONE

Environment Source	Geographic Area	Rock Unit
Tidal flat and tidal channel		
Fletcher, 1967	SE NY	Potter Hollow Fm.
Woodrow and Fletcher, 1968	SE NY	Laurens and Manorkill Fms.
McCave, 1968	SE NY	Portland Point Mbr./Moscow Fm.
Johnson and Friedman, 1969; Johnson, 1970, 1972, 1976; Friedman, 1972	SE NY	Tully Fm., clastic equivalents
Sulenski, 1969	GPO*	Bellvale Fm.
Bowen and others, 1970	EC NY	West Falls Gp.
Sutton and others, 1970	EC NY	Sonyea Gp.
Way, 1972	SE NY	Middle-Upper Devonian
Humphreys and Friedman, 1975	NC PA	Catskill Fm.
Rahmanian, 1979	C PA	Irish Valley Mbr./Catskill Fm.
Kirby, 1981	GPO	Bellvale FM.
Ehrets, 1981	EC NY	West Falls Gp.
Distributary mouth bar		
Sulenski, 1969	GPO	Bellvale Fm.
Sutton and others, 1970	EC NY	Sonyea Gp.
Friedman, 1972	SE NY	Hamilton Gp.
Mazzullo, 1973	SE NY	Hamilton Gp.
Rahmanian, 1979	C PA	Irish Valley Mbr./Catskill Fm.
Kirby, 1981	GPO	Bellvale Fm.
Beach		
Fletcher, 1967	SE NY	Potter Hollow Fm.
Krajewski and Williams, 1971	NE PA	Catskill Fm.
Mazzullo, 1973	SE NY	Hamilton Gp.
Rahmanian, 1979	C PA	Irish Valley Mbr./Catskill Fm.
Berg and others, 1981	NC PA	Catskill Fm.
Interdistributary areas		
McCave, 1968	SE NY	Portland Point Mbr./Moscow Fm.
Sutton and others, 1970	EC NY	Sonyea Gp.
Way, 1972	SE NY	Middle-Upper Devonian
Friedman, 1972	SE NY	Hamilton Gp.
Offshore bar		
Krajewski and Williams, 1971	NE PA	Catskill Fm.

*GPO - Green Pond Outlier, NJ-NY

transition facies. Unusual marginal-marine deposits are discussed by Woodrow and others (1981) and Bridge and Doser (1985).

Nonmarine facies

In the broadest sense the nonmarine facies of the Catskill coastal alluvial plain represent only two environments of deposition: meandering streams and braided streams. However, numerous variations of these environments and their associated subenvironments combine to create local complexity within the scheme of regional simplicity. These environmental variations are reflections of the previously mentioned changes in gradient, sediment supply, and subsidence.

The following generalizations apply to Catskill nonmarine facies throughout the Appalachian basin: (1) the total sequence coarsens upwards, (2) the average grain size decreases from southeast to northwest (Figure 1), (3) the maximum grain size decreases from northeast to southwest, and (4) although red color is characteristic of Catskill nonmarine facies, much of the facies is not red. Sources of information about meandering-stream facies in Catskill rocks are given in Table 2; about braided-stream facies, Table 3.

Meandering-stream facies

Delta plain. This facies comprises fining-upward cycles generally dominated by the finer-grained, overbank deposits (Figure 4a). Cycles tend to be only a few meters thick and frequently show considerable lateral continuity within large outcrops. Basal contacts are sharp and vary from nearly planar to irregular, but generally with less than 0.5 m of relief. At the base of the cycle is a very fine- to coarse-grained, gray, greenish gray or grayish red, channel sandstone. Scattered quartz pebbles commonly occur in these sandstones in the easternmost exposures, but are generally absent farther northwest. Bedding varies from planar to very low-angle crossbedding to trough crossbedding. Calcareous, lensoid, intraformational conglomerates composed of shale clasts, calcareous nodules, carbonaceous debris, quartz pebbles, and rare fish teeth or plates are common in the bases of many channel sandstones.

The sandstones grade upward into red siltstones that in turn grade up into red claystones. The fine-grained rocks are the overbank or flood-basin facies and frequently comprise more than half of the cycle. They range from fissile to structureless depending upon the amount of bioturbation. Cyclic alternations of fissile

TABLE 2. SUMMARY OF SOURCE, GEOGRAPHIC AREA, AND ROCK UNIT OF THE CATSKILL COASTAL ALLUVIAL PLAIN IN WHICH MEANDERING-STREAM FACIES HAVE BEEN IDENTIFIED

Source	Geographic Area	Rock Unit
Allen, 1965	NE PA	Catskill Fm.
Allen and Friend, 1968	NY-PA-WV	Catskill facies
Woodrow and Fletcher, 1967	SE NY-NE PA	Catskill Fm.
Buttner, 1967	SE NY	Catskill Fm.
McCave, 1968	SE NY	Moscow-Ludlowville Fm.
Johnson and Friedman, 1969	SE NY	Tully Fm.
McCave, 1969	SE NY	Catskill Fm.
Sulenski, 1969	GPO*	Bellvale Fm.
Friedman, 1972	SE NY	Catskill Fm.
Way, 1972	SE NY	Middle-Upper Devonian
Glaeser, 1974	NE PA	Duncannon Mbr./Catskill Fm.
Epstein and others, 1974	NE PA	Duncannon Mbr./Catskill Fm.
Humphreys and Friedman, 1975	NC PA	Catskill Fm.
Rahmanian, 1979	C PA	Sherman Creek Mbr./Catskill Fm.
Bridge and Dingman, 1981	SE NY	Oneonta Fm.
Berg and others, 1981	NC PA	Catskill Fm.
Ehrets, 1981	EC PA	West Falls Gp.
Bridge and Gordon, 1985	SE NY	Oneonta Fm.

*GPO - Green Pond Outlier, NJ-NY

TABLE 3. SUMMARY OF SOURCE, GEOGRAPHIC AREA, AND ROCK UNIT OF THE CATSKILL COASTAL ALLUVIAL PLAIN IN WHICH BRAIDED-STREAM FACIES HAVE BEEN IDENTIFIED

Source	Geographic Area	Rock Unit
Lucier, 1966	SE NY	Kiskatom-Kaaterskill Fms.
Buttner, 1967, 1968	SE NY	Genesee Gp.
Friedman, 1972	SE NY	Catskill Fm.
Way, 1972	SE NY	Middle-Upper Devonian
Glaeser, 1974	NE PA	Sawmill Run Mbr./Catskill Fm.
Epstein and others, 1974	NE PA	Berry Run-Clarks Ferry Mbrs./ Catskill Fm.
Humphreys and Friedman, 1975	NC PA	Catskill Fm.
Sevon and others, 1978	NE PA	Duncannon Mbr./Catskill Fm.
Rahmanian, 1979	C PA	Duncannon Mbr./Catskill Fm.
Kirby, 1981	GPO*	Skunnemunk Fm.

*GPO - Green Pond Outlier, NJ-NY

and structureless rock occur and presumably represent repeated episodes of rapid deposition during floods followed by periods of near-surface bioturbation. In situ calcareous nodules indicate soil development. Small burrows, dipnoan lungfish aestivation tubes, fish plates, rootlet traces, plant fragments and impressions, trails, and dessication cracks are common. Scattered, interbedded, thin sandstones probably represent crevasse-splay deposits. The flood-basin deposits are truncated by a sharp, erosional contact and are overlain by channel sandstones of the following cycle.

These cyclic sequences were deposited on low-gradient plains near the coast line by meandering streams with shallow channels, high sinuosity, and large flood basins (Rahmanian 1979). Subsidence and sediment supply were balanced sufficiently to result in minimal erosion of previous deposits. Figure 5a presents an idealized cycle of this facies and Figure 6a presents a typical sequence of the facies. The facies are well described by Way (1972), Epstein and others (1974), Rahmanian (1979), Bridge and Dingman (1981), and Bridge and Gordon (1985). Rahmanian presents the most completely integrated discussion of a large area, whereas Bridge and Gordon present an excellent analysis of a single outcrop.

Alluvial plain. This facies comprises fining-upward cycles (Figure 4b) dominated by the channel facies. The cycles are variable in thickness but tend to be many meters thick. The thickness and completeness of any cycle may vary laterally depending on the amount of erosion that preceded deposition of the following cycle. The bases of cycles are sharp and irregular, and may have several meters of relief. The basal unit is a gray or grayish red channel sandstone or conglomerate. Sand grains are usually medium grained or larger. Most pebbles are rounded, white quartz; although clasts of a variety of low-rank metamorphic and sedimentary rocks may also occur. These channel deposits are characterized by either trough or large-scale planar crossbedding. Grain size decreases upwards and is accompanied by a decrease in magnitude of crossbedding. Planar bedding often occurs at or near the top of the sand-size sequence.

The sandstones grade upward (Figure 4b) into red siltstones that in turn grade upward into red claystones. The finer-grained sediments represent the overbank facies and usually comprise less than a third of the cycle. The rocks are fissile to structureless depending on the amount of bioturbation that has occurred, but they tend to be less burrowed than overbank deposits of the

Figure 4. Representative views of the delta-plain, alluvial-plain, braided-stream, and miscellaneous facies of the Catskill coastal alluvial plain. a. Delta-plain, fining-upward cycles, Interstate Route 80, Centre County, PA. b. Alluvial-plain, fining-upward cycles, U.S. Route 22-322, Perry County, PA. c. Braided-stream facies, U.S. Route 209, Carbon County, PA. d. Filled, cut-off, meander channel, U.S. Route 15, Lycoming County, PA.

delta-plain facies. Dessication cracks, rootlet and plant fragment traces, burrows, and calcareous nodules occur, but less commonly than in the delta-plain facies. The overbank deposits are truncated by a sharp, commonly unconformable, erosional contact which is overlain by channel deposits of the next cycle.

These cyclic sequences were deposited by streams of moderate to high gradient and low to moderate sinuosity (Rahmanian 1979). The channels migrated freely and sometimes completely eroded an earlier deposited cycle. Figure 5b presents an idealized cycle of this facies and Figure 6b presents a typical sequence of this facies. The facies are well described by Allen and Friend (1968), Epstein and others (1974), and Rahmanian (1979).

Braided-stream facies

This facies comprises thick sequences, locally several hundred meters thick, of crossbedded sandstones and conglomerates with few interbedded siltstones and claystones and an absence of fining-upward cycles (Figure 4c). The sandstones are fine to coarse grained and, in contrast to other nonmarine facies, they are almost always gray. Granule conglomerates comprise a small part of this facies, but some horizons are laterally persistent and can be traced over a large area. Beds vary in thickness, but are rarely greater than 2 m and commonly less than 0.5 m. Lateral variability and impersistence of sandstone beds is conspicuous.

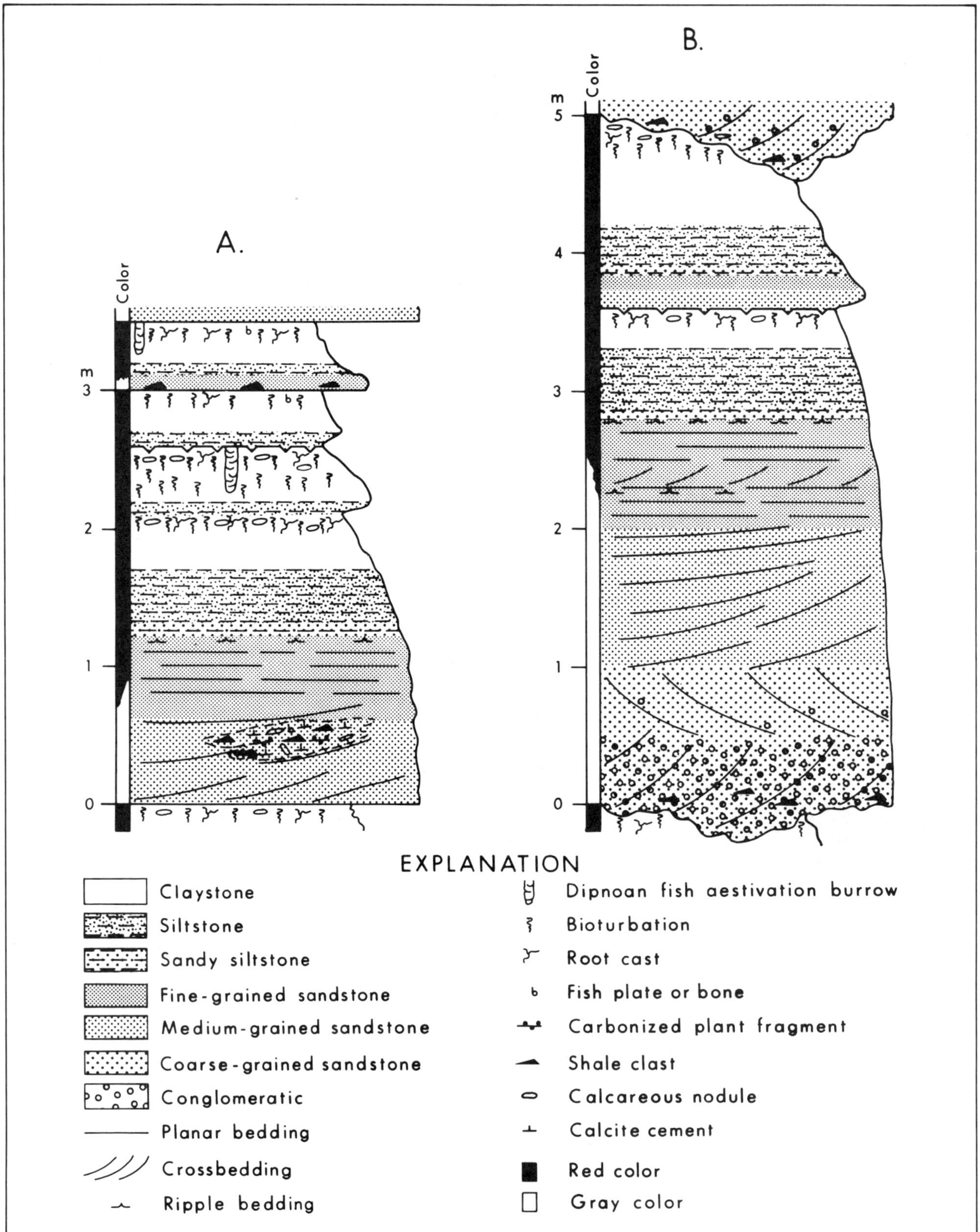

Figure 5. Idealized meandering-stream, fining-upward cycles of the Catskill coastal alluvial plain. a. Delta plain. b. Alluvial plain.

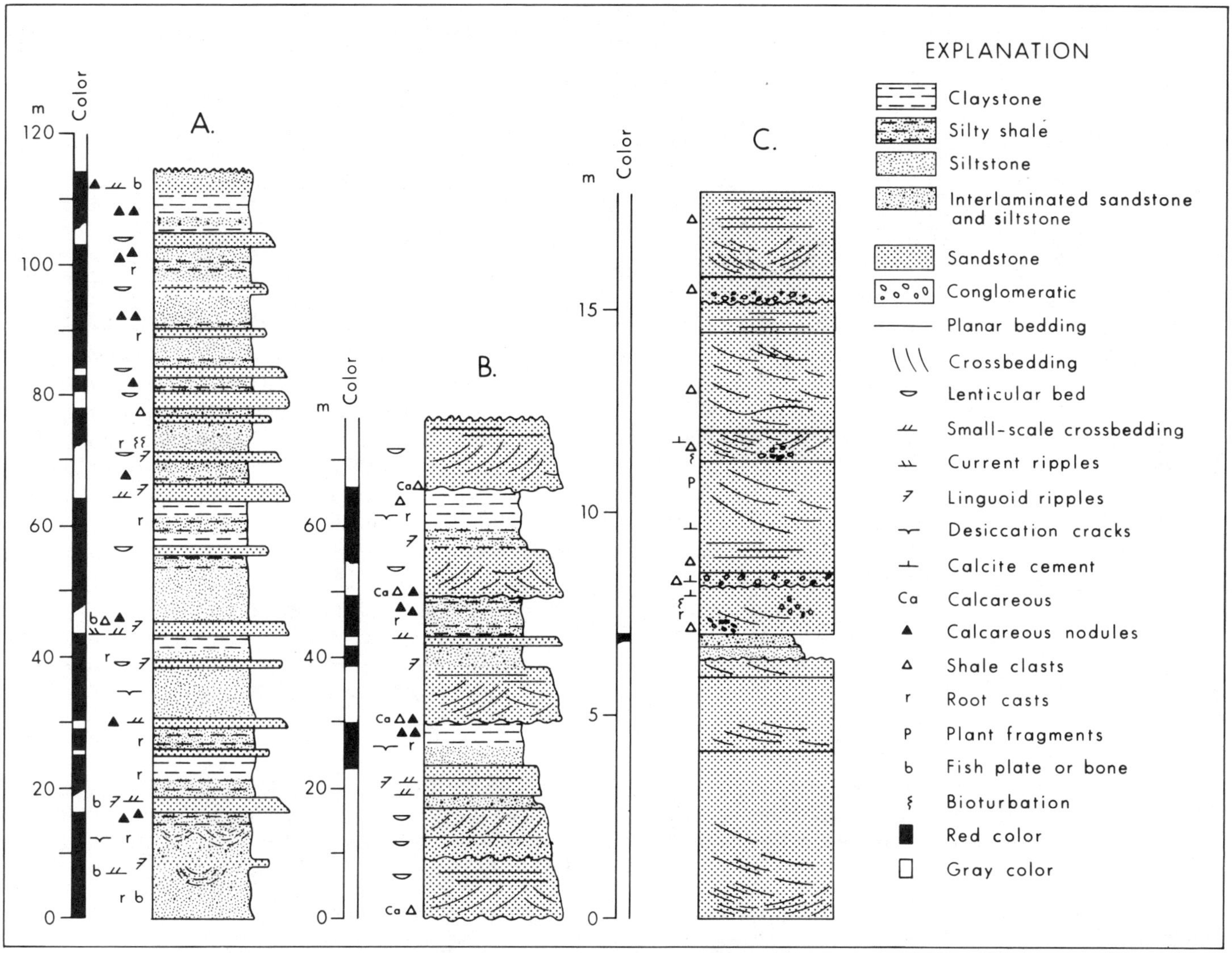

Figure 6. Representative described sections for the major nonmarine facies of the Catskill coastal alluvial plain in Pennsylvania. a. Delta-plain, meandering-stream facies, Interstate Route 80, Centre County (from Rahmanian 1979). b. Alluvial-plain, meandering-stream facies, Interstate Route 80, Centre County (from Rahmanian 1979). c. Braided-stream facies, PA Route 115, Monroe County (from Berg 1975).

Basal contacts are sharp, nearly planar, and seldom have relief of more than 0.5 m. Trough crossbedding is the dominant bedding type; although planar bedding is not uncommon. Parting and parting-step lineations are common on bedding surfaces and carbonized plant fragments occur locally. Red and gray siltstones and claystones occur locally, but they are thin, laterally discontinuous, and bounded by sharp contacts.

The braided-stream facies were deposited by moderate- to high-gradient streams with shallow but rapid discharge and large sediment load. The reason for the facies position between delta-plain and alluvial-plain meandering-stream facies and its nearly total lack of red color is not known, but may be related to a period of large-volume, sediment discharge from the Acadian Mountains. The braided-stream facies dominates all of the Catskill sequence of Pike County in northeastern Pennsylvania and in adjacent New York where no delta-plain meandering-stream facies developed. Figure 6c presents a typical sequence of this facies, and Table 3 presents sources of information about the occurrence of this facies in Catskill rocks. The facies is well described by Epstein and others (1974) and Sevon and others (in press).

Other facies

A variety of other distinctive depositional environments as well as subenvironments (Figure 4d) of the facies described existed during deposition of the Catskill nonmarine facies. Sources of information for some of these environments are summarized in

TABLE 4. MISCELLANEOUS NONMARINE ENVIRONMENTS OF DEPOSITION IDENTIFIED IN ROCKS OF THE CATSKILL COASTAL ALLUVIAL PLAIN

Environment Source	Geographic Area	Rock Unit
Alluvial fan		
Sulenski, 1969	GPO*	Skunnemunk Fm.
Kirby, 1981	GPO	Skunnemunk Fm.
Dune		
Johnson and Friedman, 1969	SE NY	Tully Fm.
Meander cut-off		
Berg and others, 1981	NC PA	Catskill Fm.

*GPO - Green Pond Outlier, NJ-NY

Table 4, but the individual facies are not discussed here. These environments have been studied in detail in specific places, but regional patterns of their occurrence have not been determined.

DEMISE OF THE CATSKILL COASTAL ALLUVIAL PLAIN AND ITS SOURCE AREA

Rocks in Maryland and Pennsylvania deposited during latest Devonian and earliest Mississippian time indicate that the following events occurred: (1) sedimentation on the Catskill coastal alluvial plain stopped, (2) erosion occurred on the alluvial plain, particularly in areas nearest to the Acadian Mountains, and was greatest at the centers of sediment-dispersal systems, (3) marine waters transgressed across at least part of the alluvial plain (Dennison and Head 1975), and (4) polymictic diamictite, a unique lithology in the Paleozoic rocks of the Appalachian basin, was deposited in the areas of greatest erosion (Sevon 1979b). The reasons for these events are not fully understood at present, but they probably resulted from tectonic activity in or near the Acadian Mountains (Sevon 1981; Thompson and Sevon 1982; Ettensohn 1985). One thing is clear however: the diverse lithologies present in the polymictic diamictite provide the final and most intimate look at the composition of the Acadian Mountains. Thereafter, throughout the remainder of the Paleozoic, sediment brought into the Appalachian basin from the east was quartz rich and was probably eroded from the most proximal parts of the former alluvial plain. The hinterland of the Acadian Mountains apparently ceased to function as a sediment source.

REFERENCES CITED

Allen, J.R.L., 1965, Fining-upwards cycles in alluvial successions: 1 Geological Society and Manchester Geological Association Geological Journal, v. 4, p. 229–246.

Allen, J.R.L. and Friend, P. F., 1968, Deposition of the Catskill facies, Appalachian region: with notes on some other Old Red Sandstone basins, *in* Klein, G. deV., ed., Late Paleozoic and Mesozoic continental sedimentation, northeastern North America: Geological Society of America Special Paper 160, p. 21–74.

Ayrton, W. G., 1963, Isopach and lithofacies map of the Upper Devonian of northeastern United States, *in* Shepps, V. C., ed., Symposium on Middle and Upper Devonian stratigraphy of Pennsylvania and adjacent states: Pennsylvania Geological Survey, 4th ser., General Geology Report 39, p. 3–6.

Barrell, J., 1913, The Upper Devonian delta of the Appalachian geosyncline: Part I. The delta and its relations to the interior sea: American Journal Science, 4th ser., v. 36, p. 429–472.

——1914, The Upper Devonian delta of the Appalachian geosyncline: Part III. The relations of the delta to Appalachia: American Journal of Science, 4th ser., v. 37, p. 225–253.

Berg, T. M., 1975, Geology and mineral resources of the Brodheadsville quadrangle, Monroe and Carbon Counties, Pennsylvania: Pennsylvania Geological Survey, 4th ser., Atlas 205a, 60 p., 2 maps.

Berg, T. M., Sevon, W. D., and Bucek, M. F., 1977, Geology and mineral resources of the Pocono Pines and Mount Pocono quadrangles, Monroe County, Pennsylvania: Pennsylvania Geological Survey, 4th ser., Atlas 204cd, 66 p., 2 maps.

Berg, T. M., Crowl, G. H., Edmunds, W. E., Luce, P. B., and others, 1981, Geology of Tioga and Bradford Counties, Pennsylvania: Guidebook, 46th Annual Field Conference of Pennsylvania Geologists, Pennsylvania Geological Survey, 162 p.

Bowen, Z. P., Sutton, R. G., and McAlester, A. L., 1970, Upper Devonian deltaic environments, *in* Heaslip, W. G., ed., Field trip guidebook: 42nd Annual Meeting, New York State Geological Assocation (Cortland), p. B-1–B-11.

Bridge, J. S. and Dingman, L. E., 1981, Depositional environment of the Oneonta Formation (Catskill Facies), near Unadilla, New York, *in* Enos, P., ed., Guidebook for field trips in south-central New York: 53rd Annual Meeting, New York State Geological Association (Binghamton), p. 255–277.

Bridge, J. S. and Droser, M., 1985, Depositional environment of an unusual limestone within the Chemung magnafacies, *in* The Catskill Delta: Geological Society of America Special Paper 201.

Bridge, J. S., and Gordon, E. A., 1985, Quantitative reconstruction of ancient river systems in the Oneonta Formation, *in* Woodrow, D. L. and Sevon, W. D., eds., The Catskill Delta: Geological Society of America Special Paper 201.

Burtner, R. L., 1964, Paleocurrent and petrographic analysis of the Catskill facies of southeastern New York and northeastern Pennsylvania [Ph.D. thesis]: Cambridge, Massachusetts, Harvard University, 222 p.

Buttner, P.J.R., 1967, Continental sequences in the proximal Genesee Group (Stop 3, Field Trip C), *in* Waines, R. H., ed., Guidebook to field trips: 39th Annual Meeting, New York State Geological Association (New Paltz), p. C24–C29.

——1968, Proximal continental rhythmic sequences in the Genesee Group (Lower Upper Devonian) of southeastern New York, *in* Klein, G. deV., ed., Late Paleozoic and Mesozoic continental sedimentation, northeastern North America: Geological Society of America Special Paper 106, p. 109–126.

Dennison, J. M., 1970, Stratigraphic divisions of Upper Devonian Greenland Gap Group ("Chemung Formation") along Allegheny Front in West Virginia, Maryland, and Highland County, Virginia: Southeastern Geology, v. 12, p. 53–82.

Dennison, J. M. and deWitt, W., Jr., 1972, Redbed zone produced by sea level drop at beginning of Devonian Cohocton age delimits Fulton and Augusta

Lobes of Catskill delta complex, *in* Dennison, J. M. and others, Stratigraphy, sedimentology, and structure of Silurian and Devonian rocks along the Allegheny Front in Bedford County, Pennsylvania, Allegheny County, Maryland, and Mineral and Grant Counties, West Virginia: Guidebook, 37th Annual Field Conference of Pennsylvania Geologists, Pennsylvania Geological Survey, p. 109–114.

Dennison, J. M. and Head, J. W., 1975, Sea level variations interpreted from the Appalachian basin Silurian and Devonian: American Journal of Science, v. 275, p. 1089–1120.

Ehrets, J. R., 1981, The West Falls Group (Upper Devonian) Catskill delta complex: stratigraphy, environments, and sedimentation, *in* Enos, P., ed., Guidebook for field trips in south-central New York: 53rd Annual Meeting, New York State Geological Association (Binghamton), p. 3–30.

Epstein, J. B., Sevon, W. D., and Glaeser, J. D., 1974, Geology and mineral resources of the Lehighton and Palmerton quadrangles, Carbon and Northampton Counties, Pennsylvania: Pennsylvania Geological Survey, 4th ser., Atlas 195cd, 460 p.

Ettensohn, F. R., 1985, The Catskill Delta complex and the Acadian Orogeny: a model, *in* The Catskill Delta: Geological Society of America Special Paper 201.

Fisher, W. L. and McGowen, J. H., 1969, Depositional systems in Wilcox Group (Eocene) of Texas and their relation to occurrence of oil and gas: American Association of Petroleum Geologists Bulletin, v. 53, p. 30–54.

Fisher, W. L., Brown, L. F., Jr., Scott, A. J., and McGowen, J. J., 1969, Delta systems in the exploration for oil and gas: University Bureau of Economic Geology Research Colloquium, 78 p.

Fletcher, F. W., 1967, Middle and Upper Devonian clastics of the Catskill Front, New York, *in* Waines, R. H., ed., Guidebook to field trips: 39th Annual Meeting, New York State Geological Association (New Paltz), p. C1–C23.

Fletcher, F. W. and Woodrow, D. L., 1970, Geology and economic resources of the Pennsylvania portion of the Milford and Port Jervis 15-minute quadrangle: Pennsylvania Geological Survey, 4th ser., Atlas 223, 64 p.

Friedman, G. M., 1972, "Sedimentary facies": products of sedimentary environments in Catskill Mountains, Mohawk Valley, and Taconic sequence, eastern New York state: Guidebook, Society of Economic Paleontologists and Mineralogists, Eastern Section, 48 p.

Friedman, G. M., and Sanders, J. E., 1982, Time-temperature-burial significance of Devonian anthracite implies former great (6.5 km) depth of burial of Catskill Mountains, New York: Geology, v. 10, p. 93–96.

Glaeser, J. D., 1974, Upper Devonian stratigraphy and sedimentary environments in northeastern Pennsylvania: Pennsylvania Geological Survey, 4th ser., General Geology Report 63, 89 p.

Humphreys, M. and Friedman, G. M., 1975, Late Devonian Catskill deltaic complex in north-central Pennsylvania, *in* Broussard, M. L., ed., Delta models for exploration: Texas, Houston Geological Society, p. 369–379.

Johnson, K. G., 1970, Transitional sedimentary facies of the Catskill deltaic system in eastern New York State, *in* Heaslip, W. G., ed., Field trip guidebook: 42nd Annual Meeting, New York State Geological Association (Cortland), p. C1-14.

——1972, Evidence for tidal origin of Late Devonian clastics in eastern New York State, U.S.A.: Proceedings, Section 6, Stratigraphy and Sedimentology, 24th International Geological Congress, Montreal, p. 285–293.

——1976, Alluvial and tidal facies of the Catskill deltaic system, *in* Johnsen, J. H., ed., Guidebook to field excursions: 48th Annual Meeting, New York State Geological Association (Poughkeepsie), p. B-8-1–26.

Johnson, K. G. and Friedman, G. M., 1969, The Tully clastic correlatives (Upper Devonian) of New York state: a model for recognition of alluvial, dune(?), tidal, nearshore (bar and lagoon), and offshore sedimentary environments in a tectonic delta complex: Journal of Sedimentary Petrology, v. 39, p. 451–485.

Kirby, M. W., 1981, Sedimentology of the Middle Devonian Bellvale and Skunnemunk Formations in the Green Pond Outlier in northern New Jersey and southeastern New York [M.S. thesis]: New Brunswick, Rutgers, The State University of New Jersey, 109 p.

Krajewski, S. and Williams, E. G., 1971, Upper Devonian flagstone from northeastern Pennsylvania: The Pennsylvania State University, College of Earth and Mineral Science Experiment Station, Special Publication 3-71, 185 p.

Kramers, J. W. and Friedman, G. M., 1971, The Centerfield Limestone and its clastic correlative (Middle Devonian, New York): A sedimentologic analysis (abs.): Geological Society of America, Abstracts with Programs, v. 3, no. 1, p. 40.

Lakatos, S. and Miller, D. S., 1983, Fission-track analysis of apatite and zircon defines a burial depth of 4 to 7 km for lowermost Upper Devonian, Catskill Mountains, New York: Geology, v. 11, p. 103–104.

Lucier, W. A., 1966, The petrology of the Middle and Upper Devonian Kiskatom and Kaaterskill sandstones: A vertical profile [Ph.D. thesis]: New York, University of Rochester, 92 p.

Mazzullo, S. J., 1973, Deltaic depositional environments in the Hamilton Group (Middle Devonian), southeastern New York state: Journal of Sedimentary Petrology, v. 43, p. 1061–1071.

McCave, I. N., 1968, Shallow and marginal marine sediments associated with the Catskill complex in the Middle Devonian of New York, *in* Klein, G. deV., ed., Late Paleozoic and Mesozoic continental sedimentation, northeastern North America: Geological Society of America Special Paper 106, p. 75–107.

——1969, The Catskill delta complex: distributaries in the Bradford subdelta (abs): Geological Society of America, Abstracts with Programs, v. 4, no. 1, p. 39.

Mencher, E., 1939, Catskill facies of New York state: Geological Society of America Bulletin, v. 50, p. 1761–1794.

Pelletier, B. R., 1958, Pocono paleocurrents in Pennsylvania and Maryland: Geological Society of America Bulletin, v. 69, p. 1033–1064.

Perry, W. J., Jr. and deWitt, W., Jr., 1977, A field guide to thin-skinned tectonics in the central Appalachians: Guidebook, post-meeting field trip no. 4, American Association of Petroleum Geologists/Society of Economic Paleontologists and Mineralogists, Annual Meeting, Washington, D.C., 54 p.

Rahmanian, V. D., 1979, Stratigraphy and sedimentology of the Upper Devonian Catskill and uppermost Trimmers Rock Formations in central Pennsylvania [Ph.D. thesis]: University Park, The Pennsylvania State University, 340 p.

Seaman, S. J., 1979, The lowermost unit of the Spechty Kopf Formation: composition and possible source areas (Senior research paper): Bryn Mawr, Pennsylvania, Bryn Mawr College, 16 p.

Sevon, W. D., 1969, The Pocono Formation in northeastern Pennsylvania: Guidebook, 34th Annual Field Conference of Pennsylvania Geologists, Pennsylvania Geological Survey, 129 p.

——1979a, Devonian sediment-dispersal systems in Pennsylvania [abs.]: Geological Society of America, Abstracts with Programs, v. 11, no. 1, p. 53.

——1979b, Polymictic diamictites in the Spechty Kopf and Rockwell Formations, *in* Dennison, J. M. and others, Devonian shales in south-central Pennsylvania and Maryland: Guidebook, 44th Annual Field Conference of Pennsylvania Geologists, Pennsylvania Geological Survey, p. 61–66; 107–110.

——1981, The Middle and Upper Devonian clastic wedge in northeastern Pennsylvania, *in* Enos, P., ed., Guidebook for field trips in south-central New York: 53rd Annual Meeting, New York State Geological Association (Binghamton), p. 31–63.

Sevon, W. D., Rose, A. W., Smith, R. C., II., and Hoff, D. T., 1978, Uranium in Carbon, Lycoming, Sullivan, and Columbia Counties, Pennsylvania: Guidebook, 43rd Annual Field Conference of Pennsylvania Geologists, Pennsylvania Geological Survey, 99 p.

Sevon, W. D., Berg, T. M., Schultz, L. D., and Crowl, G. H., in press, Geology and mineral resources of Pike County, Pennsylvania: Pennsylvania Geological Survey, 4th ser., County Report 52.

Sulenski, R. J., 1969, The paleoenvironments and fossil communities of the Bellvale Sandstone (Middle Devonian) and Skunnemunk Conglomerate (Middle Devonian?) at Skunnemunk Mountain, New York [M.A. thesis]: New York, Queens College, 138 p.

Sutton, R. G., 1963, Correlation of Upper Devonian strata in south-central New

York, *in* Shepps, V. C., ed., Symposium on Middle and Upper Devonian stratigraphy of Pennsylvania and adjacent states: Pennsylvania Geological Survey, 4th ser., General Geology Report 39, p. 87–101.

Sutton, R. G., Bowen, Z. P., and McAlester, A. L., 1970, Marine shelf environments of the Upper Devonian Sonyea Group of New York: Geological Society of America Bulletin, v. 81, p. 2975–2992.

Thompson, A. M. and Sevon, W. D., 1982, Excursion 19B: Comparative sedimentology of Paleozoic clastic wedges in the Central Appalachians, U.S.A: Field excursion guidebook, 11th International Congress on Sedimentology (Hamilton, Ontario, Canada), International Association of Sedimentologists, 136 p.

Walker, R. G., 1971, Non-deltaic depositional environments in the Catskill clastic wedge (Upper Devonian) of central Pennsylvania: Geological Society of America Bulletin, v. 82, p. 1305–1326.

Walker, R. G., 1972, Upper Devonian marine-nonmarine transition, southern Pennsylvania: Pennsylvania Geological Survey, 4th ser., General Geology Report 62, 25 p.

Walker, R. G. and Harms, J. C., 1971, The "Catskill Delta": a prograding muddy shoreline in central Pennsylvania: Journal of Geology, v. 79, p. 281–299.

Way, J. H., Jr., 1972, Depositional environmental analysis, Middle and Upper Devonian sedimentary rocks, Catskill Mountain area, New York [Ph.D. thesis]: Troy, New York, Rensselaer Polytechnic Institute, 147 p.

Willard, B., 1934, Early Chemung shore line in Pennsylvania: Geological Society of America Bulletin, v. 45, p. 897–908.

Woodrow, D. L. and Fletcher, F. W., 1968, Late Devonian paleogeography in southeastern New York and northeastern Pennsylvania, *in* Oswald, D. H., ed., International symposium on the Devonian System, Volume II: Calgary, Alberta Society of Petroleum Geologists, p. 1327–1334.

Woodrow, D. L., Enos, P., Bottjer, R. J., and Mirrero, C., 1981, Bioclastic carbonate units in the Catskill clastic wedge, *in* Enos, P., ed., Guidebook for field trips in south-central New York: 53rd Annual Meeting, New York State Geological Association, (Binghamton), p. 201–229.

Manuscript Accepted by the Society July 20, 1984

Printed in U.S.A.

Geological Society of America
Special Paper 201
1985

Catskill Delta shallow marine strata

John M. Dennison
Geology Department
University of North Carolina
Chapel Hill, North Carolina 27514

ABSTRACT

Shallow marine strata of the Devonian delta complex range from lower delta plain with some marine influence to lagoon, barrier island, shallow shelf, shelf margin, and gentle or steep slope, progressing seaward to deep-water deposits of black shales and some pelagic limestones. Shallow marine depositional history and lithofacies at any one place vary through time depending on: (1) tectonic uplift of source area and rate of erosion, (2) subsidence of the major basin, (3) sea-level changes, (4) localized basement tectonics, (5) random shifts of delta distributaries, (6) extent of delta progradation across the basin, (7) tidal amplitude, (8) longshore drift, and (9) intensity of mass wastage of the delta front.

Deposition of prodelta muds immediately above the Oriskany Sandstone and Wallbridge Discontinuity began the history of the Catskill Delta complex. The initiation of prodelta siltstones shifted southward from New York in the Cazenovia Age to Tennessee in the Bradford Age. Maximum delta lobe progradation occurred simultaneously in all delta lobes in the mid-Bradford Age. The location with maximum detrital accumulation rate in eastern outcrop belts shifted southward from New York in the Middle Devonian to west-central Virginia in the very late Devonian, to southwestern Virginia at the beginning of the Mississippian. The area uplifted east of New York appears to have been reduced from the dominant sediment-input center as the locus of Acadian maximum uplift progressed southward.

SCOPE OF PAPER

This paper emphasizes the strata deposited seaward from nonmarine, fluvial sediments and shoreward from deep-water black shales. This includes lower-delta-plain intertidal areas, the zone of wave interaction, barrier islands and lagoons, the shelf area generally below wave base except during large storms, the shelf margin, a gentle decline of slope toward deeper water and poorly oxygenated sediments, and/or turbidites at the foot of slopes and landward from basin-bottom. The deeper water sediments beyond the reach of turbidity currents or storm agitation are excluded. On the west side of the Appalachian basin some of the black shales change facies to more calcareous strata, probably representing better oxygenated conditions on the margin of the Cincinnati-Algonquin arch, but those areas are beyond the influence of the Catskill Delta. Some deep-water, pelagic micrites (Tully and Purcell Limestones from Pennsylvania to Virginia, for example) formed below wave-base when little terrigeneous detritus was entering the deeper basin, and these important time-markers are not deltaic deposits.

FACTORS AFFECTING DELTAIC SEDIMENTATION

Details of deltaic sedimentation are less understood in the southern part of the Appalachian basin than in the north, because of the longer history of studies in New York and Pennsylvania and because of less severe structural deformation in the New York and extreme eastern Pennsylvania outcrops. However, the broad outlines of sedimentary tectonics are now becoming apparent for the entire Catskill Delta complex. The complex is a series of deltas both shifting intensity and varying in position with time. In the Late Devonian these several sediment input systems coalesced into a large prograding alluvial plain which sloped generally westward or northwestward. These sediment-input sys-

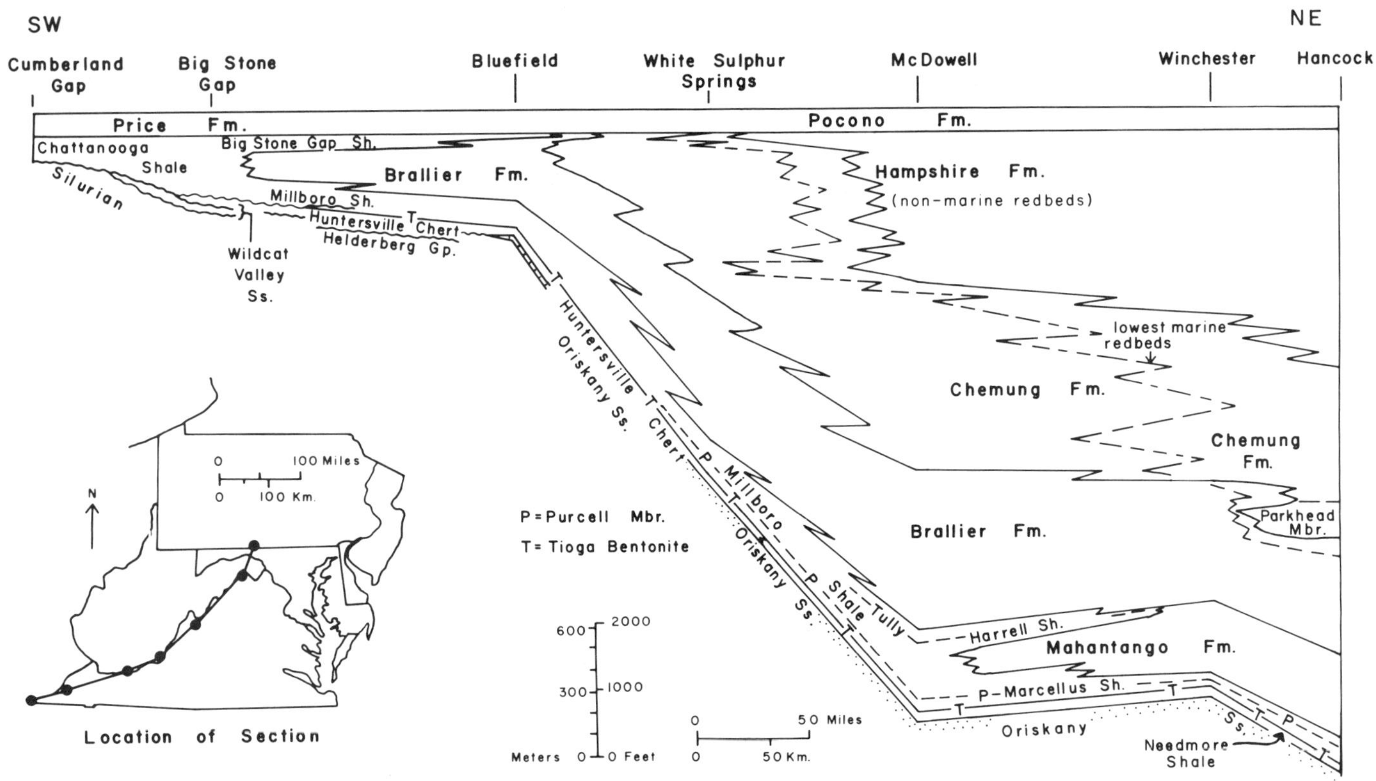

Figure 1. Northeast-southwest cross section of Catskill Delta strata in central portion of Valley and Ridge Province, showing nomenclature and generalized thickness of rock-stratigraphic units.

tems entered the Devonian sea and deposited lobate concentrations of marine strata seaward from the most powerful sites of detrital influx.

Glaeser (1979) has described a general sedimentation model for the entire Catskill Delta system in the northern part of the Appalachian basin. He is one of the few workers who has tied outcrop sedimentology to subsurface stratigraphy (Glaeser 1974). Potter and others (1982) and Woodrow (1983) also wrote sedimentologic summaries. Woodrow and Isley (1983) have presented the most recent sedimentologic analysis, working with selected outcrops in New York and Pennsylvania.

Detailed stratigraphic cross sections across New York have been understood for decades, with time relations and facies patterns quite clear (Rickard 1975). In the central latitudes of the Appalachian basin, exposures of the Catskill Delta are further complicated by folding, and faulting even becomes a problem farther south in Virginia and Tennessee. In the extreme southern Appalachians, all of the Upper Devonian is black shale of basin-bottom, presumably of deep-water origin.

Figure 1 presents facies relationships diagonally across part of the delta complex in the Valley and Ridge Province of the central Appalachians, and Figure 2 shows approximate time-rock units in the same stratigraphic cross section.

The shallow marine depositional history and lithofacies at any one place varies through time depending on:

(1) tectonic uplift of source area and rate of erosion,
(2) subsidence of the major basin,
(3) sea-level changes in the basin, perhaps reflecting global (eustatic) fluctuations,
(4) localized basement tectonics within a generally subsiding basin, producing second and third order basin details,
(5) random shifts of prograding delta distributaries and turbidite fans
(6) extent of delta progradation across the basin,
(7) tidal amplitude,
(8) longshore drift, and
(9) mass wastage of the delta front.

The effects of each of these factors will be briefly considered.

Tectonic Uplift of Source Area and Rate of Erosion

The rapidity of tectonic uplift relative to rate of erosion determines the relief of a source area and whether it is being eroded rapidly by streams with steep gradients. The volume of sediment produced depends on rate of erosion and the areal size of the source terrain. Rate of erosion also depends on rock type: the more resistant the rock, the slower the erosion rate.

Topography of the source area influences the amount of rainfall, as does the latitude. The Appalachia source area probably did not change latitude much during Catskill Delta deposition; it was probably only a few degrees south of the equator, with the trend of the Acadian highlands roughly parallel to Devonian lati-

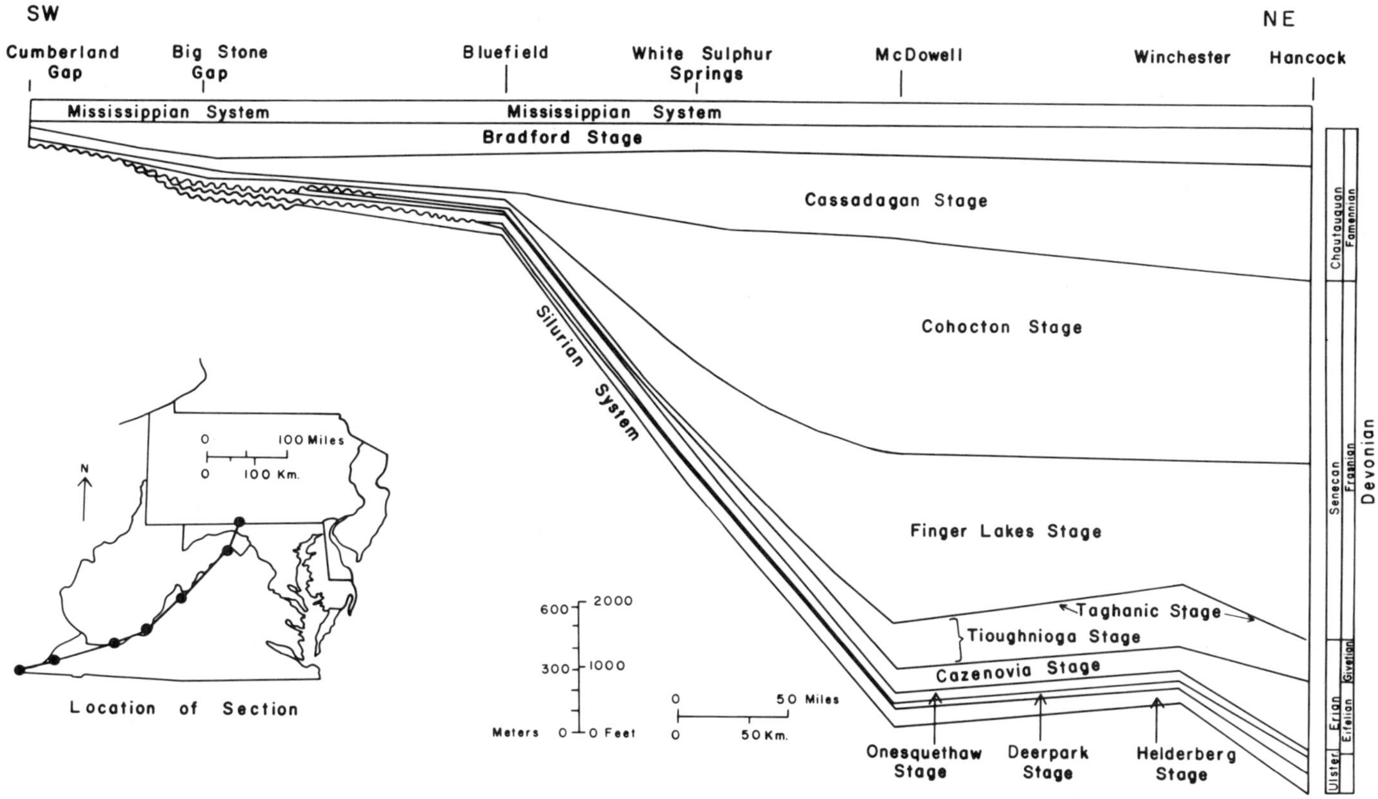

Figure 2. Northeast-southwest cross section of Catskill Delta strata in central portion of Valley and Ridge Province, showing thickness of Devonian time-stratigraphic units.

tudinal lines. Climatic fluctuations resulting from equilibrium shifts between growing and reducing mountain heights have been proposed to explain the origin of black shale tongues in the Devonian delta marine strata (Ettensohn and Barron 1981), but I doubt their mechanism because crystalline mountainous areas cannot be eroded to low elevations in the brief span of five million years or less required by their model to explain the spacing in time among the major black shale tongues.

Climate affects in other ways the rate of erosion and volume of sediment produced. In present-day North America there is more rapid erosion in the semi-arid regions of the West than in the regions with more humid climate. A humid climate protects the ground with a heavy vegetative cover, and the climate of the Catskill Delta and probably the provenance area was humid during the Middle and Late Devonian (Woodrow, Fletcher, and Ahrnsbrak, 1973). Plant spacing may not have been as dense as in modern, highly evolved floras. A humid climate promotes more thorough chemical weathering and production of kaolinitic clays than an arid climate.

Additional weathering and reworking of sediments accumulated along levees and in coastal plains produces a more mature deposit in the resulting delta than does first-cycle erosion of metamorphic or igneous mountains.

A rapidly prograding delta lobe indicates voluminous and generally rapid erosion in the source area. Thus if regional subsidence proceeds at a constant rate, timing and geographic position of the largest delta lobes in the Catskill Delta complex should indicate where the source area was yielding the most sediment and the streams with the largest discharge of water.

Subsidence of the Basin

The overall shape of the Appalachian basin was established by the Late Ordovician. During times of relatively slow detrital clastic sedimentation, such as in the Cayugan Epoch and Onesquethaw Age, the maximum water depth was along an axis trending about S25°W from Syracuse, New York to Birmingham, Alabama (Dennison 1983b). This probably reflects a tectonic grain established during the rifting of the Iapetus Ocean in Late Precambrian to Early Cambrian time (Donaldson and Shumaker, 1981; Dennison 1982a; 1982c). Figure 3 shows the position of the Rome Trough, presumed to result from this tectonism, extending eastward across Kentucky along the 38th Parallel fracture zone and deflecting northeastward across Pennsylvania. I extend the 38th Parallel fracture zone eastward into Virginia along and slightly north of the site originally proposed by Zartman and others (1967) and Heyl (1972), based on my own observations of sedimentary tectonics and in part following the LANDSAT lineament patterns published by Werner (1979). This figure also shows names of major lobes of deltaic sedimentation superimposed on a map of these early major basement fracture patterns.

Figure 3. Location of major fracture systems that developed in the Late Precambrian and Early Cambrian along with major delta lobes of the Late Devonian delta complex.

Subsequent addition of a thick terrigenous clastic wedge derived from the Acadian mountains to the east affected subsidence patterns of the basin. Isostatic loading by the heavy clastic wedge caused a greater subsidence of the sedimentation surface to the east of the original basin axis. Topography east of the original basin axis probably was built up on the sea floor by the influx of deltaic sediments adjacent to the sediment-input centers, so that the actual maximum water depth became displaced to the west of the initial basin axis as prodelta sediments moved westward.

The clastic wedge appeared earlier and was more voluminous in the northern part of the Appalachian basin than in the south, and consequently basin subsidence of the Lower Devonian strata beneath the Catskill Delta was greater in the north. The displacement of maximum water depth to the west of the general basin axis was greatest in the north, so that distal turbidite beds extend to outcrops on the east flank of the Cincinnati arch in northern Kentucky and southern Ohio (Three Lick Bed of the Ohio Shale described by Provo 1977 and by Provo and others 1978).

Basin subsidence was largely isostatic once the delta was initiated, but in the early stages it may have resulted from tectonic downwarping. It is my impression that the rate of basin subsidence from New York to Virginia was slower during the latest Devonian than during the early Late Devonian. The observed more rapid westward progradation in the latest Devonian seems consistent with a slower rate of basin subsidence as the shoreline advanced westward across the early Late Devonian site of maximum water depth toward or onto the cratonward edge of the basin.

Sea-level Changes

It is relatively easy to identify a basin-wide change in sea level during deposition of the Catskill Delta, because several delta lobes are affected simultaneously (Dennison and Head 1975). Black shale tongues seem to represent times of sea-level rise when anaerobic, deep-water conditions are superimposed onto shallower-water sediments, perhaps on a slope. This condition may result from changes in the pycnocline depth with varying chemical and water circulation conditions in the sea, but the major black shale tongues such as the Geneseo-Harrell, Rhinestreet, Pipe Creek, Dunkirk, and the Cleveland-lower Big Stone Gap strata are generally conceded to represent sea-level rises that drove the shoreline eastward. During deposition of those dark strata, corresponding coarser and nearer-shore facies also shifted eastward. Some dark shales close to the shoreline may indeed be lagoonal deposits where circulation was restricted by barrier island systems. Several marker-bed limestones in the Middle

Devonian (Centerfield-Landes, Portland Point-Pokejoy, and the northern Tully) seem to represent times when the shallower and more gently inclined shelf was flooded by sea-level rise. Then fine terrigenous clastics accumulated on the former shelf west of the delta inundated by sea-level rise (McCave 1968; Johnson and Friedman 1969; Heckel 1973). There is thus a pronounced difference between the Middle and Upper Devonian portions of the delta, with limestones usually resulting from transgressions in the Middle Devonian and black shale tongues being the most conspicuous deposits of transgressions after Tully Limestone deposition.

The pelagic limestones of the deeper-water portion of the Tully and the Purcell-Cherry Valley beds both may be basin-bottom accumulations at times of sea-level rise, if the facies patterns of the eastern Tully from New York to Virginia (Heckel 1973) can be assumed to apply to the Purcell.

The most pronounced sea-level rise caused the Taghanic onlap as the Tully sea flooded the Hamilton Group subaerial delta, while in the Midwest the rise inundated middle America and ultimately covered the Transcontinental Arch. This transgression is more than a relative deepening of the Appalachian basin (Johnson 1970) and it may have been eustatic.

Facies changes in both rock type and fauna permit recognition of lesser sea-level changes within the Cohocton Stage from New York to Virginia (McGhee and Dennison 1980). Ultimately these bands in the strata corresponding to distinct sea-level highs and lows permit tracing time with two or three times the precision possible with fossils alone.

The transgression at the end of the Devonian that produced the black Cleveland Shale tongue in Ohio and Pennsylvania and the black shale of the lower Big Stone Gap Shale in southwestern Virginia also affected both the western craton (lower Bakken strata) and the Canadian Rockies (Exshaw Shale) far beyond the Appalachian basin. The Sunbury transgression in the very early Mississippian was not as pronounced as the Cleveland Shale transgression at the end of the Devonian.

The rather abrupt and very large (perhaps 100 m) sea-level rise in the Taghanic onlap initiated a change from little turbidite deposition in the Middle Devonian to pronounced turbidite deposition in the Upper Devonian. All turbidites require a basin with bottom waters substantially below wave-base, to allow the turbidites to move in suspension toward deeper waters. Such deep basins with turbidites have originated in the Appalachians from three causes: rifting basins, localized tectonic downwarping, and major sea-level rise (Dennison 1980). The sea-level rise accompanying the Taghanic onlap was of sufficient amplitude that deep water formed over the former shallow shelf of Hamilton Group sedimentation. This permitted steep delta slopes to develop during deposition of the Portage-Brallier facies from New York to Tennessee, favoring development of turbidites. Turbidites are much less evident and voluminous in the Hamilton Group, although local tectonic basins provided a setting for some turbidite deposition from southern Pennsylvania to northeastern West Virginia (such as the distal extent of Clearville Siltstone, as described by Jolley 1983). Once turbidite deposition began in the deepened Late Devonian waters, the turbidites prograded westward and northwestward across the Appalachian basin during most of the rest of the Devonian, with isostatic subsidence caused by the entire clastic wedge maintaining a deeper area west of the shelf edge.

A drop in sea level is more difficult to identify in a prograding delta complex than a sea-level rise. The identifying criterion for a sea-level drop is an abrupt basinward shift of nearer-shore facies, affecting several delta lobes simultaneously. Three drops in sea level are apparent in the Catskill Delta complex.

The latest, and therefore the sea-level drop affecting the largest area of shallow marine beds, is a brief westward to southwestward pulse of fluvial redbeds, extending from the Cattaraugus Formation of New York across the subsurface of Pennsylvania, eastern Ohio, and West Virginia to outcrops in southwestern Virginia, all mid-Bradford in age (Dennison and Head 1975; Rickard 1975; McDonnell 1981; Dennison 1983a). This brief sea-level pulse timed the maximum progradation ever attained by the Catskill Delta facies. The extent of this shoreline is shown by line 7 in Figure 4. This sea-level drop apparently introduced turbidites into the former deep-basin shale environment of Ohio and Kentucky (Three Lick Bed), consistent with the initiation of turbidites into deep basins at a time of sea-level drop, in accord with the model proposed by the Shanmugan and Moiola (1982).

Lyke (1981) has described a turbidite fan system that produced a pronounced marine progradation at the base of the Cohocton Stage, evident in the Allegheny Front strata from southern Pennsylvania to southeastern West Virginia. It simultaneously affected the major Fulton and Augusta Lobes of the Catskill Delta complex, as well as smaller lobes intermediate in position. Locally redbed muds that washed in from the near-shore facies can be identified in these lower to middle fan deposits. Lyke agreed with the initial suggestion by Dennison and deWitt (1972) that abrupt and brief progradation of facies probably resulted from a sea-level drop. Contemporaneous strata should be studied in Virginia, central Pennsylvania, and New York to see if the sea-level drop can be detected there.

An earlier possible sea-level drop occurs in the middle of the Tioughnioga Stage and is marked by the Clearville Siltstone (as defined by Dennison and Hasson 1976) from northern Virginia to southern Pennsylvania (Hasson and Dennison 1978, 1979; Jolley 1983) and probably the Nis Hollow Siltstone of eastern Pennsylvania (Epstein and others 1974). Effects in near-shore facies are the least extensive geographically of any of these major sea-level drops, because at that time the Catskill Delta complex was relatively small (shoreline indicated by solid line 3 in Figure 4).

Sea-level drops with less amplitude may have produced the bands with marine redbeds described by Dennison (1971) in the Foreknobs Formation along the Allegheny Front from Maryland to Virginia. The bands are apparently time markers (Dennison and Head 1975), but perhaps these bandings were delimited by

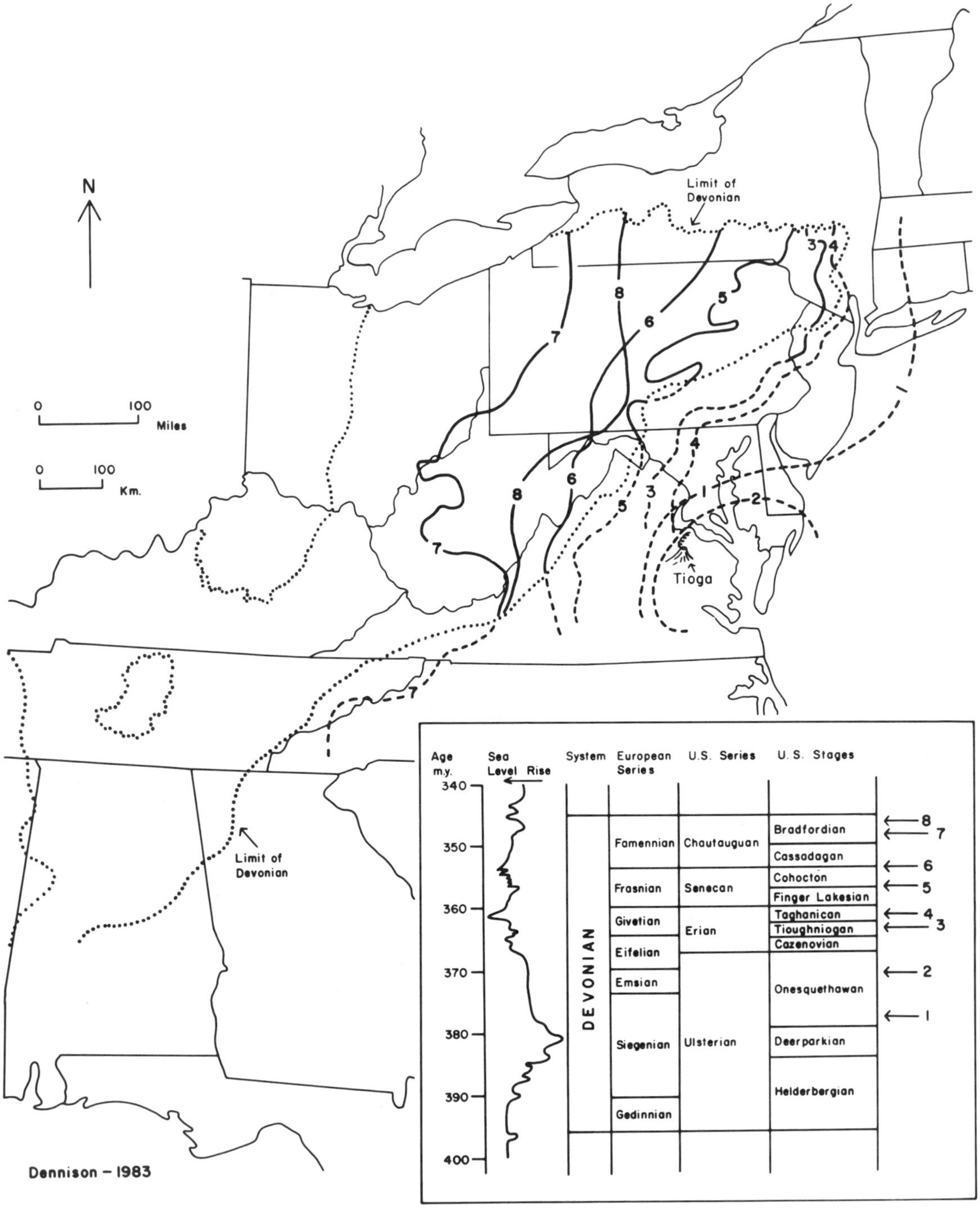

Figure 4. Shifts in shoreline position of Devonian delta lobes. Dotted line delimits extent of preserved Devonian strata. Numbered lines correspond to approximate shoreline at the times indicated by the numbers on the time scale. Note that the position of the shoreline during parts of the Devonian (where shown as dashed line) is inferred to be east of the present limit of outcrops, with the position determined by the shape and lithology of terrigenous clastic wedges. Shorelines are located relative to present-day position of Devonian strata, rather than restored to their Devonian palinspastic position.

sea-level rises superimposed on a generally prograding delta complex, rather than by an actual drop in sea level.

The summary of sea-level changes for the Devonian presented by Dennison and Head (1975) still seems to be generally valid during Catskill Delta deposition. Figure 4 portrays that sea-level curve and also maps the shifts in shoreline through time. Transgressions at times 4 and 8 are particularly evident, corresponding to the Tully Limestone and Cleveland Shale.

Localized Basement Tectonics

The effects of basement tectonics on Devonian sedimentation are only now becoming vaguely known. The following summary reflects chiefly my interpretation.

In the earliest stages of rifting which led to the Iapetus Ocean in Cambrian time, several major fracture systems crossed the region of the Appalachian basin, breaking it along major polygonal fractures which trended about N25°E and N85°E (Figure 3), as described by Dennison (1982b, 1983b). Presumably lesser fractures were somewhat guided by the fracture grain of the major blocks, and it is unlikely that the lesser fractures crossed the major fracture zones. Within the area of the Devonian delta complex the general N85°E trend is followed by what I call the 42nd Parallel, 40th Parallel, 38th Parallel, and 36th Parallel fracture zones (Figure 3). Donaldson and Shumaker (1981) recognized a cross-over zone of basement fractures intermediate between the 38th and 40th Parallel fracture zones.

I suspect that the 40th Parallel fracture zone was active intermittently from Late Precambrian through Jurassic time (Dennison 1982b), and it probably produced two effects in the Devonian. During the Middle Devonian there was probably some uplift of the block south of the fracture and east of the present Chestnut Ridge anticline which overlies the east edge of the Rome Trough. The reported facies change of Tully Limestone into a siltstone (Piotrowski and Harper 1979) is unlikely. Instead, it seems more reasonable that the absence of Tully Limestone in wells is caused by an uplifted block of Clearville Siltstone at the top of the Mahantango, and that the Tully Limestone was eroded along with about 15 meters of older shaly strata shortly after Tully deposition (Dennison 1982b). The Tully is widespread far to the east of the area in Somerset County where the Tully reportedly disappears, and the only reasonable siltstone correlative was situated to the east: the Clearville. The abrupt north border of the Fulton Lobe of the Catskill Delta (and the south border of adjacent Perry Bay) coincides with the 40th Parallel fracture, and a down-to-the-north movement along the fracture may help explain facies patterns associated with the lobe. The Catskill Formation isopach patterns trend nearly eastward across Pennsylvania just north of the 40th Parallel zone (Dennison 1982a, Fig. 15), and this too may reflect relative uplift to the south of the fracture system.

The 38th Parallel fracture zone also influenced Devonian sedimentation. The earliest manifestation of argillaceous clastics from the rising Acadian highlands in the central Appalachian basin was the Needmore Shale. The center of Needmore facies and the thickest accumulation of Needmore is at the latitude of the 38th Parallel trend. The center of Tioga volcanic eruptions that occurred at the end of Needmore deposition is also on this trend. Thirdly, the apparent shape of the western, distal end of the Catskill redbeds in western West Virginia (Figure 4) seems to be indented by the 38th Parallel trend and may represent sedimentary tectonic effects. That would be a fruitful area to study Devonian sedimentary tectonics because small vertical movements along basement fractures would produce wide shifts of shoreline in an area with exceptionally low stream gradients at the most westward progradation of the delta complex. There is evidence in Grant County, West Virginia of up-to-the-north movement of the 38th Parallel fracture zone during Mahantango and Brallier deposition (Avary and Dennison 1980). The thinned Hampshire Formation at Mouth of Seneca, West Virginia (Dennison 1971, Fig. 6) also may represent latest Devonian uplift of the 38th Parallel zone, along an east-trending arch that grew markedly in the middle Mississippian (Dennison 1983b, Fig. 52).

There were probably smaller basement blocks in addition to those major fracture zones described above. One such fracture was the cross-over zone between the 38th and 40th Parallel fractures, as described by Donaldson and Shumaker (1981) during Chemung and Catskill sedimentation.

Random Shifts of Prograding Delta Distributaries

The shape of a delta is caused by random shifts of debouching stream channels that tend to distribute the sediment evenly as the channels discharge into the sea. A weak input of sediment into the sea relative to sediment dispersal by waves and tides results in a smoothly arcuate delta, but a very strong input of sediment can produce a large birdfoot delta. No distinct birdfoot deltas have been identified in the Catskill Delta complex. Birdfoot deltas would be most expected at the time of the Cattaraugus redbeds (lines 7 on Figure 4) when progradation was strongest, or slightly later when sediments prograded onto the shelf inundated by marine transgression at the end of the Devonian (line 8 on Figure 4).

The Catskill Delta complex seems to represent the coalescing of several tectonic deltas (Johnson and Friedman 1969). The individual rivers were relatively small and birdfoot deltas probably did not form from many or any of them.

Progradation associated with random delta shifting can sometimes be separated from the effects of sea-level change. The Augusta Lobe in Virginia first appeared in the Finger Lakes Age as a distinct lower fan deposit of thickened siltstones in the Brallier Formation, with no simultaneous manifestations on the Fulton Lobe, thereby eliminating sea-level drop as the probable cause of the Back Creek Siltstone Member (Avary and Dennison 1980). Individual beds in this turbidite fan cannot be traced from outcrop to outcrop, and such a pattern is most readily explained by random shifts of distributary channels.

Lyke (1981) could not trace individual turbidite beds within submarine fans on the Augusta and Fulton Lobes, but he could trace bundles of beds across parts of the submarine fan complex.

Extent of Delta Progradation

Streams that extend their aggrading surfaces long distances from the fall line provide better opportunities in the lowered stream gradient for winnowing sediment to improve sorting and for more thorough weathering of labile minerals en route. Such streams would develop an intricate drainage pattern on a distributary system that has a broad area subject to marine influences during storms. This may help explain why the Catskill Delta shallow marine sandstones with good sorting and much intergranular permeability are principally confined to the uppermost Devonian, and why hydrocarbon production from the sandstones and siltstones of the lower part of the Catskill Delta comes from "tight sands" with fracture permeability. The Benson sand in the central Appalachian basin is generally considered a "tight sand," but locally has several percent intergranular permeability (Cheema, Donaldson, Heald, and Renton, 1977; Cheema, 1977). As the Late Devonian progressed, nearshore marine sedimentation in Pennsylvania transformed from a prograding muddy shoreline with rapid sedimentation in a rapidly subsiding basin (Walker and Harmes 1971; Walker 1971) to a delta complex that prograded onto a more slowly subsiding western edge of the Appalachian basin, with much greater opportunity for winnowing en route to the sea and after first being deposited in the sea.

Tidal Amplitude

Because the Catskill Delta formed in a cratonic seaway with no nearby connection to large basins with oceanic crust or depths, diurnal tides with large amplitudes were quite unlikely in the Appalachian basin. Furthermore, tidal amplitudes tend to be diminished in low latitudes, the accepted paleogeographic position of the Catskill Delta complex. Sedimentation studies have not located evidence of strong diurnal tides in the Catskill Delta (Walker and Harms 1975), only microtidal activity.

Storm tides may present another situation, however. The Catskill Delta was perhaps far enough south of the equator to be in a tropical hurricane path. Storm overwashes could produce substantial lower coastal plain deposits with sediment reworking on the delta margins. Storm tides could smear subaerial deposits left by the distributary system and produce geographically more extensive fining-upward cycles, washing shells inland, and occasionally flooding the delta with marine waters. Marine overwash could bring in copper chloride which formed malachite in Pennsylvania in lower delta plain deposits (Slaughter 1970; Sevon and others 1978) and geochemical copper anomalies in Virginia associated with greenish sandstones and shales in the Hampshire Formation (Lesure and Matooka 1980).

An exception to general belief in little tidal activity in the Catskill Delta is provided by the study by Rahmanian (1979, 1980). He believed that tidal processes dominated over fluvial processes during deposition of the Irish Valley Member of the Catskill Formation in the Snyder Lobe of western Centre County, Pennsylvania. The tide-dominated delta passed southward along shore to an inactive coastal plain environment (in the region of Perry Bay). He did not specify whether the tidal effects resulted from diurnal or storm tides.

It is seldom possible to find exposures of the Catskill Delta complex where individual bedding planes can be followed for a hundred or more meters. In the redbeds near the facies change from contemporaneous nonmarine to marine strata, I have seen some exposures where the strata are mostly red muds with thin sheet-sands that are quite persistent and not obviously channeled, in contrast to those Hampshire and Catskill deposits more commonly ascribed to meandering stream origin. I prefer to interpret these more evenly bedded strata as storm-tide overwash, rather than the distal edge of crevasse splays.

Longshore Drift

A general drift toward the southwest has been proposed for Devonian delta sedimentation (Willard 1939), based largely on the highly asymmetrical shape of his Snyder Lobe at the base of the Cohocton Stage in central Pennsylvania (line 5 in Figure 4). Palinspastic reconstruction of Devonian outcrops in south-central Pennsylvania restores Willard's data points in such a manner that the very pronounced asymmetry of the Snyder Lobe is partly removed, although some asymmetry of the lobe and adjacent bays is still apparent when his data points are palinspastically restored. Moderate longshore currents are still suggested. This is not readily apparent in the paleocurrent data of Rahmanian (1979, Fig. 120), with the only suggestion of longshore currents being toward the northeast at Horseshoe Curve, Pennsylvania.

The general configuration of line 7 in Figure 4 also suggests longshore drift toward the present southwest. More subsurface work is needed to determine the exact shape of that shoreline beneath the Dunkard basin, but the area with apparent southwestward drift is clearly within the Appalachian Plateau, so its shape does not need palinspastic restoration.

Contour currents can also distribute sands along the continental slope areas of modern oceans, in addition to the turbidity currents moving downslope. Contour currents seem to originate from the global circulation of large water masses associated with oceanic areas and depths as part of the global heat exchange mechanism. The restricted size and shape of the Appalachian basin probably was not conducive to development of significant contour current systems.

Woodrow and Isley (1983) have suggested that internal waves along the pycnocline might produce currents strong enough to move sand in deeper water in a direction related to the movement of the waves or even dislodge sediments on a gentle slope to generate turbidity currents. Woodrow (1983) describes sedimentary structures associated with a possible Devonian pycnocline in New York.

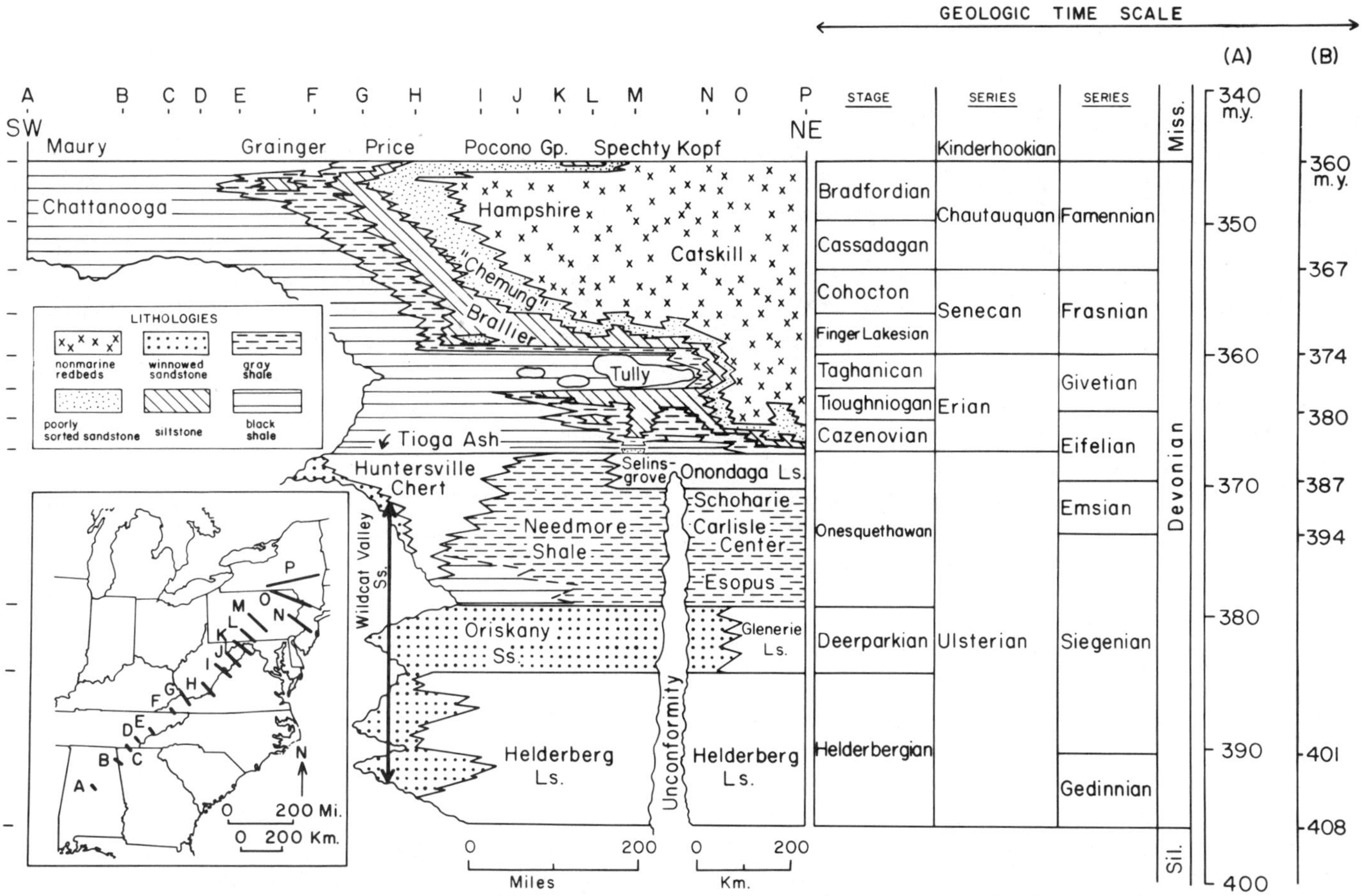

Figure 5. Stratigraphic panel showing age relationships of Catskill Delta facies in the eastern part of the Appalachian basin. For columnar strips G through P the bottom of the panel is the most eastern outcrop belt in the Valley and Ridge Province, and the panel top is at the eastern edge of the Appalachian Plateau Province. Such an inclined stratigraphic cross section permits the maximum resolution of ages of shifting marine lobes of Devonian deltas. For columnar strips A through F the panel is at the most eastern outcrop belt of the Valley and Ridge Province, closest to the source of terrigenous clastics.

Mass Wastage of the Delta Front

Basinward transport by mass wastage of parts of the subaerial and submarine deposits is an important process in modern ocean deltas and on continental slopes. This is most evident from submarine slumps and slides and from down-to-basin listric faults that cut the subaerial portions of deltas; both can be identified geomorphically. No such faulting has been clearly identified in the Catskill Delta complex.

The fault proposed by Heckel (1973) in central New York as cutting the Hamilton Group prior to Tully deposition is probably not listric because its upward movement on the west is opposite what would be expected from listric faulting of a delta prograding from the east.

Large-scale listric faulting may not have affected the Catskill Delta, either because the basin was not deep enough or steep enough to generate this sort of instability or because the Appalachian basin was a cratonic basin and unlikely to develop large unstable masses capable of slumping.

To my knowledge, none of the chaotic blocks and hummocky topography characteristic of the lower edge of submarine slides of fairly coherent sediment masses have been identified in the Catskill Delta complex, and I suspect that they did not form. Seismic profiles and closely spaced wells in the Appalachian Plateau should be searched carefully for such features, to be marked by laterally abrupt thickness variations within stratigraphic intervals tens to hundreds of meters thick.

SHIFTS IN DELTA LOBES WITH TIME

Figures 4 and 5 portray the progressive development of the Catskill Delta complex in map and stratigraphic cross section views. The published information sources used for preparing these illustrations are listed in the Appendix, and are supplemented by my own field observations. For brevity, a single description will follow the delta development in both figures. Details of Figure 4 will be referred to as time or shoreline 1 through 8, and geographic locations in Figure 5 will be designated as lettered columns A through P.

Subaerial deltas also have a submarine shallow shelf, often a delta front or slope, and distal prodelta deposits. These marine portions of deltas can outline the position and general shape of deltas, even though the subaerial delta was located east of the present limit of preservation of Devonian rocks. Thus it is possible to infer the location of shorelines which were situated east of the presently preserved Devonian strata (dashed lines in Figure 4). Figure 4 shows the approximate shoreline at eight times during the Catskill Delta development.

Figure 5 is a stratigraphic diagram of an unusual sort. Its vertical dimension is a numerical time scale in years. Numbered time scale (A) is after Harland and others (1964). Numerical ages for the geological time scale are in a state of flux. One of the newest time scales is by Harland and others (1982), shown as numbered time scale (B) in Figure 5. The time scale published by Palmer (1983) follows that of Harland and others (1982) for the Devonian. The more recent Devonian ages are about 11 to 20 million years older than those of Harland and others (1964). The facies portrayed in Figure 5 are linked by lithostratigraphic and biostratigraphic correlations with Stages in the New York stratigraphic section. These correlations form the basis for the conclusions in the present paper, independent of fluctuations of radiometric dates for the geologic time scale.

Horizontal distance in Figure 5 is portrayed along the Devonian outcrop belts of the Valley and Ridge and eastern Appalachian Plateau Provinces, from transects at positions A through P. Lithologies of terrigenous clastics are indicated by patterns. Wavy lines surrounding blank areas indicate the positions of unconformities near Swatara Gap, Pennsylvania (Inners 1975, 556) and from Virginia (within the Wildcat Valley Sandstone and the base of the Upper Devonian) southwestward across Tennessee to Alabama (at base of the Chattanooga Shale). Authigenic limestone and chert are all shown as unpatterned areas in the lithologic diagram (including small unnamed areas of lenticular Tully Limestone near column K). The dashes at the left margin of Figure 5 correspond to Stage boundaries in New York, shown at the left end of the time scale, if the reader should choose to color time lines through the lithologic diagram.

The earliest argillaceous clastic wedge began just after Oriskany Sandstone deposition, with clays appearing at the base of the Onesquethaw Stage from two sources to the east (see shoreline 1). This was just after the Wallbridge Discontinuity, a global lowering of sea level, during which the well-sorted Oriskany Sandstone formed by shoaling across the Appalachian basin (Dennison and Head 1975). The southern source region (letter K) was centered in Virginia and produced the lower Needmore Shale, and a northern source (letter O) shed clastics of the Esopus Shale. The intervening embayment is inferred from the thickness patterns of the clastic wedges (Inners 1975).

By time 2 the northern source was apparently inundated, as inferred from the lack of eastern detrital clastic input into the Onondaga Limestone sea. The Needmore source area in Virginia continued to uplift and Needmore muds radiated to the west, northwest, and north. Needmore muds coarsen upward to siltstones at Massanutten Mountain, Virginia just below the top of the Onesquethaw Stage. The Stage is terminated by the top of the middle coarse zone of Tioga eruptive rocks, a cluster of three sand-size tuffs generally occurring within a meter of strata. Grain-size and thickness studies of the tuffs place the source volcano(s) within about the size area sketched as a volcano on Figure 4 (Dennison and Textoris 1978). The volcanic complex probably extended above sea level because a subaerial eruption seems necessary to account for ash bed accumulation over 1000 km away. The main volcanic plume passed northward to about Frederick, Maryland and then arced westward toward Pittsburgh, Pennsylvania and Michigan. This wind pattern suggests that these tuffs were carried by trade winds in the Devonian southern hemisphere (Dennison and Textoris 1980) and provides an independent measure of paleolatitude, which generally confirms a Devonian equator passing through Michigan or Ohio.

Just before the Tioga middle coarse zone eruptions (about 0.3 m below the middle coarse zone in the Needmore Shale) the sea deepened abruptly, at least from Virginia to Pennsylvania, as evidenced by a change from aerobic to anaerobic muds and an abrupt decrease in grain size from silty shale to fine clay. This change suggests that the Virginia shoreline was driven eastward and that perhaps the landmass was inundated.

During early Cazenovia time Marcellus black muds spread most vigorously northward from south-central Pennsylvania or the present Maryland Piedmont and westward from a newly uplifted source area in New England as the Acadian orogeny intensified. These muds buried the sites of Onondaga, Selinsgrove, and Huntersville limestone and chert accumulation. By the end of Marcellus deposition erosion of uplifts in New England caused the earliest deltas to prograde fluvial plains westward and northward, extending from Albany, New York to northern New Jersey (Wolff 1967; Kirby 1981). This shoreline is portrayed by line 3. Evidence can be seen in the marine beds for ancestral Wyoming, Snyder, and Fulton Lobes by mid-Tioughniogan Stage (line 3 and columns L through O).

The Taghanic onlap drove the shoreline eastward (line 4 and columns J through L). The actual shoreline shift can be demonstrated in New York (Johnson and Friedman 1969), and a corresponding eastward shift of marine facies occurs in outcrops and the subsurface from New York to Virginia. There does not seem to be an orogenic slackening, but simply an eastward shoreline shift in response to the Taghanic sea-level rise.

The Augusta Lobe appeared in Virginia and West Virginia between times 4 and 5 as the turbidite fan of the Back Creek Siltstone (Avary and Dennison 1980), also shown in column I.

Line 5 shows the major delta lobes (after Willard 1934) at the base of the Cohocton Stage, when the brachiopod *Cyrtospirifer* appeared in the Appalachian basin (Greiner 1957), possibly as the Transcontinental Arch was inundated by the Late Devonian sea-level rise. Willard's recognition of the shapes of delta lobes and embayments remains a classic combination of biostratigraphy and paleogeography. *Cyrtospirifer* is a slightly offshore shallow marine facies fossil (McGhee and Sutton 1981), so its eastward disappearance as mapped by Willard in the prograding clastics probably portrays the limit of brackish water adjacent to debouching streams, rather than land itself. His mapping probably represents generally the shapes of the major delta lobes in Pennsylvania. It is important to refine this paleogeography by mapping the eastward paleoecologic loss of the oldest species, *C. chemungensis,* rather than the eastern limit of younger species. This is probably critical in delimiting Wyoming Lobe and Luzerne Bay. Other streams were probably present. Sevon (Sevon and others 1978; Sevon 1979) believes eight rivers discharged across Pennsylvania from crystalline bedrock channels in Appalachia that were rather fixed in position throughout Middle and Late Devonian time. Sevon's work may have identified more numerous streams than Willard, but Willard probably did recognize the major streams in terms of volume of terrigenous clastic input.

Lyke (1981) identified minor turbidite fans marking stream mouths associated with the Fulton and Augusta Lobes, the major input centers at time 5. The weaker turbidite fans apparently required a significant amount of time to prograde westward to the Allegheny Front because of their smaller rate of detrital influx, even though all of the turbidite fans in his informal Minnehaha member probably prograded markedly westward as a result of an abrupt sea-level drop at the Finger Lakes–Cohocton boundary.

By the end of Cohocton time (line 6) the clastic inputs of the Fulton and Augusta Lobes were dominant. Kirchgessner (1973) identified a minor sublobe at Mouth of Seneca, West Virginia. The shoreline in New York was still prograding westward, but not as vigorously as in southern Pennsylvania and Virginia.

Details of shoreline shift are unclear between lines 6 and 7, because outcrops of the shoreline strata are available only in New York, southeastern West Virginia, and adjacent Virginia. The desirable detail of well-sample descriptions has not been done in Pennsylvania and West Virginia. A series of northward-trending subsurface oil sands of this age are generally identified as barrier or offshore bar sands (Dennison 1971) because of their sorting, orientation, and stratigraphic position. The extensive network of correlations established by gamma ray logs as part of the U.S. Department of Energy Eastern Gas Shale Project (Roen and others 1978a; 1978b; West 1978; Piotrowski and Harper 1979; Piotrowski and Krajewski 1979; Schweitering 1980) identifies sand bodies fairly clearly, but does not separate fluvial from shallow shelf sandstones. Fluvial strata are commonly distinguished by red color of well cuttings of sandstones and enclosing shales, but red does not always mean nonmarine, especially if it is dull red to brownish rather than bright red.

There is no indication in Virginia and Tennessee of any delta south of the Augusta Lobe during the interval between times 6 and 7.

The shoreline at time 7 is basically that shown by Dennison and Head (1975) and represents the maximum progradation attained by the Catskill Delta complex. The shoreline that crops out in western New York and northwestern Pennsylvania as the Cattaraugus redbeds can generally be followed by the westward loss of redbeds brightly enough colored to be identified in well cuttings across Pennsylvania, Ohio, and West Virginia, cropping out again near White Sulphur Springs, West Virginia (McDonnell 1981) and traceable southward to the Roanoke area of Virginia. Brallier turbidites identified in Tennessee by Dennison and Boucot (1974) and studied by Hasson (1982) clearly delimit a Bradford-age fan indicating a Hawkins Lobe source area to the southeast in North Carolina. The general embayment in southwestern Virginia between the Hawkins and Augusta Lobes was designated Scott Bay by Hasson (1982). A minor river discharging into Scott Bay may have been present near Wytheville, but more work is needed to determine whether this was a separate input center. In southern Tennessee along the Little Tennessee River, the entire upper Devonian is represented by 7 meters of black Chattanooga Shale (Glover 1959), and there is no evidence of any Devonian delta lobes developing east of the southern Tennessee, Georgia, or Alabama outcrops.

A sea-level rise at the end of the Devonian produced a major shoreline shift, driving the edge of the land nearly 160 km east as the Penn-York Embayment reached western Maryland (Laird 1942; Dennison and Head 1975; Dennison 1983a). Black shale tongues extended eastward from the Cleveland Shale and the lower part of the Big Stone Gap Member of the Chattanooga Shale (columns E and F). The shape of shoreline 8 suggests that by the end of the Devonian the Acadian highlands of New England may have begun to wear down, with smaller volumes of sediment produced, and that the principal uplifted area was the Piedmont of Virginia.

The beginning of the Mississippian Period marks the end of the Devonian delta complex, not only because of the shift to a new Period, but because of a quite different set of tectonic conditions affecting a much larger area of the United States. The Devonian delta complex was entirely derived from uplifts of the Acadian mountains east of the Appalachian basin. In earliest Mississippian the Spechty Kopf-Pocono-Price-Grainger Formations (Figure 5) were terrigenous clastics that continued to enter the seaway from the east and southeast. The Hawkins Lobe source area disappeared, and a new prominent input center (at the word "Price" between columns G and H) appeared at the site of the latest Devonian Scott Bay, forming the Virginia-Carolina delta (Pepper and others 1954) that persisted as the most prominent stream from Appalachia during the remainder of the Paleozoic. There is no evidence that the Canadian Shield area contributed sediments to the Catskill Delta. During the early Mississippian, however, the region of the Canadian Shield was strongly uplifted, so that a major northern source was evident in the Borden delta complex extending from Illinois to Tennessee (Whitehead 1979). Pepper and others (1954) suggested that the Berea-Red Bedford delta prograded from the north across Ohio in earliest Mississippian time (just after shoreline 8). However, I accept the evidence marshalled by Lewis (1976), which indicates that the Berea-Bedford of northern Ohio represents a barrier island system foundered in prodelta muds that moved westward into Ohio. That regime was shortly followed by sediments derived from a true northern source—the Cuyahoga Formation of

Ohio—that extended across Kentucky as the Borden delta. As a further indication of increased complexity, the lower Middle Mississippian Floyd Shale of the Alabama portion of the Appalachian basin seems to be derived from a newly risen southern or southwestern Ouachita source area (Thomas 1972). The change from Devonian to Mississippian tectonics is very marked, and in many ways heralded even more complex Pennsylvanian tectonics.

PROGRESSION OF THE ACADIAN OROGENY

Once a stream system initiated a delta lobe in the Middle and Late Devonian, that lobe seemed to persist, indicating that the stream was firmly notched through major water gaps in the source area. Streams progressively appeared from north to south, as shown in the sediments of the eastern margin of the Appalachian basin. Does this imply a progressive uplift to the south (in present compass directions) that developed clastic wedges along the southwest-trending eastern margin of the Valley and Ridge Province? Do the wedges result from a gradual westward approach of an already present south-trending continent or microcontinent, compared to the southwest trend of the outcrop belts? Or do the wedges represent the clockwise rotation of a microcontinent located east of the central and southern Appalachian basin?

Examination of the volume of clastic wedges does not answer these questions clearly. If a series of major rivers approximately equal in size existed along the west slope of an Acadian highland reaching from New York to North Carolina, terrigenous clastic wedges should first indicate vague sources to the east of the present margin of Devonian preservation, and then become more specific distributary centers as the elongate crystalline landmass approached. This is not the case. Instead, each delta in the Middle and Late Devonian appears as a small area of siltstone introduced into the eastern margin of the black shales of the seaway. This suggests the new appearance of specific streams draining newly uplifted areas, rather than the gradual, oblique approach of long established streams draining an already developed large landmass.

The answer to these questions can best come from sedimentary petrography. A single, fairly uniform uplift along the entire length of the Acadian highlands that eroded to produce the Catskill Delta complex should produce evidence of simultaneous unroofing of progressively deeper metamorphic zones within the source area. There is limited, possibly inadequate evidence that this unroofing was not simultaneous, but rather that the northern area uplifted in Middle Devonian time, with a progression of uplift southward in the Late Devonian.

In the Catskill-Pocono-Green Pond-Skunnemunk region (column O), the earliest uplift shed detritus from greenschist terrain of the old Appalachia, followed by stronger input of reworked sediments as the wave of folding progressed westward in the Middle Devonian (Kirby 1981). Unroofing of higher metamorphic grades or igneous rocks waited until the Late Devonian.

Between columns I to K, in studies of the Foreknobs Formation (Cohocton and Cassadaga Stages) along the Allegheny Front, Kirchgessner (1973) found petrographically that the uplift was progressing southward in the source terrain for the Fulton Lobe, Grant Bay, and Augusta Lobe. Feldspar content increased upward, and clasts from greenschist facies decreased upward along the 160 km outcrop trend, as unroofing progressed. The Fulton Lobe drained from a provenance with a hotter thermal history (more feldspar in the metamorphic and possibly intrusive rocks) than equivalent beds in the Augusta Lobe.

Needless to say, these petrographic studies are inadequate to test clearly the hypothesis of uplift progressing southward, but they are consistent with that hypothesis. A series of petrographic traverses throughout stratigraphic sections at the positions of columns F through P in Figure 5 is needed to characterize the changes through time of the shallow marine sandstones, as the clearest indicator of changes in the source area east of the Appalachian basin. Paleocurrent studies of the marine beds of the Catskill Delta indicate northwestward to westward grain transport (McIver 1961; Potter and others 1982). It is desirable for sampling to remain in the nearshore region for two reasons. Time resolution is much more precise from marine fauna than in nonmarine beds. The thick coating of hematite in many sandstones of the nonmarine facies would make mineralogic interpretation much more difficult than in the ocean-winnowed facies.

There is a strong suggestion from the mapped shorelines of Figure 4 that the maximum uplift (and hence supply of the largest volume of detritus) did progress southward during the Acadian orogeny. The most rapidly accumulating preserved deltas were in the north during the Middle Devonian (times 3 and 4); there was a large volume of nonmarine beds in New York and New Jersey, compared with a probably smaller volume of marine detrital clastics in the Snyder and Fulton Lobes. By time 5 (at base of Cohocton Stage), the Wyoming, Snyder, and Fulton Lobes were all very pronounced, with the Augusta Lobe appearing shortly before, and the Devonian delta lobe near Albany seemingly weakened. By time 6 the Augusta Lobe had become the most rapidly growing lobe, with moderately voluminous accumulation in a combined Fulton and Snyder Lobe. Not enough work has been done in Pennsylvania to delimit clearly the separate effects of the Fulton and Snyder Lobes at that time, although the work of Glaeser (1979) and Rahmanian (1980) would suggest that both lobes were fairly prominent. By mid-Bradford time (line 7) a much extended Augusta Lobe was reaching far beyond the more slowly prograding shoreline of the northern lobes, and a tiny point-source of a weak Hawkins Lobe had just appeared in the south. The transgression at the end of the Devonian (line 8) shows a large Penn-York Embayment masking the Fulton and Snyder Lobes. This embayment appears as a zone of siltstone at the top of the Devonian between columns K and L. At that time there was an indistinct lobe west of the earlier lobes near Kingston, New York and Wyoming County, Pennsylvania. The same transgression obliterated evidence of the Hawkins Lobe (between columns E and F), and probably the small source area to the east in North Carolina was inundated. At the beginning of the Mississippian Period the very voluminous Virginia-Carolina delta centered near the south border of West Virginia (at the word "Price" in Figure 5) completed the southward progression of centers of greatest uplift.

In summary, the strata contain evidence that the uplift of the Appalachia source area progressed southward during the Acadian orogeny, and that the northern uplifted area was progressively reduced in elevation as the center of maximum uplift moved southward. This suggests some sort of rotational or sequential motion of colliding continental plates during the Acadian orogeny. Such a motion places specific constraints on models of Devonian plate behavior.

The apparent north-south progression of uplift of the Devonian delta source areas may not reflect a true succession of topographic development, because many modern rivers do not debouch into the sea closest to the highlands in the source area of the river. It is significant that such a progression is consistent with 20-30 degrees of clockwise rotation of the Traveler block in New England during the Acadian orogeny, as proposed by Spariosu and Kent (1983) from paleomagnetic evidence. The extrusives of the Traveler Felsite are paleontologically dated as probably early Onesquethaw age, corresponding to the clastic wedge of the Espopus-Carlisle Center-Schoharie Formations shown in Figures 4 and 5 of this paper. The rotated Acadian highlands described by Spariosu and Kent (1983) may have extended south as far as North Carolina, and the turbidite siltstones of the Hawkins Lobe may represent the final arrival of the southern tip of an uplifting Acadian block close enough to the Appalachian basin that a detrital clastic wedge appears in the most eastern Devonian outcrop belts of northeastern Tennessee.

Osberg (1978) has also pointed out that the anticlines produced in New England by the orogeny become more closely spaced and increasingly overthrust to the south from Maine to Massachusetts, indicating more intensive compression southward, perhaps as a result of this rotational movement. Indications of the New England rotated block have not been specifically identified south of Long Island Sound, but the evidence of this paper suggests that its uplift effects may have extended as far south as the Carolinas.

ACKNOWLEDGMENTS

Support for this research came in part from University of North Carolina Research Grant VP446.

Appendix—Documentation for Figures 4 and 5

Sources of information used in preparing Figures 4 and 5 are listed from south to north, according to the lettered sequence of column locations on the map in Figure 5. Subsurface information sources are listed for the Appalachian Plateau to the northwest of those lines. These published sources were supplemented by my field observations beginning in 1958.

Column A
Glover (1959), Kiefer (1970)

Column B
Glover (1959), Kiefer (1970)

Column C
Glover (1959)

Column D
Glover (1959)

Column E
Glover (1959)

Between Columns E and F
Dennison and Boucot (1974), Hasson (1982), Whitehead (1979)

Column F
Kepferle and others (1981), Roen and others (1964), Whitehead (1979)

Column G
Dennison (1961), Englund (1968), Whitehead (1979), Woodward (1943)

Column H
Dennison (1961), Dennison and Wheeler (1975), Lyke (1981), McDonnell (1981), Woodward (1943)
Subsurface information
Martens (1939, 1945)

Column I
Avary and Dennison (1981), Dennison (1961, 1971), Dennison and deWitt (1972), Dennison and Wheeler (1975), Hasson and Dennison (1978), McGhee and Dennison (1980), Lyke (1981), Woodward (1943)
Subsurface information
Martens (1939, 1945)

Column J
Avary and Dennison (1980), Dennison (1961, 1971), Dennison deWitt (1972), Dennison and Wheeler (1975), Hasson and Dennison (1978), Jolley (1983), Lyke (1981), McGhee and Dennison (1980), McIver (1961, 1970), Meckel (1970), Woodward (1943)
Subsurface information
Martens (1939, 1945)

Column K
Dennison (1961, 1971, 1982a), Dennison and deWitt (1972), Dennison and Hasson (1976), Dennison and Wheeler (1975), Hasson and Dennison (1978, 1979), Inners (1975), Jolley (1983), Laird (1942), Lyke (1981), McIver (1961, 1970), Meckel (1970), Willard (1939), Woodward (1943)
Subsurface information
Fettke (1961)

Column L
Dennison (1961, 1982a), Dennison and deWitt (1972), Dennison and Hasson (1976), Ellison (1965), Frakes (1967), Hasson and Dennison (1978, 1979), Inners (1975), Jolley (1983), Kaiser (1972), Laird (1942), Lyke (1981), McIver (1961, 1970), Meckel (1970), Willard (1939).
Subsurface information
Fettke (1961), Piotrowski and Harper (1979), Piotrowski and Krajewski (1979)

Column M
Dennison (1982a), Ellison (1965), Faill and others (1978), Frakes (1967), Glaeser (1974), Inners (1975), Kaiser (1972), McIver (1961, 1970), Meckel (1970), Rahmanian (1979), Walker (1971), Walker and Harmes (1971, 1975), Willard (1939)
Subsurface information
Fettke (1961), Piotrowski and Harper (1979)

Between Columns M and N
Dennison (1982a), Epstein and others (1974), Frakes (1974),

Glaeser (1974), Inners (1975), Willard (1939)

Column N

Burtner (1963, 1965), Dennison (1982a), Fletcher and Woodrow (1970), Glaeser (1974), Inners (1975), Willard (1939)

Subsurface information

Fettke (1961), Glaeser (1974)

Column O

Burtner (1963, 1965), Dennison (1982a), Glaeser (1974), Kirby (1981), McCave (1968), Mencher (1939), Rickard (1975), Wolff (1967), Woodrow and Fletcher (1968)

Column P

Burtner (1963, 1965), Dennison (1982a), Heckel (1973), Johnson and Friedman (1969), McCave (1968), Mencher (1939), Rickard (1975), Wolff (1967)

REFERENCES CITED

Avary, K. L., and Dennison, J. M., 1980, Back Creek Siltstone Member of Devonian Brallier Formation in Virginia and West Virginia: Southeastern Geology, v. 21, p. 121–153.

Burtner, R. L., 1963, Sediment dispersal patterns within the Catskill facies of southeastern New York and northeastern Pennsylvania: Pennsylvania Geological Survey, 4th ser., General Geology Report G 39, p. 7–23.

—— 1965, Paleocurrent and petrographic analysis of the Catskill facies of southeastern New York and northeastern Pennsylvania [Ph.D. thesis]: Cambridge, Harvard University, 222 p.

Cheema, M., 1977, Sedimentation and gas production of the Upper Devonian Benson sand in north-central West Virginia: A model for exogeosynclinal mid-fan turbidites of a delta complex [Ph.D. thesis]: Morgantown, West Virginia University, 117 p.

Cheema, M. R., Donaldson, A. C., Heald, M. T., and Renton, J. J., 1977, Gas-producing submarine-fan deposits of Catskill delta complex in north-central West Virginia: American Association of Petroleum Geologists Bulletin, v. 61, p. 775–776.

Dennison, J. M., 1961, Stratigraphy of Onesquethaw Stage of Devonian in West Virginia and bordering states: West Virginia Geological Survey Bulletin 22, 87 p.

—— 1971, Petroleum related to Middle and Upper Devonian deltaic facies in central Appalachians: American Association of Petroleum Geologists Bulletin, v. 55, p. 1179–1193.

—— 1980, Causes of Appalachian basin turbidites: Geological Society of America Abstracts with Programs, v. 12, no. 2, p. 30.

—— 1982a, Uranium favorability of nonmarine and marginal marine strata of Late Precambrian and Paleozoic age in Ohio, Pennsylvania, New Jersey, and New York: U.S. Department of Energy, National Uranium Resource Evaluation Report GJBX-(50)82, 254 p.

—— 1982b, Geologic history of the 40th Parallel lineament in Pennsylvania: Morgantown, Appalachian Basin Industrial Associates, Program-Spring Meeting, April 15 and 16, 1982 (Knoxville, Tennessee), 19 p.

—— 1982c, Geologic history of the Appalachian basin keel line: Morgantown, Appalachian Basin Industrial Associates, Program-Fall Meeting, October 14 and 15, 1982 (Blacksburg, Virginia), 25 p.

—— 1983a, Latest Devonian shoreline trend in the Appalachian basin: Morgantown, Appalachian Basin Industrial Associates, Program-Spring Meeting, April 7 and 8, 1983 (Research Triangle Park, North Carolina), p. 38–62.

—— 1983b, Sedimentary tectonics of the Appalachian basin: Dallas, Dallas Geological Society, 129 p. (Notes for short course, April 16 and 17, 1983 meeting of American Association of Petroleum Geologists).

Dennison, J. M., and Boucot, A. J., 1974, Little War Gap at Clinch Mountain provides standard reference section for Silurian Clinch Sandstone and most nearly complete Devonian section in eastern Tennessee: Southeastern Geology, v. 16, p. 79–101.

Dennison, J. M., and deWitt, W., Jr., 1972, Redbed zone produced by sea level drop at beginning of Devonian Cohocton Age delimits Fulton and Augusta Lobes of Catskill delta complex (Pennsylvania Geological Survey, 4th ser., Guidebook, 37th Annual Field Conference of Pennsylvania Geologists), p. 109–114.

Dennison, J. M., and Hasson, K. O., 1976, Stratigraphic cross section of Hamilton Group (Devonian) and adjacent strata along south border of Pennsylvania: American Association of Petroleum Geologists Bulletin, v. 60, p. 278–287.

Dennison, J. M., and Head, J. W., 1975, Sealevel variations interpreted from the Appalachian basin Silurian and Devonian: American Journal of Science, v. 275, p. 1089–1120.

Dennison, J. M., and Textoris, D. A., 1978, Tioga Bentonite time-marker associated with Devonian shales in Appalachian basin, *in* Schott, G. L., Overbey, W. K., Jr., Hunt, A. E., and Komar, C. A., eds., Proceedings First Eastern Gas Shales Symposium, October 17–19, 1977, Morgantown, West Virginia: U.S. Department of Energy, Morgantown Energy Research Center, Publication MERC/SP-77/5, p. 166–182.

—— 1980, Middle Devonian wind direction for the North American plate determined from the Tioga Bentonite: 26th Congrès Géologique International, Résumés (Abstracts), v. 1, p. 223.

Dennison, J. M., and Wheeler, W. H., 1975, Stratigraphy of Precambrian through Cretaceous strata of probable fluvial origin in southeastern United States and their potential as uranium host rocks: Southeastern Geology, Special Publication 5, 210 p.

Donaldson, A. C. and Shumaker, R. C., 1981, Late Paleozoic molasse of central Appalachians, *in* Maill, A. D., ed., Sedimentation and tectonics in alluvial basins: Geological Association of Canada Special Paper 23, p. 99–124.

Ellison, R. L., 1965, Stratigraphy and paleontology of the Mahantango Formation in south-central Pennsylvania: Pennsylvania Geological Survey, 4th ser., General Geology Report G 48, 298 p.

Englund, K. J., 1968, Geologic map of the Bramwell Quadrangle, West Virginia–Virginia: U.S. Geological Survey Geologic Quadrangle Map GQ-745.

Epstein, J. B., Sevon, W. D., and Glaeser, J. D., 1974, Geology and mineral resources of the Lehighton and Palmerton quadrangles, Carbon and Northampton Counties, Pennsylvania: Pennsylvania Topographic and Geologic Survey, 4th ser., Atlas 195cd, 460 p.

Ettensohn, F. R., and Barron, L. S., 1981, Depositional model for the Devonian-Mississippian black-shale sequence of North America: a tectonic-climatic approach: U.S. Department of Energy, Morgantown Energy Technology Center Document 12040-2, 80 p.

Faill, R. T., Hoskins, D. M., and Wells, R. B., 1978, Middle Devonian stratigraphy in central Pennsylvania—a revision: Pennsylvania Geological Survey, 4th ser., General Geology Report 70, 28 p.

Fettke, C. R., 1961, Well-sample descriptions in northwestern Pennsylvania and adjacent states: Pennsylvania Topographic and Geologic Survey, 4th ser., Bulletin M 40, 691 p.

Fletcher, F. W., and Woodrow, D. L., 1970, Geology and mineral resources of the Pennsylvania portion of the Milford and Port Jervis fifteen-minute quadrangles: Pennsylvania Geological Survey, 4th ser., Atlas 223, 64 p.

Frakes, L. A., 1967, Stratigraphy of the Devonian Trimmers Rock in eastern Pennsylvania: Pennsylvania Geological Survey, 4th ser., General Geology Report G 51, 208 p.

Glaeser, J. D., 1974, Upper Devonian stratigraphy and sedimentary environments in northeastern Pennsylvania: Pennsylvania Geological Survey, 4th ser., General Geology Report 63, 89 p.

—— 1979, Catskill delta slope sediments in the central Appalachian basin: source deposits and reservoir deposits, *in* Doyle, L. J. and Pilkey, O. H., eds., Geology of continental slopes: Society of Economic Paleontologists and Mineralogists Special Publication 27, p. 343–357.

Glover, L., 1959, Stratigraphy and uranium content of the Chattanooga Shale in northeastern Alabama, northwestern Georgia, and eastern Tennessee: U.S. Geological Survey Bulletin 1087-E, p. 133–168.

Greiner, H., 1957, *"Spirifer disjunctus"*: its evolution and paleoecology in the Catskill Delta: Peabody Museum of Natural History Bulletin 11, 75 p. + 13 plates.

Harland, W. B., Smith, A. G., and Wilcox, B., 1964, The Phanerozoic time scale:

Geological Society of London Journal, v. 120s, 458 p.

Harland, W. B., Cox, A. V., Llewellyn, P. G., Pickton, C.A.G., Smith, A. G., and Walters, R., 1982, A geologic time scale: Cambridge, Cambridge University Press, 131 p.

Hasson, K. O., 1982, Stratigraphy of the Chattanooga Shale, northeastern Tennessee: Southeastern Geology, v. 23, p. 171–185.

Hasson, K. O., and Dennison, J. M., 1978, Stratigraphy of the Devonian Harrell and Millboro Shales in parts of Pennsylvania, Maryland, West Virginia, and Virginia: U.S. Department of Energy Research and Development Administration, Project Final Report for Contract #EY-77-C-21-8153, 124 p. (Open-File Report US DOE ESGP 110.)

—— 1979, Devonian shale stratigraphy between Perry Bay and the Fulton Lobe, south-central Pennsylvania and Maryland: Pennsylvania Topographic and Geologic Survey, 4th ser., Guidebook 44th Annual Field Conference of Pennsylvania Geologists, p. 1–17.

Heckel, P. H., 1973, Nature, origin, and significance of the Tully Limestone: Geological Society of America Special Paper 138, 244 p.

Heyl, A. V., 1972, The 38th Parallel lineament and its relationship to ore deposits: Economic Geology, v. 67, p. 879–894.

Inners, J. D., 1975, The stratigraphy and paleontology of the Onesquethaw Stage in Pennsylvania and adjacent states [Ph.D. thesis]: Amherst, University of Massachusetts, 666 p.

Johnson, J. G., 1970, Taghanic onlap and the end of North American Devonian provinciality: Geological Society of America Bulletin, v. 81, p. 2077–2105.

Johnson, K. G., and Friedman, G. M., 1969, The Tully clastic correlatives (Upper Devonian) of New York State—a model for recognition of alluvial, dune (?) tidal, nearshore (bar and lagoon), and offshore sedimentary environments in a tectonic delta complex: Journal of Sedimentary Petrology, v. 39, p. 451–485.

Jolley, R. M., 1983, The Clearville Siltstone member of the Middle Devonian Mahantango Formation in parts of Pennsylvania, Maryland, West Virginia, and Virginia [M.S. thesis]: Chapel Hill, University of North Carolina, 192 p.

Kaiser, W. R., 1972, Delta cycles in the Middle Devonian of central Pennsylvania [Ph.D. thesis]: Baltimore, Johns Hopkins University, 183 p.

Kepferle, R. C., Potter, P. E., Pryor, W. A., and Stenbeck, T. V., 1981, Stratigraphy of the Chattanooga Shale (Upper Devonian and Lower Mississippian) in vicinity of Big Stone Gap, Wise County, Virginia: U.S. Geological Survey Bulletin 1499, 20 p.

Kiefer, J. D., 1970, Pre-Chattanooga Devonian stratigraphy of Alabama and northwest Georgia [Ph.D. thesis]: Urbana, University of Illinois, 175 p.

Kirby, W. M., 1981, Sedimentology of the Middle Devonian Bellvale and Skunnemunk Formations in the Green Pond outlier in northern New Jersey and southeastern New York [M.S. thesis]: New Brunswick, Rutgers University, 109 p.

Kirchgessner, D. A., 1973, Sedimentology and petrology of Upper Devonian Greenland Gap Group along the Allegheny Front, Virginia, West Virginia and Maryland [Ph.D. thesis], Chapel Hill, University of North Carolina, 93 p.

Laird, W. M., 1942, The stratigraphy of the Upper Devonian and Lower Mississippian of southwestern Pennsylvania [Ph.D. thesis]: Cincinnati, University of Cincinnati, 244 p.

Lesure, F. G., and Matooka, J. M., 1980, Possible stratiform copper occurrence Ramseys Draft Wilderness study area, Augusta County, Virginia: Southeastern Geology, v. 21, p. 227–238.

Lewis, T. L., 1976, Late Devonian and Early Mississippian paleoenvironments, northern Ohio: Geological Society of America Abstracts with Programs, v. 8, p. 220.

Lyke, W. L., 1981, The Minnehaha member of the Upper Devonian Scherr and Brallier Formations of the central Appalachians [M.S. thesis]: Chapel Hill, University of North Carolina, 120 p.

Martens, J.H.C., 1939, Petrography and correlation of deep-well sections in West Virginia and adjacent states: West Virginia Geological Survey, v. 11, 255 p.

—— 1945, Well-sample records: West Virginia Geological Survey, v. 17, 889 p.

McCave, I. N., 1968, Shallow and marginal marine sediments associated with the Catskill complex in the Middle Devonian of New York, *in* Klein, G. deV., ed., Late Paleozoic and Mesozoic continental sedimentation, northeastern North America: Geological Society of America Special Paper 106, p. 75–107.

McDonnell, D. E., 1981, Limits of the Devonian Hampshire Formation in parts of West Virginia and Virginia [M.S. thesis]: Chapel Hill, University of North Carolina, 128 p.

McGhee, G. R., Jr., and Dennison, J. M., 1980, Late Devonian chronostratigraphic correlations between the central Appalachian Allegheny Front and central and western New York: Southeastern Geology, v. 21, p. 279–286.

McGhee, G. R., Jr., and Sutton, R. G., 1981, Late Devonian marine ecology and zoogeography of the central Appalachians and New York: Lethaia, v. 14, p. 27–43.

McIver, N. L., 1961, Upper Devonian marine sedimentation in the central Appalachians [Ph.D. thesis]: Baltimore, Johns Hopkins University, 327 p.

—— 1970, Appalachian turbidites, *in* Fisher, G. W., Pettijohn, F. J., Reed, J. C., Jr., and Weaver, K. N., eds., Studies of Appalachian geology, central and southern: New York, Interscience Publishers, p. 69–81.

Meckel, L. D., 1970, Paleozoic alluvial deposition in the central Appalachians: A summary, *in* Fisher, G. W., Pettijohn, F. J., Reed, J. C., Jr., and Weaver, K. N., eds., Studies of Appalachian geology: central and southern: New York, Interstate Publishers, p. 49–67.

Mencher, E., 1939, Catskill facies of New York State: Geological Society of America Bulletin, v. 50, p. 1761–1794.

Osberg, P. H., 1978, Synthesis of the geology of the northern Appalachians, U.S.A.: Geological Survey of Canada Paper 78-13, p. 137–147.

Palmer, A. R., compiler, 1983, The Decade of North American Geology geologic time scale: Geology, v. 11, p. 503–504.

Pepper, J. F., deWitt, W., Jr., and Demarest, D. F., 1954, Geology of the Bedford Shale and Berea Sandstone in the Appalachian basin: U.S. Geological Survey Professional Paper 259, 111 p.

Piotrowski, R. G., and Harper, J. A., 1979, Black shale and sandstone facies of the Devonian "Catskill" clastic wedge in the subsurface of western Pennsylvania: U.S. Department of Energy, Morgantown Energy Technology Center, EGSP Series No. 13.

Piotrowski, R. G., and Krajewski, S. A., 1979, Preliminary stratigraphic cross section (A_1–A_4) showing radioactive black shale zones and sandstones in the Middle and Upper Devonian, western Pennsylvania: U.S. Department of Energy, Morgantown Energy Technology Center, EGSP Series No. 9.

Potter, P. E., Maynard, J. B., and Pryor, W. A., 1982, Appalachian gas bearing Devonian shales: statement and discussions: Oil and Gas Journal, v. 80, p. 290–318.

Provo, L. J., 1977, Stratigraphy and sedimentology of radioactive Devonian-Mississippian shales of the central Appalachians: U.S. Department of Energy Report GJBX37(77), 102 p.

Provo, L. J., Kepferle, R. C., and Potter, P. E., 1978, Division of Ohio black shale in eastern Kentucky: American Association of Petroleum Geologists Bulletin, v. 62, p. 1703–1713.

Rahmanian, V. D., 1979, Stratigraphy and sedimentology of the Upper Devonian Catskill and uppermost Trimmers Rock Formations in central Pennsylvania [Ph.D. thesis]: State College, Pennsylvania State University, 340 p.

—— 1980, Upper Devonian tide-dominated deltaic-interdeltaic sedimentation in west-central Pennsylvania: A sedimentologic model for distribution of petroleum sandstone reservoir types: American Association of Petroleum Geologists Bulletin, v. 64, p. 1288–1289.

Rickard, L. V., 1975, Correlation of the Silurian and Devonian rocks in New York State: New York State Museum and Science Service Map and Chart Series No. 24.

Roen, J. B., Miller, R. L., and Huddle, J. W., 1964, The Chattanooga Shale (Devonian and Mississippian) in the vicinity of Big Stone Gap, Virginia: U.S. Geological Survey Professional Paper 501-B, p. B43–B48.

Roen, J. B., Wallace, L. G., and deWitt, W., Jr., 1978a, Preliminary stratigraphic cross section showing radioactive zones of the Devonian black shales in eastern Ohio and west-central Pennsylvania: U.S. Geological Survey Oil and Gas Investigation Chart OC-82.

—— 1978b, Preliminary stratigraphic cross section showing radioactive zones in the Devonian black shales in the central part of the Appalachian basin: U.S. Geological Survey Oil and Gas Investigation Chart OC-87.

Schweitering, J. F., 1980, Preliminary cross sections (A′-B′-D′) of Middle and Upper Devonian in West Virginia: U.S. Department of Energy MET-C/EGSP no. 211.

Sevon, W. D., 1979, Polymictic diamictites in the Specty Kopf and Rockwell Formations: Pennsylvania Geological Survey, 4th ser., Guidebook 44th Annual Field Conference of Pennsylvania Geologists, p. 61–66, 107–110.

Sevon, W. D., Rose, A. W., Smith, R. C., II, and Hoff, D. T., 1978, Uranium in Carbon, Lycoming, Sullivan, and Columbia Counties, Pennsylvania: Pennsylvania Topographic and Geologic Survey, Guidebook 43rd Annual Field Conference of Pennsylvania Geologists, 99 p.

Shanmugan, G. and Moiola, R. J., 1982, Eustatic control of turbidites and winnowed turbidites: Geology, v. 10, p. 231–235.

Slaughter, J., 1970, Genesis of stratiform copper deposits of the Catskill Formation of northern Pennsylvania [M.S. thesis]: New Brunswick, Rutgers University, 67 p.

Spariosu, D. J. and Kent, D. V., 1983, Paleomagnetism of the Lower Devonian Traveler Felsite and the Acadian orogeny in the New England Appalachians: Geological Society of America Bulletin, v. 94, p. 1319–1328.

Thomas, W. A., 1972, Mississippian stratigraphy of Alabama: Alabama Geological Survey Monograph 12, 121 p.

Walker, R. G., 1971, Nondeltaic depositional environments in the Catskill clastic wedge (Upper Devonian) of central Pennsylvania: Geological Society of America Bulletin, v. 82, p. 1305–1326.

Walker, R. G., and Harms, J. C., 1971, The "Catskill Delta"—a prograding muddy shoreline in central Pennsylvania: Journal of Geology, v. 79, p. 381–399.

—— 1975, Shorelines of weak tidal activity: Upper Devonian Catskill Formation—central Pennsylvania, *in* Ginsburg, R. N., ed., Tidal deposits: New York, Springer-Verlag, p. 103–108.

Werner, E., 1979, The 38th Parallel lineament in West Virginia, *in* Podwysocki, M. H. and Earle, J. L., eds., Proceedings of the Second International Conference on Basement Tectonics: Denver, Basement Tectonics Committee, Inc., p. 406–412.

West, M., 1978, Preliminary stratigraphic cross section showing radioactive zones in the Devonian black shales in the eastern part of the Appalachian basin: U.S. Geological Survey Oil and Gas Investigation Chart OC-86.

Whitehead, N. H., III, 1979, Lithostratigraphy of the Grainger-Price Formation, northeastern Tennessee and southwestern Virginia, and depositional history of the Lower Mississippian of east-central U.S. [Ph.D. thesis]: Chapel Hill, University of North Carolina, 214 p.

Willard, B., 1934, Early Chemung shore line in Pennsylvania: Geological Society of America Bulletin, v. 45, p. 897–907.

—— 1939, Middle and Upper Devonian, *in* Willard, B., Swartz, F. M., and Cleaves, A. B., The Devonian of Pennsylvania: Pennsylvania Geological Survey, 4th ser., General Geology Report 19, p. 131–481.

Wolff, M. P., 1967, Deltaic sedimentation of the Middle Devonian Marcellus Formation in southeastern New York [Ph.D. thesis]: Ithaca, Cornell University, 215 p.

Woodrow, D. L., 1983, Topography and sedimentary processes in an epicontinental sea, *in* Stanley, D. J., ed., The shelfbreak; Critical interface on continental margins: Society of Economic Paleontologists and Mineralogists Special Publication No. 33, p. 159–166.

Woodrow, D. L., and Fletcher, F. W., 1968, Late Devonian paleogeography in southeastern New York and northeastern Pennsylvania, *in* Oswald, D. H., ed., International symposium on the Devonian System: Alberta Society of Petroleum Geologists, v. 2, p. 1327–1334.

Woodrow, D. L., Fletcher, F. W., and Ahrnsbrak, W. F., 1973, Paleogeography and paleoclimate at the deposition sites of the Devonian Catskill and Old Red facies: Geological Society of America Bulletin, v. 84, p. 3051–3063.

Woodrow, D. L., and Isley, A. M., 1983, Facies, topography, and sedimentary processes in the Catskill sea (Devonian), New York and Pennsylvania: Geological Society of America Bulletin, v. 94, p. 459–470.

Woodward, H. P., 1943, Devonian System of West Virginia: West Virginia Geological Survey, v. 15, 655 p.

Zartman, R. E., Brock, M., Heyl, A. V., and Thomas, H. H., 1967, K-Ar and Rb-Sr ages of some alkalic intrusive rocks from central and eastern United States: American Journal of Science, v. 265, p. 848–870.

Manuscript Accepted by the Society July 12, 1984

Printed in U.S.A.

Geological Society of America
Special Paper 201
1985

Upper Devonian turbidite sequence, central and southern Appalachian basin: Contrasts with submarine fan deposits

Paul D. Lundegard*
Neil D. Samuels*
Wayne A. Pryor
H. N. Fisk Laboratory of Sedimentology
Department of Geology
University of Cincinnati
Cincinnati, Ohio 45221

ABSTRACT

The Upper Devonian turbidite sequence in the central and southern Appalachian basin reaches 1400 m in thickness and consists of siltstone turbidites interbedded with mudstone, claystone, and shale. Facies and paleocurrent analyses, based on approximately 6100 m of detailed measured section at 35 localities and 700 measurements of directional structures, indicate major differences between the Upper Devonian sequence and submarine fan and ancient flysch deposits.

Compared to flysch and submarine fan deposits, the Upper Devonian turbidites are finer grained and thinner bedded, reflecting the smaller size and lower maximum velocity of the turbidity currents. The stratigraphic transition from turbidites to overlying deltaic rocks shows a different succession from that expected for proximal deposits of submarine fans associated with fan channels and canyons. This transition in the Upper Devonian deposits is gradual and marked by upward-thickening and -coarsening sequences of evenly-bedded turbidites and lenticular beds of crossbedded sandstone. Higher energy mass flow deposits such as debris flows, massive and pebbly sandstone, and conglomerates are notably absent.

Paleocurrent data are remarkably uniform, both areally and vertically through the stratigraphic section. They indicate a uniform dispersal pattern, transverse to the basin axis, for nearly 600 km along depositional strike. This suggests turbidity currents had multiple point sources along the basin margin and that the paleoslope had little topographic relief on it. Dip-oriented, short-lived turbidite lobes were built and abandoned as the source of turbidity currents shifted. Migrating delta distributaries, instead of fixed feeder channels, probably supplied sediment to the slope.

INTRODUCTION

Most ancient turbidite sequences studied to date are of the Alpine flysch type (McBride 1962; Mutti 1974; Niem 1976) or the continental borderland type (Sullwold 1960; Nilsen and Simoni 1973; Webb 1981). Few examples of turbidites in deltaic settings, such as the Upper Devonian rocks of the Appalachian basin, have been adequately described (Link 1975; Kepferle 1978). Evidence is emerging, however, that existing turbidite facies models are in some cases inapplicable to deltaic turbidite deposits (Walker 1978). Our paper documents some of these discrepancies and suggests a depositional model for Upper Devonian turbidites in the central and southern Appalachian basin.

Upper Devonian turbidites in the central and southern Appalachian basin occur in a regressive sequence of fine-grained turbidites, mudstone, claystone, and shale. Within the study area, the section containing turbidites has a maximum thickness of

*Present addresses: Lundegard, Department of Geological Sciences, University of Texas at Austin, Austin, Texas 78712; Samuels, Amoco Production Company, P.O. Box 3092, Houston, Texas 77253.

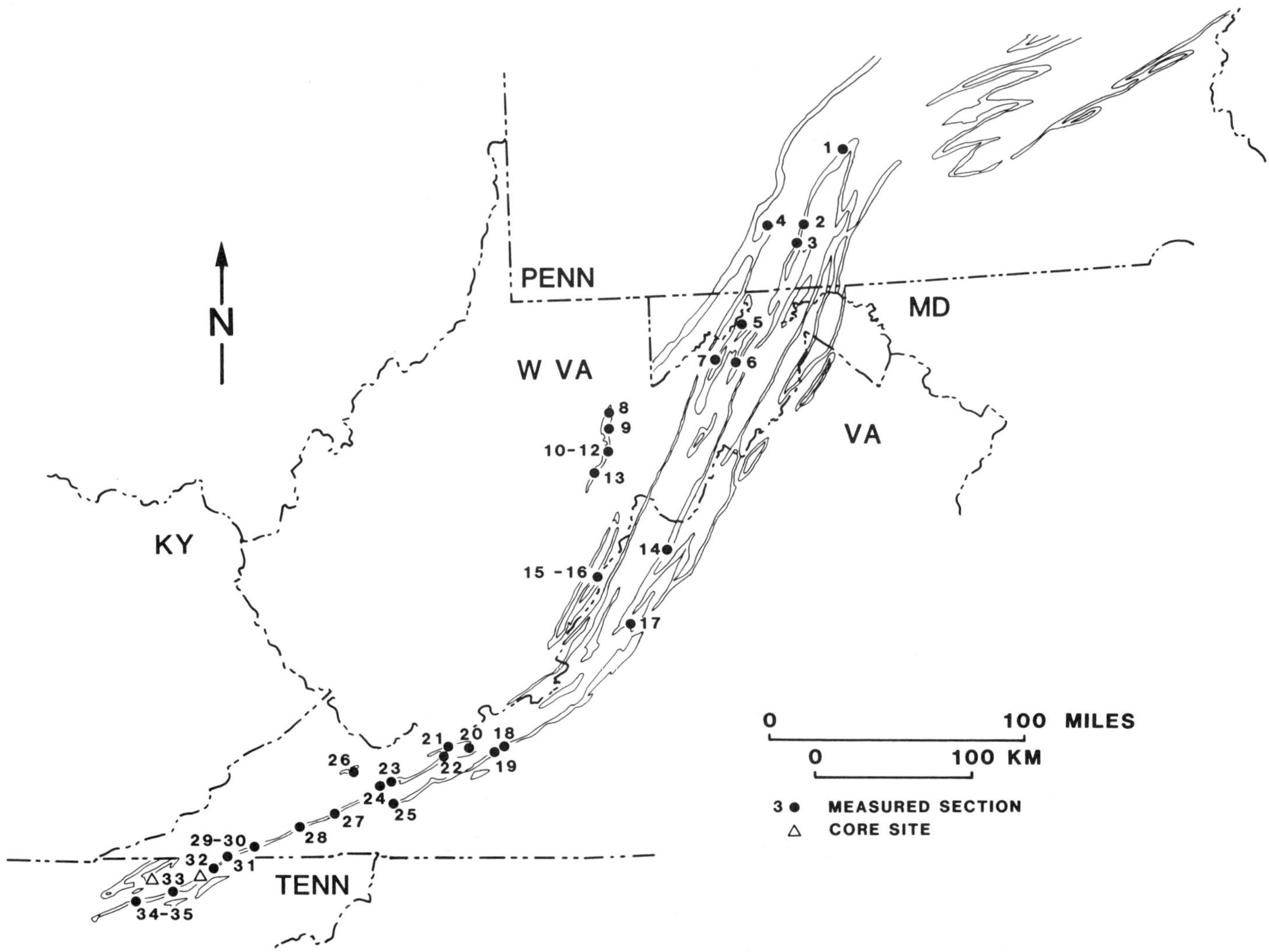

Figure 1. Map of study area showing outcrop of Upper Devonian turbidites, measured sections, and selected drill cores in eastern Tennessee that were also studied (eastern core is TDG-DOE-4; western core is TDG-DOE-3). For detailed descriptions of measured sections see Lundegard and others (1980).

about 1400 m along the Allegheny Front in West Virginia and crops out mainly in the Valley and Ridge Province from south-central Pennsylvania to southwestern Virginia (Fig. 1). Upper Devonian turbidites and their stratigraphically adjacent and equivalent strata were examined throughout most of their outcrop in south-central Pennsylvania, Maryland, West Virginia, Virginia, and eastern Tennessee (Figure 1). Turbidites are also present in the sequence of Devonian rocks in northeastern Pennsylvania and western and central New York. (Sutton 1959; Walker and Sutton 1967). Thirty-five sections, totaling over 6100 m, were measured and observations made at about 50 other localities where sections were not measured (for descriptions of sections see Lundegard 1979; Samuels 1979; Lundegard and others 1980). Strata are exposed in a series of gently plunging anticlines and synclines that strike N-NE. Beds generally dip less than 45 degrees, but locally are overturned in the vicinity of thrust faults. Depositional strike was N-S. (Potter and others 1979) and because of its oblique relationship to structural strike, both onshore-offshore and lateral variations are observable along the outcrop belt.

PREVIOUS WORK

Sedimentological analyses of the Upper Devonian turbidite sequence in the central and southern Appalachian basin have been focused in central Pennsylvania. Both Frakes (1967) and Walker (1971) interpreted the nature of the Late Devonian paleoslope in this area. Woodrow and Isley (1983) studied the Upper Devonian turbidite sequence at one outcrop in northeastern Pennsylvania and three outcrops in south-central New York. Walker (1967) also made a detailed study of turbidite sedimentary structures at two outcrops in Maryland. In the only regional study of Upper Devonian turbidites prior to this one,

		SOUTH-CENTRAL PENNSYLVANIA	EASTERN WEST VIRGINIA	WESTERN VIRGINIA	EASTERN TENNESSEE
MISS.	LOWER	ROCKWELL FM	POCONO FM	PRICE FM	GRAINGER FM
DEVONIAN	UPPER	HAMPSHIRE FM	HAMPSHIRE FM	"CHEMUNG" FM	CHATTANOOGA SHALE: BIG STONE GAP SHALE MBR
	UPPER (GREENLAND GAP GRP)	FOREKNOBS FM	FOREKNOBS FM		
	UPPER (GREENLAND GAP GRP)	SCHERR FM	SCHERR FM	BRALLIER FM	CHATTANOOGA SHALE: GRAY SILTY SHALE UNIT
	UPPER	BRALLIER FM	BRALLIER FM		
	UPPER	HARRELL SH	HARRELL SH	MILLBORO SH	CHATTANOOGA SHALE: LOWER BLACK SHALE UNIT
	MIDDLE	MAHANTANGO FM	MAHANTANGO FM		
	MIDDLE	MARCELLUS SH	MARCELLUS SH		

Figure 2. Stratigraphic nomenclature in the different parts of the study area. The Tully Limestone is present locally above the Mahantango Formation.

McIver (1961, 1970) collected a large number of paleocurrent observations and described the sedimentary features of the turbidites in the central Appalachians. Relevant stratigraphic investigations include the works of Dennison (1970), Dennison and Boucot (1974), and Avary and Dennison (1978). This paper is a summary of the sedimentologic studies of Upper Devonian rocks by Samuels (1979) and Lundegard (1979) in the central and southern Appalachians, respectively.

STRATIGRAPHY

Upper Devonian turbidites in the central and southern Appalachians occur mainly in the Brallier Formation and the Scherr Formation (Figure 2), the basal formation of the Greenland Gap Group (Dennison 1970). These two formations are sedimentologically similar and have conformable, gradational contacts with adjacent stratigraphic units. The Brallier and Scherr formations are overlain by non-turbidite marine sandstones and mudrocks of the "Chemung" and Foreknobs formations, respectively. Following the recommendation of Dennison (1970), the name "Chemung" is used in quotations because it has been abandoned as a formal stratigraphic term in south-central New York where it was first proposed. Throughout the study area the Brallier Formation is underlain by dark-colored mudrocks of either the Harrell Shale or Millboro Shale (Figure 2). In eastern Tennessee the unit of "Brallier lithology" described by Dennison and Boucot (1974) is equivalent to the informal "middle gray silty shale" unit of the Chattanooga Shale identified by Oliver and others (1969), whose terminology will be used here.

The Upper Devonian turbidite sequence is regressive (Figure 3). In general, the Brallier Formation becomes younger from east to west, a direction that is perpendicular to the paleoshoreline. In south-central Pennsylvania it is entirely Fingerlakeian in age, whereas in southwest Virginia it ranges from late Fingerlakeian to early Cassadagan in age (Oliver and others 1969). Furthermore, the gray silty shale unit of the Chattanooga Shale in eastern Tennessee is entirely Cassadagan in age (Oliver and others 1969). Thus, the Brallier Formation also becomes younger from northeast to southwest.

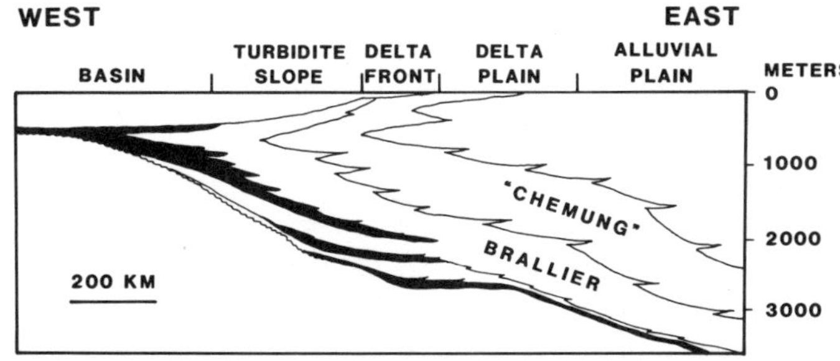

Figure 3. Schematic cross-section of Upper Devonian rocks showing position of turbidites in regressive sequence. Black areas indicate black shale units. Data from diverse sources including Oliver and others (1969).

PALEOCURRENTS

Approximately 700 directional structures were measured in Upper Devonian turbidites and related rocks in Virginia, West Virginia, and eastern Tennessee. Of these, 560, or 80 percent are sole marks including flute and groove molds, prod and bounce marks, and ridge and furrow structures. The remainder are mostly ripple marks but also include elongate wood fragments, parting lineations, and cross bedding. These data are distributed over eighty-five 7½-minute quadrangles at an average density of eight observations per quadrangle. Approximately 90 percent of the data are from the Brallier Formation, the rest are from the "Chemung" Formation, the Scherr Formation, and the Chattanooga Shale (east Tennessee data). Many more observations of sole mark orientations were made than other directional structures because they are more accurate indicators of turbidite flow patterns (Potter and Pettijohn 1977; 172) and are easier to measure.

The orientation of directional features in folded strata were tilt corrected in the field whenever possible as recommended by Dott (1974). Otherwise they were tilt corrected in the lab. Directional structures in severely deformed strata were not measured. Vector means rather than arithmetic means are used on all maps and profiles, because the former are a much better measure of central tendency (Potter and Pettijohn 1977; 376). The VECMEN program written and modified from several sources by James M. Parks of Lehigh University was used for the determination of vector means.

Regionally, sole marks on Brallier Formation siltstones show a strikingly homogeneous westerly trend (Figure 4), approximately perpendicular to the N-S depositional strike (Denni-

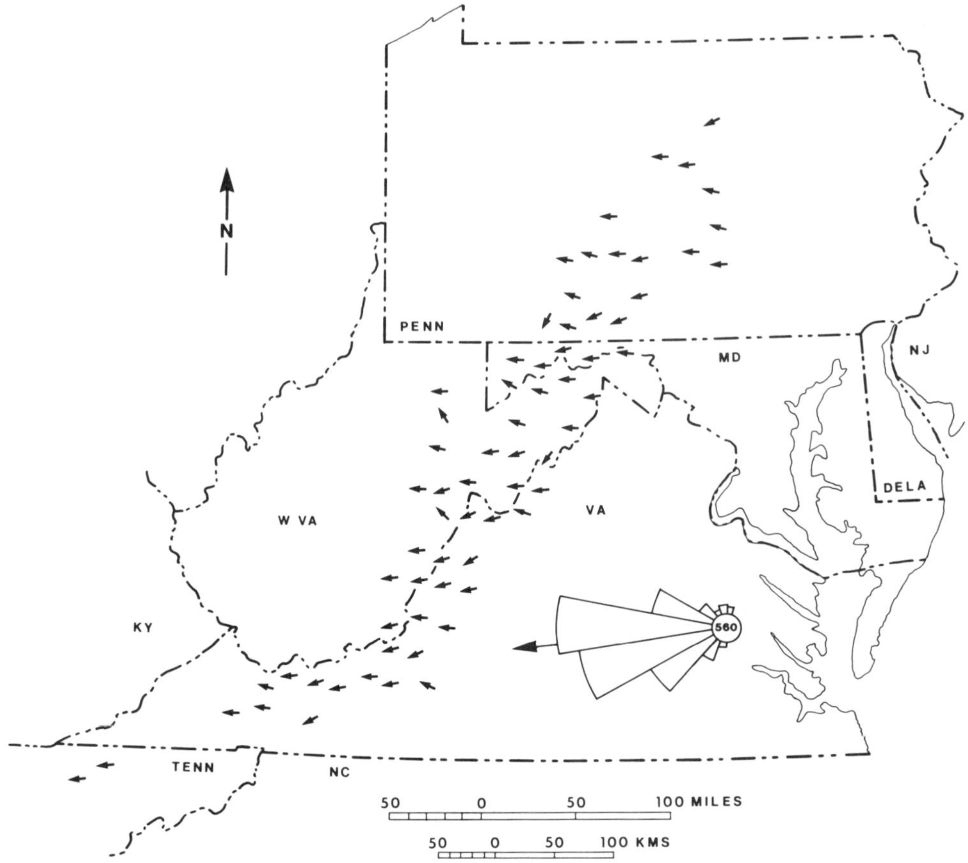

Figure 4. Paleocurrent map of Upper Devonian turbidites in the Appalachian basin. Data are from sole marks and are averaged in 15-minute quadrangles. Note the consistent westerly trend for nearly 600 km along depositional strike. The data in Pennsylvania and Maryland are from McIver (1961). Rose diagram is for all sole mark data (N = 560). The vector mean is 263° and the standard deviation is 32°.

son and Hasson 1976; 278; Potter and others 1979), and transverse to the basin axis. This trend shows little variation from central Pennsylvania to eastern Tennessee, a distance of approximately 600 km along depositional strike. In addition, Upper Devonian turbidite paleocurrents are very similar in New York state (Sutton 1959; McIver 1970; Potter and others 1979), indicating the same dispersal pattern persisted for nearly 750 km along depositional strike. Indications of longitudinal flow (parallel with the basin axis), are virtually absent. When all of the data from sole marks in Virginia, West Virginia, and Tennessee are plotted on a single rose diagram (Figure 4), a remarkably distinct unimodal distribution results (standard deviation equals 32°), demonstrating that even over very large areas the paleocurrents show little variation in direction.

Four vertical profiles of turbidite directional features were made through thick stratigraphic sections to see if divergent flow patterns associated with overlapping submarine fans were recognizable. However, both distal turbidites near the bottom of the section and proximal turbidites near the top show overall uniformity and an absence of distinct divergent trends (Figure 5), suggesting that the stratigraphic section was not constructed by the overlapping of submarine fans with radial dispersal patterns.

A review of the literature shows that longitudinal transport (i.e. parallel to the axis of the basin) is typical of turbidite systems (Potter and Pettijohn 1977; 332). It is, therefore, interesting that in the Appalachian basin, an elongate geosynclinal trough, Upper Devonian turbidites show almost exclusively transverse paleocurrents (perpendicular to the axis of the basin) whereas Ordovician turbidites show longitudinal paleocurrents (McBride 1962; McIver 1970). There are several implications of such a dispersal pattern.

The transverse paleocurrent pattern is evidence that little sediment was input longitudinally. The question arises, however, as to why the transverse turbidity currents were not deflected longitudinally. Several possible explanations exist, but the one we favor is that the sides of the depositional basin had gentle slopes and therefore the basin lacked a sharply defined axis. This interpretation is supported by sedimentologic evidence discussed later.

It is also possible that turbidity currents simply did not reach the axis of the basin. This idea is not appealing because it requires either deposition of all the turbidites on the side of the basin or that the basin had a terraced profile (Contescu 1974: Fig. 12) which subsurface data do not support. Cheema (1977; 67), however, favored this latter explanation for the absence of evidence of

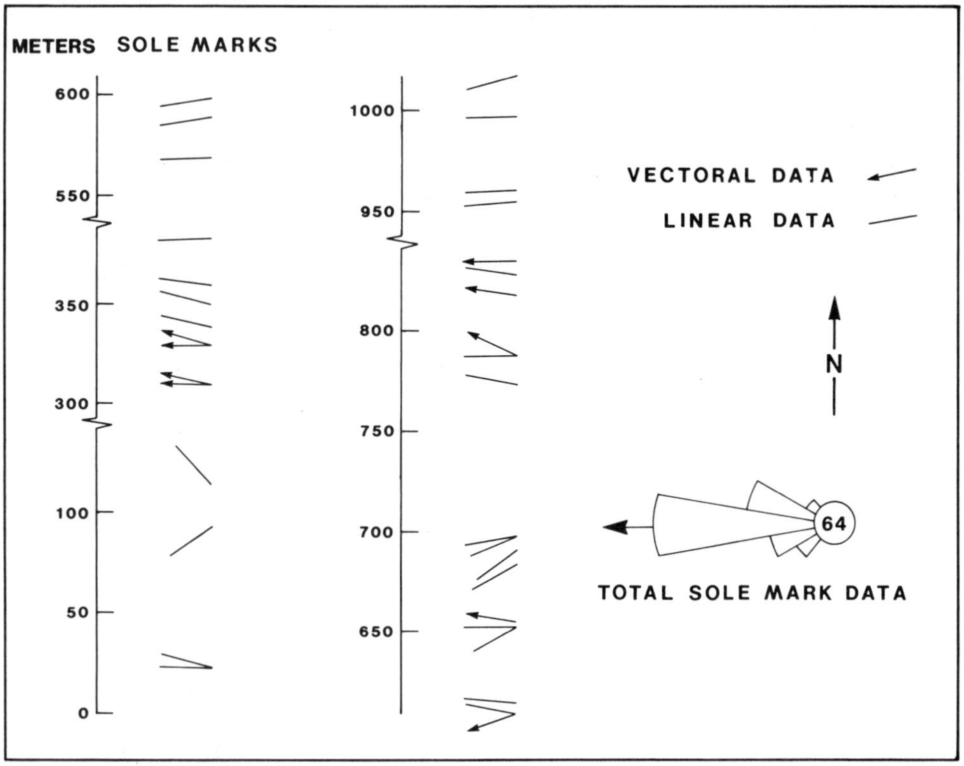

Figure 5. Representative vertical profile of paleocurrents in Brallier Formation (Section 18, Figure 1). Note overall uniformity. On rose diagram all data are treated as vectoral because no east-oriented structures were observed.

turbidity current flow parallel with the basin axis in Upper Devonian strata in the subsurface of north-central West Virginia. It is also possible that turbidity currents reached the basin axis and were deflected but were too weak to erode the bottom and form sole marks. The orientation of wood fragments and ripples in the upper parts of turbidites also indicate westerly transport and therefore do not support this idea (Lundegard 1979; 68). It also seems unlikely that this explanation could be valid for all the turbidites.

Most turbidite sequences show far greater regional variability in their paleocurrents than does the Upper Devonian sequence in the Appalachian basin. Examples include the Ordovician of the Appalachian basin (McBride 1962), the Ordovician-Silurian sequence of Arctic Canada (Trettin 1970), the Mississippian of the Marathon basin (McBride 1970), the Cretaceous of the eastern Carpathians (Contescu 1974), and the Upper Mississippian of the Ouachita basin (Morris 1974). Several of these examples show both a fairly homogeneous longitudinal pattern of paleocurrents and a more variable transverse pattern (Trettin 1970; Contescu 1974; Morris 1974). The Ordovician of the Appalachian basin shows two opposing longitudinal paleocurrent trends (McBride 1962). Very few turbidite deposits show a homogeneous unimodal paleocurrent pattern, and in those that do the flow is predominantly longitudinal (ten Haaf 1959; Scott 1966). In other words, whereas a few turbidite sequences do show homogeneous paleocurrent patterns, it is unusual for them to show a homogeneous *transverse* paleocurrent pattern. The only other examples of this that we know of are the Borden Formation (Mississippian) of Kentucky (Moore and Clarke 1970; Fig. 3; Kepferle 1978; Fig. 7; Maynard and Lauffenburger 1978; Fig. 4) and the Difunta Group (Late Cretaceous to Paleocene) of northeastern Mexico (McBride and others 1975). Interestingly, both rock units are delta-associated turbidite deposits like the Devonian in the Appalachian basin.

The areal homogeneity of paleocurrent patterns for more than 600 km along depositional strike strongly suggests that turbidity currents had multiple point sources of origin. Had there been only a few long-lived sources, distinct dispersal patterns radiating from these points would be evident (Sullwold 1960). Also, if the deposits at a given place on the sea floor were formed by the coalescence of a few, large, long-lived submarine fans, divergent paleocurrent patterns associated with each fan ought to be seen in vertical profiles through the deposit. However, profiles through the Upper Devonian sequence show only small, random divergence in paleocurrents (Figure 5). Therefore, we believe that the Upper Devonian paleoslope was one of little topographic relief on which dip-oriented, short-lived turbidite lobes were built and abandoned as the loci of turbidity currents shifted.

FACIES

We have distinguished five lithofacies of the Brallier Forma-

TABLE 1. SUMMARY OF FACIES

Facies	Lithology and Thickness	SLST + SS/ Shale Ratio	SLST or SS Bed Thickness (Mode and Range)	Sedimentary Structures
Delta front	Fine sandstone to siltstone and olive gray mudstone. 10's to 100's m	Moderate to very high	M 30 cm R 10 cm - 2 m	Lenticular to irregular bedding, cross bedding, fossils. Coarsening- and thickening-upward sequences
Turbidite slope	Siltstone and olive gray mudstone or shale 3 - 30 m	Low to high	M 6 cm R 1 - 30 cm	Even, persistent beds with base-truncated Bouma sequences (Tbcde, Tcde) common, tops of beds commonly rippled, sole marks very common. Upward-thickening and -coarsening sequences in upper Brallier Formation
Siltstone bundle	Siltstone to very-fine sandstone with minor shale 5 - 22 m	High to very high	M 20 cm R 2 cm - 1.5 m	Even, sharply defined beds; structureless non-graded beds common; Ta, Tab, and Tabc Bouma sequences; shale clasts and bed amalgamation common; wood fragments common; sole marks generally uncommon; rare load structures. Mostly upward-thinning and -fining sequences
Interlobe slope	Olive gray mudstone and minor shale 3 - 60 m	Zero to very low	M 3 cm R 1 - 10 cm	Mudstone: distinct horizontal burrows, 3-5 mm wide, some like *Pteridichnites biseriatus*; *Ambocoelia* sp.; Siltstoned Tcde and Tde Bouma sequences
Lobe margin	Olive gray to dark gray, silt-laminated shale, and minor yellowish gray, burrowed claystone 15 cm - 210 m	Zero to very low	Less than 3 cm, if present Silt Laminae 1 - 2 mm thick are very common	Shale: even, persistent silt laminae are very common; siderite layers up to 2 cm thick are locally common. Claystone: distinct to indistince horizontal burrows less than 2 mm wide
Basinal	Black shale Up to 30 m	Zero to very low	M 3 cm R 1.5 - 10 cm	Shale: dominantly clayshale with textural lamination; silt laminae less than 2 mm thick are locally common. Siltstone: if present, beds generally have Tce Bouma sequences and rippled tops

tion and adjacent strata by gross rock type, siltstone + sandstone/shale ratio, bedding thickness and character, and physical and biogenic structures. The interpretive names assigned to these facies are: delta-front, turbidite-slope, interlobe-slope, lobe-margin, and basinal. A siltstone-bundle sub-facies of the turbidite-slope facies is also recognized. The salient features of the facies are summarized in Table 1 and briefly discussed below. Mudrock terminology suggested by Lundegard and Samuels (1980) is used.

Delta-front Facies

Delta-front facies, which thin distally to the west and south, comprise the uppermost Brallier Formation and lower "Chemung" Formation and are conformably overlain by non-marine red beds of the Hampshire Formation in the northern part of the study area. High siltstone + sandstone/shale ratio, thick beds, lenticular bedding (Figure 6), low-angle trough crossbedding (Figure 7), and abundant and diverse marine fossils in the sandstone beds of this facies (Table 1) indicate a shallow-marine, current-agitated environment. Fossils include abundant crinoids and brachiopods, and fewer molluscs. Paleocurrent data collected from this facies in Pennsylvania by McIver (1970) indicate offshore (westerly) transport. Repetitive upward-thickening and -coarsening sequences are common and typically grade upward from plane-laminated siltstone beds a few cm thick to cross-bedded, fossiliferous sandstone beds a few tens of cm thick. Individual sequences are a few m in thickness but may be stacked within larger sequences tens of m thick (Figure 8). In bedding style and sedimentary structures the vertical sequences resemble those produced by migrating shelf bars (Spearing 1975) as well as prograding deltas (Coleman 1976). However, we believe the latter process better explains the offshore paleocurrents, absence

Figure 6. Delta-front facies. Note high siltstone + sandstone/shale ratio and thick beds. Upper Brallier Formation (Section 18, Figure 1). Mattock handle (right) is 1 m long.

Figure 7. Trough crossbedding in delta-front facies. Same location as Figure 6. Hammer (center-right) is 30 cm long.

of hummocky cross stratification, and upward gradation into non-marine rocks.

Turbidite-slope Facies

Most of the turbidites of the Brallier Formation occur in the turbidite-slope facies and most of the paleocurrent data were collected from the abundant flute and groove molds on its siltstone beds. Thin bedding, moderate to high siltstone + sandstone/shale ratio, and base-truncated Bouma sequences in the siltstone beds of the turbidite-slope facies (Table 1 and Figure 9) indicate that small, low-velocity turbidity currents frequently interrupted the more or less continuous hemipelagic sedimentation of mud. The delta-front facies overlies and is gradational with the turbidite-slope facies that also thins to the west (Figure 8). The turbidite-slope facies is commonly recurrent in thick stratigraphic sections and shows variable siltstone + sandstone/shale ratios. Bed thickness and siltstone + sandstone/shale ratio generally increase up section, indicating an increase in proximity to the source. The increase in proximity was accompanied by an increase in oxygenation of the water column as indicated by the occurrence of a few vertical feeding burrows in the siltstone beds.

In the upper part of the Brallier Formation upward-thickening and -coarsening sequences are common (Figure 8), some of which grade from turbidites at the base to crossbedded sandstones of the delta-front facies at the top (Figure 8). This relationship indicates progradation of the source of turbidity currents. Elsewhere in the Brallier Formation megasequences are poorly developed in turbidite-slope facies. Evidence of channeling was not observed in the turbidite-slope facies. Infauna were locally capable of reworking the mud deposited between turbidites, suggesting that bottom conditions were generally dysaerobic. Sedimentary structures in the siltstone beds of the turbidite-slope facies are typical of "distal" turbidites (Table 1), although the gradual vertical transition to delta-front facies indicates some were deposited proximally. We believe the predominance of base-truncated Bouma sequences in the siltstone beds are more indicative of the small size and low-velocity of turbidity currents rather than proximity to the source because beds with sedimentary structures typical of classic proximal turbidites are rare throughout the stratigraphic section.

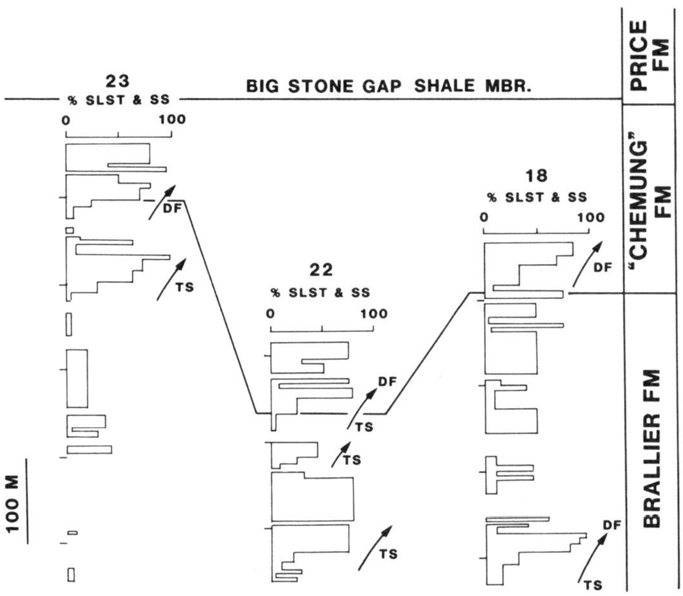

Figure 8. Cross-section through Sections 18, 22, 23 (Figure 1) showing percent-siltstone and sandstone beds. Note that the stratigraphic transition from turbidites of the Brallier Formation to top-set sandstones of the "Chemung" Formation is gradual and characterized by upward-thickening and -coarsening sequences (arrows) of turbidite-slope (TS) and delta-front (DF) facies. Note also that thick slope mudstone units between the turbidite sequence and overlying top-set sequence are absent. Datum for section is the middle of the Big Stone Gap Shale Member of the Price Formation (Bartlett 1974).

Siltstone-bundle Sub-facies

Thickly-bedded bundles of siltstone turbidites comprise a small proportion of the total Brallier Formation but record the occasional occurrence of relatively large, erosive turbidity currents (Figure 10). They are slightly more abundant in the lower half of the section. The features of seven siltstone bundles described in bed-by-bed detail are summarized in Table 2. Individual bundles are 3 to 23 m thick and their lateral extent, parallel to depositional strike (estimated by tracing their topographic expression as linear ridges and correcting for divergence of structural and depositional strike), is at most a few km and typically is less than 1 km. Rapid deposition from relatively large, high-flow intensity turbidity currents is indicated by the thick siltstone beds, high siltstone + sandstone/shale ratio, top-truncated Bouma sequences, bed amalgamation, and the occurrence of shale clasts (Tables 1 and 2). Sole marks are rare in the siltstone bundles (Section 25 is an exception), however, the capacity of the turbidity currents to erode the bottom is demon-

Figure 9. Turbidite-slope facies. Note even, persistent siltstone beds. Brallier Formation (Section 18, Figure 1). Mattock handle (lower left) is 1 m long.

Figure 10. Turbidites of the siltstone-bundle sub-facies. Note high siltstone + sandstone/shale ratio and thick, even beds. Brallier Formation, (Section 19, Figure 1). Hammer (lower left) is 30 cm long.

TABLE 2. FEATURES OF SILTSTONE-BUNDLES

Feature	Section							
	17	19	20	22	21	25	33	Avg.
Thickness (m)	12	11	8	7	7	12	10	13
Percent siltstone	70	62	67	56	61	67	52	64
Number of siltstone beds	47	49	30	26	26	46	43	42
Mean siltstone bed thickness (cm)	21	26	20	17	16	22	15	21
Mean shale bed thickness (cm)	9	16	6	14	11	11	14	12
ABC proximity index (%)*	..	69	70	75	61	74	54	67
Vertical sequence	FU	FU CU	FU CU	FU	FU	NT	FU	..

Note: For location of sections see Figure 1.
*After Walker (1967).
FU = fining- and thinning-upward. CU = coarsening- and thickening-upward. NT = no trend. Note that some sections show both fining- and coarsening-upward sequences.

strated by abundant shale clasts up to 25 cm long. The shale clasts in the siltstone beds are lithologically similar to the interbedded shales. Siltstone bundles commonly show upward-thinning and -fining sequences (Table 2 and Figure 11) that can result either from channel filling (Walker 1978; 953–956), or by the gradual shifting of active turbidite deposition to another part of the slope. The lateral dimensions of siltstone bundles in the Brallier Formation are comparable to those of gullies on prodelta slopes (Shepard 1955) and small, mid-fan channels (Normark 1978; 912). However, channel-margin features were not observed. Available evidence, therefore, suggests that the siltstone bundles are either shallow channel deposits or simply higher energy turbidites deposited in an area that was gradually abandoned.

Interlobe-slope Facies

Pervasive bioturbation of the mudstones, sparse body fossils, and rarity of thin turbidite siltstone beds in the interlobe-slope facies (Table 1 and Figure 12) suggest slow deposition of hemipelagic mud on a dysaerobic sea floor. Shells of the brachiopod *Ambocoelia* sp., are locally common in the mudstones. Because *Ambocoelia* are thought to have been epifaunal sessile organisms (McGhee 1975; 177) bioturbation was probably the work of soft-bodied infauna. The mudstones have low organic carbon contents, generally less than 0.3 weight percent (Lundegard 1979; 42–46). Turbidite sedimentation was limited to a few thin siltstone beds with T_{cde} or T_{de} Bouma sequences. Repeated occurrences of mudstone units 3 to 60 m in thickness between turbidite-slope or siltstone-bundle facies could be caused either by lateral shifting of environments or repeated episodes of facies progradation and retreat. We therefore interpret these mudstones as slope sediments that were deposited on the slope lateral to and basinward of the areas of active turbidite sedimentation. Thick mudstone units between the turbidite sequence and overlying delta-front facies are absent (Figure 8), indicating that a turbidity-current bypass zone on the upper slope was not present in this region as suggested by Walker (1971) for Upper Devonian rocks in the Susquehanna Valley in Pennsylvania.

Lobe-margin Facies

The lobe-margin facies consists of bioturbated claystone and shale with abundant silt laminae (Figure 13). This facies predominates in distal outcrops in southwestern Virginia in the lower part of the Brallier Formation and in eastern Tennessee in the "middle gray silty shale" unit of the Chattanooga Shale. The silt laminae in this facies show many of the same features as Holocene and Pleistocene turbidite silt laminae (Stow and Shanmugam 1980), including graded individual laminae, graded-laminated units,

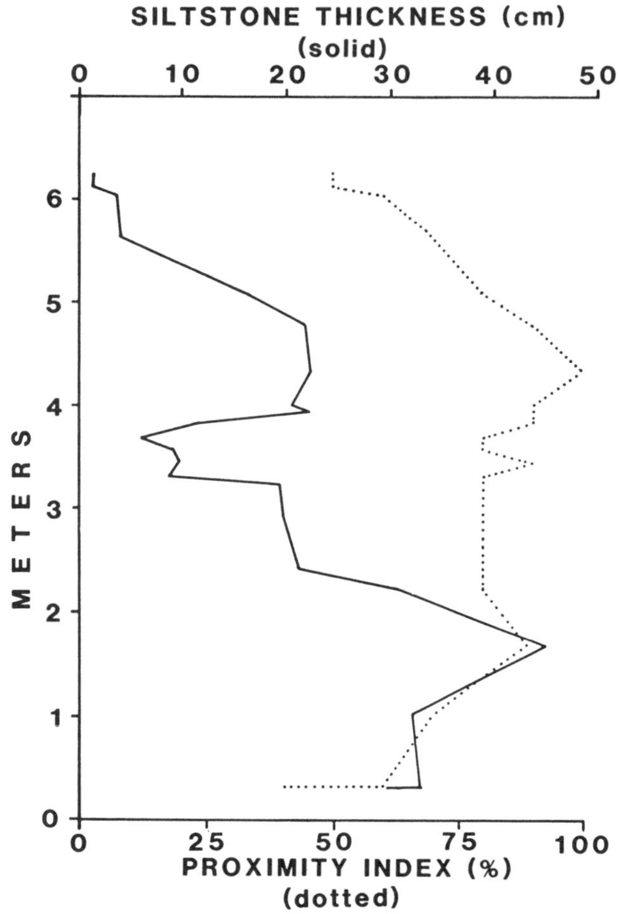

Figure 11. Vertical profile of a siltstone bundle. Five-bed moving average. Proximity index calculated as suggested by Walker (1967). High proximity index valves indicate a predominance of turbidites beginning with Bouma "a" units. Brallier Formation (Section 21, Figure 1).

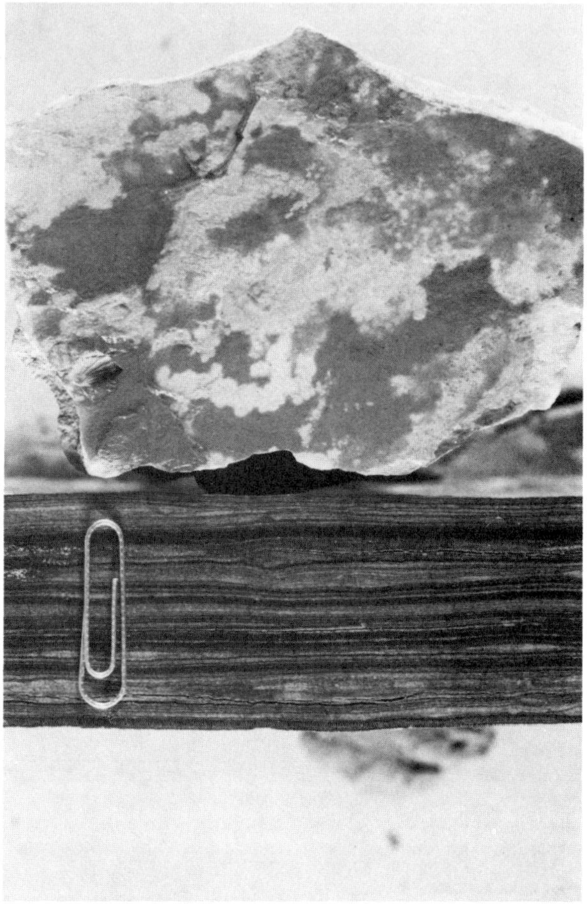

Figure 13. Lobe-margin facies. Top: bioturbated claystone with yellowish gray burrow mottles. Plan view of a bedding surface. Perpendicular to bedding, the mottles appear as thin "biolaminae" and streaks. Bottom: cross-sectional view of shale with abundant silt laminae interpreted as turbidites.

Figure 12. Interlobe-slope facies. Note absence of siltstone beds. Brallier Formation (Section 18, Figure 1). Mattock handle (left) is 1 m long.

sharp lower contacts, gradational upper contacts, fading ripples, and load casts. Based on these similarities, we believe the silt laminae in the lobe-margin facies are turbidites. The near absence of siltstone beds and the finer grain size of the shales and claystones suggest that this facies was deposited in areas farther from active turbidite sedimentation than the interlobe-slope facies. The burrows in the claystones of the lobe margin facies (Figure 13) are smaller than in the mudstones of the interlobe slope facies, possibly indicating that the dissolved oxygen concentration was lower where the claystones were deposited (Raff and Raff 1970). Thus, the lobe margin facies was probably deposited in a less oxygenated (deeper water), more distal environment than the interlobe slope facies.

Basinal Facies

Basinal black shales with minor siltstone turbidites and locally common silt laminae, like those in the lobe-margin facies (Table 1), occur in the lower part of the Brallier Formation and in the Millboro Shale in distal outcrops in southwestern Virginia. Subsurface data further indicate that this facies interfingers with the turbidite facies and thickens to the west (Figure 3). Fourteen black shale samples from the eastern Appalachian basin average 1 weight-percent organic carbon (Organic carbon content of the basinal shales generally increases to the west and may locally exceed 10 weight percent; Potter and others 1980). As indicated by their high organic carbon content, well-developed textural lamination, and absence of bioturbation, these black shales probably were deposited as hemipelagic mud and organic matter in a stagnant, quiet-water environment.

DEPOSITIONAL MODEL-CONTRASTS WITH SUBMARINE FAN MODELS

Turbidite facies models have largely been based on studies of modern submarine fans (Nelson and Kulm 1973; Normark 1978) and turbidite sequences in either Alpine flysch (McBride 1962; Mutti 1974; Niem 1976) or continental borderland settings (Sullwold 1960; Nilsen and Simoni 1978; Webb 1981). We found, however, that current turbidite facies models are inapplicable to delta-associated turbidite deposits such as the Upper Devonian sequence in the Appalachian basin.

As shown in schematic vertical sequences through a typical submarine fan and the Upper Devonian turbidite sequence (Figures 14 and 15), the Upper Devonian sequence is finer grained and thinner bedded than most modern submarine fan and ancient flysch deposits. Grain size is largely in the silt to very fine sand range and siltstone and sandstone bed thickness averages less than 10 cm. The entire stratigraphic sequence from basinal shales to non-turbidite deltaic sandstones and mudrocks includes only the classic turbidites member of the resedimented, coarse, clastic family (Walker 1978; 933). Important differences also exist in the proximal facies, especially in the stratigraphic transition from turbidites to non-turbidites. In the Upper Devonian sequence this

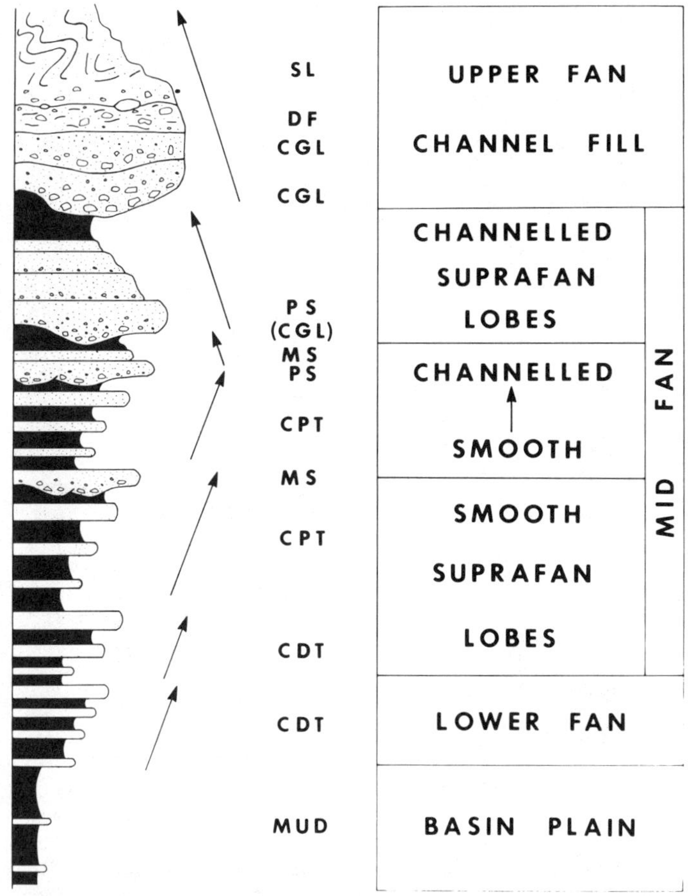

Figure 14. Schematic stratigraphic sequence produced by a prograding submarine fan. Inclined arrows indicate upward-thickening and -coarsening sequences (inclined to the right), or upward-thinning and -fining sequences (inclined to the left); CDT, classic distal turbidites; CPT, classic proximal turbidites; MS, massive sandstones; PS, pebbly sandstones; CGL, conglomerate; DF, debris flows; and SL, slumps. Modified from Walker (1978, Figure 14).

transition is gradual and characterized by upward-thickening sequences of classic turbidites of the turbidite-slope facies, and crossbedded traction deposits of the delta-front facies (Figure 15). Contrasting with this are the channeled, upward-thinning sequences of massive and pebbly sandstones, conglomerates, debris flows deposits, and slumps that characterize the proximal deposits of modern submarine fans (Figure 14). Therefore, the grain size, bed thickness, and facies of the Upper Devonian deposits indicate that turbidity currents were smaller and had lower maximum velocities than those of most modern submarine fans and ancient flysch sequences.

Furthermore, the facies relationships and the homogeneous transverse paleocurrent pattern indicate that the Late Devonian slope of the Appalachian basin was not fed by submarine canyons or upper-fan channels (Figure 16). Instead, sediment was probably spread directly onto the slope. The absence of canyon and channeled upper-fan facies indicates that the slope was previously unsculptured by rivers, as some modern shelf margins were dur-

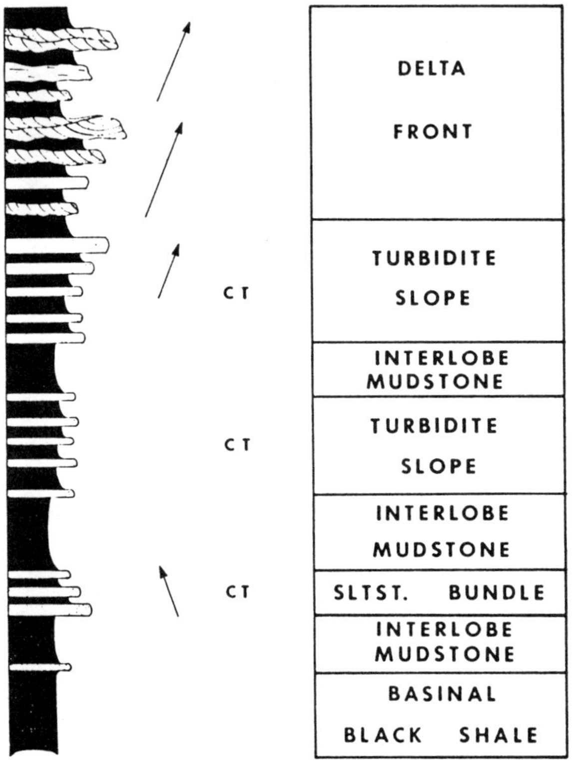

Figure 15. Schematic representation of the Upper Devonian stratigraphic sequence and its interpreted facies. Inclined arrows indicate upward-thickening and -coarsening (inclined to the right), or upward-thinning and -fining sequences (inclined to the left); CT, classic turbidites. Total thickness is approximately 1000 m. Note that the sequence contains only the classic turbidites member of the resedimented coarse clastic family. In addition, the stratigraphic transition from turbidites to overlying deltaic rocks is gradual and differs in its facies succession from the deposits of the proximal parts of submarine fans (Figure 14).

ing the Pleistocene, or that turbidity currents were too weak or scattered in their occurrence to incise the slope significantly. The predominance of upward-thickening sequences in the proximal turbidite slope facies and the continuity of bedding also suggest turbidity currents were generally unchanneled. Only the siltstone-bundle facies shows any evidence of channelling (upward-thinning and -fining sequences) and this evidence is inconclusive.

The uniformity of paleocurrents for hundreds of kilometers along depositional strike and vertically through the stratigraphic sequence suggests that the location of sediment sources for turbidity currents shifted frequently, because fixed sources would eventually result in submarine fan development and consequent radial divergence of turbidity currents. This shifting of turbidity current sources is to be expected in delta-associated environments because of the frequent diversion of deltas and distributaries. Holocene delta lobes of the Mississippi River have shifted position about every 1000 years along approximately 175 km of coastline (Coleman 1976; 26). Individual distributaries shift position even more frequently. The routine diversion of the deltas and distributaries of several rivers feeding the eastern margin of the basin could easily account for the observed paleocurrent pattern. Because turbidity current sources were not fixed for long periods of time, submarine fans did not form. Thus, we favor changing locations of turbidity current sources, probably as a result of frequent river diversion, as an explanation for the uniformity of paleocurrents, stratigraphic alternation of turbidite-slope and interlobe-slope facies, and the repetitive upward-thickening and -coarsening sequences of turbidite-slope and delta-front facies. The same mechanism was preferred by Walker and Harms (1971; 395) as an explanation for repetitive sequences of coastal facies (their Irish Valley motif) in the Upper Devonian of central Pennsylvania.

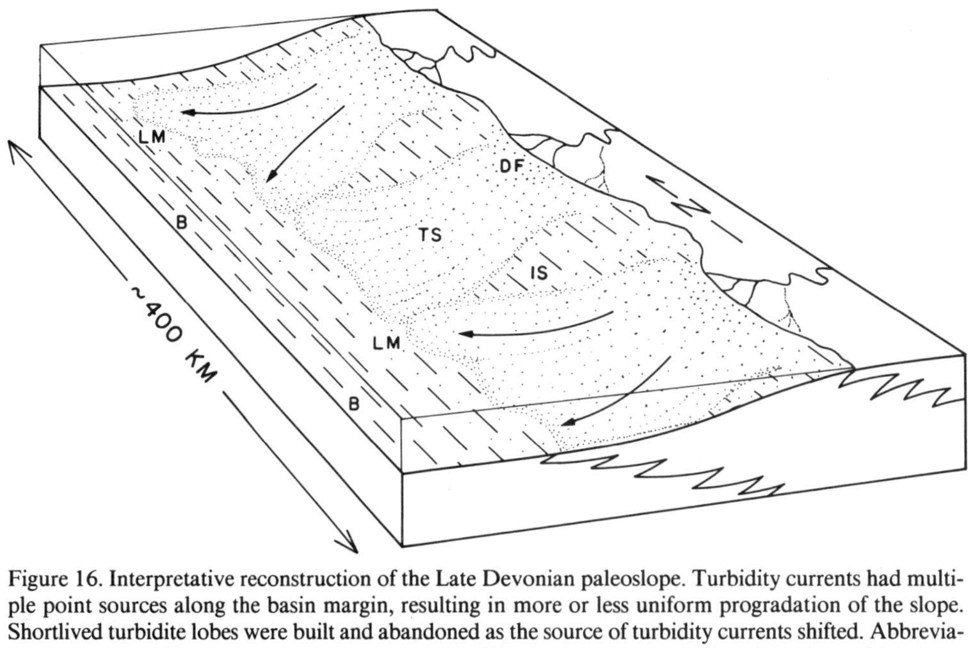

Figure 16. Interpretative reconstruction of the Late Devonian paleoslope. Turbidity currents had multiple point sources along the basin margin, resulting in more or less uniform progradation of the slope. Shortlived turbidite lobes were built and abandoned as the source of turbidity currents shifted. Abbreviations of facies discussed in text: DF, delta-front; TS, turbidite-slope; IS, interlobe-slope; LM, lobe-margin; and B, basinal.

The shifting of turbidity current sources resulted in the formation of an apron of turbidites (turbidite-slope facies) along the Late Devonian slope and base-of-slope, and more or less uniform progradation of the slope (Figure 16). Bioturbated mudstone (interlobe-slope facies) accumulated by hemipelagic sedimentation on the slope lateral to areas of active turbidite deposition. At the base of the slope, hemipelagic, bioturbated claystone and shale with turbidite silt laminae (lobe-margin facies) where deposited. The slope environment was generally dysaerobic and inhabited only by soft-bodied organisms. Black shales (basinal facies) accumulated by hemipelagic sedimentation of mud and organic matter basinward of the turbidites in quiet, stagnant water.

The apparent absence of major slumping, so conspicuous in other turbidite sequences, suggests that the Late Devonian paleoslope was probably very gentle. Slumping of mud slopes as low as 0.6 degrees is very common on the Mississippi River prodelta (Shepard 1955). Also, because the deposits of modern lower-fan environments are sedimentologically similar to those of the Upper Devonian sequence, one can infer that their slopes are similar. If the Late Devonian paleoslope was steeper than modern lower-fan slopes, the turbidites should show different sedimentation styles because of the increased velocity and erosive capacity of the currents. Modern lower-fan gradients range from about 0.06 degrees to 0.15 degrees (Nelson and others 1970; 277, 284).

Water depth in the basin receiving the turbidites is difficult to estimate. Using Klein's (1974) stratigraphic method of estimating water depth in deltaic sequences the water could not have been deeper than 950 m. This value is based on a stratigraphic thickness of 915 m for the rocks between the basinal shales and the topset deltaic sandstones, and 20 m of water overlying these topset deposits. It is considered a maximum value because it does not allow for penecontemporaneous subsidence. At present no better estimate is possible because widespread marker beds have not been recognized in the Brallier Formation. For comparison, Klein's method was applied to the overlying Lower Mississippian deltaic deposits, which also include turbidites. Using the measured sections of Bartlett (1974) in southwestern Virginia and Hasson (1972) in eastern Tennessee, water depths were calculated and range from 140 to 210 m. One expects water depths to be shallower during the later phase of basin filling represented by these Lower Mississippian clastics. The topography of the Borden delta in southern Illinois suggests that the Upper Devonian black shales in that area were deposited in at least 183 m of water (Lineback 1966; 22). It is probable that the Late Devonian sea was shallower in cratonal areas, such as Illinois, than in the Appalachian basin.

The cause or initiating mechanism of the turbidity currents is unknown. Turbidity currents are commonly thought to be initiated by submarine slumping of sediments on a slope, and indeed some turbidite sequences have major slump strucatures in their proximal facies (Cline 1970; McBride 1970). However, we have observed no such features in the study area although Walker (1971) observed some slump structures in the Upper Devonian sequence of central Pennsylvania. Where evidence for slumping is absent, alternative initiating mechanisms such as storm surge and high river discharge should be considered. The relatively thin and uniform thickness of turbidites in the Upper Devonian sequence indicates that a uniformly low intensity mechanism was responsible for triggering the turbidity currents.

The topography and sedimentology of the Late Devonian paleoslope that we infer for the central and southern Appalachian basin may also be valid beyond our study area to the north. Woodrow and Isley's (1983) and Walker and Sutton's (1967) measured sections of the Upper Devonian turbidite sequence in south-central New York and north-eastern Pennsylvania are very similar to ours. Woodrow and Isley (1983) also believe the paleoslope was unchanneled and that turbidites had a low maximum velocity.

This study confirms Walker's (1978) observation that submarine-fan models are inapplicable to delta-associated turbidite deposits. Differences in the dispersal pattern and facies sequence are largely explainable in terms of how sediment is supplied to the slope, that is by shifting sediment sources rather than fixed feeder channels. The gentle depositional slope was also important because it inhibited channeling and allowed turbidites to be deposited near their source. Descriptions of other delta-associated turbidite sequences (Allen 1960; Walker 1966; Moore and Clarke 1970; Link 1975; Kepferle 1978; Maynard and Lauffenburger 1978) suggest that there may be a continuum between low flow intensity sequences such as the Upper Devonian of the Appalachian basin, and high flow intensity sequences such as the Namurian Shale Grit of northern England. The slope and tectonics of the depositional basin are probably controlling factors.

ACKNOWLEDGMENTS

Drs. Paul Edwin Potter and J. Barry Maynard of the University of Cincinnati were instrumental in the completion of this project. A special thanks goes to Roy Kepferle whose wisdom and wit are appreciated more than he knows. Dr. Earle F. McBride, Bob Blodgett, and Johnny C. Wilson of the University of Texas critically reviewed the manuscript. Joan Wolff kindly assisted in the drafting. Funding for this study came from the Department of Energy contract ERDA No. EY-76-C-05-5205 ORD 5201-4 and is gratefully acknowledged.

REFERENCES CITED

Allen, J.R.L., 1960, The Mam Tor Sandstones: a "turbidite" facies of the Namurian deltas of Derbyshire, England: Journal of Sedimentary Petrology, v. 30, p. 193–208.

Avary, K. L., and Dennison, J. M., 1978, Distinctive turbidite siltstone in Devonian Brallier Formation of West Virginia and Virginia (abstract): American Association of Petroleum Geologists Bulletin, v. 62, p. 492.

Bartlett, D. S., 1974, Anatomy of the Lower Mississippian delta of southwest Virginia [Ph.D. thesis]: University of Tennessee, 373 p.

Cheema, M., 1977, Sedimentation and gas production of the Upper Devonian Benson sand in north-central West Virginia—a model for exogeosynclinal mid-fan turbidites off a delta complex [Ph.D. thesis]: West Virginia University, 117 p.

Cline, L. M., 1970, Sedimentary features of Late Paleozoic flysch, Ouachita Mountains, Oklahoma, *in* Lajoie, J., ed., Flysch sedimentology of North America: Geological Association of Canada Special Paper No. 7, p. 85–101.

Coleman, J. M., 1976, Deltas: processes of deposition and models for exploration: Continuing Education Publishing Company, Inc., 102 p.

Contescu, L. R., 1974, Geologic history and paleogeography of eastern Carpathians, example of Alpine geosyncline evolution: American Associated of Petroleum Geologists Buletin, v. 58, p. 2436–2476.

Dennison, J. M., 1970, Stratigraphic divisions of the Upper Devonian Greenland Gap Group ("Chemung Formation") along the Allegheny Front in West Virginia, Maryland, and Highland County, Virginia: Southeastern Geology, v. 12, p. 53–82.

Dennison, J. M., and Boucot, A. J., 1974, Little War Gap at Clinch Mountain provides standard reference section for Silurian Clinch Sandstone and most nearly complete Devonian section in eastern Tennessee: Southeastern Geology, v. 16, p. 79–101.

Dennison, J. M., and Hasson, K. O., 1976, Stratigraphic cross-section of Hamilton Group (Devonian) and adjacent strata along south border of Pennsylvanian: American Association of Petroleum Geologists Bulletin, v. 60, p. 278–287.

Dott, R. J., Jr., 1974, Paleocurrent analysis of severely deformed flysch-type strata—a case study from South Georgia Island: Journal of Sedimentary Petrology, v. 44, p. 1166–1173.

Frakes, L. A., 1967, Stratigraphy of the Devonian Trimmers Rock in eastern Pennsylvania: Pennsylvania Geological Survey, 4th Series, Bulletin G51, 208 p.

Hasson, K. O., 1972, Lithostratigraphy of the Grainger Formation (Mississippian) in northeast Tennessee [Ph.D. thesis]: University of Tennessee, 143 p.

Kepferle, R. C., 1978, Prodelta turbidite fan apron in Borden Formation (Mississippian), Kentucky and Indiana, *in* Stanley, D. J., and Kelling, G., eds., Sedimentation in submarine canyons, fans, and trenches: Pennsylvania, Dowden Hutchinson and Ross, p. 224–238.

Klein, G. deV., 1974, Estimating water depths from analysis of barrier island and deltaic sedimentary sequences: Geology, v. 2, p. 409–412.

Lineback, J. A., 1966, Deep-water sediments adjacent to the Borden Siltstone (Mississippian) delta in southern Illinois: Illinois State Geological Survey Circular 401, 48 p.

Link, M. H., 1975, Matalija Sandstone—a transition from deep water turbidite to shallow marine deposition in the Eocene of California: Journal of Sedimentary Petrology, v. 45, p. 63–78.

Lundegard, P. D., 1979, Sedimentology and Petrology of a prodeltaic turbidite system—The Brallier Formation (Upper Devonian), western Virgina and adjacent areas [M.S. thesis]: University of Cincinnati, 296 p.

Lundegard, P. D., and Samuels, N. D., 1980, Field classification of fine-grained sedimentary rocks: Journal of Sedimentary Petrology, v. 50, p. 781–786.

Lundegard, P. D., Samuel, N. D., and Pryor, W. A., 1980, Sedimentology, petrology, and gas potential of the Brallier Formation—Upper Devonian turbidite facies of the central and southern Appalachians: U.S. Department of Energy, DOE/METC/5201-5, 220 p.

McBride, E. F., 1962, Flysch and associated beds of the Martinsburg Formation (Ordovician), central Appalachians: Journal of Sedimentary Petrology, v. 32, p. 39–91.

—— 1970, Flysch sedimentation in the Marathon Region, Texas, *in* Lajoie, J., ed., Flysch sedimentology of North America: Geological Association of Canada Special Paper No. 7, p. 67–84.

McBride, E. F., Weidie, A. E., and Wolleben, J. A., 1975, Deltaic and associated deposits of Difunta Group (Late Cretaceous to Paleocene), Parras and La Popa basins, northeastern Mexico, *in* Broussard, M. L., ed., Deltas, models for exploration: Houston Geological Society, p. 485–522.

McGhee, G. R., Jr., 1975, Late Devonian benthic marine communities of the central Appalachian Allegheny Front: Lethaia, v. 9, p. 111–136.

McIver, N. L., 1961, Upper Devonian marine sedimentation of the central Appalachians [Ph.D. thesis]: Johns Hopkins University, 530 p.

—— 1970, Appalachian turbidites, *in* Fisher, G. W., Pettijohn, F. L., Reed, J. C., Jr., Weaver, K. N., eds., Studies of Appalachians geology—central and southern: New York, Interscience Publishing, p. 69–81.

Maynard, J. Barry, and Lauffenburger, S. K., 1978, A marcasite layer in prodelta turbidites of the Borden Formation (Mississippian) in eastern Kentucky: Southeastern Geology, v. 20, p. 47–58.

Moore, B. R., and Clarke, M. K., 1970, The significance of a turbidite sequence in the Borden Formation (Mississippian) of eastern Kentucky and southern Ohio, *in* Lajoie, J., ed., Flysch sedimentology of North America: Geological Association of Canada Special Paper No. 7, p. 211–218.

Morris, R. C., 1974, Sedimentary and tectonic history of the Ouachita Mountains, *in* Dickenson, W. R., ed., Tectonics and sedimentation: Society of Economic Paleontologists and Mineralogists Special Publication 22, p. 120–142.

Mutti, E., 1974, Examples of ancient deep-sea fan deposits from Circum-Mediterranean geosynclines, *in* Modern and ancient geosynclinal sedimentation: Society of Economic Paleontologists and Mineralogists Special Publication 19, p. 92–105.

Nelson, C. H., Carlson, P. R., Byrne, J. V., and Alpha, T. R., 1970, Development of the Astoria Canyon-Fan physiography and comparison with similar systems: Marine Geology, v. 8, p. 259–291.

Nelson, C. H., and Kulm, V., 1973, Submarine fans and channels: Society of Economic Paleontologists and Mineralogists, short course lecture notes, Anaheim, Turbidites and deep water sedimentation, p. 39–78.

Niem, A. R., 1976, Patterns of flysch deposition and deep-sea fans in the lower Stanley Group (Mississippian), Ouachita Mountains, Oklahoma and Arkansas: Journal of Sedimentary Petrology, v. 46, p. 633–646.

Nilsen, T. H., and Simoni, T. R., 1973, Deep-sea fan paleocurrent patterns of the Eocene Butano Sandstone, Santa Cruz Mountains, California: U.S. Geological Survey Journal Research, v. 1, p. 439–452.

Normark, W. R., 1978, Fan valleys, channels, and depositional lobes on modern submarine fans: characters for recognition of sandy turbidite environments: American Association of Petroleum Geologists Bulletin, v. 62, p. 912–931.

Oliver, W. A., Jr., deWitt, W., Jr., Dennison, J. M., Hoskins, D. M., and Huddle, J. W., 1969, Correlation of Devonian rock units in the Appalachian basin: U.S. Geological Survey, Oil and Gas Investigations Chart OC-64.

Potter, P. E., Maynard, J. B., and Pryor, W. A., 1980, Final report of special geological, geochemical, and petrological studies of the Devonian shales in the Appalachian basin: Final report to U.S. Dept. of Energy by the Univ. of Cincinnati, Dept. of Geology, Contract DE-AC21-76MCO5201, 86 p.

Potter, P. E., and Pettijohn, F. J., 1977, Paleocurrents and basin analysis: New York, Springer-Verlag, 452 p.

Potter, P. E., Pryor, W. A., Lundegard, P. D., Samuels, N. D., and Maynard, J. Barry, 1979, Devonian paleocurrents in the Appalachian basin: U.S. Department of Energy, METC/CR-79/22, 60 p.

Raff, R. A., and Raff, E. C., 1970, Respiratory mechanisms and the fossil record: Nature, v. 228, p. 1003–1005.

Samuels, N. D., 1979, The sedimentology and petrology of the Brallier Formation—Upper Devonian turbidite slope facies of the central Appalachians [M.S. thesis]: University of Cincinnati, 153 p.

Scott, K. M., 1966, Sedimentology and dispersal pattern of a Cretaceous flysch sequence, Patagonian Andes, southern Chile: American Association of Petroleum Geologists Bulletin, v. 50, p. 72–107.

Shepard, F. P., 1955, Delta front valleys bordering the Mississippi distributaries: Geological Society of America Bulletin, v. 66, p. 1489–1498.

Spearing, D. R., 1975, Shallow marine sands: Society of Economic Paleontologists and Mineralogists Short Course 2, p. 103–132.

Stow, D.A.V., and Shanmugam, G., 1980, Sequence of structures of in fine-grained turbidites: comparison of Recent deep-sea and ancient flysch sediments: Sedimentary Geology, v. 25, p. 23–42.

Sullwold, H. H., Jr., 1960, Tarzana Fan, deep submarine fan of Late Miocene age, Los Angeles County California: American Association of Petroleum Geologists Bulletin, v. 44, p. 433–457.

Sutton, R. G., 1959, Use of flute casts in stratigraphic correlation: American Association of Petroleum Geologists Bulletin, v. 43, p. 230–237.

ten Haaf, E., 1959, Graded beds of the northern Appennines [Ph.D. thesis]: Rijks University of Groningen, 102 p.

Trettin, H. P., 1970, Ordovician-Silurian flysch sedimentation in the axial trough of the Franklinian geosyncline, northeastern Ellesmere Island, Arctic Canada, *in* Lajoie, J., ed., Flysch sedimentology in North America: Geological Association of Canada Special Paper No. 7, p. 13–36.

Walker, R. G., 1966, Shale Grit and Grindslow Shales: transition from turbidite to shallow water sediments in the Upper Carboniferous of Northern England: Journal of Sedimentary Petrology, v. 36, p. 90–114.

——1967, Turbidite sedimentary structures and their relationship to proximal and distal depositional environments: Journal of Sedimentary Petrology, v. 37, p. 25–43.

——1971, Nondeltaic depositional environments in the Catskill clastic wedge (Devonian) of central Pennsylvania: Geological Society of America Bulletin, v. 82, p. 1305–1326.

——1978, Deep-water sandstone facies and ancient submarine fans: models for exploration for stratigraphic traps: American Association Petroleum Geologists Bulletin, v. 62, p. 119–158.

Walker, R. G., and Harms, J. C., 1971, The "Catskill Delta": a prograding muddy shoreline in central Pennsylvania: Journal of Geology, v. 79, p. 381–399.

Walker, R. G., and Sutton, R. G., 1967, Quantitative analysis of turbidites in the Upper Devonian Sonyea Group, New York: Journal of Sedimentary Petrology, v. 37, p. 1012–1022.

Woodrow, D. L., and Isley, A. M., 1983, Facies, topography, and sedimentary processes in the Catskill Sea (Devonian), New York and Pennsylvania: Geological Society of America Bulletin, v. 94, p. 459–470.

Webb, G. W., 1981, Stevens and earlier Miocene turbidite sandstones southern San Joaquin Valley, California: American Association of Petroleum Geologists Bulletin, v. 65, p. 438–465.

MANUSCRIPT ACCEPTED BY THE SOCIETY JULY 20, 1984

Printed in U.S.A.

Geological Society of America
Special Paper 201
1985

Devonian continental ecosystems and habitats

Richard Beerbower
Department of Geological Sciences
State University of New York at Binghamton
Binghamton, New York 13901

ABSTRACT

During the Devonian, continental ecosystems underwent major transformations including increases in primary and secondary production and in trophic differentiation and specialization. These are consequent on adaptations in vascular plants which increased resource acquisition and assimilation and reduced the effects of disturbance. The adaptations were probably initiated in relatively stable, rich, low disturbance habitats; their development involved a positive feedback with habitat characteristics so that improvements in utilization of resources and regulation of disturbance actually added to the availability of resources and moderated disturbance. Progressive development of this feedback system produced significant changes in the physical characteristics of continental habitats through the course of the Devonian. In particular, by late Devonian, the rate of chemical weathering must have been higher, and the rates of runoff, throughflow, and sheet and rill erosion lower with a concomitant reduction in variations in stream discharge. These changes would have had major consequences for alluvio-deltaic sedimentation, especially for determination of channel regimes and of depositional and early diagenetic patterns in overbank environments. Detailed studies of Catskill sediments can help to elucidate the extent of ecosystem changes.

The paper from which the abstract derives is published in: Tiffney, R. F. (ed.), in press, Geological Factors and the Evolution of Plants, Yale University Press, New Haven, Conn.

Printed in U.S.A.

Geological Society of America
Special Paper 201
1985

The flora of the Catskill clastic wedge

Harlan P. Banks
Section of Plant Biology
Cornell University
Ithaca, New York 14853

James D. Grierson
Patricia M. Bonamo
Department of Biological Sciences
State University of New York at Binghamton
Binghamton, New York 13901

ABSTRACT

In the last 25 years, many paleobotanists and palynologists have focused on the paleoflora of the Catskill clastic wedge, to take advantage of the sequence of abundant floral remains. In this interim a considerable body of paleobotanical information has accrued. It is clear now that these fossil plants are not a localized paleoflora. At the generic level many of them are to be found in Europe, Siberia, China, Australia and, less commonly, in Africa. Most of these genera are of the plant lineages that give rise to the dominant floras of the Carboniferous. Unlike most animal megafossils, it is the form and position of the reproductive organs of the plant that have been primarily used in placing the plant in its rightful position in major evolutionary groupings. Yet plant megafossils are mostly found as fragments of the vegetative plant, as leaves, stems and occasionally cellular anatomy. Because of this, it has been only recently that a tentative scheme of sequential megafossil assemblages in the Siluro-Devonian has appeared to augment earlier palynological schemes. Several prolific localities are discussed and one example of exceptional preservation is illustrated. These details show the kind and extent of information obtainable from the Catskill delta deposits, which then give clues to depositional environments and climatic conditions at the time of deposition.

INTRODUCTION

Prior to 1960 detailed descriptions of the plants of the Catskill clastic wedge were sparse. A number of papers by Arnold (e.g., 1939), one by Banks (1944), and one by Beck (1957) presaged subsequent studies but they did not warrant broad generalizations. Since 1960 an imposing array of data has appeared as a result of careful monitoring of the blasting at newly opened quarries, road cuts, and dam sites, followed by mass collecting when densely fossiliferous lenses were exposed. Intensive study of the new collections using an ever-expanding battery of techniques has yielded insights previously thought to be unavailable. This work of the last two decades has been carried out primarily by students of plant morphology and anatomy who were unspecialized in geology. Space prohibits discussion of extensive collections of land plants from the marine portion of the clastic wedge. Many of the specimens can be identified with genera known in greater detail from the terrestrial strata.

RELATIONSHIP OF THE CATSKILL FLORA TO THAT OF THE REST OF THE WORLD

As a result of the work of many investigators, we can say that the flora of the Catskill deposits is by no means a localized one; it is clearly part of a uniform worldwide flora. If it were not for the exciting evolutionary story it reveals and the unfolding of this comprehensive world picture, one might call the Devonian flora monotonous.

In terms of genera, and often species as well, one can follow the Catskill Devonian flora from the Central Appalachians northward through Maine (Andrews et al., 1977), eastern Canada (Banks et al., 1975; Gensel, 1982; Hueber, 1983), Ellesmere Island (Andrews et al., 1965), thence westward across the Canadian Arctic Islands (Fry, *in* Tozer, 1956; Hueber, 1972) to the Rockies in Alberta (Scheckler, 1978), southward to Wyoming (Dorf, 1933), and finally to an outlier in northern Arizona (Canright, 1970) along the Salt River. Westward from Alberta, it is found in Yukon Territory (Scheckler and Bassinger, 1983) and Alaska (Churkin et al., 1969; Mamay, 1962). Back to the east, the flora is found in Greenland (Pedersen, 1976), Svalbard (Spitsbergen: Høeg, 1942), Bear Island (Nathorst, 1902), Norway, Great Britain, Western Europe, central Europe, across southern Russia to the Saian-Altai Mountain Range in Siberia (see Boreau, 1967; Chaloner and Sheerin, 1979; Banks, 1980). Finally, comparable material is currently being described from the People's Republic of China (Li and Tsai, 1978). Nor is this flora confined to the Northern Hemisphere. It is becoming increasingly clear that several genera are found in both Australia and the Northern Hemisphere (Fairon-Demaret, 1974; Hueber, 1983). Even the limited evidence from Antarctica tells the same story (Grindley, Mildenhall and Schopf, 1980). Africa and South America are much less well known, but evidence from palynology (Richardson and Ioannides, 1973) and from a few poorly preserved megafossils (Klitsch, Lejal-Nicol and Massa, 1973) indicates that North Africa shares the same flora.

EVOLUTIONARY DEVELOPMENT OF THE FLORA

Both evolutionary aspects and stratigraphic occurrences of the Siluro-Devonian early land flora have been markedly clarified during the last two decades. For many years the first land plants were referred to as psilophytes (Psilopsida; Psilophytales). But as early as 1954, Leclercq pointed out that reproductive structures of the psilophytes were found in two distinct positions, terminal (Fig. 1a) and lateral (Fig. 1b). Ten years later Hueber (1964) showed that the shape and mode of dehiscence of these two sporangial types were different. In 1967 Hueber and Banks demonstrated that the vascular anatomy of the plants was distinct as well. Then Banks (1968) proposed new subdivisions of vascular plants to replace the old Psilophytales which had become a taxonomic repository for the bulk of the Early Devonian flora. He erected as new subdivisions of Tracheophyta: Rhyniophytina, Zosterophyllophytina and Trimerophytina.

Rhyniophytina (Fig. 1a) includes the first vascular land plants which appeared in Wenlock time (Edwards et al. 1983) and continued into Siegenian or Emsian. They are characterized by leafless, rootless stems that branched dichotomously and bore single, terminal, globose to fusiform reproductive bodies (sporangia).

In Gedinnian time a second group of plants, the Zosterophyllophytina (Fig. 1b) appeared (see Edwards, 1972). They too had leafless stems, but probably bore an occasional single (adventitious) root. Their sporangia were borne laterally along the stem, were reniform, and dehisced distally along the convex margin. Some members of the group persisted into early Late Devonian.

Mid-Siegenian saw the advent of two more groups (see papers cited for Assemblage Zone III in Banks, 1980). Plants in Trimerophytina (Fig. 1c) shared leafless stems and terminal sporangia with rhyniophytes, but several of their characteristics indicated a more complex evolutionary level. Trimerophytes were more robust, perhaps reaching a meter or more in height. Single plants branched profusely dichotomously, helically and alternately. Their sporangia were borne in large terminal clusters of 64 to 256. Anatomically, they were much more complex also. The widely known genus *Psilophyton* (Fig. 1c) represents trimerophytes. Some of the group may have persisted into the Frasnian.

Also appearing in Mid-Siegenian time were lycopods (Lycophytina; Fig. 1d), a group that has continued to the present, giving it the longest fossil record known among vascular land plants. Lycopods shared features with zosterophylls. They were more complex in their possession of true leaves which bore single sporangia on their upper surface. Sporangia in both groups were reniform and dehisced distally. Anatomically, the two groups shared many similarities. Lycopods represent an evolutionary level more complex than zosterophylls, and the two groups seem always to have been distinct from all other vascular plants.

The four groups introduced above had appeared in the Early Devonian prior to the commencement of deposition of the Catskill clastic wedge. They initiated two evolutionary series that expanded markedly in Mid-Devonian time, along with other, lesser groups that apparently failed to contribute to the two predominant series.

Middle and Upper Devonian strata record the continuation of the Early Devonian adaptive radiation. The two most exciting and thoroughly studied groups are the lycopods and the progymnosperms. Lycopods were diverse, with small simple leaves or leaves which were variously forked. They evolved the arborescent habit, e.g., *Lepidosigillaria* (the Naples tree, Fig. 2b). During Carboniferous far more diversity appeared, including the giant lepidodendrids of the coal swamps. The latter disappeared in Permian time but trailing, herbaceous lycopods have persisted with relatively minor change to the present.

Progymnosperms, a group erected by Beck (1960), were plants with reproductive processes comparable to those of ferns, but with wood comparable to the seed-bearing gymnosperms. They represent a series of evolutionary levels more complex than trimerophytes. Appearing first in the Eifelian, they diversified steadily from then on into the Mississippian (Fig. 2c-f, 3b-d).

BIOSTRATIGRAPHIC IMPLICATIONS OF THE CATSKILL FLORA

Arber (1921) separated the Devonian into an Early Devonian *Psilophyton*-Flora and a Late Devonian *Archaeopteris*-Flora. Since then, many more plants have been found, and newer techniques have provided important data. In addition, a better knowledge of the stratigraphic position of the plants has contrib-

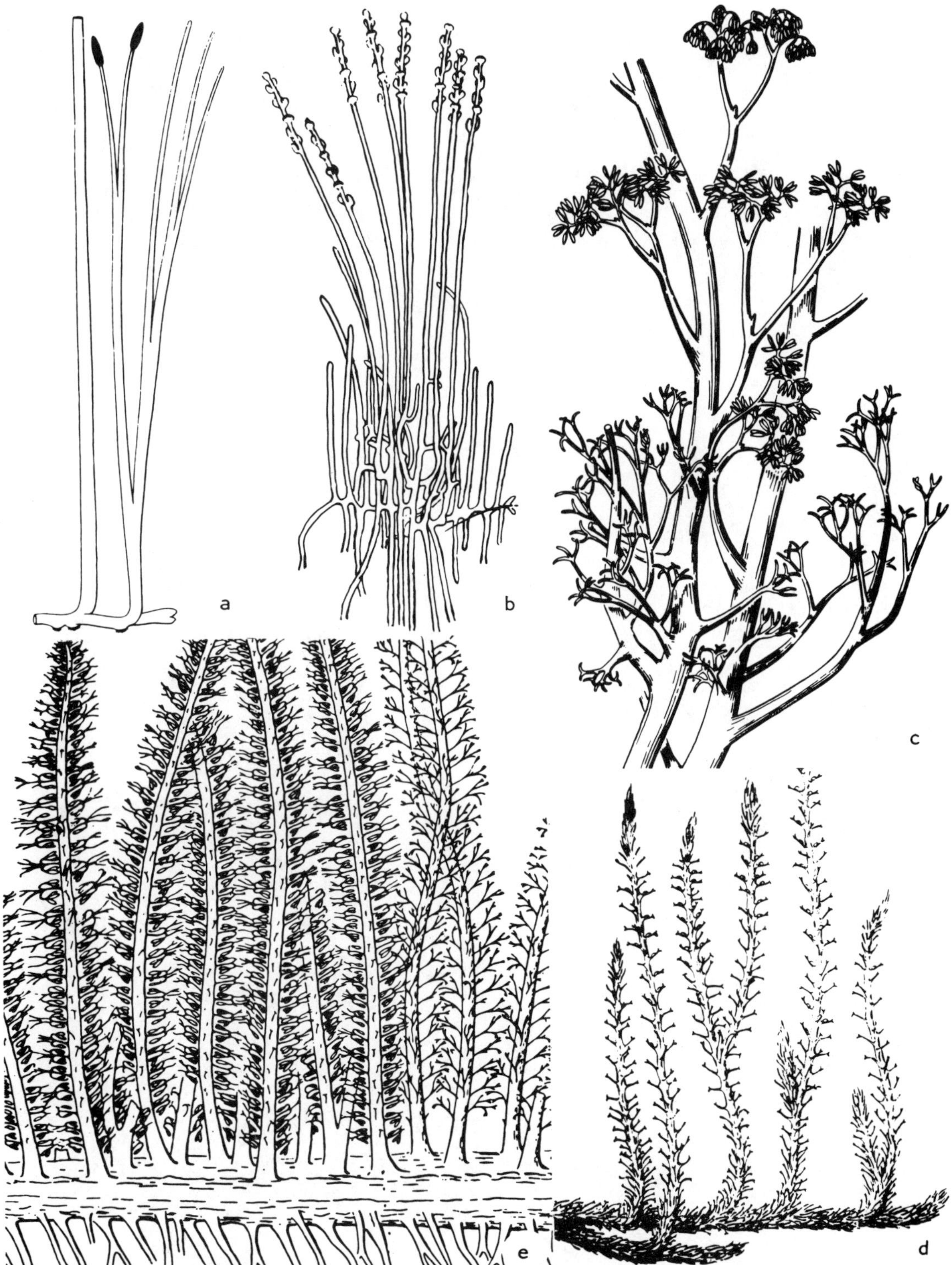

Figure 1. a. *Rhynia,* a rhyniophyte with horizontal stem and erect branches, to 20-25 cm high. Note terminal sporangia, dichotomous branching, lack of leaves and lack of roots. b. *Zosterophyllum,* a zosterophyllophyte. Erect branches to 20-25 cm high. Note the peculiar branching near the base, lateral sporangia, and absence of leaves and roots. c. *Psilophyton,* a trimerophyte. Note the abundant and variable branching, recurved tips on some branches, masses of terminal sporangia on the fertile branches, and the absence of leaves. Aerial axis to 90 cm high. d. *Protolepidodendron,* a lycopod with trailing stems and erect branches 20-25 cm tall. Note the small leaves, some of which have sporangia on their upper surfaces. Roots occur only occasionally; none shown. e. *Hyenia,* a horsetail with trailing stem, erect branches to 20-25 cm, small leaves and occasionally branched roots. Note the complicated sporangium-bearing branches. Fig. 1a, b, d, e after Banks, 1970; 1c after Banks, Leclercq and Hueber, 1975.

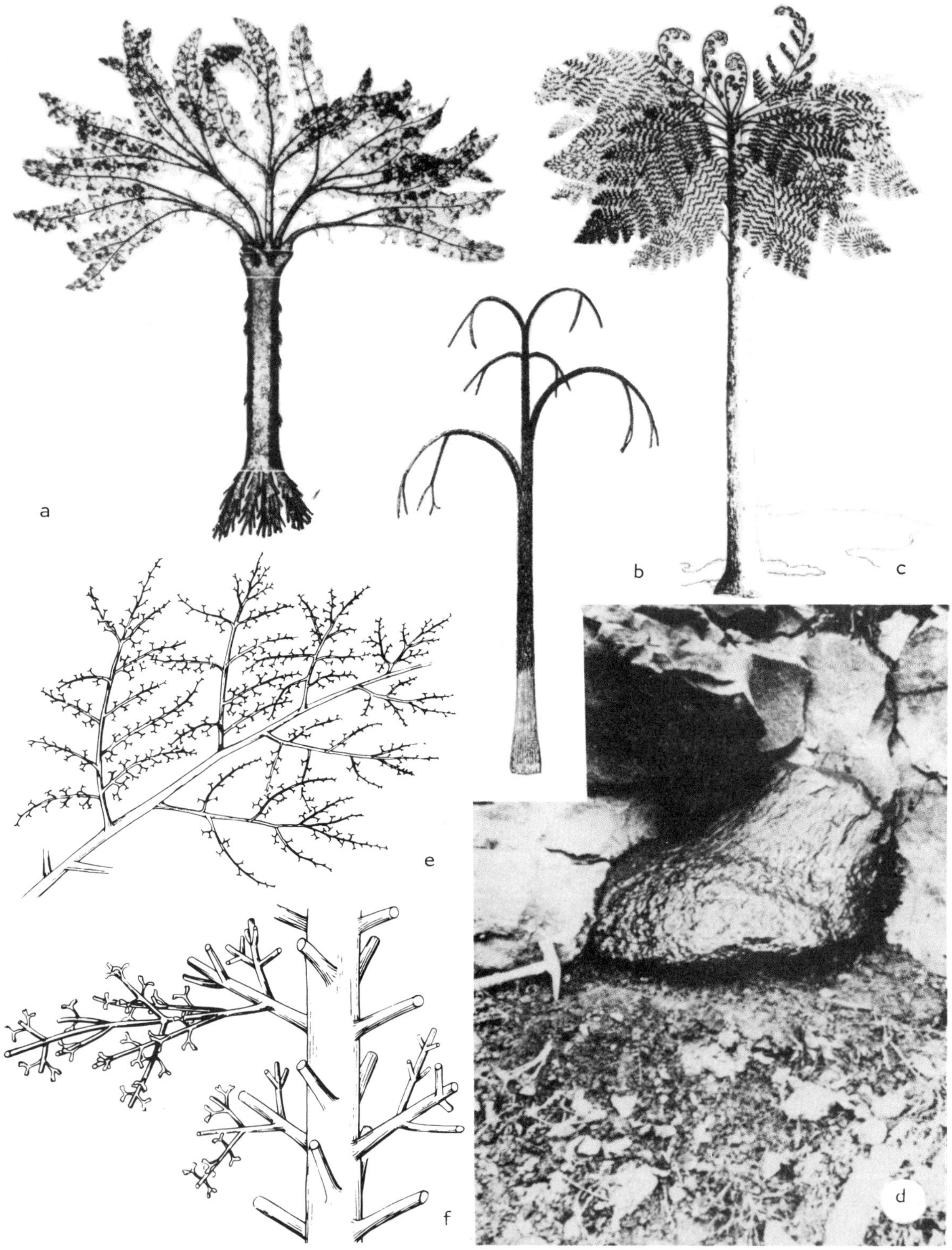

Figure 2. a. *Pseudosporochnus,* a cladoxylalean. Stem erect, ultimate branchlets of the large branch system were precursors of leaves. Plant to 27 dm tall. b. *Lepidosigillaria,* an arborescent lycopod to 49 dm. The stem was abundantly covered by small leaves. c. *Eospermatopteris,* might be a progymnosperm. This is Goldring's reconstruction for which we have evidence only of isolated stump casts (Fig. 2d) and unattached branch systems. Some specimens of associated branches bore sporangia; none bore seeds. Plants may have been 90 dm tall. d. *Eospermatopteris,* a stump in situ ca 5 dm in height. e. *Aneurophyton,* a large branch system extending to 9 dm. The ultimate branchlets may be precursors of leaves. The internal structure and sporangia show this genus to be a progymnosperm. Goldring thought it was borne by *Eospermatopteris,* solely on the basis of association. f. *Tetraxylopteris,* a branch system of a progymnosperm with unwebbed (non-laminar) leaves; it may have reached 3 dm in height as a shrub. Fig. 2a, b, e, f after Banks, 1970; 2c, d after Goldring, 1924.

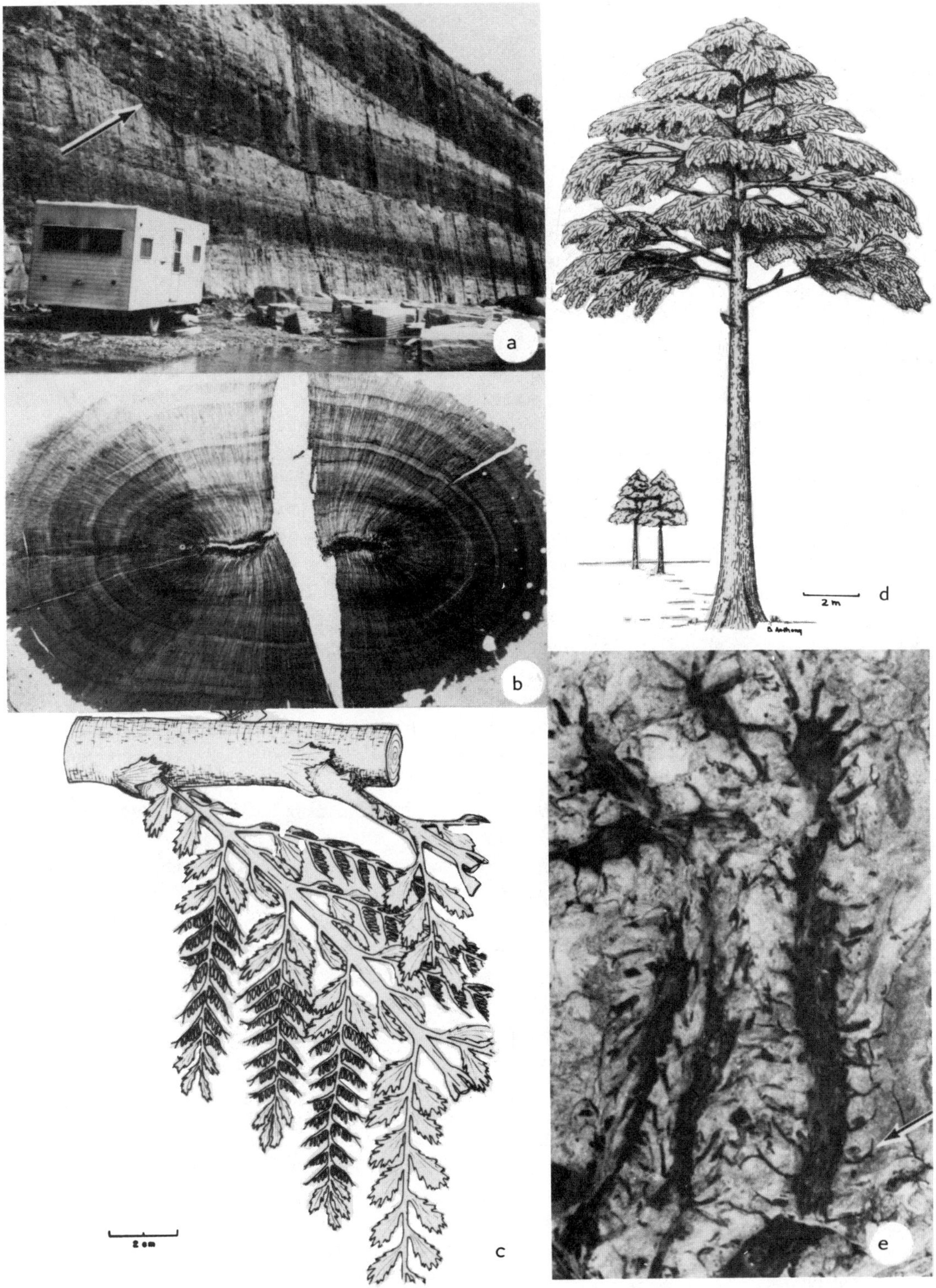

Figure 3. a. Photograph of one wall of the overflow channel at Cannonsville Reservoir (N.Y.) showing a siltstone lens (arrow) that is filled with well-preserved leafy branches of *Archaeopteris.* b. Transverse section of Arnold's *Callixylon erianum,* presumably a large branch of *Archaeopteris,* with several obvious growth layers. The central pith, extending here from left to right, has been crushed during preservation. x 2.1. c. Beck's (1962) reconstruction of a portion of one leafy branch of *Archaeopteris.* Both vegetative (sterile) and fertile leaves are shown. Note the expanded laminae of the sterile leaves (webbed). d. Beck's (1962) reconstruction of a plant of *Archaeopteris (Archaeopteris + Callixylon)* to 18 m in height. e. *Leclercqia complexa,* a lycopod. The leafy stems are abundant in the matrix. Note the helical arrangement of the leaves (top of right stem) and the leaves bearing sporangia on their upper surfaces (arrow). x. 1.8.

uted fundamentally to our understanding of evolutionary trends. Significant collections of land plants are found only in terrestrial strata. The precise correlation of these beds with securely dated marine strata is often lacking. Palynology has contributed notably to stratigraphy since spores are often distributed widely in both marine and terrestrial strata. In the former, they can be used to build a stratigraphic sequence for which marine invertebrates provide reliable indices. Terrestrial sequences can then be matched against them for dating purposes. An excellent analysis of this use of plant spores is given by McGregor (1979).

Many workers have wanted more precise subdivisions of Devonian time. Russian workers have used terrestrial plants as indices for the age of continental strata, e.g., Petrosian in Oswald (1968). Recently, Chaloner and Sheerin (1979) produced an excellent summary of changes in Devonian plants through time, of stratigraphic ranges of many genera, and of evolutionary relationships among them.

Banks (1980) suggested a seven-fold division (Table 1) of the stages from Pridolian through post-Famennian Devonian, on the basis of time of first appearance of widespread new floras. Individual genera may cross the boundaries of the Generic Assemblage Zones, but it is the time of abundance of a given flora that sets the limits. Examination of Table 1 will show that the lowest three zones correspond to three of the early plant groups discussed above. *Cooksonia* is a rhyniophyte (Fig. 1a), *Zosterophyllum* (Fig. 1b) a zosterophyll, and *Psilophyton* (Fig. 1c) a trimerophyte. *Hyenia* (Fig. 1e) is an articulate (horsetail) and, while not abundant, is restricted to Eifelian strata. *Svalbardia,* with dissected leaves, might be a precursor to *Archaeopteris* (Fig. 3b-d) and presumably is a progymnosperm. *Rhacophyton* is an early representative of coenopterid ferns, an extinct group that was prominent in the Pennsylvanian. Genera that occur in each of these zones are given in Banks (1980), and one should consult Chaloner and Sheerin (1979) as well. A source of considerable detail and good illustrations of these plants is Boureau (1967). These assemblage zones provide a clue to the possible age of a collection made in undated terrestrial strata. Either the collection must be good enough to permit ready identification or, if it is fragmentary, it must be studied sufficiently to permit an accurate description of whatever tissues or organs are present. The time when macerated specimens, ruined by transport, can be considered worthwhile specimens of plant fossils is long past, despite continued submission of such material for identification. The tentative nature of these zones is but a challenge to continue the work that has led to their erection.

Precise definition of stratigraphic divisions is made difficult because relatively few Devonian genera are known as whole plants. Some are known only from external vegetative appearance, some only from a portion of their internal anatomy, some solely from reproductive structures. However, it requires all of these parts to reconstruct a whole plant. Organic connection of parts or identity of specific characters must be established with certainty before a reconstruction can be attempted. For example, for years *Callixylon* (Fig. 3b), found only as silicified, calcified, or pyritized stems with gymnosperm-like wood, or as logs with no structure, was regarded as the stem of a gymnosperm-like plant. It was common around the globe in the Late Devonian. In a similar way, *Archaeopteris* (Fig. 3c) was known as the foliage of an Upper Devonian free-sporing fern for an even longer time and it was often found in association with *Callixylon.* Only in 1960 did Beck find an axis with *Callixylon* anatomy (wood) bearing a branch with *Archaeopteris* foliage and attached reproductive parts. This showed that the whole plant was neither a gymnosperm, as supposed for *Callixylon* wood, nor a fern, as supposed for *Archaeopteris* foliage, but a hitherto unknown combination of these characteristics. Beck's striking discovery led him to suggest a new group of vascular plants, Progymnospermopsida, for the new plant. *Archaeopteris* is now the name for the whole organism (Fig. 3d) although *Callixylon* continues to be used as a form genus for isolated pieces of the stem. Other genera, showing this combination of characters, have been found and added to the group, e.g., *Aneurophyton, Tetraxylopteris* and *Rellimia.*

A major problem in using such a series of assemblage zones is the error which arises from attempting to identify fragmentary material. As one example, a small branch of a progymnosperm such as *Aneurophyton* or *Tetraxylopteris* (Fig. 2e, f) could easily be mistaken for a *Psilophyton* (Fig. 1c) branch. Incorrectly identified, one would arrive at a dating of Zone III when in fact the specimen belonged to Zone V or VI.

Nonetheless, certain specialized characteristics can be used to supplement taxonomic and stratigraphic conclusions. Without repeating details published elsewhere (see Chaloner and Sheerin, 1979; Banks, 1980; this paper), some examples are: specimens preserved with secondary xylem (wood) cannot be older than Eifelian (Zone IV), those showing the arborescent habit are not older than Givetian (Zone V), those with large leafy branch systems (e.g., *Archaeopteris,* Fig. 3c, d) are not older than Frasnian (Zone VI), those with spore tetrads consisting of three aborted and one functional spore are not older than Frasnian (Zone VI), and specimens of seeds are not older than Mid-Famennian (Zone VIII).

EXCEPTIONAL EXAMPLE: OPTIMUM PRESERVATION + PRECISE TECHNIQUES = OPTIMUM INSIGHTS

Specimens of *Leclercqia complexa* Banks, Bonamo and Grierson 1972, 3.5-7.0 mm in diameter and up to 46 cm in length were found densely filling a black, fine-grained mudstone at Brown Mountain along Schoharie Creek in the Panther Mountain Formation north of the Schoharie (Gilboa) Reservoir. The plants are preserved predominantly as compressions in which the originally terete stems are now flattened, but still retain considerable coaly cellular structure. The dichotomizing stems are densely covered by leaves borne in a low helix (Fig. 3e). The leaves fork and some bear adaxial sporangia (Fig. 3e, arrow). The stems are so abundant that it is often difficult to decipher the real characteristics without using the appropriate techniques. The dense and

TABLE 1. COMPARISON OF TENTATIVE MEGAFOSSIL GENERIC ASSEMBLAGE-ZONES IN DEVONIAN STRATA WITH PALYNOLOGICAL ZONES SUGGESTED BY MC GREGOR (1977) AND RICHARDSON (1974) BASED ON FIRST APPEARANCES OF THE TAXA (FROM BANKS, 1980)

Series or Stage	Banks, 1980	McGregor, 1977	Richardson, 1974
Post Fammenian Tn 1b Tn 1a	?		V. nitidus
Famennian	Rhacophyton Assemblage-Zone VII		V. pusillites S. lepidophytus L. cristifer
Frasnian	Archaeopteris Assemblage-Zone VI		optivus-bullatus
Givetian	Svalbardia Assemblage-Zone V		Triangulatus Densosporites devonicus
Eifelian	Hyenia Assemblage-Zone IV	devonicus-orcadensis velata-langii	Rhabdosporites langii Acinosporites acanthomammillatus
Emsian — Upper		annulatus-lindlarensis / Grandispora sextantii	Calyptosporites biornatus-proteus
Emsian — Lower	Psilophyton Assemblage-Zone III	caperatus-emsiensis	Emphanisporites annulatus
Siegenian			Dibolisporites cf. gibberosus
Gedinnian	Zosterophyllum Assemblage-Zone II	micrornatus-proteus	Emphanisporites micrornatus Streelispora newportensis
Downtonian	Cooksonia Assemblage-Zone I	chulus-?vermiculata	Synorisporites tripapillatus

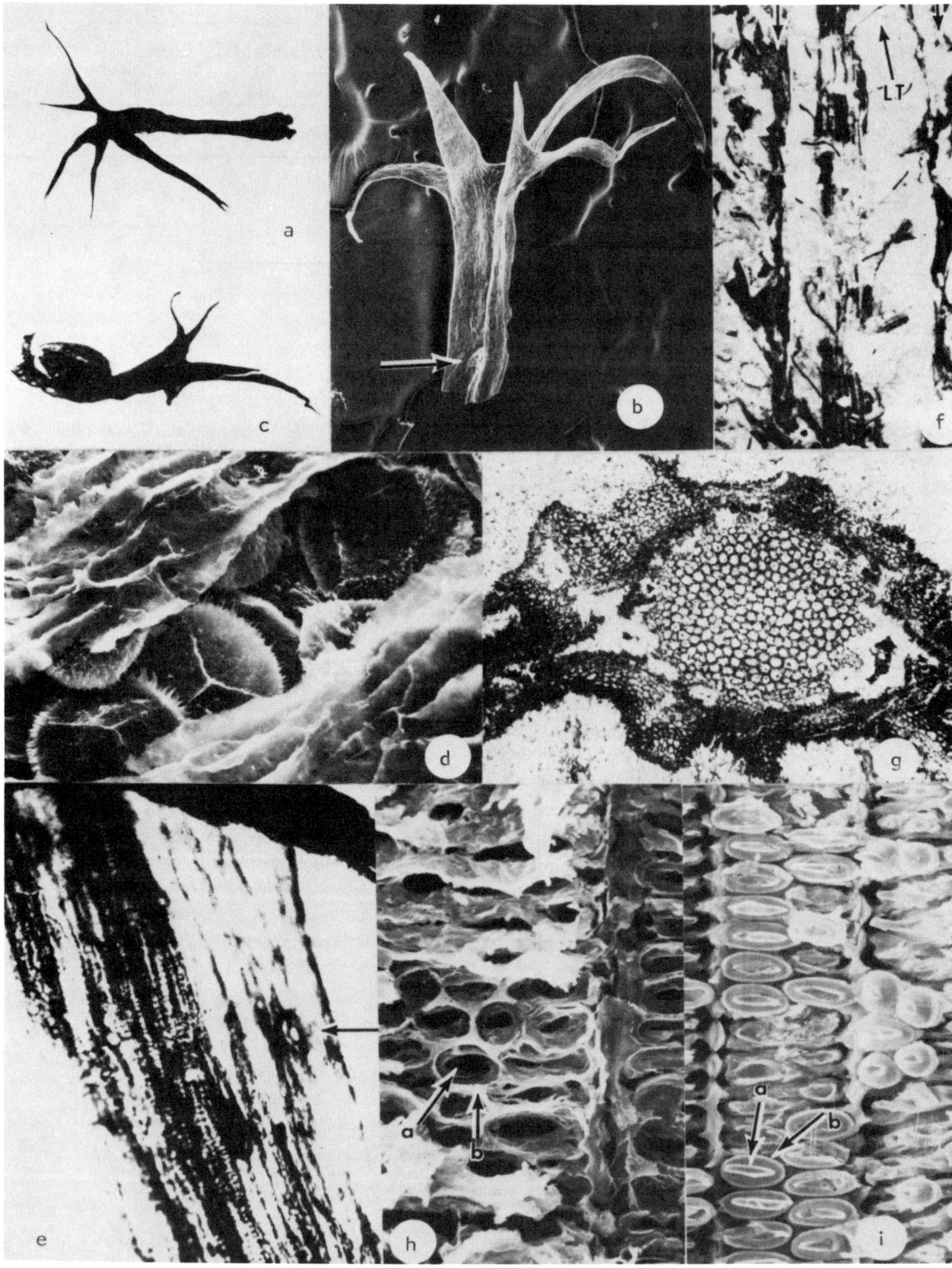

Figure 4. *Leclercqia complexa,* a trailing lycopod. Longest specimen 5 dm. Upcurving branches may have reached a height of 20 cm. a. A complete five-parted leaf macerated out of the matrix. Note the middle segment is longest and recurved abaxially. x 9. b. A broken leaf seen with SEM. The forking (distal) portion is complete but the proximal section from the stem attachment level to about mid-leaf is missing (cf. Fig. 4a). Note the tiny ligule (arrow) on the adaxial (upper) surface. x 30. c. An almost complete fertile leaf, similar to the vegetative leaf but bearing one sporangium on its adaxial surface. x 9. d. A crack in an attached sporangium (like that in 4c) seen with SEM shows crowded spores still held within the sporangial cuticle. Note trilete marks and spiny ornament. x 500. e. Oxidation of the middle recurved segment of a leaf reveals the central strand of coaly tracheids and a stoma (arrow). x 128. f. Partial transfer of a stem reveals the vasculature in saggital (longitudinal) section. The two vertical

forking leaves, for example, are so interlaced that it is impossible to determine whether they fork only once, or more than once.

Dissolution of the matrix in hydrofluoric acid shows the real morphology of the complex leaves. They are laminar with five tips (Fig. 4a, b). Oxidation of the coaly matter of the leaves reveals more detail. Each leaf has a single vein throughout its length (Fig. 4e) and stomata on its surface (Fig. 4e, arrow). The vein conducted food and water and the stomata facilitated gas exchange.

Veins in leaves are continuous with the central vascular column of the stem, and the strand of cells (leaf trace) that connects the two can be demonstrated in *Leclercqia* (Fig. 4f, arrow LT). This is an unusual discovery.

Frequently short lengths of stems are three-dimensionally preserved. This results from crystallization of pyrite (cellular permineralization) within the cells whose walls retain organic matter and the original shape of the cells. Transverse sections viewed with a metallurgical microscope reveal a fluted cylinder of xylem (typical for lycopods) and some of the surrounding tissues (Fig. 4g). Longitudinal sections demonstrate the patterns of pitting on the tracheids (Fig. 4h, i). The development of special etching and maceration techniques enabled Grierson (1976) to separate the pyritized lumens (Fig. 4i) of the tracheids from their coalified walls (Fig. 4h). The two portions were then studied by light microscopy and SEM and compared with data from compression specimens whose tracheids consisted only of coaly matter. This provided a unique opportunity to synthesize data obtained from the several modes of preservation and methods of analysis and to reassess interpretations. Compare the SEM view of tracheids in which only the coaly wall remains (Fig. 4h) with that in which the wall has been etched away leaving only the pyritized lumen and pit cavity casts (Fig. 4i).

Sporangium-bearing leaves (Fig. 4c) suitably oxidized, or viewed with SEM, reveal an oval pad of tissue to which sporangia are attached. Oxidation of sporangia permits study of their mode of dehiscence and observation of stomata on their surface. The latter character suggests thick-walled, primitive sporangia. Spores are abundant in the sporangia (Fig. 4d) and are of one kind (homospory). They have two kinds of ornamentation, spinose and biform, and they illustrate various stages in maturation (Bonamo and Grierson, 1973).

The excellent preservation also allowed the discovery, with the SEM, of a tiny organ called a ligule (Fig. 4b, arrow). Ligules are tongue-like outgrowths on the upper surface of both vegetative and fertile leaves which were previously thought to occur only on heterosporous lycopods. This occurrence of a ligule in *Leclercqia* is the oldest recorded documentation of a ligule in the fossil record, and the first report of a ligule occurring on a homosporous lycopod (Grierson and Bonamo, 1979).

The preservation of *Leclercqia* is indeed extraordinary. But it is only through the use of a suite of techniques that we were able to reveal the morphological, anatomical, and reproductive structures to a degree that is unusual for such an early land plant. The delicate, five-forked leaves were often intact; sporangia were retained on the fertile leaves; spores still filled the sporangia and showed their developmental stages from small, young, thin-walled groups of tetrads to larger, mature spores with typical spiny ornament; and the delicate leaf traces were still connected from the stem stele out into the complex leaves. Certainly these plants were not transported prior to or during fossilization. They must have lived close to areas of quiet water either in swamps or pond edges, or possibly along the levees of streams. In any case, to be so well preserved, they had to have been buried in situ.

EXAMPLES OF SOME PROLIFIC LOCALITIES AND THEIR POSSIBLE DEPOSITIONAL ENVIRONMENTS

Of the many localities for fossil plants in the Catskill clastic wedge of eastern New York, we have chosen six (Table 2), which illustrate a range of possible habitats around ponds, along streams, and close to shorelines.

Locality 1 is the site of the well-known Gilboa Forest described by Goldring (1924, 1927). Goldring reported three levels of stumps (sandstone casts, Fig. 2d), each resting in situ on mudstones from 15 to 60 cm in thickness and buried in a thick, coarse sandstone. Trunks had broken off at heights of 6 to 12 dm and were rarely preserved in the vicinity of the stumps. Goldring mentions (1924) the discovery of a few flattened stems (trunks?) 9, 12 and 37 dm in length, but their precise locality is not given. Strap-like, unbranched roots 7-25 mm in diameter radiated from 15 cm to 28 dm out into the mudstone from the swollen bases of the stumps. The situation suggests an area populated by numerous arborescent plants that were suddenly partially buried by an influx of sand. Eighteen intact stumps were found in one area measuring 4.5 m^2. The plants subsequently died and the stems rotted or were washed away. The rotted stumps were filled when more sand was washed in.

arrows indicate the outer stem surface. Within is the ribbed xylem strand from which arise the delicate leaf traces (arrow LT) that go into the leaves. x 6. g. Transverse section of a pyritized stem showing the central fluted xylem strand surrounded by outer cortical tissue. x 27. h. Longitudinal SEM view of tracheids where only the coaly secondary wall remains. Here the pits in the wall appear as oval openings (black aperture, arrow a) surrounded by broad sunken pit borders (grey area, arrow b). x 500. i. Longitudinal SEM view of tracheids where all wall material has been etched away leaving only the pyrite filling the cells. Thus we have a replica of the space system of the cells; a reverse image of that seen in Fig. 4h. The fine pyrite also fills the minute pit cavities of the walls circumscribed by their pit borders. Before treatment, the small whitish ovals here (arrow a) filled areas like the black apertures in Fig. 4h (arrow a), and the larger oval around them (arrow b) fit into the limits of the grey border of Fig. 4h (arrow b). x 500.

TABLE 2. SIX SELECTED LOCALITIES WITH LARGE ACCUMULATIONS OF FOSSIL PLANT REMAINS

Locality	Location	Rock Unit	Flora
1. Schoharie Reservoir Gilboa, N.Y. Riverside Quarry W. side Schoharie Cr. 1/5 mile S. of bridge.	Gilboa, N.Y. 7 1/2' Quad. sheet 6168 IV NW. AMS Series V 821, 1945, approx. 2,231, 400 yds. N. by 869,300 yds. E.	Continental Moscow Formation ? = Kiskatom Givetian	Eospermatopteris (Goldring, 1924) Aneurophyton (Serlin and Banks, 1978) Caulopteris (Goldring, 1924) Protolepidodendron (Grierson and Banks, 1963) Archaeosigillaria (Fairon-Demaret and Banks, 1978) Ibyka (Skog and Banks, 1973) Prosseria (Banks, 1966) Pseudosporochnus (Schuchman, 1969) Lepidosigillaria (In Cornell University cf. Hyenia Paleobotanical Collection)
2. Base of Brown Mt. site of hydroelectric Pumped Storage Project on Schoharie Creek.	Gilboa, N.Y. 7 1/2' Quad. sheet 6168 IV NW 1945, approx. 890,000 ft. N. by 469,000 ft. E.	Panther Mt. Formation Lower Givetian	Leclercqia (Banks, Bonamo and Grierson, 1972; Bonamo and Grierson, 1973; Grierson, 1976; Grierson and Bonamo, 1979) Rellimia (Bonamo, 1977) Haskinsia (Grierson and Banks, 1983) Centipedes, various arachnids, a mite, and a possible insect (Shear, Bonamo, Grierson, Rolfe, Smith, and Norton, 1984)
3. Quarry along N.Y. State Route 145 2 miles NW of Cairo, N.Y., Lens 9 m. across, 1 m. thick.	Freehold, N.Y. 7 1/2' Quad. sheet 6168 I SE AMS Ser. V 821, 1945, approx. 905, 700 yds. E by 2,221,000 yds. N.	Kiskatom Formation Givetian	Short lengths of permineralized axes of: Rhacophyton (Matten, 1968; 1973; 1974; 1975) Cladoxylon " " " " " Stenokoleos " " " " " Reimannia " " " " " Cairoa " " " " " Triloboxylon " " " " " Actinoxylon " " " " " Ostracodes and fish scales
4. County road metal quarry on N flank of South Mt. 1.1 mi. W of Schoharie-Greene Co. line on W Durham to Conesville Rd.	Livingstonville, N.Y. 7 1/2' Quad. sheet 6168 IV NE, AMS Series V 821. 1945, approx. 2,230,250 yds. N by 885,500 yds. E.	Oneonta Formation Lower Frasnian	Pallaviciniites (Hueber, 1961) Sawdonia (Hueber and Grierson, 1961) Serrulacaulis (Hueber and Banks, 1979) Colpodexylon (Grierson and Banks, 1963) Pseudosporochnus (Hueber, 1960; 1971) Cladoxylon " " " Astralocaulis " " " Stenokoleos (Matten and Banks, 1969) Archaeopteris (Carluccio, Hueber and Banks, 1966) Callixylon (Hueber, 1960) Eospermatopteris (Hueber, 1960) Arachnids (Kjellesvig-Waering, Posthumous MS submitted)
5. Abandoned roadside Quarry on N. side of N.Y. Rte. 97 in W. bank Fish Cabin Creek, Pond Eddy.	Milford, PA.-N.Y.-N.J. 15' Quad. N 4115 W 7445/15 1915	Delaware River Formation Frasnian	Colpodexylon (Banks, 1944) Aneurophyton (Serlin and Banks, 1978) Eospermatopteris " " " " Eddya (Beck, 1967) Svalbardia (Matten, 1981)
6. Cannonsville Reservoir esp. lens in channel below spillway.	Deposit, N.Y. 15' Quad. Cut not mapped, at bend of West Branch of Delaware River, BM 1008; now Stilesville Weir: inaccessible.	Beers Hill Shale Formation Lower Frasnian	Archaeopteris spp. (Carluccio, Hueber and Banks, 1966) Callixylon

Coarse sandstones, like those enclosing the stumps, also preserve other plant parts. Goldring found branch systems (as in Fig. 2e), which she called fronds, in the sandstone with the stumps. Some were up to 9 dm in length. In shale, interbedded with the sandstones, she found organs she thought were seeds. She proceeded to assemble all of these parts into one reconstruction (Fig. 2c) which she named *Eospermatopteris,* the dawn seed fern. The seeds have been shown to contain spores, hence are not seeds, and the branches have never been found attached to the stems. Today the branches are assigned to *Aneurophyton.* Nevertheless, the stumps do indicate that some kind of tree grew at Gilboa. Also found in the sandstones with *Eospermatopteris* and *Aneurophyton* were lycopods (as Fig. 1d), horsetails (as Fig. 1e) and cladoxylaleans (as Fig. 2a) listed under Locality 1 (Table 2). Thus the forest in the vicinity of Gilboa probably consisted of tree-like forms, bushes, and ground cover. All of the plants are extinct, all lacked large webbed leaves, and all reproduced by spores as ferns do, not by seeds.

Locality 1 presents a picture of in situ, broken, rooted stumps, mixed with other relatively poorly preserved broken plant parts which have obviously been washed about, all embedded in a coarse sandstone. This indicates deposition by fast-moving water bringing in sand which abraded the plant parts, but sometimes allowed internal anatomical permineralization by pyrite (e.g., *Protolepidodendron, Archaeosigillaria,* Grierson and Banks, 1963). This fast-moving water was probably flood water. The sandstone entombing the trunks occurs just above a mudstone layer which is in turn supported by a coarse sandstone. It is likely that the stumps were growing on a levee (indicated by the lower sandstone) which was undisturbed long enough to allow the accumulation of a soil layer (represented by the mudstone) and time enough for the growth of the trees. The successive vertical levels of stumps support these hypotheses of successive floods separated in time by a period of soil accumulation and tree growth. The origin of the allochthonous part of the vegetation can only be speculative; it could have been growing among the trees, or it could have been washed in with the flood waters.

Well-preserved fossil plants often occur in lenses of limited horizontal and vertical extent, sometimes as few as $9m^2$ in area. At these sites the rocks are richly filled with plant parts. This abundance suggests a rich and varied vegetation growing in the vicinity of ponds and/or streams that were then buried by an influx of mud. The parts are preserved as compressions (Fig. 3e, 4a-c) and petrifactions (Fig. 4f, g, h, i) or some combination of the two (Fig. 4e) which together give considerable histological detail.

Locality 2 (Brown Mountain Gilboa) is the site at which *Leclercqia complexa* was collected. As we discussed previously, its excellent preservation indicates in situ burial of a dense ground cover of plants (Fig. 3e, 4). At the same locality, other abundant remains (e.g., *Rellimia*; Table 2) give evidence of the presence of taller plants of shrubby habit, but without shading leaves.

Locality 4 (South Mountain) yields plant remains of similar preservation to those in Locality 2 because in both areas one can find comparable mudstone lenses representing in situ burial. At both localities, however, other lithologies occur where the plant parts are less well preserved and more abraded.

Localities 2 and 4 give evidence of both autochthonous and allochthonous deposition. Whether these sites were near streams or ponds is difficult to determine from the plant evidence, but they were certainly wholly terrestrial deposits.

Locality 3 (Cairo), on the other hand, is characterized only by broken, cellularly permineralized plant parts. These were evidently deposited away from their site of growth, but they provide botanical information which adds significantly to our evolutionary evidence.

The best preserved material at Localities 5 (Pond Eddy) and 6 (Cannonsville) is found in blue siltstones. Strata at Pond Eddy are characterized by fining upward cycles (Fletcher and Woodrow, 1970), and the plants were deposited along Devonian streams. At Cannonsville one finds a variety of strata, most of which include plant remains. Gray and blue-gray sandstones are predominant. They are interbedded with flags and thin layers of shale or siltstone. Here and there one finds lenses of siltstones (Fig. 3a) that are filled with remarkably well-preserved specimens of *Archaeopteris* (Fig. 3c). Both sterile and fertile leaves are abundant and the delicate venation of the leaves is intact (Beck, 1964; Carluccio, Hueber and Banks, 1966). We interpret these specimens to be parts of trees that grew around a body of water into which they fell and were preserved. There is no indication that they were transported prior to burial. Sandstones are filled with much less well-preserved specimens that probably were transported prior to burial. Some massive sandstones are literally filled with pyritized axes (*Callixylon*) of *Archaeopteris*. The size of these axes varies, but some were obviously segments of large trunks. Animal fossils in the Cannonsville area are rare. Beck (1964) reported finding some trilobites. Carluccio, Hueber and Banks (1966) found a few branchiopods of the genus *Estheria*. This admixture of marine and less obviously marine animals, the absence of any other marine invertebrates, the tremendous abundance of land plants, and the presence of untransported fossils preserved in a body of standing water suggest to us that deposition occurred on low land, fed by streams, so close to the shoreline that some marine life was washed in.

The obvious relationship between our plant localities and bodies of water or streams is in agreement with the conclusions of Woodrow, Fletcher and Ahrnsbrak (1973) that the vegetation was concentrated in these areas. We cannot refute their concept of a low water table and generally unfavorable conditions (hot and dry during at least part of the year) for plant growth in the interfluves on the coastal plain. We do point out, however, that the rarity of fossil plant parts in the red beds can be predicted and is not evidence for the absence of plants on that portion of the landscape. Oxidation, as evidenced by the red color, destroys the plants before there is opportunity for fossilization. The rare plant parts we have seen in red beds are simply impressions (e.g., small branches of *Archaeopteris*). No coaly matter remains. The botanical value of such fossils is negligible. In stark contrast to the red beds are our dark-colored siltstone and mudstone lenses described above that represent former ponds, swamps or streams. There, the plant parts sank to the bottom or were buried in a flood. In either case, burial was rapid and the parts were preserved under anoxic (reducing) conditions. The significance to paleobotany of locating and studying such deposits cannot be overemphasized, nor can the problem of locating them in the first place. It requires constant vigilance in the field to watch all large cuts into the strata of the Catskill clastic wedge.

THE HABIT OF THE PLANTS; ITS EVOLUTIONARY AND ECOLOGICAL SIGNIFICANCE

Any attempt to reconstruct the landscape of the Devonian from the plant evidence should involve patterns of sedimentation and, as we mentioned above, most paleobotanists working on the Catskill clastic wedge have neither been sedimentologists, nor have they involved sedimentologists. Much valuable information is thus lacking. Edwards (1980) wrote an excellent summary of much of the available ecological and sedimentological data and their bearing on ancient landscapes. Scott (1980) has given a useful introduction to the problems of Devonian plant ecology, and we offer a few comments based on our experience in New York state.

The several genera discussed in this account display a variety of growth forms that require a brief summary to aid visualization of the assemblages at our six localities. This is found in Table 3.

The plant evidence demonstrates that Silurian and Early Devonian time saw the introduction of an array of plant types destined to play distinct roles later in the Devonian. Zosterophylls and lycopods were trailing, herbaceous, presumably perennial

TABLE 3. LIST OF GENERA INCLUDED IN THE SIX LOCALITIES OF THIS REPORT; THEIR BOTANICAL AFFINITIES AND THEIR POSSIBLE HABIT OF GROWTH

Genus	Botanical Affinity	Habit of Growth
Pallaviciniites	liverwort	flat on moist soil, green
Sawdonia Serrulacaulis	zosterophylls (Fig. 1b)	ground cover; not over 3 dm high, no leaves, stem green
Archaeosigillaria Haskinsia Colpodexylon Leclercqia (Fig. 3e, 4) Protolepidodendron (Fig. 1d)	lycopods	ground cover; trailing, arching up and recurving; small leaves; leaves and stems green
Lepidosigillaria (Fig 2b)		arborescent, to 5 m; small, green leaves.
Hyenia (Fig. 1e)	probable horsetail	low ground cover; small leaves; branches and leaves green
Prosseria	probable horsetail	stiff plant, ? to 3 dm
Rhacophyton	Coenopteridales, an extinct group, common in Carboniferous	erect, much branched, large branch systems with unwebbed (non-laminar) leaves; a bush
Pseudosporochnus (Fig. 2a) Cladoxylon	Cladoxylales, an extinct, mostly Pennsylvanian group	erect, to 3 m; large branch systems without laminar leaves; bush
Stenokoleos	affinity unknown	small bush
Aneurophyton (Fig. 2e)	progymnosperm	large, to 9 dm, branch systems; ultimate branchlets with unwebbed (non-laminar) leaves.
Eospermatopteris (Fig. 2c-d)	possible progymnosperm	stumps perhaps of trees 9 m high that could have borne branches like Aneurophyton.
Caulopteris	affinity unknown	bases of branches similar to Aneurophyton
Rellimia; Tetraxylopteris (Fig. 2f) Triloboxylon Reimannia; Cairoa	progymnosperms probable progymnosperm possible progymnosperms	probably all are parts of bushy or shrubby plants, much branched, leaves unwebbed (non laminar) (e.g., Fig. 2e, f)
Svalbardia; Actinoxylon Archaeopteris (+Callixylon) Eddya	probable progymnosperms progymnosperm possible progymnosperm	shrubs or trees; highly dissected leaves. tree to 18 m; true laminar leaves (Fig. 3b-d). Eddya may be a sporeling of Archaeopteris

plants that served as ground cover (Fig. 1b, d). Rhyniophytes and trimerophytes gradually established profuse branching (Fig. 1a, c). Advanced trimerophytes may have reached three feet in height. The aerial parts of trimerophytes were simply a system of branches that diminished in size distally (Fig. 1c). Leaves were lacking. Photosynthesis proceeded in the stems which must have been green.

Middle Devonian saw the first progymnosperms capable of increasing in diameter by the production of secondary xylem (wood, Fig. 3b). This, plus the production of complex lateral branching systems (Fig. 2e, f), meant a whole new level (height) of plants, producing a canopy. The ultimate parts of the branch systems gradually became modified until, in *Archaeopteris,* they were true webbed leaves (Fig. 3c). These, in turn, meant the production of shade. The height of the tallest progymnosperms is unknown, but *Eospermatopteris* (Fig. 2c) may have reached 30 feet, and *Archaeopteris* (Fig. 3d) 60 or more.

It is frequently stated (e.g., Woodrow et al., 1973) that there was no grass in the Devonian to stabilize the soil. Grass, a flowering plant, did not appear until Tertiary time. However, lycopods had become abundant by Mid-Devonian, and in some deposits they are so profuse that the rock is almost exclusively plant remains (Fig. 3e). Why, if lycopods covered the ground as densely as we think, do sedimentologists feel that the soil was not bound? We feel the answer may lie in the roots produced by lycopods. The so-called fern allies, the lycopods and the horsetails (Fig. 1e), grew by horizontal or gently arching and recurving stems that produced some erect, aerial branches. From the stems occasional (adventitious), little-branched roots penetrated the soil. The plants had no dense root system like that of modern grasses. Quite possibly, flood waters had little difficulty dislodging the loosely attached club mosses and horsetails. So far as we know, true, much-branched root systems may not have appeared until Frasnian time in plants like *Archaeopteris.*

The emerging picture of abundant plant remains clustered around sources of water is more than merely one of plants living near likely sites of fossilization. The early plants were, in fact, tied to the wet banks of streams or ponds by their life histories and the structure of their *gametophyte* plant bodies. The plant parts documented by fossils are overwhelmingly portions of the sporophyte plant body. This body is covered by a cuticle and supported by a vascular or water system. Both of these parts are highly resistant to decay and are the parts most often remaining in fossils. The sporophyte plant produces the cuticularized spores used by palynologists in stratigraphic studies. We know from living plants that the spores of such free-sporing plants grow into very small (no larger than 3 cm), non-cuticularized, non-vascularized gametophyte plant bodies completely independent of the larger, more resistant sporophyte. This gametophyte has no protection (such as the cuticle) from desiccation, nor any means (such as roots) to obtain more than surface water, or to transport water (such as tracheids). They are thus only able to exist and survive in generally moist environments. The gametophyte must live long enough to produce the gametes. The sperm must swim to the egg to produce the fertilized egg which will grow into the sporophyte plant. The time required from germination of the spore to mature gametophyte plant capable of producing eggs and sperms varies considerably among living plants. In most ferns and horsetails, this takes only a few months because the gametophytes are surface living and photosynthetic. In some ferns and some lycopods, this may take many years because their gametophytes are underground and saprophytic. They grow so slowly because their only nutrition comes from a symbiotic interaction with endophytic soil fungi. Although a sporophyte plant may be physically capable of resisting dry periods, it can live only where the environment favors growth of its delicate gametophyte. The living flora provides many such appropriate niches because of the dense shade provided by gymnosperm and angiosperm forest trees. The story is considerably different in Early Devonian. There, the early sporophytes were small leafless plants (Fig. 1a, b) which provided no shade. The only possible habitat for the delicate gametophytes, and thus the sporophytes they produced, was along streams or beside bodies of water.

We have no way of knowing which type of gametophyte was produced by the early plants. Underground gametophytes are presumed to be the primitive type, according to evidence from living plants. A gametophyte found in the Rhynie chert (Remy and Remy, 1980) might have been surface living. If the early gametophytes were of the surface-living type, they would correlate with the idea of alternating wet/dry periods (as suggested by Woodrow et al., 1973) because they could grow and produce sporophytes in a short time period (months). In this case the gametophyte could mature and produce gametes and a subsequent sporophyte, all during the wet season. The sporophyte would then be the growing stage during the drier season.

As the plants grew denser (e.g., *Leclercqia,* Fig. 3e) and extended farther away from the water sites by means of their horizontal growth (e.g., *Protolepidodendron,* Fig. 1d), new gametophytes would be produced farther from the immediate vicinity of the water source, following the shady microniches provided by the ground cover of the dense sporophytes. However, no area significantly distant from water could be populated by plants until the advent of highly branched, bushy plants and then the taller trees with laminar, shade-producing leaves such as *Archaeopteris.*

Along with the development of the tree habit was a switch from homospory to heterospory. Spores from homosporous plants germinate into gametophytes bearing both eggs and sperms, while spores from heterosporous plants produce gametophytes with eggs or sperms, not both. Hence the sperm that fertilizes an egg from a heterosporous plant is unlikely to have the same genetic makeup as the egg it fertilizes. Thus heterospory introduced considerably greater genetic recombination.

The next step in the evolution of the land flora was a giant step, the development of the seed habit. The seed is a highly efficient packaging of the female gametophyte and its egg, protected by one or more enveloping, water-resistant layers. The seed is retained on the sporophyte plant and connected to it by vascular tissue. The gametophyte and its egg are supplied by the sporophyte with water and nourishment. This new structure, the seed, was important because it isolated the vulnerable female gametophyte from the vagaries of the physical environment.

Concomitant changes in the male gametophyte involved retention of sperm within a light-weight, water-resistant pollen grain and the development of mechanisms for the transfer of the sperm to the enclosed egg.

The fertilized egg, and then the developing sporophyte, are contained in the seed and supplied with stored food. Not only is the gametophyte protected, but the difficulties of the early establishment phases for the developing sporophyte are diminished because of the ready availability of this stored food. When it is finally shed from the sporophyte, the seed is often capable of remaining dormant for long periods until conditions are right for the development of the sporophyte with all of its adaptations to obtaining, transporting and retaining water.

This revolution in plants, the development of the seed habit, is functionally comparable to the revolution in vertebrate animals, the development of the amniotic egg.

These complicated evolutionary developments opened the way for the colonization of the whole of the land because some plants no longer had a vulnerable water-requiring period. As the seed plants colonized greater and greater land areas in subsequent geologic Periods, they shaded more and more niches and thus provided new environments for those plants that still retained the delicate, free-living, water-dependent gametophytes.

HIGHLIGHTS OF THE EVOLUTION OF DEVONIAN FLORAS SIGNIFICANT TO ECOLOGY

1. Evolution of low-growing lycopods (Siegenian) and horsetails (Eifelian).

2. Evolution of secondary growth (production of wood) in Eifelian, leading to the arborescent habit.

3. Evolution of profuse branching by Trimerophytina (Emsian) continued by progymnosperms in Eifelian and onward.

4. Flattening and webbing of ultimate branchlets, resulting in large, leafy branch systems capable of shading.

5. Evolution in reproductive biology from the production of one kind of spore (homospory) to two kinds (heterospory) in

Emsian-Eifelian time and then to the seed habit in Famennian time.

These changes markedly affected the Devonian landscape as they resulted in: ground cover (1), tiers or stories of vegetation (2), creation of shade (3, 4), the possibility of greater genetic variation (5), and the possibility of a resting stage (5) capable of surviving unfavorable conditions and obviating the constant need for water.

CLIMATIC IMPLICATIONS OF THE FLORA

Woodrow et al. (1973) discuss the climate during deposition of the Catskill clastic wedge. They suggest a tropical wet and dry or savanna climate for the central coastal plain on which the plants were growing. The one reliable feature of plants that we can use to check their suggestion is the presence or absence of growth layers in the secondary xylem (wood) of shrubby or arborescent species. Arnold (1930) reviewed the North American literature on growth "rings" in Devonian plants. Some workers found them present in *Callixylon,* some found them absent. Arnold himself counted five complete "rings" and several incomplete ones in *C. erianum* which he illustrated by a diagram. We illustrate his type specimen (Fig. 3b) on which the so-called "rings" are obvious. Study of the stem along a radius from the pith outward shows the first-formed tracheids to be uniform in radial diameter. Subsequent tracheids are narrower until the narrowest are reached. At this point growth ceases. When growth is resumed, the tracheids have a distinctly greater radial diameter. Subsequently, the pattern of narrowing reappears. What one sees as a "ring" is simply the band(s) of narrowest cells produced just prior to cessation of seasonal growth, whatever the length of the season. We prefer the term growth layer to "annual ring" because the new cells produced constitute a layer over the entire plant, not simply one ring of cells, and because we do not know that the alternation is annual. There could be two such rings per year, or none in a given year. Arnold (1931) illustrated growth layers in *Callixlon newberryi.* Matten and Banks (1967) illustrated them in *Tetraxylopteris.* Chaloner and Creber (1973) have recently reviewed the current literature on growth layers and climates.

Seasonal growth does not distinguish between a summer/winter or a wet/dry alternation. Larger cells can be formed during either a summer or a wet period, smaller cells as that season draws to a close. On the assumption that the Catskill landmass lay in a tropical zone, our evidence favors the tropical wet and dry alternation discussed by Woodrow et al. (1973). If at any time in this latitude or at a latitude closer to the equator rainfall was equal in its distribution through a year, growth layers would be lacking because cell sizes would be uniform. This situation could have obtained over the northern part of the Devonian landmass (Woodrow et al., 1973).

CO-EVOLUTION: ARTHROPODS, PLANTS, FUNGI

Kevan, Chaloner and Savile (1975) discussed in detail the arthropods and fungi associated with the plants in the Rhynie chert (Emsian). They considered that food-seeking arthropods wounded the plants by chewing the surface and/or sucking juices but that most of the arthropods were carnivorous. Their analysis of the injuries to the plants included evidence of division of cells around the wound, suggesting a repair activity by the living plant. Banks (1981) has since reported extensive periderm (cork-like tissue) formation underneath surface wounds in the stem of *Psilophyton* (Emsian) from the Gaspé Peninsula. Surface wounding such as could have been caused by chewing animals stimulates inner cells to commence dividing. The result is radially aligned files of cells underneath the obviously damaged superficial cells. This new peridermal tissue decreases damage caused by evaporation of water from an exposed surface. *Psilophyton,* approximately the same age as the Rhynie chert plants, is not accompanied by chewing invertebrates, but arthropods are found in New York at Localities 2 and 4 (Brown Mountain and South Mountain). There, however, the plants associated with the arthropods have not yet demonstrated wounding. Thus the Rhynie chert is unique in preserving both the animals and the wounded plants.

Locality 2, Brown Mountain along Schoharie Creek, provides a striking demonstration of the quality of preservation found in sediments formed from anoxic muds (reducing conditions). We discussed earlier the importance of techniques to extract data from the rich cache of the lycopod *Leclercqia* found at this locality. During the dissolution of samples in hydrofluoric acid, whole leaves, sporangia, spores, and other parts of *Leclercqia* floated free. Bonamo and Grierson found that an abundance of fossilized invertebrates also floated free from the matrix. This constitutes the largest terrestrial fauna yet found in American Devonian rocks. According to Shear et al. (1984), the fauna consists of "two or more trigonotarbids, two centipedes, one mite, possible amblypigid, one spider, undetermined arachnids, possible insects." A second example of the co-existence of plants and animals is found at Locality 4 where Hueber isolated abdomens of arthropods during acid maceration of the matrix. These were sent to Kjellesvig-Waering some years ago and will be published in a posthumous manuscript submitted to the Paleontological Research Institution (Ithaca, New York).

The presence of fungal mycelia in the plants of the Rhynie chert led Kevan et al. (1975) to consider that growth of fungi in plant tissues was a means of protection of fungi from desiccation. Interest in the fungi in the Rhynie chert recalls Arnold's (1931) illustration of fungal mycelium in cells of the wood of *Callixylon.* Stubblefield and Banks (1983) have found oomycetous fungi in aerial axes of *Psilophyton dawsonii.* They were unable to determine whether the fungi were parasitic, saprophytic, or perhaps participated in some symbiotic relationship with the plants.

THE FLORA OF THE CATSKILL CLASTIC WEDGE IN PENNSYLVANIA AND WEST VIRGINIA

Detailed studies of the Catskill flora south of New York

state are less numerous. In Pennsylvania, Arnold (1935, 1939) early recognized rich fossiliferous layers, Beck (1964) reported that *Archaeopteris* was abundant and predominant in New York and well into Pennsylvania, and Pettitt and Beck (1968) described a new Late Devonian seed *Archaeosperma.* This was the earliest-known seed at that time. Brauer (1980, 1981) described the morphology of the cones of *Barinophyton* and two types of spores in each sporangium. This is unusual for either fossil or living plants.

Work in West Virginia is equally significant. The study of *Rhacophyton* by Andrews and Phillips (1968) and by Cornet, Phillips and Andrews (1976), of *Archaeopteris* by Phillips, Andrews and Gensel (1972), and of the earliest appearance of seeds by Gillespie, Rothwell and Scheckler (1981) are examples. Gillespie et al. (1981) and Scheckler (1982) are utilizing sedimentological investigations in conjunction with plant studies to attempt to reconstruct the Famennian landscape. Preliminary conclusions indicate that the Famennian swamps might be the forerunners of the coal-bearing formations of Carboniferous age.

CONCLUSIONS

1. Plants of the Catskill clastic wedge were found also around the globe.
2. Plants were concentrated around ponds, along levees and near the shoreline.
3. Plants evolved continuously from Pridolian onward.
4. Plants were evolving in concert with arthropods and fungi.
5. Plants were regularly evolving new morphologies resulting in greater adaptive possibilities:
   a. arborescence,
   b. then vertical tiers of vegetation—shading,
   c. reproductive biology: first homospory, then heterospory and finally the seed habit appeared.
6. Correlation of sites of deposition of fossil plants with sedimentary patterns is underway and will lead to better understanding of paleoecology.
7. Emphasis is placed on the necessity of locating deposits of well-preserved plants that can yield truly essential data and on the use of techniques to extract these data.

REFERENCES CITED

Andrews, H. N., Kasper, A. E., Forbes, W. H., Gensel, P. G., and Chaloner, W. G., 1977, Early Devonian flora of the Trout Valley Formation of Northern Maine: Review of Palaeobotany and Palynology, v. 23, p. 255–285.

——and Phillips, T. L., 1968, *Rhacophyton* from the Upper Devonian of West Virginia: Journal of the Linnean Society (Botany), v. 61, p. 37–64.

——, ——and Radforth, N. W., 1965, Paleobotanical studies in Arctic Canada. I. *Archaeopteris* from Ellesmere Island: Canadian Journal of Botany, v. 43, p. 545–556.

Arber, E.A.N., 1921, Devonian Floras: Cambridge University Press, 100 p.

Arnold, C. A., 1930, The genus *Callixylon* from the Upper Devonian of Central and Western New York: Papers of the Michigan Academy of Science, Arts and Letters, v. XI, p. 1–50.

——1931, On *Callixylon newberryi* (Dawson) Elkins et Wieland: Contributions from the Museum of Paleontology, University of Michigan, v. III, p. 207–232.

——1935, On seed like structures associated with *Archaeopteris* from the Upper Devonian of Northern Pennsylvania: Contributions from the Museum of Paleontology, University of Michigan, v. IV, p. 283–286.

——1939, Observations on fossil plants from the Devonian of Eastern North America. IV. Plant Remains from the Catskill Delta Deposits of Northern Pennsylvania and Southern New York: Contributions from the Museum of Paleontology, University of Michigan, v. V, p. 271–314.

Banks, H. P., 1944, A new Devonian lycopod genus from southeastern New York: American Journal of Botany, v. 31, p. 650–659.

——1966, Devonian flora of New York State: The Empire State Geogram, v. 4, p. 10–24.

——1968, The early history of land plants, *in* Drake, E. T., ed., Evolution and Environment: New Haven and London, Yale University Press, p.73–107.

——1970, Evolution and Plants of the Past: Belmont, California, Wadsworth Publishing Co., Inc., 170 p.

——1980, Floral assemblages in the Siluro-Devonian, *in* Dilcher, D. L., and Taylor, T. N., eds., Biostratigraphy of Fossil Plants: Stroudsburg, Pa., p. 1–24.

——1981, Peridermal activity (wound repair) in an Early Devonian (Emsian) trimerophyte from the Gaspé Peninsula, Canada: The Paleobotanist, v. 28–29, p. 20–25.

——Bonamo, P. M., and Grierson, J. D., 1972, *Leclercqia complexa* gen. et sp. nov., a new lycopod from the late Middle Devonian of Eastern New York: Review of Palaeobotany and Palynology, v. 14, p. 19–40.

——Leclercq, S., and Hueber, F. M., 1975, Anatomy and morphology of *Psilophyton dawsonii,* sp. n. from the late Lower Devonian of Quebec (Gaspé) and Ontario, Canada: Palaeontographica Americana, v. 8, p. 75–127.

Beck, C. B., 1957, *Tetraxylopteris schmidtii* gen. et sp. nov., a probable pteridosperm precursor from the Devonian of New York: American Journal of Botany, v. 44, p. 350–367.

——1960, The identity of *Archaeopteris* and *Callixylon:* Brittonia, v. 12, p. 351–368.

——1962, Reconstructions of *Archaeopteris* and further consideration of its phylogenetic position: American Journal of Botany, v. 49, p. 373–382.

——1964, Predominance of *Archaeopteris* in Upper Devonian flora of Western Catskills and adjacent Pennsylvania: Botanical Gazette, v. 125, p. 126–128.

——1967, *Eddya sullivanensis,* gen. et sp. nov., a plant of gymnospermic morphology from the Upper Devonian of New York: Palaeontographica B, v. 121, p. 1–22.

Bonamo, P. M., 1977, *Rellimia thomsonii* (Progymnospermopsida) from the Middle Devonian of New York State: American Journal of Botany, v. 64, p. 1272–1285.

——and Grierson, J. D., 1973, Sporophylls, sporangia and spores of *Leclercqia complexa* [Abs.]: American Journal of Botany, v. 60 (supp. to No. 4), p. 16.

Boureau, E., ed., Traité de Paléobotanique, v. 2: Paris, Masson et Cie, Eds., 845 p.

Brauer, D. F., 1980, *Barinophyton citrulliforme* (Barinophytales Incertae Sedis, Barinophytaceae) from the Upper Devonian of Pennsylvania: American Journal of Botany, v. 67, p. 1186–1206.

——1981, Heterosporous barinophytacean plants from the Upper Devonian of North America and a discussion of the possible affinities of the Barinophytaceae: Review of Palaeobotany and Palynology, v. 33, p. 347–362.

Canright, J. E., 1970, Spores and associated macrofossils from the Devonian of Arizona: Geoscience and Man, v. 1, p. 83–88.

Carluccio, L. M., Hueber, F. M., and Banks, H. P., 1966, *Archaeopteris macilenta,* anatomy and morphology of its frond: American Journal of Botany, v. 53, p. 719–730.

Chaloner, W. G., and Creber, G. T., 1973, Growth rings in fossil woods as evidence of past climates, *in* Tarling, D. H., and Runcorn, S. K., eds., Implications of Continental Drift to the Earth Sciences: Nato Advanced Study Inst., v. 1, Academic Press, London and New York, p. 425–437.

——and Sheerin, A., 1979, Devonian macrofloras, *in* House, M. R., and others, eds., The Devonian System: London, the Palaeontological Association, p. 145–161.

Churkin, M. Jr., Eberlein, G. D., Hueber, F. M., and Mamay, S. H., 1969, Lower Devonian land plants from graptolitic shale in south-eastern Alaska: Palaeontology, v. 12, p. 559–573.

Cornet, B., Phillips, T. L., and Andrews, H. N., 1976, The morphology and variation in *Rhacophyton ceratangium* from the Upper Devonian and its bearing on frond evolution: Palaeontographica B, v. 158, p. 105–129.

Dorf, E., 1933, A new occurrence of the oldest known terrestrial vegetation, from Beartooth Butte, Wyoming: Botanical Gazette, v. XCV, p. 240–257.

Edwards, D., 1972, A *Zosterophyllum* fructification from the Lower Old Red Sandstone of Scotland: Review of Palaeobotany and Palynology, v. 14, p. 77–83.

—— 1980, Early Land Plants, *in* Panchen, A. L., ed., The Terrestrial Environment and the Origin of Land Vertebrates: London and New York, Academic Press, p. 55–88.

—— Feehan, J., and Smith, D. G., 1983, A late Wenlock flora from County Tipperary, Ireland: Botanical Journal of the Linnean Society, v. 86, p. 19–36

Fairon-Demaret, M., 1974, Nouveaux specimens du genre *Leclercqia* Banks, H. P., Bonamo, P. M. et Grierson, J. D., 1972 du Givetien (?) du Queensland (Australie): Bulletin Institute Royale Sciences Naturelles belgique, v. 50, p. 1–4.

—— and Banks, H. P., 1978, Leaves of *Archaeosigillaria vanuxemii,* a Devonian lycopod from New York: American Journal of Botany, v. 65, p. 246–249.

Fletcher, F. W., and Woodrow, D. L., 1970, Geology and economic resources of the Pennsylvania portion of the Milford and Port Jervis 15-minute quadrangles: Pennsylvania Geological Survey Atlas, v. 223, p. 1–64.

Fry, W. L., 1956, Identification of plants, *in* Tozer, E. T., Geological Reconnaissance, Prince Patrick, Eglinton, and Western Melville Islands, Arctic Archipelago, Northwest Territories: Geological Survey of Canada, Paper 55c, p. 16–17.

Gensel, P. G., 1982, A new species of *Zosterophyllum* from the Early Devonian of New Brunswick: American Journal of Botany, v. 69, p. 651–669.

Gillespie, W. H., Rothwell, G. W., and Scheckler, S. E., 1981, The earliest seeds: Nature, v. 293, p. 1–3.

Goldring, W., 1924, The Upper Devonian forest of seed ferns in Eastern New York: New York State Museum, Bulletin 251, p. 5–92.

—— 1927, The oldest known petrified forest: The Scientific Monthly, v. XXIV, p. 514–529.

Grierson, J. D., 1976, *Leclercqia complexa* (Lycopsida, Middle Devonian): its anatomy, and the interpretation of pyrite petrifactions: American Journal of Botany, v. 63, p. 1184–1202.

—— and Banks, H. P., 1963, Lycopods of the Devonian of New York State: Palaeontographica Americana, v. IV, p. 220–295.

—— and —— 1983, A new genus of lycopods from the Devonian of New York State: Botanical Journal of the Linnean Society, v. 86, p. 81–101.

—— and Bonamo, P. M., 1979, *Leclercqia complexa:* earliest ligulate lycopod (Middle Devonian): American Journal of Botany, v. 66, p. 474–476.

Grindley, G. W., Mildenhall, D. C., and Schopf, J. M., 1980, A mid-Late Devonian flora from the Ruppert Coast, Marie Byrd Land, West Antarctica: Royal Society of New Zealand Journal, v. 10, p. 271–285.

Høeg, O. A., 1942, The Downtonian and Devonian flora of Spitsbergen: Norges Svalbard-Og Ishavs-Undersøkelser, Skrifter 83, p. 1–228.

Hueber, F. M., 1960, Contributions to the fossil flora of the Onteora "Red Beds" (Upper Devonian) in New York State [Ph.D. thesis]: Ithaca, N.Y., Cornell University, 212 p.

—— 1961, *Hepaticites devonicus,* a new fossil liverwort from the Devonian of New York: Annals of the Missouri Botanical Garden, v. 48, p. 125–132.

—— 1964, The psilophytes and their relationship to the origin of ferns: The Torrey Botanical Club, v. 21, p. 5–9.

—— 1971, *Astralocaulis:* a new name for *Schizopodium* Harris, and *Sawdonia ornata*: a new name for *Psilophyton princeps* var. *ornatum*: Taxon, v. 20, p. 640–642.

—— 1972, Early Devonian land plants from Bathurst Island, District of Franklin: Geological Survey of Canada: Paper 71–28, p. 1–17.

—— 1983, A new species of *Baragwanathia* from the Sextant Formation (Emsian) Northern Ontario, Canada: Botanical Journal of the Linnean Society, v. 86, p. 57–79.

—— and Banks, H. P., 1967, *Psilophyton princeps:* the search for organic connection: Taxon, v. 16, p. 81–85.

—— and —— 1979, *Serrulacaulis furcatus* gen. et sp. nov., a new zosterophyll from the lower Upper Devonian of New York State: Review of Palaeobotany and Palynology, v. 28, p. 169–189.

—— and Grierson, J. D., 1961, On the occurrence of *Psilophyton princeps* in the early Upper Devonian of New York: American Journal of Botany, v. 48, p. 473–479.

Kevan, P. G., Chaloner, W. G., and Saville, D.B.O., 1975, Interrelationships of early terrestrial arthropods and plants: Palaeontology, v. 18, p. 391–417.

Kjellesvig-Waering, E., in press, A restudy of the fossil Scorpionida of the World: Palaeontographica Americana, Ithaca, N.Y.

Klitzsch, E., Lejal-Nicol, A., and Massa, D., 1973, Le Siluro-Dévonien à Psilophytes et Lycophytes du bassin de Mourzouk (Libye): Academie des Sciences Paris, Compte Rendus, v. 277, p. 2465–2467.

Leclercq, S., 1954, Are the psilophytes a starting or a resulting point?: Svensk Botanisk Tidskrift, v. 48, p. 301–315.

Li, H. H., and Tsai, C. Y., 1978, Devonian floras of China: Papers for the International Symposium on the Devonian System, 1978, No. 3, Nanking Institute of Geology and Palaeontology, Academia Sinica, p. 1–15.

Mamay, S. H., 1962, Occurrence of *Pseudobornia* Nathorst in Alaska: The Palaeobotanist, v. 11, p. 19–22.

Matten, L. C., 1968, *Actinoxylon banksii* gen. et sp. nov.: a progymnosperm from the Middle Devonian of New York: American Journal of Botany, v. 55, p. 773–782.

—— 1973, The Cairo flora (Givetian) from Eastern New York. I. *Reimannia,* terete axes and *Cairoa lamaneckii* gen. et sp. n.: American Journal of Botany, v. 60, p. 619–630.

—— 1974, The Givetian flora from Cairo, New York: *Rhacophyton, Triloboxylon* and *Cladoxylon*: Botanical Journal of the Linnean Society, v. 68, p. 303–318.

—— 1975, Additions to the Givetian Cairo flora from Eastern New York: Torrey Botanical Club Bulletin, v. 102, p. 45–52.

—— 1981, *Svalbardia banksii* sp. nov. from the Upper Devonian (Frasnian) of New York State: American Journal of Botany, v. 68, p. 1383–1391.

—— and Banks, H. P., 1967, Relationship between the Devonian progymnosperm genera *Sphenoxylon* and *Tetraxylopteris*: Torrey Botanical Club Bulletin, v. 94, p. 321–333.

—— and —— 1969, *Stenokoleos bifidus* sp. n. in the Upper Devonian of New York State: American Journal of Botany, v. 56, p. 880–891.

McGregor, D. C., 1977, Lower and Middle Devonian spores of eastern Gaspé, Canada II. Biostratigraphy: Palaeontographica B, v. 163, p. 111–142.

—— 1979, Spores in Devonian stratigraphical correlation, *in* House, M. R., and others, eds., The Devonian System: London, The Palaeontological Association, p. 163–184.

Nathorst, A. G., 1902, Zur Oberdevonischen Flora der Bären-Insel: Kongl. Svenska Vetenskaps-Akademiens Handlingar, v. 36, p. 1–60.

Pedersen, K. R., 1976, Fossil floras of Greenland, *in* Geology of Greenland, Escher and Watt, eds.: Geological Survey of Greenland, Copenhagen, p. 519–535.

Petrosian, N. M., 1968, Stratigraphic importance of the Devonian flora of the U.S.S.R., *in* Oswald, D. H., ed.: International Symposium on the Devonian System, Calgary, 1967, Alberta Society of Petroleum Geologists, v. 2, p. 579–586.

Pettitt, J. M., and Beck, C. B., 1968, *Archaeosperma arnoldii*—a cupulate seed from the Upper Devonian of North America: Contributions from the Museum of Paleontology, University of Michigan, v. 22, p. 139–154.

Phillips, T. L., Andrews, H. N., and Gensel, P. G., 1972, Two heterosporous species of *Archaeopteris* from the Upper Devonian of West Virginia: Palaeontographica B, v. 139, p. 47–71.

Remy, W., and Remy, R., 1980, *Lyonophyton rhyniensis* nov. gen. et nov. spec., ein Gametophyt aus dem Chert von Rhynie (Unterdevon, Schottland): Argumenta Palaeobotanica, v. 6, p. 37–72.

Richardson, J. B., 1974, The stratigraphic utilization of some Silurian and Devonian miospore species in the northern hemisphere: an attempt at a synthesis: International Symposium Belgian Micropaleontological Limits, Pub. No. 9, (Namur 1974), p. 1–13.

——and Ioannides, N., 1973, Silurian palynomorphs from the Tanezzuft and Acacus Formations, Tripolitania, North Africa: Micropaleontology, v. 19, p. 257–307.

Scheckler, S. E., 1978, Ontogeny of progymnosperm: II. Shoots of Upper Devonian Archaeopteridales: Canadian Journal of Botany, v. 56, p. 3136–3170.

——1982, Effects of paleotopography on the Late Devonian flora and the origin of coal swamps in Southern Larussia [Abs.]: Journal of Paleontology, v. 57, Supplement to No. 2, p. 24.

——and Basinger, J. F., 1983, Late Devonian *Archaeopteris fissilis* from northernmost ancient Laurentia, Eastern Yukon, Canada [Abs.]: American Journal of Botany, v. 70, no. 5, pt. 2, p. 77–78.

Schuchman, P. G., 1969, *Pseudosporochnus* in the Middle Devonian of New York State [M.S. thesis]: Ithaca, Cornell University, 70 p.

Scott, A. C., 1980, The ecology of some Upper Palaeozoic floras, *in* Panchen, A. L., ed., The Terrestrial Environment and the Origin of Land Vertebrates: London and New York, The Academic Press, p. 87–115.

Serlin, B. S., and Banks, H. P., 1978, Morphology and anatomy of *Aneurophyton,* a progymnosperm from the Late Devonian of New York: Palaeontographica Americana, v. VIII, no. 51, p. 343–359.

Shear, W. A., Bonamo, P. M., Grierson, J. D., Rolfe, W.D.I., Smith, E. L., and Norton, R. A., 1984, Early Land Animals in North America: Science, v. 224, p. 492–494.

Skog, J. E., and Banks, H. P., 1973, *Ibyka amphikoma,* gen. et sp. n., a new protoarticulate precursor from the late Middle Devonian of New York State: American Journal of Botany, v. 60, p. 366–380.

Stubblefield, S., and Banks, H. P., 1983, Fungal remains in the Devonian trimerophyte *Psilophyton dawsonii*: American Journal of Botany, v. 70, p. 1258–1261.

Woodrow, D. L., Fletcher, F. W., and Ahrnsbrak, W. F., 1973, Paleogeography and paleoclimate at the deposition sites of the Devonian Catskill and Old Red Facies: Geological Society of America Bulletin, v. 84, p. 3051–3064.

MANUSCRIPT ACCEPTED BY THE SOCIETY JULY 20, 1984

Printed in U.S.A.

Geological Society of America
Special Paper 201
1985

Unusual marginal-marine lithofacies from the Upper Devonian Catskill clastic wedge

John S. Bridge
Mary L. Droser
Department of Geological Sciences
State University of New York at Binghamton
Binghamton, New York 13901

ABSTRACT

Upper Devonian marginal-marine deposits exposed at Ashcraft Quarry, northernmost Pennsylvania, are unusual in that they contain limestone in addition to the sandstone and shale which is prevalent in the Catskill clastic wedge. A 3 m. thick *lower unit* is a lateral-accretion deposit, composed mainly of planar and cross-stratified sandstones with subordinate wavy-flaser bedding. Erosion surfaces beneath sandstones are overlain by intraformational breccias containing transported crinoid, brachiopod, bivalve and plant remains. Paleocurrents are unidirectional westward, but current ripples rarely indicate bidirectional paleoflow.

The 3.5-5.5 m. *limestone unit,* comprising skeletal grainstone interbedded with calcareous sandstone, fines up or coarsens up from an extensive erosional base, and shows lateral-accretion bedding. The lime-grainstone contains abraded fragments of crinoids, brachiopods, bivalves, gastropods, and fish bones; also ankerite concretions, shale chips and plant remains. The whole unit is mainly large-scale cross-stratified, with bidirectional paleocurrents; and its top surface is marked by sandwaves, interfering wave and current ripples, and abundant burrows.

The *upper unit* comprises interbedded sandstones, siltstones and shales. Sheet-like and channel-filling sandstones have basal skeletal lags, large- and small-scale cross-stratification, planar stratification, and hummocky cross-stratification. Finer-grained strata have wavy-lenticular bedding (with wave and current ripple marks), concretions, and abundant burrows. Throughout the exposure faunal diversity is low relative to coeval marine-shelf facies, and trace fossils belong mainly to the *Skolithos* ichnofacies.

The depositional environment of the lower unit, the limestone unit, and immediately adjacent beds, is laterally migrating sand bars adjacent to curved tidal channels with strong tidal-current asymmetry, probably in an estuary with marginal intertidal flats. Overlying deposits were introduced by periodic unidirectional currents and reworked by waves, possibly in a brackish coastal bay.

INTRODUCTION

Sedimentological studies of the marginal-marine facies of the Upper Devonian Catskill clastic wedge in New York state and Pennsylvania have suggested a range of different types of coastline (e.g., Allen and Friend, 1968; Johnson and Friedman, 1969; Sutton and others, 1970; Walker and Harms, 1971; Woodrow and Isley, 1983; see also references in Sevon, 1981). Both deltaic and nondeltaic coastlines have been claimed, but the relative importance of wave, tidal and river currents from different localities remains unclear. In this study we intend to throw further light on this problem by detailing an example of marginal-marine deposits exposed at Ashcraft Quarry in northernmost Pennsylvania (Figs. 1 and 2). The deposits are unusual in that they contain limestone in addition to the usual sandstone and shale. Indeed, somewhat similar carbonate-rich beds are only rarely seen else-

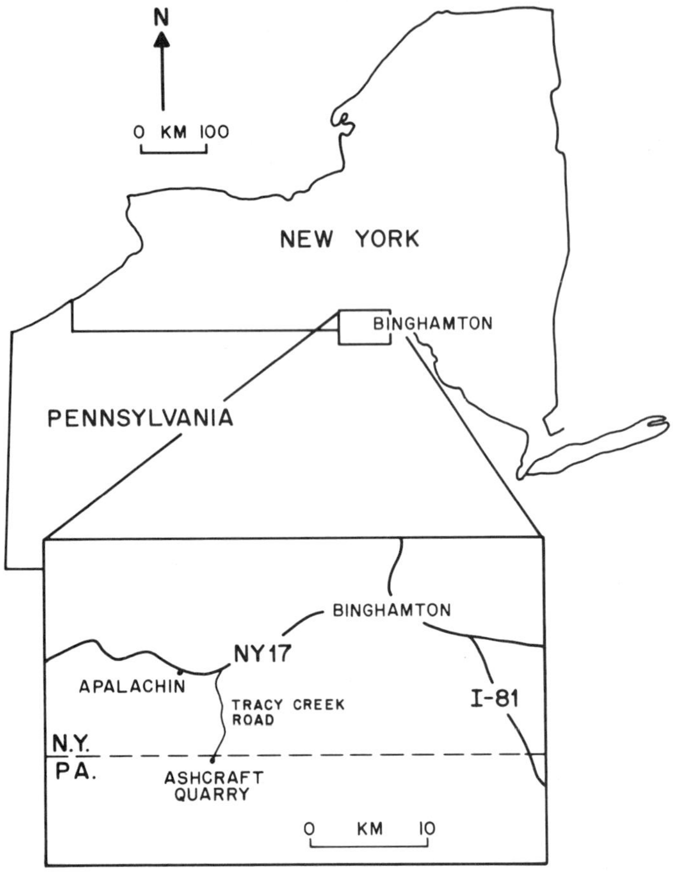

Figure 1. Location of the Ashcraft Quarry.

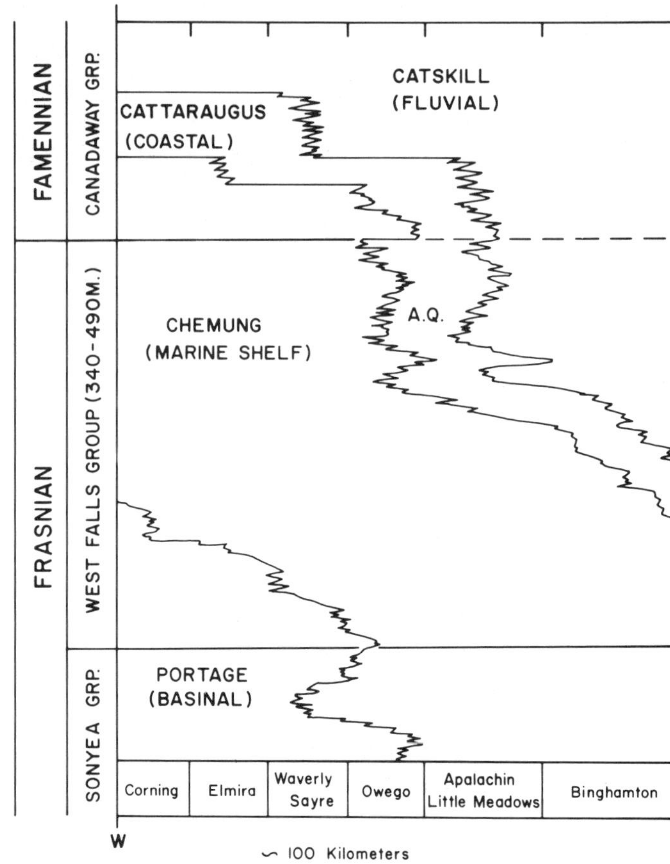

Figure 2. Tentative stratigraphic position of the Ashcraft Quarry (A.Q.), within the context of the major magnafacies of the Catskill clastic wedge (after Rickard, 1975).

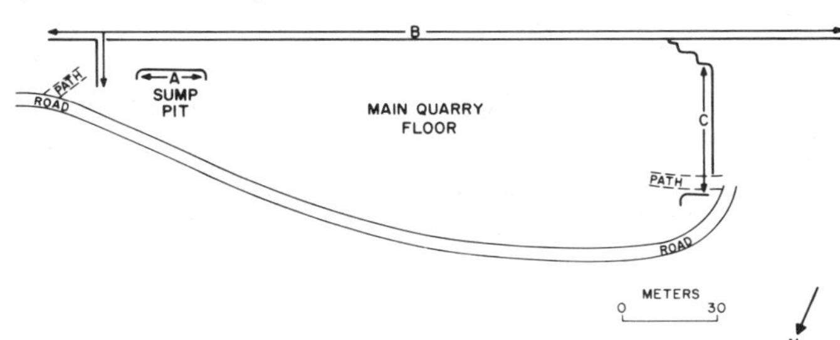

Figure 3. Plan view of the Ashcraft Quarry, showing location of main quarry faces studied. Note North azimuth.

where within the Catskill clastic wedge (Rickard, 1975; Woodrow and others, 1981). Previous explanations of the origin of the Ashcraft limestone have been an offshore shell mound (Krajewski and Cuffey, 1976a, b), a break-point bar (Bowen, 1978), a land-tied accretion lobe or spit, and a barrier bar (Woodrow and others, 1981). The following detailed study of the sedimentology and paleontology of the limestone and associated strata should resolve this inconsistency as well as contribute to the regional paleogeographic reconstruction.

DESCRIPTION OF LITHOFACIES

Organisation

The rocks are exposed sporadically on two quarry-floor levels, and in variously oriented quarry faces which are degraded enough to allow three-dimensional study (Fig. 3). Two-dimensional representations of the major lithological units and large-scale bedding geometry in the main quarry faces were constructed from photomosaics (Fig. 4). Details of the lithofacies are shown in a series of spatially correlated vertical logs (Fig. 5). Three major divisions are readily apparent in Figure 4. A *lower unit* (~3 m. thick) of sandstones with subordinate shales is exposed in the sump pit (face A) and sporadically above and below the main quarry floor. This is overlain by the *limestone-calcareous sandstone unit* (3.5-5.5 meters thick), exposed in the lower part of face B and in face C. Its top surface was examined on the upper quarry level adjacent to face C. It has been suggested that this unit is a north-south elongated "mound" (Krajewski and Cuffey, 1976a, b; Woodrow and others, 1981), but as this geometry is based on only four boreholes it must be considered highly speculative. The *upper unit* (up to ~10 m thick) of interbedded sandstone, siltstone and shale is exposed mainly in the upper part of face B.

The beds within each unit (Fig. 5) are normally arranged

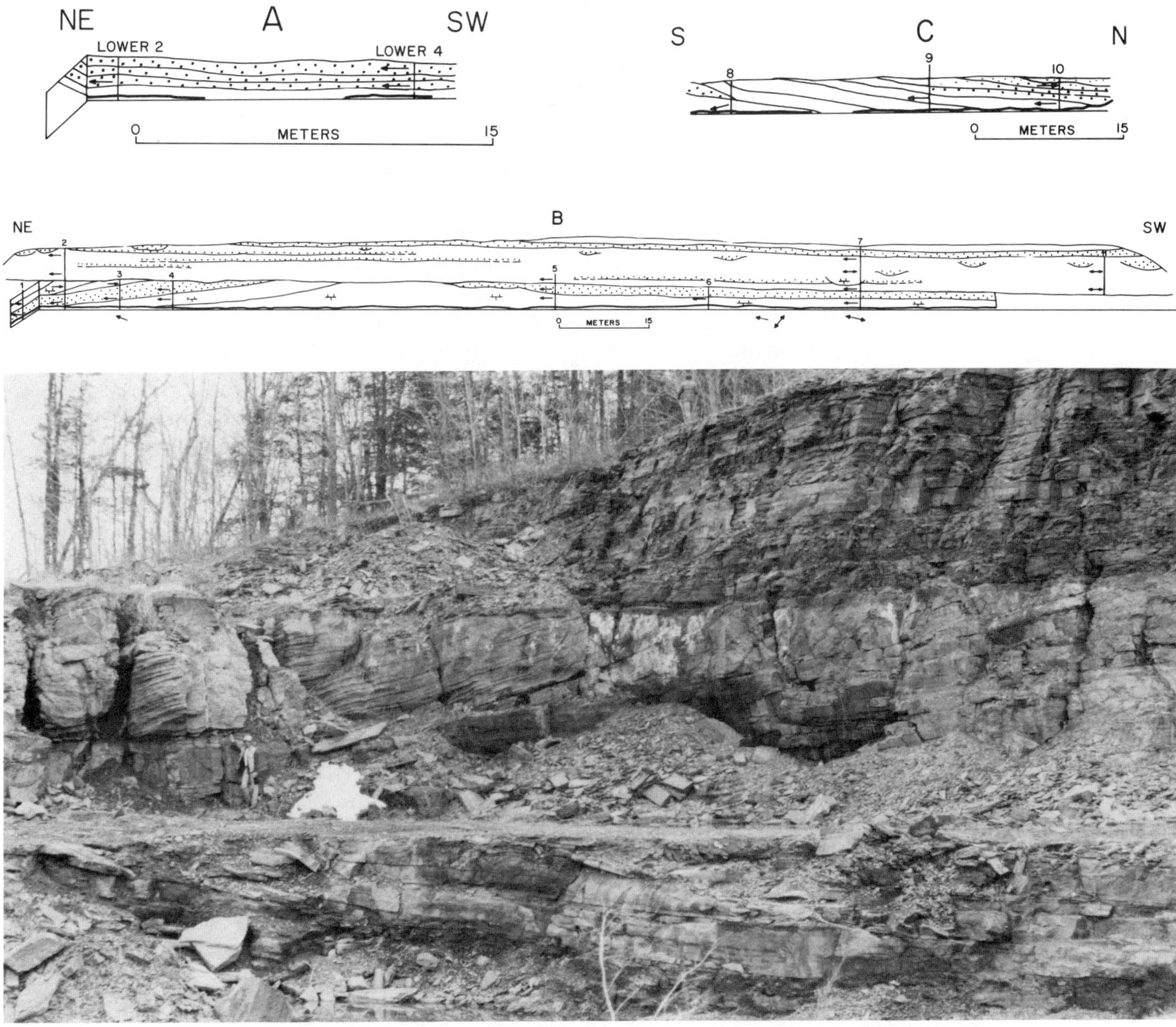

D

Figure 4. Two-dimensional sections showing major lithologies and bedding features exposed in faces A, B, and C shown in Figure 3. Stipple represents major sandstone beds, brick ornament represents lime-grainstone, and the remainder is interbedded sandstone and shale. Positions of measured sections are indicated by numbered vertical lines, and paleocurrent azimuths are also shown. Photograph (D) shows the northeast end of the quarry. Lateral-accretion deposits of the *lower unit* in the sump pit are overlain by upward-coarsening *limestone-calcareous sandstone unit* with large-scale planar cross-stratification, overlain sharply by sheet-like sandstones and shales of the *upper unit.*

into decimeter-scale bedsets with erosive bases and vertical and lateral variations in composition, texture, sedimentary structure and fossils (i.e., lithofacies). The three distinctive units constitute a larger-scale spatial variation in lithofacies. These different scales of lithofacies variability are described below, within the context of the three major divisions.

Lower unit

Planar- and cross-stratified gray sandstones are dominant, but interbedded sandstones and dark-gray shales (heterolithic bedding) also occur (Fig. 5A). The sandstones are fine to very fine grained, and occur normally as 40-50 cm. thick upward-fining bedsets. Basal erosion surfaces of bedsets are ornamented with tool marks and load casts, and are overlain by intraformational breccias consisting primarily of shale chips, crinoid, brachiopod, and rarely bivalve fragments and carbonised plant remains. The breccias are normally a few centimeters thick and extend laterally for meters.

Large-scale, trough cross-stratification, mostly in 10-20 cm.

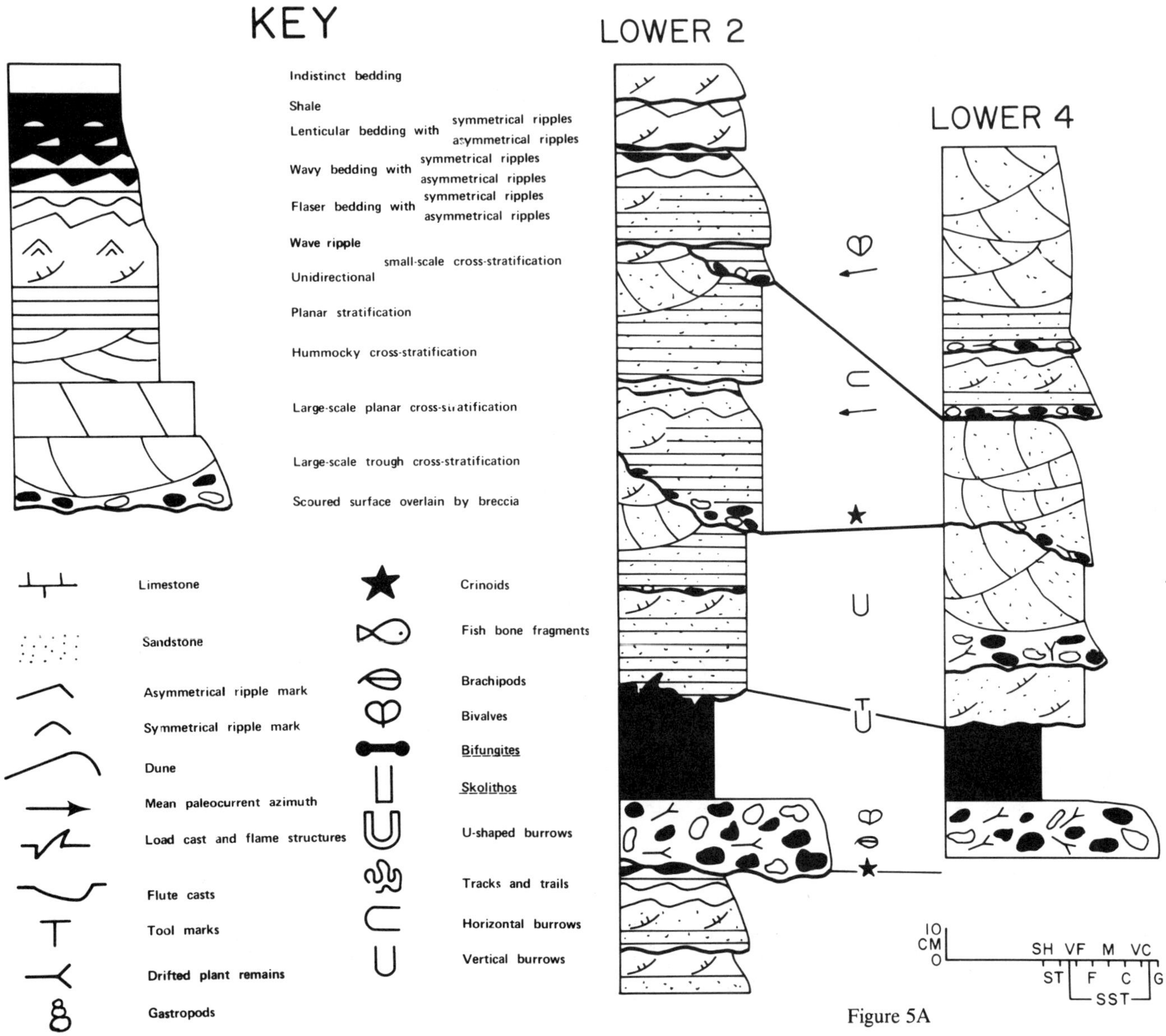

Figure 5A

Figure 5 (this and following pages). Detailed vertical sections, as indicated on Figure 4, subdivided as (A) lower unit (in sump pit), (B) limestone unit, (C) upper unit. Symbols explained in key. Note large-scale cross-sets drawn approximately to scale.

thick sets, and planar stratification with parting lineation, dominate the thicker bedsets. Small-scale cross-stratification occurs in the upper parts of such bedsets and comprises the thinner bedsets entirely. Climbing-ripple types are common and correspond to type A of Jopling and Walker (1968), the erosional-stoss type of Ashley and others (1982). Flaser and wavy bedding occurs at the top of some bedsets, the shale drapes preserving asymmetrical, sinuous and lingoid, ripple marks. Lateral transitions between sedimentary structures are as common as the vertical transitions. Cross-stratification and tool marks indicate a unidirectional paleocurrent azimuth, ranging from 255° to 270°. In the sump pit the major bedding surfaces are horizontal when viewed parallel to this paleocurrent direction, but dip 9° to the south when viewed normal to paleocurrent direction (Fig. 4D); this is lateral-accretion bedding. *Planolites* and ?*Sagittichnus* are abundant in convex hyporelief on bases of beds (Fig. 8). Apart from the fossiliferous breccias, the bivalve *Grammysia* (Fig. 9A) occurs rarely in these sandstones.

The uppermost sandstone of this unit forms the main quarry floor; its top surface is marked by asymmetrical ripple marks (paleocurrent azimuths 290° and 115°) and interfering symmetrical ripple marks (crestlines oriented 100° and 10°). Trace fossils are abundant on this surface and include *Arenicolites,* retrusive *Diplocraterion, Planolites, Paleophycus, ?Astersoma,* and numerous other tracks, trails and burrows of various sizes and orientations (Fig. 8A, C, D). In places this uppermost sandstone fines

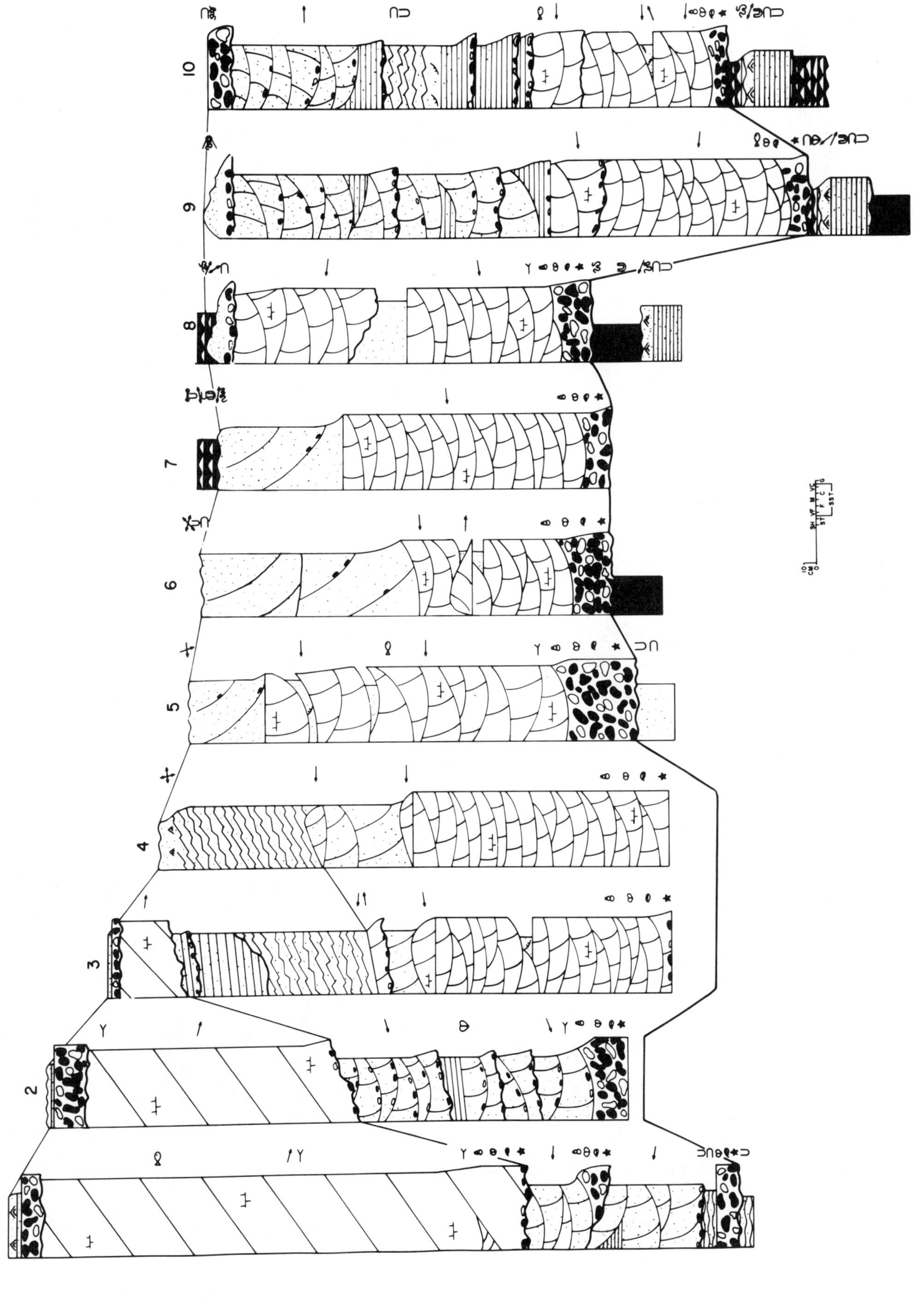

Figure 5B

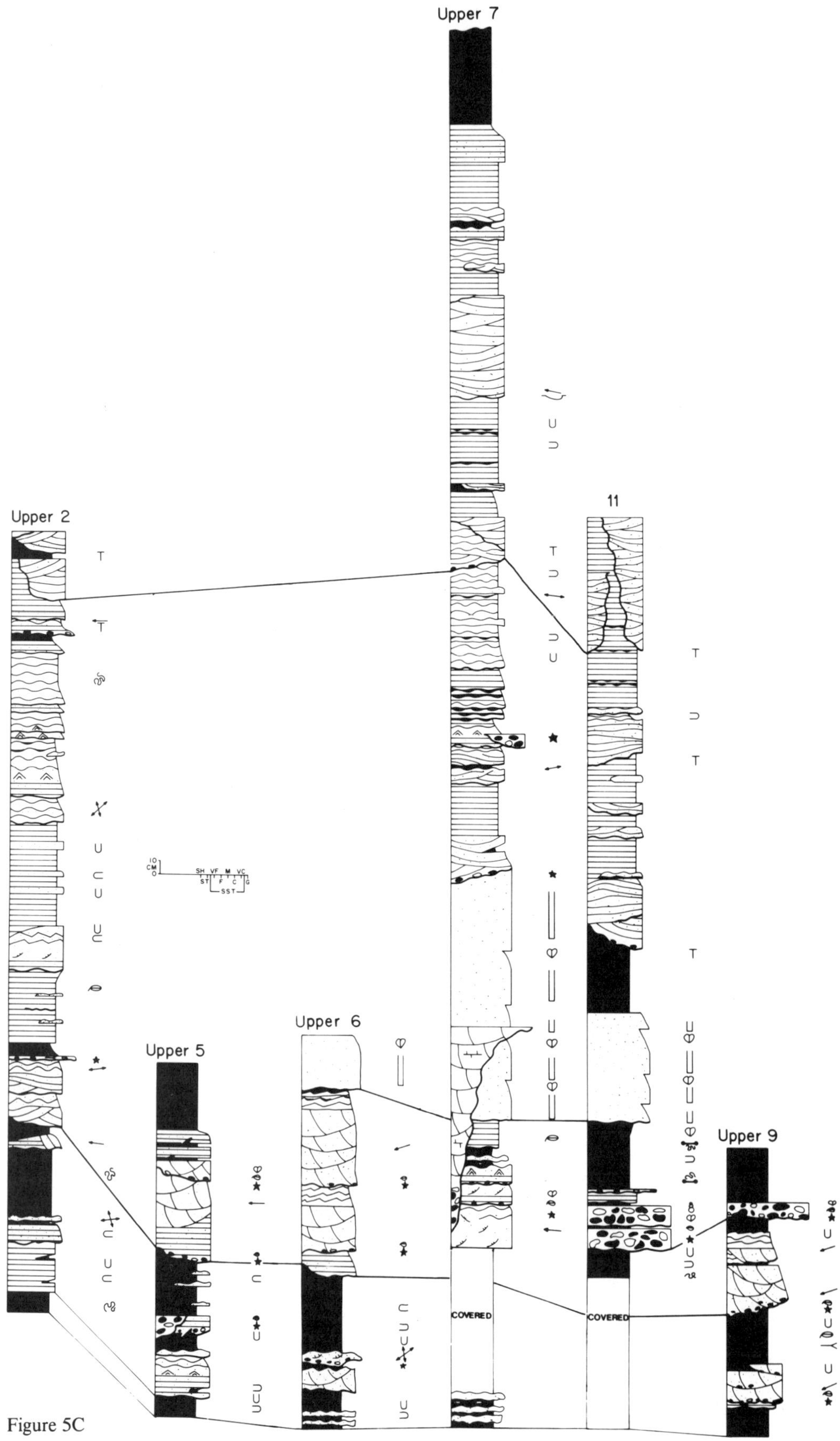
Upper 7
Upper 2
11
Upper 5
Upper 6
Upper 9
COVERED
COVERED
SH VF M VC
ST F C G
SST

Figure 5C

Figure 6. Structures within the *limestone unit.* (A) Basal erosion surface overlain by intraformational breccia near section 6. Note load casts in underlying beds. 10 cm. scale. (B) Complete upward-fining limestone-calcareous sandstone unit near section 6, showing basal erosion surface and sandwaves on sharp upper surface. Top of meter-stick is on dune-form migrating to left, whereas large-scale cross-stratification is generally directed to the right. (C) Large-scale cross-stratified lime-grainstone with finer sandstone drapes. Some drapes show asymmetrical ripple marks (and cross-lamination) migrating to left, whereas large-scale cross-strata are directed to right. 10 cm. scale. (D) Large-scale cross-stratified lime-grainstone showing bundles of relatively finer-grained cross-strata in a single cross-set (below meter-stick). (E) West end of quarry (face c) showing lateral-accretion surfaces in limestone unit dipping to right. Scale is 1 m. (F) Asymmetrical sandwave with superimposed wave-ripple marks, from top of limestone unit at west end of quarry. Scale is 1 m.

upward into shale, or into flaser-wavy bedding with asymmetrical and symmetrical ripple marks. Trace fossils in these beds include *Rhizocorallium* (Fig. 8E) in addition to those occurring in the underlying sandstone. *Arenicolites* are particularly abundant, with burrows almost always truncated by the overlying limestone unit.

Limestone-calcareous sandstone unit

This unit is everywhere bounded beneath by an erosion surface with local relief up to 30 cm. Overlying this surface is up to 50 cm. of intraformational breccia (Figs. 5B, 6A), composed predominantly of shale clasts, ankerite nodules, and shell fragments. The major lithology is gradational from light-gray, quartzose, skeletal lime grainstone (in places rudstone) to calcareous quartz arenite. The grainstone consists primarily of highly abraded shell fragments (including brachiopods, crinoids, bivalves, gastropods, fish plates and rare trilobites), fine-medium grained quartz, ankerite nodules, shale fragments, and carbonised and pyritised plant fragments (mainly stems). The cement is sparry calcite, and many shell fragments are preserved within micrite envelopes as calcite-filled molds or neomorphic calcite. The gray ankerite ($Ca_{1.0}\,Mg_{0.75}\,Fe_{0.25}\,(CO_3)_2$) nodules weather to yellow-brown iron oxide. Nodules consist of micritic to mm-sized carbonate crystals that coarsen toward the margins in a crudely concentric pattern, enclose quartz grains that are more angular and finer grained than those in the surrounding grainstone, and rarely show alveolar structure. They are rounded and elliptical (average long dimension ~2 cm.), and their maximum projection surfaces commonly parallel local bedding surfaces. The nodules and shell fragments are thus clearly reworked (Woodrow and others, 1981).

Throughout most of the large quarry face B, this unit generally fines upward through successive bedsets, from skeletal grainstone to calcareous quartz arenite (Figs. 5B, 6B). Large-scale, trough cross-stratification, in decimeter-scale sets, is the dominant internal structure. Cross-strata within the grainstone have distinctive, finer-grained quartz arenite drapes, in places with asymmetrical ripple marks directed up the cross-stratal surfaces (Fig. 6C). Successive bundles of relatively thinner (finer-grained) and thicker (coarser-grained) cross-strata can rarely be seen when a single cross-set is traced in the downcurrent direction (Fig. 6D). Planar-stratification and small-scale cross-stratification are locally interbedded with the large-scale cross-strata. Trough cross-sets generally indicate a westerly paleocurrent; however, rarely, small- and large-scale cross-sets and dune forms (Fig. 6B) indicate an eastward paleocurrent.

At the east end of the quarry, the trough cross-stratified lime grainstone thins eastward and becomes overlain by wavy-flaser bedded quartz arenite with dominantly westward-directed, herringbone cross-lamination (Figs. 4B, 5B, sections 3 and 4). Bioturbation is primarily by *Arenicolites.* These heterolithic beds lie on a 9° depositional slope and grade laterally eastwards and down-dip into large-scale trough cross-stratified sandstone, still with a westward paleocurrent. Truncating and overlying this sandstone are large-scale, wedge-shaped, planar cross-sets up to 3.7 m. thick, with an *eastward* paleocurrent, in lime-grainstone (Figs. 4B, 4D, 5B). In places, the cross-strata are deformed by synsedimentary faults and flexures. Rarely, large-scale (8-10 cm. thick) and small-scale (~1 cm. thick) cross-sets occur within these larger planar cross-strata, indicating flow up the cross-stratal surfaces. Thus, an overall coarsening upward occurs in this unit at the east end of the quarry.

At the west end of the quarry, face C exposes bedsets dipping at 6° to the north (i.e., lateral-accretion bedding, Figs. 4C, 6E). These lateral-accretion deposits show a northerly increase in sandstone relative to lime-grainstone (Fig. 5B, sections 8 through 10), paleocurrents depositing the sandstones were dominantly eastward directed, and planar-stratification becomes more abundant. Sequences of bioturbated flaser bedding with asymmetrical ripple marks, and bimodal paleocurrents, occur locally between the grainstone and sandstone.

The upper boundary of this unit is sharp, in places erosional (Fig. 5B). At the west end of the quarry, symmetrical and asymmetrical sandwaves are preserved on this top surface (Fig. 5B, 6F). Symmetrical forms range up to ~10 cm. high with average wavelength of a meter. Asymmetrical forms are up to 20 cm. high and 2 m. long, and indicate an easterly (110°) paleoflow. Superimposed on these sandwaves are asymmetrical ripple marks (75° paleoflow) and symmetrical (including interfering) ripple marks which are variously oriented 200°, 212°, 150°. Tracks, trails, and horizontal and vertical burrows of various sizes are abundant on this sandwave surface. Long horizontal traces (up to 30 cm. long and 5 mm. wide) are abundant and commonly parallel the paleflow on sandwave crests. *Chondrites* (Fig. 8F), *Beaconites* (Fig. 8B) and the inorganic *Aristophycus* structure (Seilacher, 1982, Fig. 8G) are noteworthy occurrences.

Invertebrate fossils within the limestone unit were difficult to identify because they have been abraded and diagenetically altered. However, recognisable bivalves include *Paleoneilo* (Krajewski and Cuffey, 1976a), *Grammysia* (Fig. 9A), *Cypricardella* (Fig. 9B), also the gastropod *Bellerophon,* and brachiopods *Cupularostrum* ("*Camarotoechia*") and *Tylothyris(?)* (Fig. 9C). Plants include *Lycopsida* and *Archeopteris,* and fish plates have been tentatively identified as *Dinichthys.* Notably absent are bryozoans and corals.

Upper Unit

Immediately overlying the lime-grainstone/calcareous arenite unit in places is a sequence of very fine-grained, gray sandstones, dark-gray siltstones and shales, occurring mainly as wavy and lenticular bedding with interfering symmetrical ripple marks (Fig. 5, sections 7 and 8; Fig. 7A, B, C). Planar-lamination is present at the base of some of the thicker bedsets. Trace fossils within these facies include *Arenicolites,* retrusive *Diplocraterion, Planolites, Paleophycus, Rhizocorallium,* the dumbell variety of

Bifungites (Fig. 8H), and numerous other vertical and horizontal burrows, tracks and trails. Relatively thick, interbedded shales contain the double-arrow variety of *Bifungites* (Fig. 8I) and other small sinuous traces concentrated within 1-2 cm. thick zones, but are otherwise not bioturbated. Yellow-brown globular calcareous concretions up to 1.5 cm. in diameter occur in these shales.

In other places cm.-dm. thick, upward-fining, sandstone-bedsets occur immediately above the limestone, with large- and small-scale cross-stratification, planar-stratification, and burrowed flaser-wavy bedding with asymmetrical ripple marks (Fig. 5, sections 6 and 9; Fig. 7D). These facies are similar to those occurring in the "lower unit." Basal erosion surfaces have less than 1 cm. relief and show load casts, longitudinal ridge marks, flute marks, various tool marks, and obstacle marks (Fig. 7E); they are normally overlain by a thin, basal breccia of crinoid, brachiopod and plant fragments. Trough cross-stratified limestone beds may be associated with the coarser-grained facies, occurring as sheets and rarely in a channel fill (1 m. deep and 3 m. wide). Large-scale, trough cross-beds are directed 275° - 295°. Unusual structures are wrinkle mark ridges up to 1 mm. high, which generally curve to make "honeycomb-like" structures rather than run parallel to each other (Reineck and Singh, 1973; *Kinneya?* of Seilacher, 1982). Similar marks have been documented by Goldring (1971) in the *Diplocraterion Yoyo* facies of the Devonian Baggy Beds (England). Trace fossils are abundant throughout this sequence, including *Planolites, Lockeia* (Fig. 8J), and arthropod tracks and trails.

In the western part of the quarry, above the lowermost 2 m. of the upper unit, is a characteristic zone of fine- to medium-grained sandstones in 5 to 30 cm. thick bedsets, with internal structures completely destroyed by *Skolithos* burrows (Figs. 5C, 8K). Individual bedsets commonly fine upward into bioturbated, very fine-grained sandstone or siltstone with few well-defined burrows (rare, large, horizontal burrows), a notable paucity of *Skolithos,* and commonly clusters of the bivalve *Goniofera* (Fig. 9D), articulated but not in life position. The *Skolithos-Goniofera* sequence is repeated vertically up to five times over almost 2 m., and the bedsets extend laterally on the order of tens of meters.

The uppermost part of the upper unit comprises light olive-gray, very-fine grained sandstone/coarse siltstone interbedded with subordinate gray shale. The sandstones occur as erosively based, cm.-dm. thick bedsets with sheet-like, lenticular or channel-filling geometry (Figs. 4B, 5C). Basal erosion surfaces show tool marks, flutes and load casts, and are locally overlain by intraformational breccias containing plant, crinoid and brachiopod debris. Internal structure within the sandstones is primarily hummocky cross-stratification: concave upward "synforms" (10-30 cm. high, 1-2 m. across) are dominant over hummocks, and the 5-10 mm. thick laminae parallel erosion surfaces (Fig. 7G). A distinctive horizon of hummocky cross-strata in steep-sided channel fills (Figs. 4B, 7H) contrasts with a 1 m. thick hummocky cross-stratified sandstone sheet (approximately 1 m. above it) which extends throughout the quarry. In most places hummocky cross-stratified bedsets are lenticular. Planar-stratified sandstone or siltstone may directly overlie, or grade laterally from, hummocky cross-stratification, and can also be the dominant structure within a bedset. Other common bedsets comprise wave-ripple lamination grading upward into flaser, wavy, or lenticular bedding with symmetrical ripple marks, and locally into shale. Planar-stratification is present at the bases of some of these bedsets. Tool marks on the base of the channel fills are directed 225°-275° and flute marks on the base of the thick sheet sandstone are directed 290°. Symmetrical ripple marks have crests oriented approximately E-W or N-S. Trace fossils are not common. *Planolites* occur rarely throughout, along with other horizontal and vertical burrows. Trace fossils found in the float from the upper unit include *Rusophycus* (Fig. 8L), *Scalarituba* (Fig. 8M), and a bed-junction preservation (Simpson, 1957) of *Chondrites* (Fig. 8N). Also found in the float from the upper unit are syneresis cracks (Fig. 7I) and concentric injections around a vertical burrow (Fig. 7J).

Invertebrate fossils occur throughout the upper unit within lenticular and sheet-like "coquinites" (intraformational breccias), most commonly associated with the planar and cross-stratified sandstones in the lower parts of this unit. The "coquinites" are size sorted (maximum shell length is ~4 cm.) with a very fine to fine-grained sandstone matrix; however, abrasion was minimal. Over half of the fossil material is disarticulated and articulated crinoid stems: articulated crinoid stems are commonly preferentially oriented parallel and perpendicular to paleocurrent direction. The shelly fossils occur as molds and casts, normally oriented convex upward. The most abundant shelly fossil is *Cupularostrum,* and a few coquinite lenses are composed entirely of this brachiopod plus plant debris. The brachiopods *Chonetes, Tylothyris,* and *Platyrachella* occur in approximately equal abundances, together constituting less than half of the shelly fossils. The bivalves *Cypricardella* and *Grammysia* occur less commonly than the brachiopods within the coquinites.

Individual, disarticulated and unabraded shelly fossils also occur outside the coquinites within very fine and fine-grained sandstones. *Grammysia* (up to 8 cm. long) occurs patchily as molds or casts with convex-up orientation. *Cypricardella* and *Tylothyris(?)* occur rarely. As mentioned earlier, *Goniofera* is the only (articulated) bivalve found in the siltstones.

INTERPRETATION OF DEPOSITIONAL ENVIRONMENT

A coastal environment of deposition is suggested by the association of (1) complex interaction of unidirectional, wave and tidal currents; (2) abundance of land-derived plant remains; (3) abundant trace fossils of the *Skolithos* and *Cruziana* ichnofacies; and (4) abundance of shelly invertebrate fossils but low species diversity compared with coeval fully marine deposits (cf. Bowen and others, 1974; Thayer, 1974; McGhee and Sutton, 1981). This evidence is explored in more detail below.

A B C D E F

G

I

H

J

Figure 7 (this and facing page). Structures within *upper unit.* (A) Wavy bedding with symmetrical ripple marks and bioturbated shales. 10 cm. scale (B) Bioturbated wavy-flaser bedding. 5 cm. scale. (C) Finely laminated sandstone and shale (lenticular bedding). 3 cm. scale. (D) Bedset with large-scale, trough cross-stratification overlain by flaser bedding with asymmetrical ripple marks. Flow towards observer. (E) Obstacle marks associated with shells and tool marks on base of wavy-flaser bedded sandstone. Flow left to right. 10 cm. scale. (F) Wrinkle marks on top of ripple-marked sandstone bed. (G) Hummocky cross-stratified sandstone with channelled base and sharp top, near upper section 2. (H) Complete upper unit at westernmost part of quarry showing hummocky cross-stratified sandstone in a channel fill and (above and left) as a sheet. (I) Syneresis cracks (2 mm. high) on base of sandstone bed, corresponding roughly to incomplete orthogonal class (Allen, 1982). (J) Concentric injections around vertical burrow. 2 cm. scale.

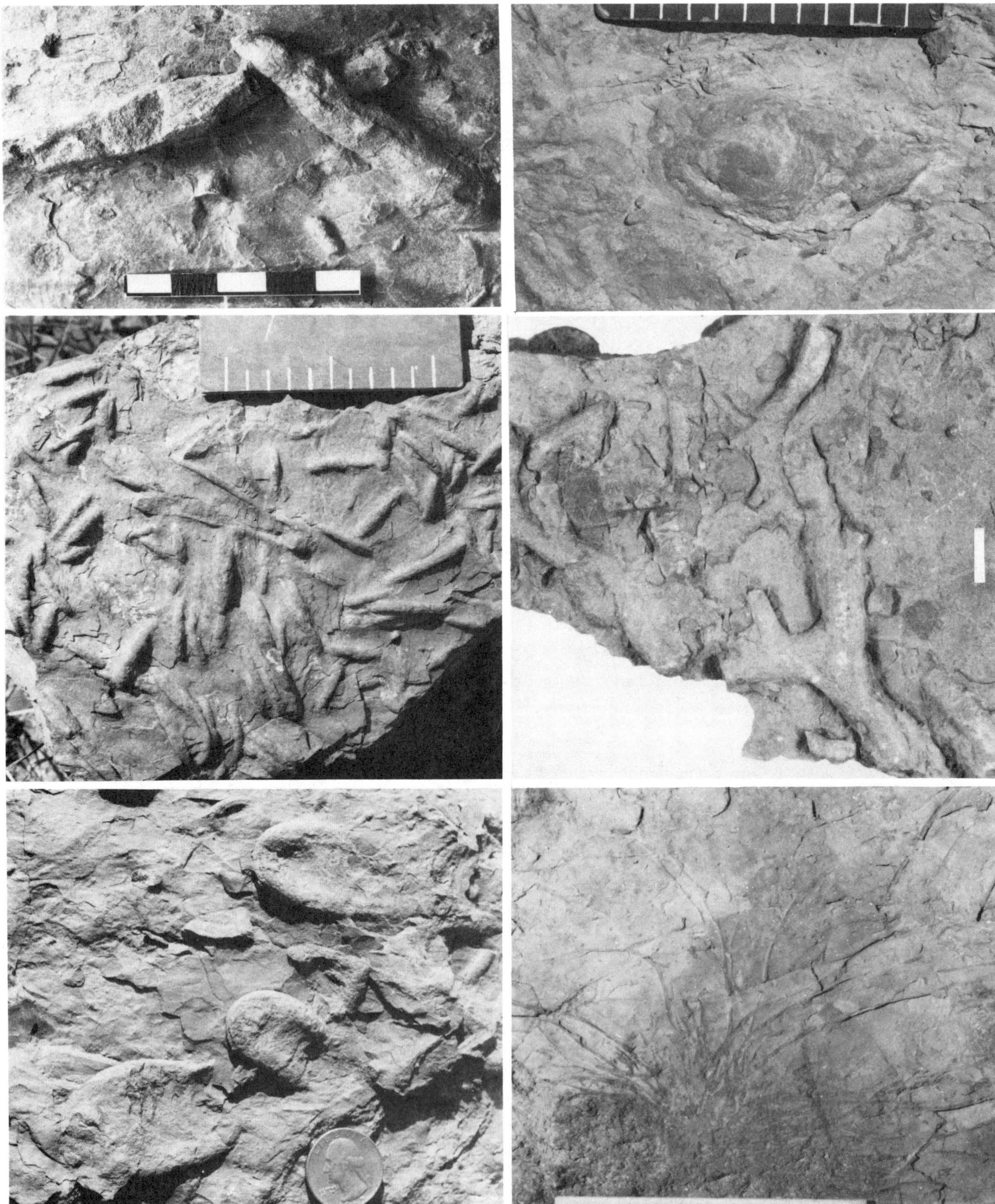

A B C D E F

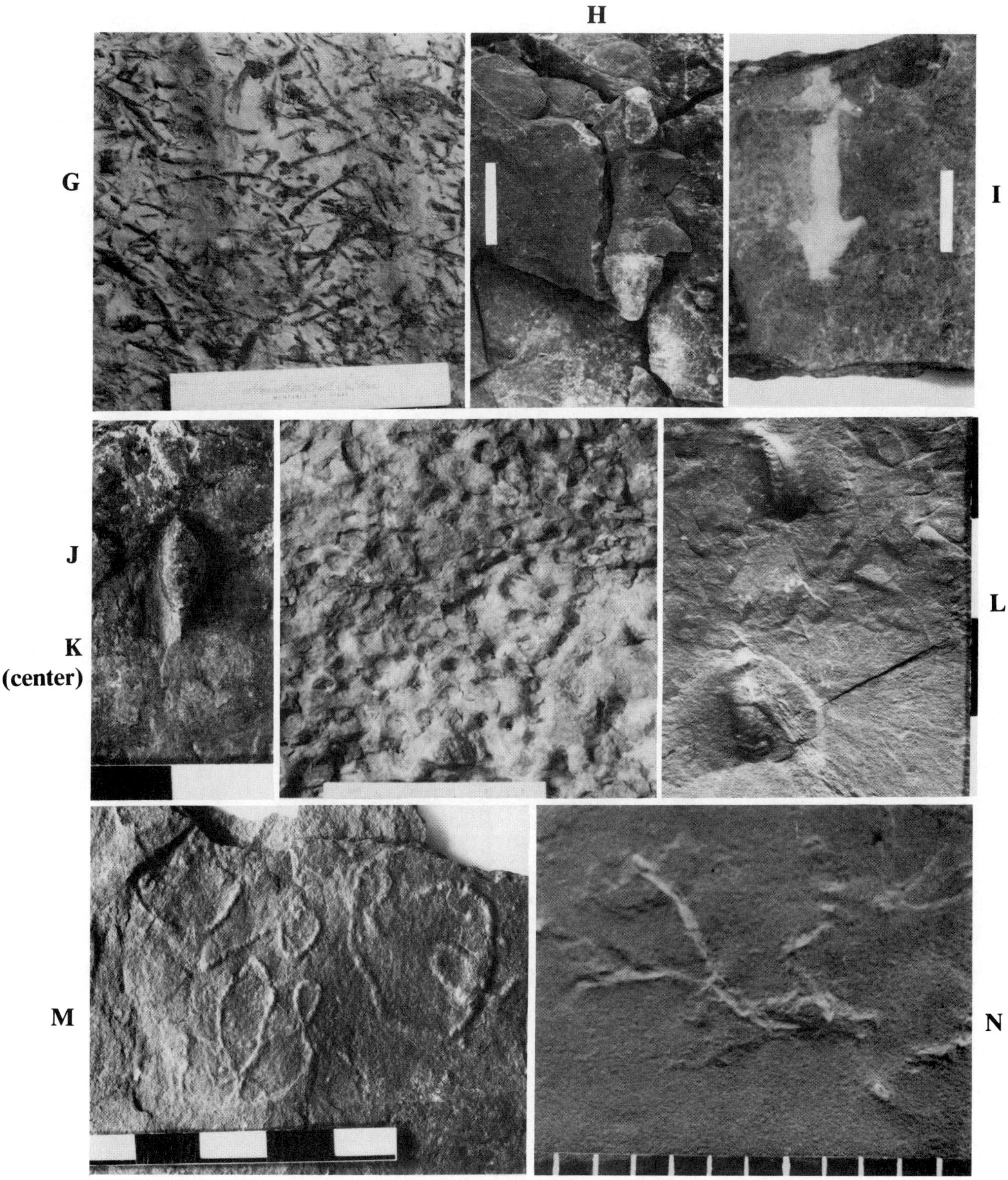

Figure 8 (this and facing page). Trace fossils. (A) Planolites in convex hyporelief. 3 cm. scale. (B) *Beaconites* on bedding surface. 10 cm. scale. (C) *Arenicolites* in convex hyporelief. 10 cm. scale. (D) *Paleophycus* in convex hyporelief. 1 cm. scale. (E) Basal bedding surface with *Rhizocorallium.* Coin is 2.3 cm. across. (F) Top surface of limestone unit with *Chondrites.* 31 cm. scale. (G) Wave-rippled top of limestone unit with *Aristophycus* and various burrow types. 15 cm. scale. (H) Bedding surface with noncompacted *Bifungites.*1 cm. scale. (I) Bedding surface with arrow-head variety of *Bifungites.* 1 cm. scale. (J) *Lockeia* in convex hyporelief. 2 cm. scale. (K) Bedding surface with *Skolithos* burrows. 5 cm. scale. (L) *Rusophycus* in convex hyporelief. 4 cm. scale. (M) *Scalaritubia* in convex hyporelief. 3 cm. scale. (N) Bed-junction preservation of *Chondrites* from hummocky cross-stratified sandstone. 10 cm. scale.

Figure 9. Body fossils. (A) *Grammysia.* 5 cm. scale. (B) *Cypricadella.* 4 cm. scale. (C) *c.f. Tylothyris.* 5 cm. scale. (D) *Goniofera.* 5 cm. scale.

Physical paleoenvironment

Large-scale, trough cross-stratification and associated planar-stratification are produced by sinuous-crested dunes and upper-stage plane beds, respectively. Small-scale cross-stratification is produced by asymmetrical current ripples. The bedsets containing these structures in the lower unit, and immediately above the limestone unit, record unsteady westerly (seaward) directed currents which were initially erosive and then decreased in power with time during deposition. The power of these depositing currents also varied spatially.

Heterolithic bedding within, and immediately adjacent to, the limestone unit records the alternation of deposition from bedload and suspended load in a relatively low-powered, unsteady flow. Types with small-scale cross-stratification and asymmetrical ripple marks with bimodal paleocurrents are of tidal origin. The thin mud drapes in flaser bedding probably resulted from deposition during a single tidal slack water period (Reineck and Wunderlich, 1968; Terwindt and Breusers, 1972, 1982; Terwindt, 1981; Hawley, 1982). Mud layers thicker than about 0.5 cm. cannot be deposited during a single tidal slack water period and may have accumulated over a series of (neap?) slack water periods (Terwindt and Breusers, 1972; Terwindt, 1981) or during storm wave deposition (McCave 1970, 1971; Hawley, 1981). Indeed, the wave-ripple laminated types of heterolithic facies probably result from storm-wave deposition. Fining upward within bedsets may be due to seasonal changes in wave currents, neap-spring tidal current variations (e.g., Terwindt, 1981), or seasonally variable unidirectional currents.

Large-scale cross-stratification in the limestone unit records flow unsteadiness on various scales, and bimodal flow directions, both attributable to tidal currents. The relatively fine-grained sandstone drapes with current ripples directed up the foreset slope suggest the alternation of a dominant (ebb) tidal current and a weaker (flood) tidal current. The downcurrent variation in cross-stratal thickness and grain size within individual cross-sets may represent neap-spring tidal bundles (cf. Visser, 1980; Terwindt, 1981; Boersma and Terwindt, 1981; Allen, 1982; Van Den Berg, 1982; Nio and others, 1983; Allen and Homewood, 1984); however, it was not possible to count individual ebb-flood couplets to prove this. Flow unsteadiness of a longer (seasonal?) time scale is evidenced by the upward-fining nature of some bedsets, with small-scale cross-strata formed at the lower flow powers associated with current ripples. There is also clearly spatial variability in flow power as indicated by lateral transitions from large to small-scale cross-stratification and planar-stratification. Large-scale (flood-oriented), planar cross-strata in wedge-shaped units are due to advancement of asymmetrical sandwaves into deepening water. The associated smaller-scale cross-strata within the planar cross-strata are caused by ripple and dune migration up the avalanche face of the sandwave during the opposing (ebb) tidal flow. Strong tidal current asymmetry is suggested by the dominance of either flood- or ebb-current directions in different parts of the limestone unit. Asymmetrical dune forms agree with this interpretation, although rare symmetrical sandwaves indicate ebb and flood currents locally of equal strength and duration. Wave activity on the top surface of the limestone unit is rarely suggested by symmetrical ripples.

Lateral-accretion bedding within the limestone (and the lower unit) with facies fining-up or coarsening-up from an extensive basal erosion surface, and prevalent planar- or large-scale cross-stratification, strongly suggest laterally migrating subtidal channel bars adjacent to curved tidal channels. These may be comparable to modern estuarine channel bars (Oomkens and Terwindt, 1960; Land and Hoyt, 1966; Visher and Howard, 1974; Howard and others, 1975; Greer, 1975; Barwis, 1978; Nio and others, 1981) or tidal inlet deposits (deRaaf and Boersma, 1971; Kumar and Sanders, 1974, 1975; Hubbard and others, 1979; Hayes, 1980). Because there is a continuum between these two types of deposits, they are difficult to distinguish (Terwindt, 1971; deRaaf and Boersma, 1971; Barwis, 1978); however, associated strata and presence or absence of fluvial input should be distinctive.

Modern tidal inlets are associated with barrier beaches, back barrier lagoons, tidal flats and washovers (Kumar, 1973; Kumar and Sanders, 1974; Davies and others, 1971; Bridges, 1976; Heward, 1981). Although the deposits from these subenvironments have varying degrees of preservability in transgressive or regressive sequences, the absence of facies indicative of washovers or beaches mitigates against the inlet origin. Furthermore, tidal inlet deposits are expected to overlie either lagoonal or marine shelf deposits, or other inlet deposits showing the same sense of lateral accretion (e.g., Donselaar and Nio, 1982). The dominance of westward- (offshore) directed paleocurrents in these units, and opposing lateral-accretion directions (N and S), suggests, instead, superimposed estuarine channel bars with marginal intertidal flats and appreciable fluvial input. The grain-size difference between the lower and limestone units can be explained by variation in local hydraulics and sediment availability (e.g., Howard and others, 1975). Currents attained their maximum velocities and turbulence intensities during deposition of the lime-grainstone, judging by the relatively coarse grain size and lack of mud. In fact, the unusual composition of this facies is directly related to the availability of large grains, which were probably reworked from other associated facies. The absence of mud and bioturbation reflects the persistent mobility of the substrate in a subtidal setting.

Bedform geometry in estuaries is complex; however, channel deposits are dominated by stratification formed by migrating dunes and sandwaves (Land and Hoyt, 1966; Klein, 1970; deRaaf and Boersma, 1971; Boothroyd and Hubbard, 1975; Nio and others, 1981j). Strong time-velocity tidal asymmetry is also well known in estuaries (Price, 1963; Postma, 1967; Klein, 1970; Greer, 1975; Barwis, 1978; Hayes, 1980; Boersma and Terwindt, 1981). Here the topographically higher parts of bars are commonly flood-dominated while lower parts are ebb-dominated. The dominance of ebb-oriented structures in the basal and limestone units suggests estuarine deposition with fluvial input strengthening the ebb flow. Despite the strong ebb dominance, evidence of tidal flood flows is present in both large-scale cross-sets, sandwaves and heterolithic bedding. The multioriented ripple marks between sandwave crests on the top surface of the carbonate unit indicate subaerial exposure, with water ponded in troughs being acted upon by wind (cf. Land and Hoyt, 1966; Elliott and Gardiner, 1981; and others).

Facies above the limestone unit suggest an upward transition away from sedimentation dominated by unidirectional and tidal currents to that dominated by wave currents. A coastal embayment with marginal tidal flats (possibly interdistributary) is proposed. Heterolithic bedding immediately above the limestone is suggestive of intertidal flats. It is also similar to that documented in Mississippi delta embayments where sand is deposited during times of river flood or storm waves, and mud is deposited from suspension during waning flows (Moore and Scruton, 1957; Coleman and others, 1964). Intermittent subaerial exposure may be reflected by wrinkle marks and other "dewatering" structures. Extensive bioturbation occurs when sedimentation rate is low, and fine plant debris is common along bedding planes in fine-grained sediments. The carbonate concretions may be analogous to the siderite concretions found in modern coastal marshes and bays (Ho and Coleman, 1969; Pye, 1981; Postma, 1982), and their alveolar structure is associated with plant debris (i.e., rhizocretions; Klappa, 1980). Commonly, crevasse channels and splays cutting into fine-grained deposits show large-scale cross-strata grading into small-scale cross-stratified sandstone, as seen in the lower parts of the upper unit (e.g., Coleman and others, 1964; Allen, 1965, 1970; Arndorfer, 1973; Elliott, 1974, 1978). The limestone-filled channel may be such a crevasse channel fill, or a tidal channel cutting through intertidal flat deposits.

The presence of hummocky cross-stratification, planar-stratification, wave-ripple lamination and heterolithic bedding with symmetrical ripple marks in the uppermost part of the upper unit, and the general lack of biogenic structures, suggests rapid deposition in a wave-dominated area. Hummocky cross-stratification is generally thought to be formed by deposition under waves between fair weather and storm-wave base, and is most commonly associated with a shoreface-shelf environment (Goldring and Bridges, 1973; Harms and others, 1975, 1982; Hamblin and Walker, 1979; Bourgeois, 1980; Brenchley and Newall, 1982; Mount, 1982; Hunter and Clifton, 1982; Dott and Bourgeois, 1982). However, it has also been associated with intertidal and estuarine deposits (Campbell and Oaks, 1973) and deltaic deposits (Dott and Bird, 1979). Maximum near-bed orbital velocities are possibly intermediate between those which form two-dimensional wave ripples and planar beds (Harms and others, 1982; Dott and Bourgeois, 1982). In this unit, planar stratification is probably formed by wave-induced currents, by virtue of its association with hummocky cross-stratification; however, deposition of some of these strata on unidirectional upper-stage plane beds cannot be ruled out. Wave-ripple laminated bedsets are formed at lower kinetic-energy levels. Tool and flute marks on the bases of bedsets persistently give a westerly directed flow. It appears that westerly-directed, unidirectional, channelised or sheetflood currents periodically caused erosion and introduction of sediment, which was normally acted upon by wave-induced currents during deposition. Each complete bedset therefore represents a major flood or storm event or season. Similar facies have been described by Dott and Bird (1979), Hamblin and Walker (1979), and Brenchley and Newall (1982). Hummocky cross-stratification has not been recognised in modern coastal bays (due to sampling problems?) although Fisk and others (1954) described wave reworking of crevasse sands in bays.

Paleoecology

The clearly transported and sorted fossil assemblages within the limestone and the "coquinites" are discussed separately from the essentially untransported shelly fossils and trace fossils. Within the "coquinites" of both upper and lower units, the lack of abrasion and commonly articulated crinoid stems suggests little bedload transport. Thus, although not a "natural" population, the fossil assemblage is considered indicative of the paleo-community in the general nearshore area. The limestone has a similar faunal composition, and hence derivation; however, the highly abraded biogenic fragments suggest more prolonged bedload transport within the depositional area (by tidal currents). The plant remains were, of course, land-derived. The "coquinite" fauna is similar to others found in Upper Devonian shoreline deposits of New York (Harrington, 1970, 1978; Sutton and others, 1970; Bowen and others, 1974; Thayer, 1974; McGhee, 1976; McGhee and Sutton, 1981), and most closely resembles the *Cypricardella* community of the Glen Aubrey Formation (Sonyea Group), interpreted by Bowen and others (1974) as dominantly immobile epifaunal suspension feeders living on relatively stable substrates in nearshore areas. However, the infaunal, mobile *Cypricardella* was interpreted as adapted to areas of unstable substrates (Bowen and others, 1974). The occurrence in an upper unit "coquinite" of the brachiopod *Cupularostrum* with plant debris only supports the idea that *Cupularostrum* may have lived attached to floating plants (Bowen and others, 1974).

Shelly invertebrates occurring outside the limestone and "coquinites" are dominated by the bivalves *Goniofera* and *Grammysia,* showing little evidence of transport. The low faunal diversity and dominance of bivalves over brachiopods within this assemblage is similar to other Upper Devonian nearshore communities in New York (see references above), and may reflect the eurytypic nature of these bivalves (Thayer, 1974), and thus their ability to adjust to environmental fluctuations in salinity, temperature, and sedimentation rate (Bretsky, 1969; Harrington, 1970; Sutton and others, 1970). The assemblage most closely resembles the *Grammysia* benthic community from the Smethport nearshore deposits of the Genesee Group (Harrington, 1970, 1978), which also has abundant vertical burrows attributable to infaunal suspension feeders. *Grammysia* was probably a semi-infaunal or epifaunal, immobile, suspension feeder (Bambach, 1971), whereas *Goniofera* was apparently an immobile, almost completely infaunal suspension feeder (Stanley, 1972). The rare occurrence in an upper-unit "coquinite" of unabraded *Grammysia* with cm.-long articulated crinoid stems reflects little transport, and supports the presence of locally dense crinoid colonies within the *Grammysia* benthic community (Harrington, 1970, 1978).

The dominant trace fossil assemblage corresponds to Seilacher's (1967) shallow-water, *Skolithos (Glossifungites)* ichnofacies, but some traces in the upper unit belong to the *Cruziana* ichnofacies. Many of these trace fossils (e.g., *Arenicolites, Scalartubia*) have been interpreted as resulting from the activity of worms or worm-like organisms under rapidly fluctuating environmental conditions (Hantzschel, 1975; and others). The sandy substrate *Skolithos* and its firm muddy substrate counterpart *Rhizocorallium* (Seilacher, 1978) reflect suspension feeding in turbulent water (Siemers, 1970; Frey, 1978; Miller, 1979; Goldring and Langenstrassen, 1979; and others). *Diplocraterion* and *Arenicolites* similarly reflect turbulent water (Goldring, 1971; Fursich, 1975; and others). Periods of negligible water motion are indicated by other trace fossils. The *Bifungites* trace-making organism was probably a sedentary infaunal filter feeder living in a shallow, nearshore environment with low current (kinetic) energy and possibly brackish water (Gutschick and Lamborn, 1975). *Chondrites, Palaeophycus* and *Planolites,* interpreted as deposit feeders (Hantzschel, 1975, Pemberton and Risk, 1982b, Simpson, 1957), are also probably indicative of a "quiet" environment (Pemberton and Risk, 1982b). Similar trace fossil assemblages have been found by Pemberton and Risk (1982a, b) who suggest that the "quiet" water organisms are the "resident community" and that periodically deposited sand (during storms or floods, for instance) is bioturbated by opportunistic tube builders. Both the trace fossil and *in situ* invertebrate assemblages are similar to those occurring in modern interdistributary bays (Coleman and others, 1964) and intertidal flats; these assemblages are primarily mud-inhabiting molluscs and burrowing worms. The particularly intense bioturbation above the limestone unit, and the *Skolithos* beds, suggest local hiatuses in deposition within this sequence.

CONCLUSION

Sedimentological and paleontological evidence points strongly towards an estuarine-brackish coastal bay origin for the sedimentary sequence exposed at Ashcraft Quarry. The limestone and subjacent beds were deposited as estuarine channel bars, with intertidal flat deposits immediately overlying the limestone. The somewhat unusual occurrence of the lime-grainstone is due to deposition of relatively coarse grains (of which invertebrate shells were the major source) by currents of locally increased power. The uppermost beds appear to have been introduced by unidirectional freshwater currents and reworked by wave currents within a coastal embayment. The overall sequence could be explained by abandonment and subsidence of a tide-dominated delta distributary and subsequent establishment of an interdistributary bay. Such a marine transgression would be of local significance only. An alternative origin might be a progressive marine transgression over a nondeltaic estuarine coastal plain, even though the Upper Devonian sequence in this region is generally considered to be regressive. The importance of tidal currents is consistent with the relatively large tidal ranges expected in partly enclosed epeiric seas; therefore, it is surprising that reports of tidal facies from the Catskill clastic wedge are not more common. Clearly more detailed studies of Upper Devonian coastal facies are required throughout the region.

ACKNOWLEDGMENT

We wish to express our gratitude to Kirk Ashcraft for allowing us free access to the quarry.

REFERENCES CITED

Allen, J.R.L., 1965, Late Quaternary Niger delta and adjacent areas: sedimentary environments and lithofacies: American Association of Petroleum Geologists Bulletin, v. 44, p. 547–600.

——1970, Sediments of the Modern Niger delta, *in* Morgan, J. R., ed., Deltaic sedimentation, modern and ancient: Society of Economic Paleontologists and Mineralogists Special Publication 15, p. 138–151.

——1982, Mud drapes in sand-wave deposits: a physical model with application to the Folkestone Beds (early Cretaceous, southeast England): Philosophical Transactions of the Royal Society of London A, v. 306, p. 29–345.

——1982, Sedimentary Structures: their character and physical basis, v. 1 and 2: Amsterdam, Elsevier, 593 and 653 p.

Allen, J.R.L., and Friend, P. F., 1968, Deposition of the Catskill facies, Appalachian region: with notes on some other Old Red Sandstone basins, *in* Klein, G. deV., ed., Late Paleozoic and Mesozoic Continental Sedimentation, northeastern North America: Geological Society of America Special Publication 106, p. 21–74.

Allen, P. A., and Homewood, P., 1984, Evolution and mechanics of a Miocene tidal sandwave: Sedimentology, v. 31, p. 63–81.

Arndorfer, D. J., 1973, Discharge patterns in two crevasses in the Mississippi River delta: Marine Geology, v. 15, p. 269–287.

Ashley, G. M., Southard, J. B., and Boothroyd, J. C., 1982, Deposition of climbing-ripple beds: A flume simulation: Sedimentology, v. 29, p. 67–79.

Bambach, R. K., 1971, Adaptations in *Grammysia Obliqua:* Lethaia, v. 4, p. 169–183.

Barwis, J. H., 1978, Sedimentology of some South Carolina tidal-creek point bars and a comparison with their fluvial counterparts, *in* Miall, A. D., ed., Fluvial Sedimentology: Canadian Society of Petroleum Geologists Memoir 5, p. 129–160.

Barwis, J. H., and Makurath, J. H., 1978, Recognition of ancient tidal inlet sequences: an example from the Upper Silurian Keyser Limestone in Virginia: Sedimentology, v. 25, p. 61–82.

Boersma, J. R., and Terwindt, J.H.J., 1981, Neap-spring tide sequences of intertidal shoal deposits in a mesotidal estuary: Sedimentology, v. 28, p. 151–170.

Boothroyd, J. C., and Hubbard, D. K., 1975, Genesis of bedforms in mesotidal estuaries, *in* Cronin, L. E., ed., Estuarine research, volume II, geology and engineering: New York, Academic Press, p. 217–234.

Bourgeois, J., 1980, A transgressional shelf sequence exhibiting hummocky stratification: the Cape Sebastian Sandstone (Upper Cretaceous), Southwestern Oregon: Journal of Sedimentary Petrology, v. 50, p. 681–702.

Bowen, D. W., 1978, A depositional model for the Ashcraft quarry, [honors thesis]: Geneva, New York, Hobart and William Smith Colleges, 47 p.

Bowen, Z. P., Rhoads, D. C., and McAlester, A. L., 1974, Marine benthic communities in the Upper Devonian of New York: Lethaia, v. 2, p. 93–120.

Brenchley, P. J., and Newall, G., 1982, Storm-influenced inner-shelf sand lobes in the Caradoc (Ordovician) of Shropshire, England: Journal of Sedimentary Petrology, v. 52, no. 4, p. 1257–1270.

Bretsky, P., 1969, Central Appalachian Late Ordovician communities: Geological Society of America Bulletin, v. 46, p. 305–342.

Bridges, P. H., 1976, Lower Silurian transgressive barrier islands, southwest Wales: Sedimentology, v. 23, p. 347–362.

Campbell, C. V., and Oaks, R. Q., Jr., 1973, Estuarine sandstone filling tidal scours, Lower Cretaceous Fall River Formation, Wyoming: Journal of Sedimentary Petrology, v. 43, p. 765–778.

Coleman, J. M., Gagliano, S. M., and Webb, J. E., 1964, Minor sedimentary structures in a prograding distributary: Marine Geology, v. 1, p. 240–258.

Davies, D. K., Ethridge, F. G., and Berg, R. R., 1971, Recognition of barrier environments: American Association of Petroleum Geologists Bulletin, v. 55, p. 550–565.

Donselaar, M. E., and Nio, S. D., 1982, An Eocene tidal inlet/washover type barrier island complex in the south Pyrenean marginal basin, Spain: Geologie en Mijnbouw, v. 61, p. 343–354.

Dott, R. H., Jr., and Bird, K. J., 1979, Sand transport through channels across an Eocene shelf and slope in southwestern Oregon, U.S.A., *in* Doyle, L. J., and Pilkey, O. H., eds., Geology of continental slopes: Society of Economic Paleontologists and Mineralogists Special Publication 27, p. 327–342.

Dott, R. H., Jr., and Bourgeois, J., 1982, Hummocky Stratification: Significance of its variable bedding sequence: Geological Society of America Bulletin, v. 93, p. 663–680.

Elliott, T., 1974, Interdistributary bay sequences and their genesis: Sedimentology, v. 21, p. 611–622.

——1978, Deltas, *in* Reading, H. G., ed., Sedimentary environments and facies: New York, Elsevier, p. 97–142.

Elliott, T., and Gardiner, A. R., 1981, Tidal current bedforms in the Loughor Estuary, *in* Nio, S. D., and others, eds., Holocene marine sedimentation in the North Sea basin: International Association of Sedimentologists Special Publication No. 5, Oxford, Blackwell, p. 51–64.

Fisk, H. N., McFarlan, E., Jr., Kolb, C. R., and Wilbert, L. J., Jr., 1954, Sedimentary framework of the modern Mississippi delta: Journal of Sedimentary Petrology, v. 24, p. 76–99.

Frey, R. W., 1978, Behavioral and ecological implications of trace fossils, *in* Basan, P. B., ed., Trace fossil concepts: Society of Economic Paleontologists Short Course Number 2, Oklahoma City, p. 43–66.

Fursich, F. T., 1975, Trace fossils as environmental indicators in the Corallian of England and Normandy: Lethaia, v. 8, p. 151–172.

Goldring, R., 1971, Shallow-water sedimentation: Geological Society of London Memoir 5, 80 p.

Goldring, R., and Bridges, P. H., 1973, Sublittoral sheet sandstones: Journal of Sedimentary Petrology, v. 43, p. 736–747.

Goldring, R., and Langenstrassen, F., 1979, Open Shelf and near-shore clastic facies in the Devonian, *in* House, M. R., and others, eds., The Devonian System: Special Papers in Paleontology, no. 23, p. 81–97.

Greer, S. A., 1975, Estuaries of the Georgia coast, U.S.A.: Sedimentology and biology—Part III, sandbody geometry and sedimentary facies at the estuary-marine transition zone, Ossabaw Sound, Georgia: a stratigraphic model: Senckenbergiana Maritima, v. 7, p. 105–135.

Gutschick, R. C., and Lamborn, R., 1975, *Bifungites,* trace fossils from Devonian-Mississippian rocks of Pennsylvania and Montana, U.S.A.: Palaeogeography, Palaeoclimatology, Palaeoecology, v. 18, p. 193–212.

Hamblin, A. P., and Walker, R. G., 1979, Storm-deposited shallow marine deposits: the Fernie-Kootenay (Jurassic) transition, southern Rocky Mountains: Canadian Journal of Earth Science, v. 16, p. 1673–1690.

Hantzschel, W., 1975, Trace fossils and Problematica (second edition), *in* Teichert, C., ed., Treatise on invertebrate paleontology, Part W, Miscellanea, Supplement 1: Boulder, Colorado, Geological Society of America (and Univ. of Kansas Press), 269 p.

Harms, J. C., Southard, J. B., Spearing, D. R., and Walker, R. G., 1975, Depositional environments as interpreted from primary sedimentary structures and stratification sequences: Society of Economic Paleontologists and Mineralogists Short Course No. 2, Dallas, 161 p.

Harms, J. C., Southard, J. B., and Walker, R. G., 1982, Structures and Sequences in clastic rocks: Society of Economic Paleontologists and Mineralogists Short Course No. 9, Calgary, 249 p.

Harrington, J. W., 1970, Benthic communities of the Genesee Group (Upper Devonian): New York State Geological Association 42nd Annual Meeting, p. A1–A18.

——1978, Biostratigraphy and paleocology of the Upper Devonian Ithaca Formation near Cortland, N.Y.: New York State Geological Association 50th Annual Meeting, Guidebook, p. 344–353.

Hawley, N., 1981, Flume experiments on the origin of flaser bedding: Sedimentology, v. 28, p. 699–712.

——1982, Reply to discussion on flume experiments on the origin of flaser bedding: Sedimentology, v. 29, p. 906–907.

Hayes, M. O., 1980, General morphology and sediment patterns in tidal inlets: Sedimentary Geology, v. 26, p. 139–156.

Heward, A. P., 1981, A review of wave-dominated clastic shoreline deposits: Earth Science Reviews, v. 17, p. 223–276.

Ho, C., and Coleman, J. M., 1969, Consolidation and cementation of recent sediments in the Atchafalaya basin: Geological Society of America Bulletin, v. 80, p. 183–192.

Howard, J. D., Elders, C. A., and Heinbokel, J. F., 1975, Estuaries of the Georgia Coast, U.S.A. Sedimentology and biology—Part V, Animal-sediment relationships in estuarine point bar deposits, Ugeechee River-Ossabauw Sound, Georgia: Senckenbergiana Maritima, v. 7, p. 181–203.

Hubbard, D. K., Oertel, G., and Nummedal, D., 1979, The role of waves and tidal currents in the development of tidal-inlet sedimentary structures and sand body geometry: examples from North Carolina, South Carolina, and Georgia: Journal of Sedimentary Petrology, v. 49, p. 1073–1092.

Hunter, R. E., and Clifton, H. E., 1982, Cyclic deposits and hummocky cross-stratification of probable storm origin in Upper Cretaceous rocks of the Cape Sebastian area, Southwestern Oregon: Journal of Sedimentary Petrology, v. 52, p. 127–143.

Johnson, K. G., and Friedman, G. M., 1969, The Tully clastic correlatives (Upper Devonian) of New York State: a model for recognition of alluvial, dune (?), tidal, nearshore (bar and lagoon), and offshore sedimentary environments, in a tectonic delta complex: Journal of Sedimentary Petrology, v. 39, p. 451–485.

Jopling, A. V., and Walker, R. G., 1968, Morphology and origin of ripple-drift cross-lamination, with examples from the Pleistocene of Massachusetts: Journal of Sedimentary Petrology, v. 40, p. 1095–1127.

Klappa, C. F., 1980, Rhizoliths in terrestrial carbonates: classification, recognition, genesis and significance: Sedimentology, v. 27, p. 613–629.

Klein, G. deV., 1970, Depositional and dispersal dynamics of intertidal sand bars: Journal of Sedimentary Petrology, v. 40, p. 1095–1127.

Krajewski, S. A., and Cuffey, R. J., 1976a, The Ashcraft-quarry limestone mound, an Upper Devonian shell-bank in northeastern Pennsylvania: Geological Society of America Abstracts with Programs, v. 8, p. 212.

——1976b, The Ashcraft limestone mound: Pennsylvania Geology, v. 7, p. 9–12.

Kumar, N., 1973, Modern and ancient barrier sediments: new interpretation based on stratal sequence in inlet-filling sands and on recognition of the nearshore storm deposits: Annals of The New York Academy of Sciences, v. 220, p. 245–340.

Kumar, N., and Sanders, J. E., 1974, Inlet sequence: A vertical succession of sedimentary structures and textures created by the lateral migration of tidal inlets: Sedimentology, v. 21, p. 491–532.

——1975, Inlet sequence formed by the migration of Fire Island Inlet, Long Island, New York, *in* Ginsburg, R. N., ed., Tidal deposits: New York, Springer-Verlag, p. 75–85.

Land, L. S., and Hoyt, J. H., 1966, Sedimentation in a meandering estuary: Sedimentology, v. 6, p. 191–207.

McCave, I. N., 1970, Deposition of fine-grained suspended sediments from tidal currents: Journal of Geophysical Research, v. 75, p. 4151–4159.

——1971, Mud layers and deposition from tidal currents: discussion of "Tidal origin of a PreCambrian quartzite—the lower fine-grained quartzite (Middle Dalradian) of Islay, Scotland": Journal of Sedimentary Petrology, v. 41, p. 1147–1148.

McGhee, G. R., Jr., and Sutton, R. G., 1981, Late Devonian marine ecology and zoogeography of the central Appalachians and New York: Lethaia, v. 14, p. 27–43.

Miller, M. F., 1979, Paleoenvironmental distribution of trace fossils in the Catskill deltaic complex, New York State: Palaeogeography, Palaeclimatology, Palaeoecology, v. 28, p. 117–141.

Moore, D. G., and Scruton, P. C., 1957, Minor internal structures of some recent unconsolidated sediments: American Association of Petroleum Geologists, v. 41, p. 2723–2751.

Mount, J., 1982, Storm-surge-ebb origin of hummocky cross-stratified units of the Andrews Mountain Member, Campito Formation (Lower Cambrian), White-Inyo Mountains, eastern California: Journal of Sedimentary Petrology, v. 52, p. 941–958.

Nio, S. D., Shuttenhelm, R.T.E., and Weering, Tj, C. E. van, 1981, Holocene marine sedimentation in the North Sea basin: International Association of Sedimentologists Special Publication No. 5: Blackwell, Oxford, 515 p.

Nio, S. D., Siegenthaler, C., and Yang, C. S., 1983, Megaripple cross-bedding as a tool for the reconstruction of the palaeo-hydraulics in a Holocene subtidal environment, S. W. Netherlands: Geologie en Mijnbouw, v. 62, p. 499–510.

Oomkens, E., and Terwindt, J.H.J., 1960, Inshore estuarine sediments in the Haringvliet (Netherlands): Geologie en Mijnbouw, v. 39, p. 701–710.

Pemberton, S. G., and Risk, M. J., 1982a, Opportunistic ichnofossils from the Silurian Thorold Sandstone of Southern Ontario: 17th International Sedimentological Congress, Hamilton, Ontario, Canada, Abstracts with Programs.

——1982b, Middle Silurian trace fossils in the Hamilton, Ontario region: their identification, abundance, and significance: Northeastern Geology, v. 2, p. 98–104.

Postma, D., 1982, Pyrite and siderite formation in brackish and freshwater swamp sediments: American Journal of Science, v. 282, p. 1151–1183.

Postma, H., 1967, Rates of sediment accumulation in modern estuaries, *in* Lauff, G. H., ed., Estuaries: Publication No. 83, American Association for the Advancement of Science, p. 158–179.

Price, W. A., 1963, Patterns of flow and channelling in tidal inlets: Journal of Sedimentary Petrology, v. 33, p. 279–290.

Pye, K., 1981, Marshrock formed by iron sulphide and siderite cementation in saltmarsh sediments: Nature, v. 294, p. 650–652.

Raaf, J.F.M. de, and Boersma, J. R., 1971, Tidal deposits and their sedimentary structures: Geologie en Mijnbouw, v. 50, p. 479–504.

Reineck, H. E., and Singh, I. B., 1973, Depositional sedimentary environments: Berlin, Springer-Verlag, 439 p.

Reineck, H. E., and Wunderlich, F., 1968, Classification and origin of flaser and lenticular bedding: Sedimentology, v. 11, p. 99–104.

Rickard, L. V., 1975, Correlation of the Silurian and Devonian rocks of New York State: New York State Museum and Science Service, Map and chart series no. 24, 16 p.

Seilacher, A., 1967, Bathymetry of trace fossils: Marine Geology, v. 5, p. 413–428.

—— 1978, Use of trace fossil assemblages for recognizing depositional environments, *in* Basan, P. B., ed., Trace fossil concepts: Society of Economic Paleontologists short course number 2, Oklahoma City, p. 167–181.

—— 1982, Distinctive features of sandy tempestites, *in* Einsele, G., and Seilacher, A., eds., Cyclic and Event Stratification: Springer-Verlag, p. 333–349.

Sevon, W. D., 1981, The Middle and Upper Devonian clastic wedge in northeastern Pennsylvania: New York State Geological Association, 53rd Annual Meeting Guidebook, p. 31–65.

Siemers, C. T., 1970, Facies distribution of trace fossils in a deltaic environmental complex: upper part of the Dakota Formation (Upper Cretaceous), central Kansas: Geological Society of America Abstracts with Programs, v. 2, p. 683–684.

Simpson, S., 1957, On the trace fossil *Chondrites:* Geological Society of London Quarterly Journal, v. 112, p. 478–496.

Stanley, S. M., 1972, Functional morphology and evolution of byssally attached bivalve mollusks: Journal of Paleontology, v. 46, p. 165–212.

Sutton, R. G., Bowen, Z. P., and McAlester, A. L., 1970, Marine Shelf environments of the Upper Devonian Sonyea Group of New York: Geological Society of America Bulletin, v. 81, p. 2975–2992.

Thayer, C. W., 1974, Marine paleoecology in the Upper Devonian of New York: Lethaia, v. 7, p. 121–155.

Terwindt, J.H.J., 1971, Lithofacies of inshore estuarine and tidal inlet deposits: Geologie en Mijnbouw, v. 3, p. 515–526.

—— 1981, Origin and sequences of sedimentary structures in inshore mesotidal deposits of the North Sea, *in* Nio, S. D. and others, eds., Holocene marine sedimentation in the North Sea basin: International Association of Sedimentologists Special Publication No. 5, Oxford, Blackwell, p. 51–64.

Terwindt, J.H.J., and Breusers, H.N.C., 1982, Discussion: Flume experiments on the origin of flaser bedding: Sedimentology, v. 29, p. 903–907.

Van Den Berg, J. H., 1982, Migration of large-scale bedforms and preservation of crossbedded sets in highly accretional parts of tidal channels in the Oosterschelde, S.W. Netherlands: Geologie en Mijnbouw, v. 61, p. 253–263.

Visher, G. S., and Howard, J. D., 1974, Dynamic relationship between hydraulics and sedimentation in the Altamaha Estuary: Journal of Sedimentary Petrology, v. 44, p. 502–521.

Visser, M. J., 1980, Neap-spring cycles reflected in Holocene sub-tidal large-scale bedform deposits: a preliminary note: Geology, v. 8, p. 543–546.

Walker, R. G., and Harms, J. C., 1971, The Catskill delta: a prograding muddy shoreline in central Pennsylvania: Journal of Geology, v. 79, p. 381–399.

Woodrow, D. L., Enos, P., Bottjer, R. J., and Minero, C., 1981, Bioclastic carbonate units in the Catskill clastic wedge: New York State Geological Association, 53rd Annual Meeting, Guidebook, p. 201–225.

Woodrow, D. L., and Isley, A. M., 1983, Facies, topography, and sedimentary processes in the Catskill Sea [Devonian], New York and Pennsylvania: Geological Society of American Bulletin, v. 94, p. 459–470.

MANUSCRIPT ACCEPTED BY THE SOCIETY JULY 1984

Printed in U.S.A.

Geological Society of America
Special Paper 201
1985

Quantitative interpretation of ancient river systems in the Oneonta Formation, Catskill Magnafacies

John S. Bridge
Elizabeth A. Gordon
Department of Geological Sciences
State University of New York at Binghamton
Binghamton, New York 13901

ABSTRACT

The Oneonta Formation (Catskill Magnafacies) in south-central New York is composed of two lithofacies associations: (1) Medium- to very fine-grained, cross- and planar-stratified sandstone bodies, with bedsets (lithofacies) arranged into one or (usually) more erosively based storeys. Individual storeys generally have upward-fining lithofacies, also lateral-accretion bedding and channel fills: (2) Interbedded mudstones and erosively based sandstones with a diversity of primary sedimentary structures, calcareous concretions, plant remains and trace fossils. Upward-fining bedsets (lithofacies) are sheet-like or channel-filling, and are arranged in meter-scale lithofacies sequences.

The sandstone bodies are interpreted as deposits of laterally migrating and aggrading single-channel (perennial) rivers. Vegetated point-bar tops were subjected to both sheet floods and chute-channel formation. Minor low-flow deposition occurred on bar surfaces. Quantitative reconstruction of bankfull channel geometry and hydraulics gives channel widths, mean depths and slopes of approximately 60 m., 2.5 m., and 10^{-4} respectively: low reconstructed bend sinuosities (1.1 to 1.2) are supported by paleocurrent data, and help to explain the dominantly coarse-grained channel fills associated with chute cut-off. Sinuosity changes during lateral migration are documented quantitatively.

The sandstone-mudstone lithofacies association is interpreted as overbank flood deposits, on levees, crevasse channels and splays, and flood basins. Plant and faunal activity, and soil-forming processes, were abundant. Periodic channel-belt diversions (avulsions) caused the meter-scale lithofacies sequences in this association, also the cyclicity of the two lithofacies associations. If avulsion frequency averaged about once per 10^3 years, floodplain deposition rates are estimated at about 2×10^{-3} m/year.

INTRODUCTION

The Upper Devonian Catskill Magnafacies is generally accepted as alluvial in origin (e.g., Shepps, 1963; Allen and Friend, 1968; Johnson and Friedman, 1969; Woodrow and others, 1973; Rickard, 1975; Sevon, 1981); however, much more refined description and interpretation of these important rocks are needed before they are fully understood. This is now possible as a result of the recent increase in our knowledge of alluvial sedimentary processes. The information presented here is part of a major study which aims to: (1) describe in detail the Catskill Magnafacies in New York state and interpret it by close comparison with modern sedimentary facies; (2) quantitatively reconstruct paleochannel geometry and hydraulics, deposition rates, tectonic setting, and the nature of large-scale channel diversions; and (3) explain how and why these environmental parameters changed in space and time during the Upper Devonian regression. Such analysis should contribute to a more thorough understanding of the evolution of ancient alluvium, and towards the quantitative link between depositional environment and lithofacies (see Bridge and Diemer, 1983, for example).

We studied four large roadside outcrops in south-central

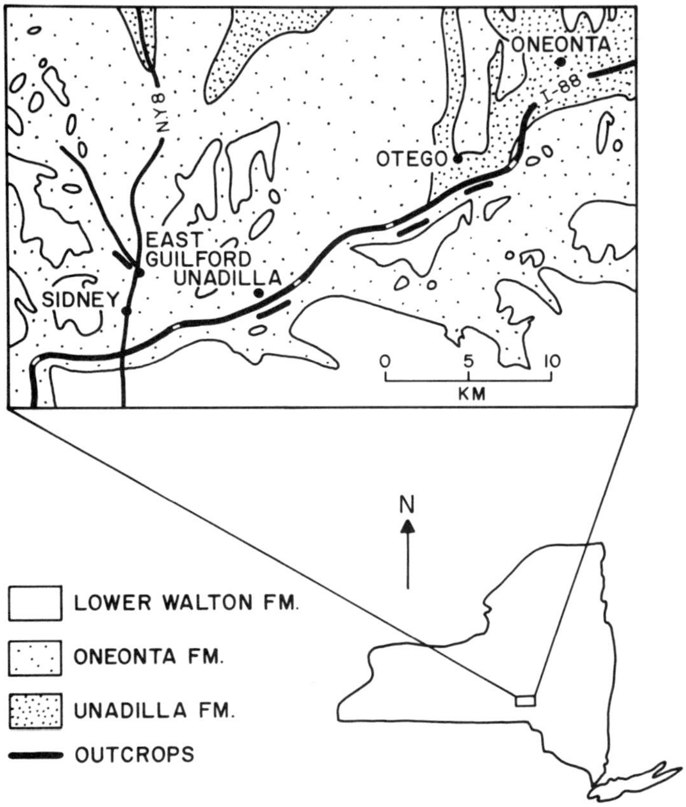

Figure 1. Location map of study area.

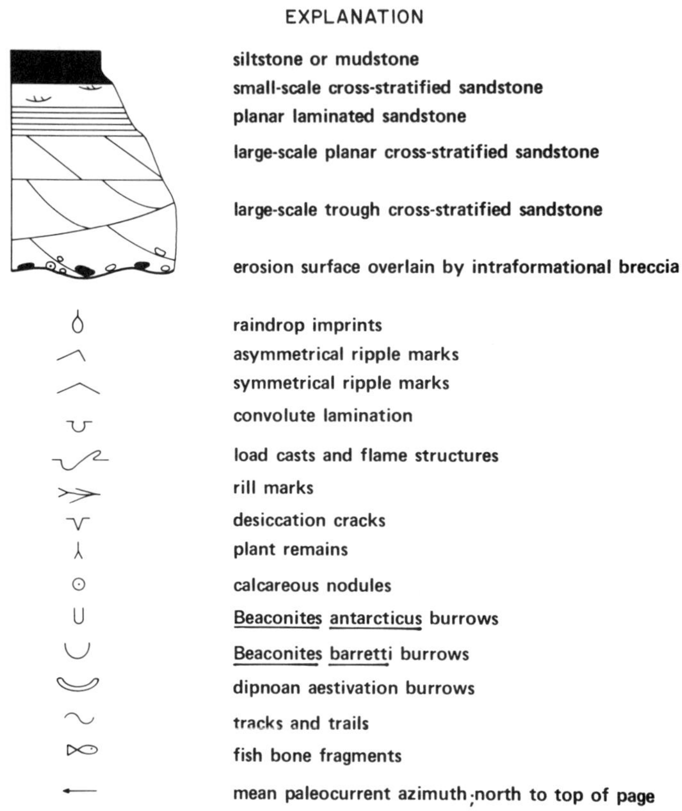

Legend for Figures 2, 3, 4, 5, 6.

New York (Fig. 1): near East Guilford, Unadilla, east of Unadilla and Otego. All exposures are of the Oneonta Formation (Fletcher, 1963; Rickard, 1975). The postulated Frasnian age of this formation is supported by miospore data (Alfred Traverse, personal communication); however, more detailed information is not yet available. The Oneonta Formation in this area has been previously interpreted as deposits of a near-coastal alluvial plain traversed by meandering rivers (Woodrow and Fletcher, 1967; Johnson and Friedman, 1969; Bridge and Dingman, 1981), although Dunne (1980) and Dunne and Boothroyd (1981) suggested a distal-fan braided stream origin for these same outcrops. The rocks can be described conveniently in terms of two lithofacies associations: (1) a sandstone-dominant lithofacies association, each occurrence of which is referred to as a sandstone body (Fig. 2); (2) an association of interbedded sandstone, siltstone, and mudstone lithofacies (Fig. 3). Scale diagrams with sedimentological logs of three of the outcrops are shown in Figure 4, 5, and 6. The "Unadilla" outcrop is detailed in Bridge and Dingman (1981).

DESCRIPTION OF THE SANDSTONE BODIES

Texture and composition

The sandstone bodies consist mainly of medium- to very fine-grained sandstones, ranging in color from gray to grayish-red. Intraformational breccias with fine- to medium-grained sandstone matrix are common locally, in most cases overlying erosion surfaces. The gravel-sized fragments (up to decimeters across) are of angular to subrounded sandstone, siltstone and mudstone; but abraded calcrete nodules, carbonized or pyritized plant remains, and fish bones are also common. Discontinuous layers of gray and green siltstone occur locally, particularly in the uppermost parts of sandstone bodies and in channel fills.

Primary sedimentary structures

Large-scale, trough cross-stratification is the dominant internal structure; planar cross-sets are rare. Cross-sets are typically between 0.2 and 0.8 m. thick. Planar-lamination and parting lineation are common and transitional to cross-strata in places. Small-scale planar- and trough cross-stratification occurs in the finest grain sizes. Straight and sinuous-crested asymmetrical ripples are preserved locally, as are some straight-crested symmetrical ripples. Flutes and gutter casts occur on the erosion surfaces overlying mudstone. In one instance, rills were developed in a siltstone layer near the top of a sandstone body, flowing normal to mean paleocurrent direction and down a local depositional slope (Fig. 7B). Raindrop imprints and desiccation cracks occur rarely in siltstone layers near the tops of sandstone bodies.

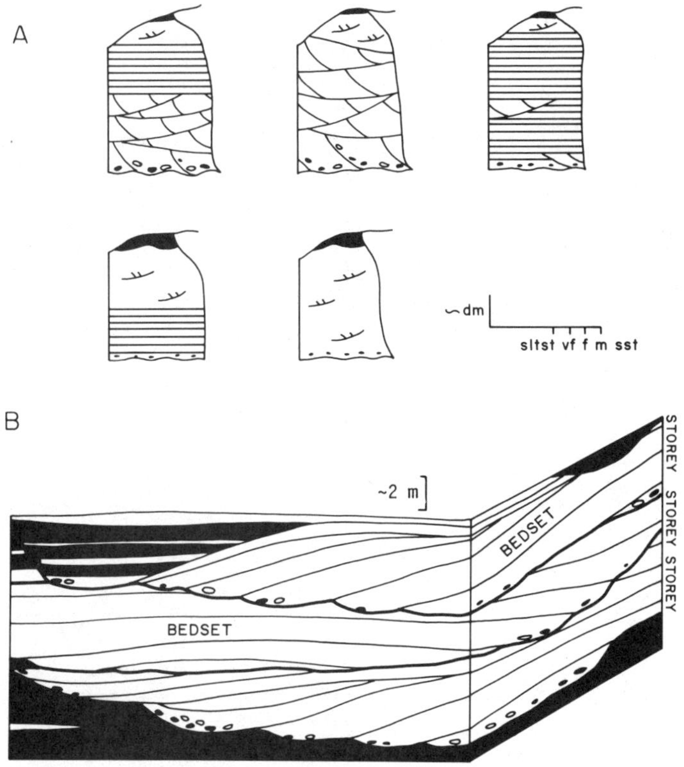

Figure 2. (A) Typical lithofacies (bedset) types in sandstone bodies (excluding fine-grained channel fills). See Legend. Bedset and cross-set thicknesses are approximate; burrows and plant casts not shown. (B) Schematic large-scale lithofacies geometry. Major erosion surfaces (thick lines) underlie each storey; minor erosion surfaces (thin lines) underlie each bedset.

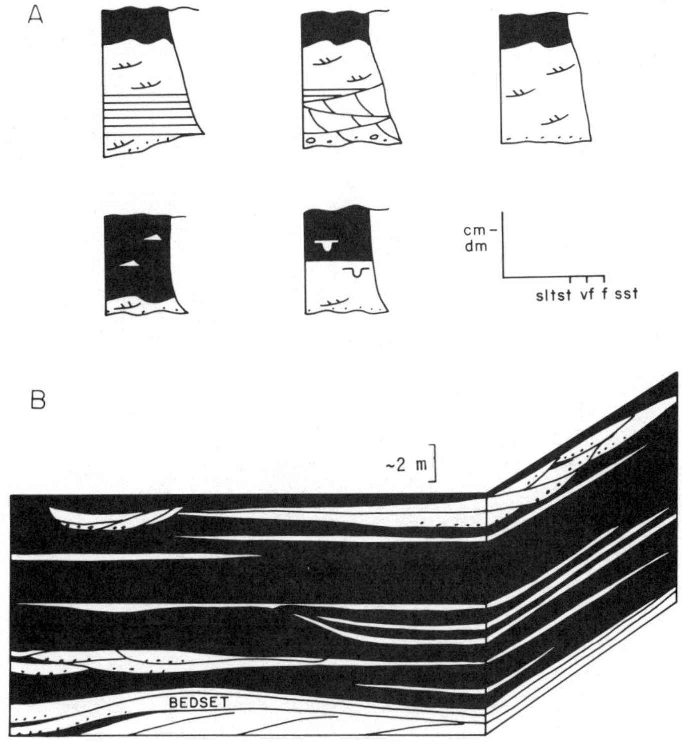

Figure 3. (A) Typical lithofacies (bedset) types in sandstone-mudstone lithofacies association. See Legend. Bedset and cross-set thicknesses are approximate; burrows and plant casts not shown. (B) Schematic large-scale lithofacies geometry, showing transition from underlying sandstone body.

Biogenic structures

Siltstone casts of *in situ* and transported plant roots and stems, about 1-3 cm. across and decimeters in length, are abundant in the topmost parts (a meter or so) of each sandstone body (and in the overlying facies association). *Beaconites antarcticus* and *Beaconites barretti* burrows (Gevers and others, 1971; Bradshaw, 1981) also occur in the topmost parts of the sandstone bodies (Fig. 8A-C). The larger *B. barretti* burrows are similar to those found in comparable rocks in Britain (Allen and Williams, 1981), Ireland (Bridge and others, 1980; Graham and Pollard, 1982), and Antarctica (Gevers and others, 1971; Bradshaw, 1981), although extensive horizontal and inclined tubes are not found. The European burrows are assigned to *Beaconites antarcticus,* but Bradshaw (1981) distinguishes such large burrow types as *Beaconites barretti.* Detailed description of these trace fossil occurrences is in preparation.

Lithofacies and their organization

Lithofacies are defined in this study as bedsets with distinctive textures and sedimentary structures (Fig. 2). Bedsets are commonly decimeters thick, but vary in thickness laterally. Each bedset is underlain by a relatively minor erosion surface and variably developed intraformational breccia. In most cases, mean grain size fines upward and sedimentary structures also vary vertically (Fig. 2).

Lithofacies (bedsets) are grouped together into "storeys." Storeys are up to 6 m. thick, and are underlain by major erosion surfaces (extending laterally for hundreds of meters and with local relief up to a meter) with locally developed intraformational breccia (Figs. 4, 5, 6). In sections roughly normal to local paleocurrent direction, the bedset boundaries (minor erosion surfaces) within storeys are inclined up to 10° from the major basal erosion surface, and extend from the base to the top of the storey. In places, this "lateral-accretion bedding" increases in inclination laterally, particularly adjacent to channel fills (e.g., Figs. 4, 9D), and also "climbs" in association with a rising basal erosion surface (Figs. 4, 5). Sections roughly parallel to local paleocurrent direction show bedset boundaries roughly parallel to the basal erosion surface, but small inclinations may also be present (topmost storey of the East Guilford outcrop, Figs. 6 and 9B). Each sandstone body is normally composed of more than one storey (i.e., multistoreyed).

Individual storeys generally show upward-fining of mean grain size and a decrease upward in scale of cross-stratification;

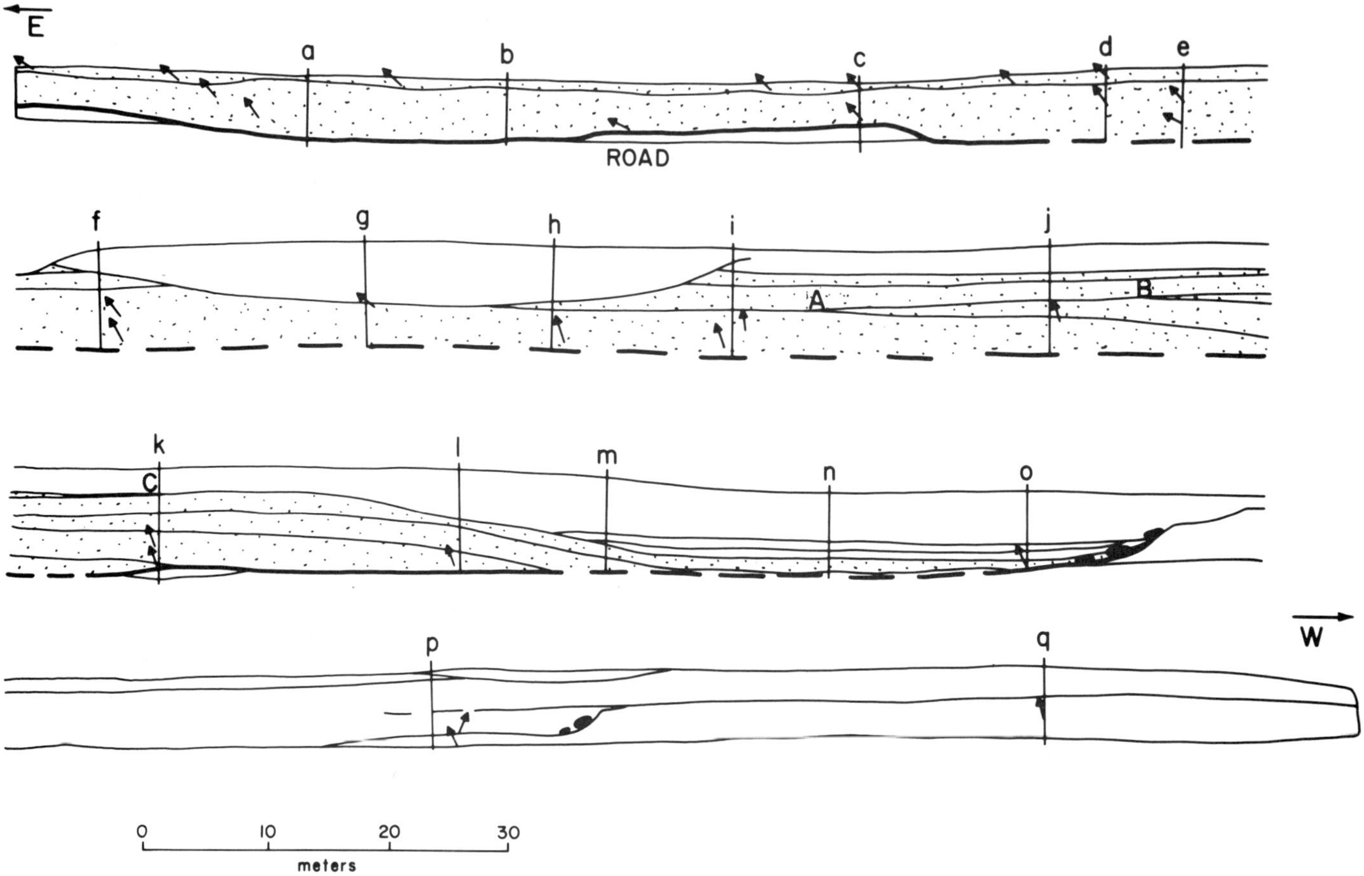

Figure 4. (A) Scale diagram of the Otego outcrop showing major bedding features and lithofacies associations. The diagram was constructed from a photomosaic, and because of distortion in the photographs, the vertical scale is inaccurate towards the top of the diagram. True vertical thicknesses are shown in the detailed measured sections (B), the positions of which are marked by letters. Stippled areas represent the sandstone bodies, with some important interbedded shales unstippled. Major erosion surfaces are marked by heavy lines (dashed where approximate); important minor erosion surfaces are marked by thinner lines. Paleocurrent azimuths (north to top of page) are shown by arrows. The azimuth of the outcrop is N 100° E.

lithofacies dominated by planar-lamination can occur virtually anywhere in such a vertical sequence. However, coarsening-upward sequences, and those showing little vertical change in grain size and sedimentary structure, also occur (Figs. 4, 5, 6). Storeys may in fact change laterally from upward-coarsening to upward-fining (Fig. 4). Where lateral-accretion bedding occurs, an inclined bedset that extends from the base to the top of a storey will necessarily vary in lithofacies laterally along its length if the storey shows an overall upward-fining or coarsening.

Some storeys terminate laterally in channel fills. Fine-grained channel fills show steep cut-banks, along which rest large siltstone blocks. The channel-filling bedsets are horizontally bedded and show overlapping disconformity with the adjacent inclined lateral-accretion surfaces and cut-banks (Figs. 4 and 9D); the bedsets are similar in lithofacies to the other finer-grained lithofacies association. Coarse-grained channel fills also show steep cut-banks overlain in places by coarse breccias; however, these margins are more complex than the fine-grained channel fills (Figs. 5 and 9E). Coarse-grained channel fills have concave-upward bedset boundaries, and may even show lateral-accretion bedding (Bridge and Dingman, 1981). The lithofacies are generally similar to those in the adjacent sandstone storeys, but are dominated by the fine-grained types. Finally, the topmost parts of some storeys are marked by smaller channels, and sheets of sandstone that truncate the tops of lateral accretion bedding (Figs. 4, 6, and 9F; Bridge and Dingman, 1981).

Paleocurrents

Mean paleocurrent directions were obtained from large-scale cross-stratification and parting lineations on planar laminae. Within any storey, paleocurrents are unidirectional and show little variation either vertically (generally less than 15°) or laterally (generally less than 40°). More paleocurrent variability occurs near the top of storeys, associated mainly with small-scale cross-strata, ripple marks, and isolated large-scale planar cross-

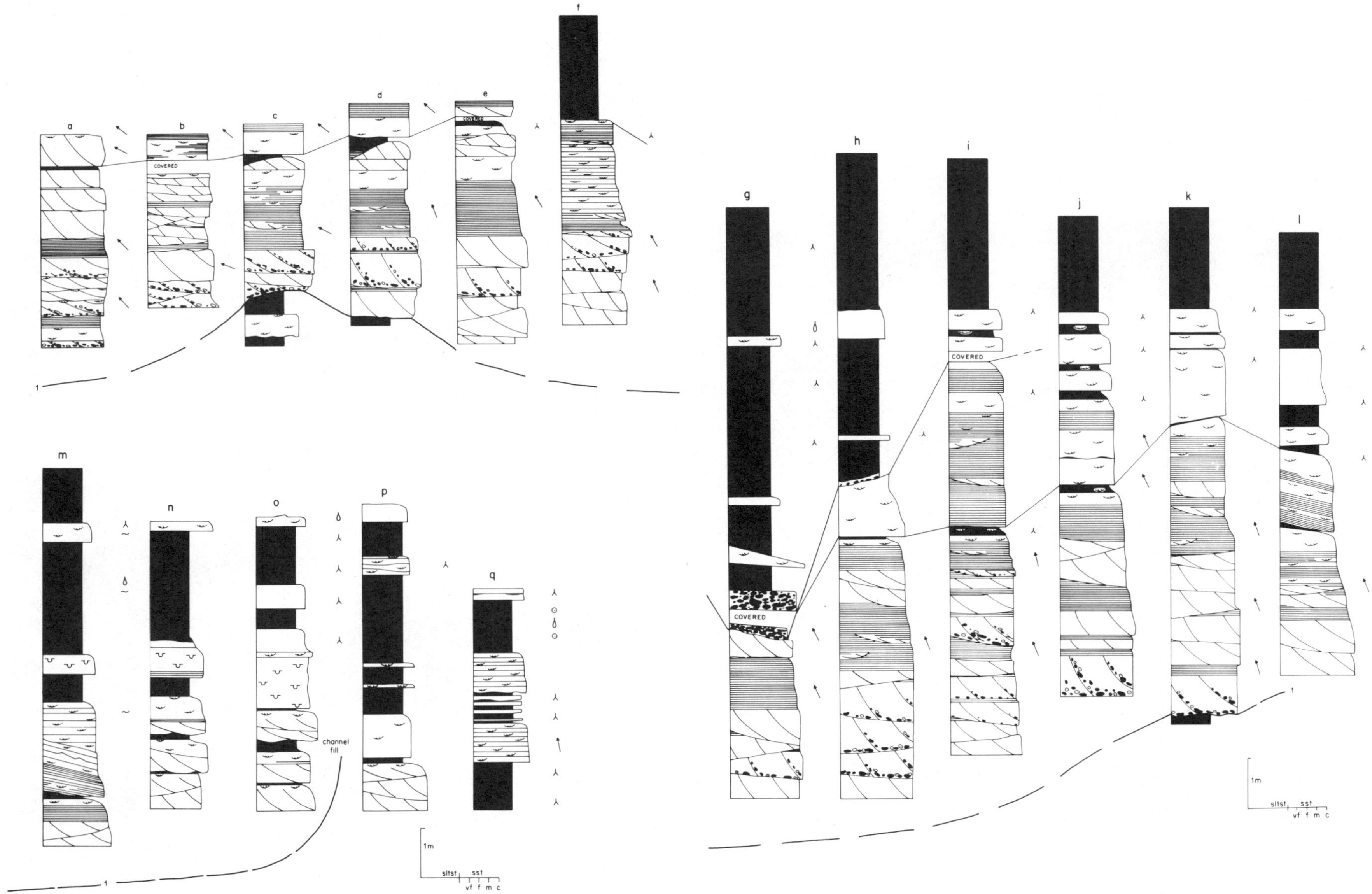

Figure 4. (B) Detailed vertical sections of the Otego outcrop. Paleocurrent azimuths (north to top of page) and additional sedimentary properties are shown to right of each graphic log. See Legend for explanation of symbols.

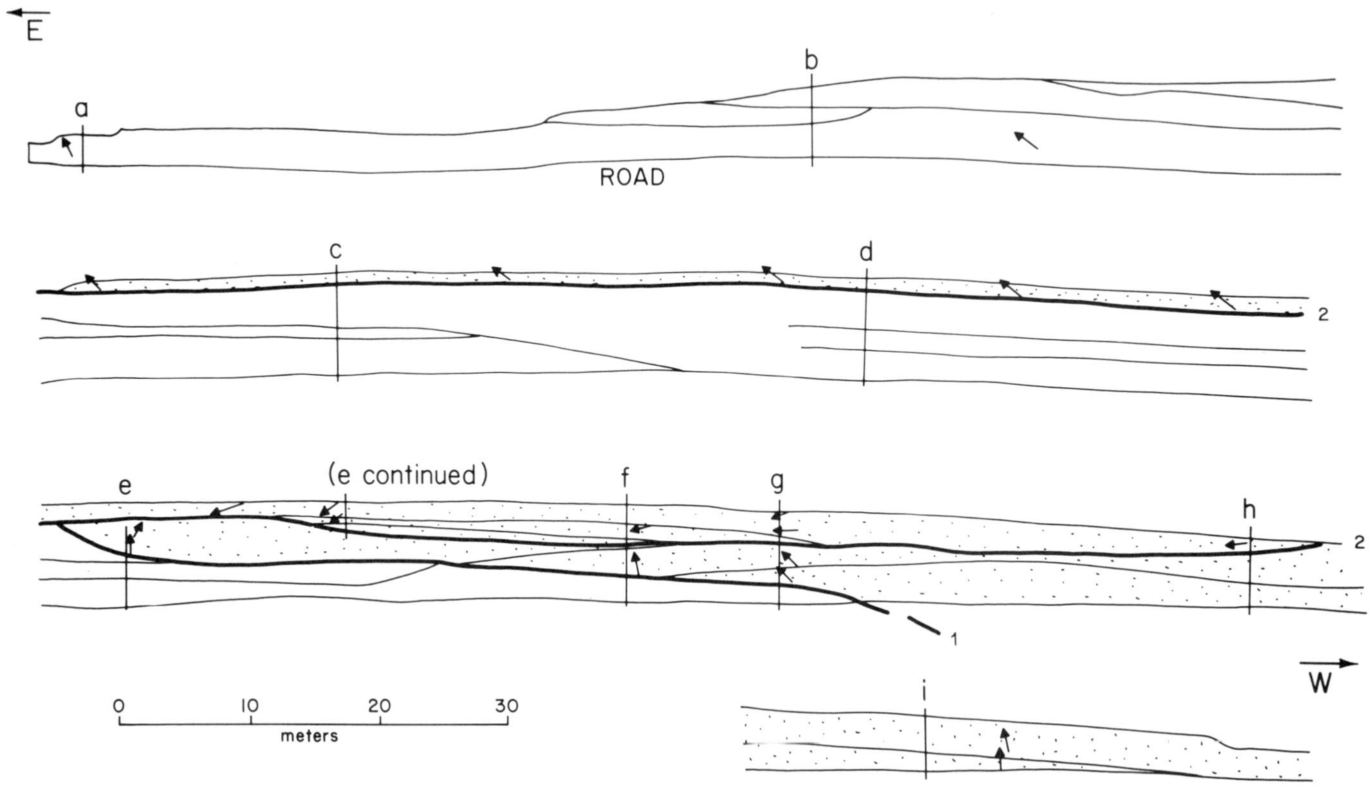

Figure 5A. Scale diagram of the East Unadilla outcrop. The outcrop azimuth is N 80° E. See Figure 4 caption for explanation. Numbers refer to basal erosion surfaces of individual storeys.

strata. Mean paleocurrent directions normally vary between storeys, mostly within an 80° range (N 260° E to N 340° E.)

Separate storeys in a sandstone body are distinguished most easily using the differing mean paleocurrent azimuths and/or changing orientation of bedset boundaries. However, if adjacent storeys have very similar mean paleocurrent directions, it may be difficult to distinguish the storeys, particularly in sections parallel to paleocurrent direction where bedset boundaries are roughly parallel, and not necessarily distinct from the major basal erosion surface.

DESCRIPTION OF THE SANDSTONE-MUDSTONE LITHOFACIES ASSOCIATION

Texture and composition

Interbedded siltstone, mudstone, and fine- to very fine-grained sandstones comprise this lithofacies association. Intraformational breccia is locally common along erosion surfaces. The siltstones and mudstones are mainly grayish-red, but also gray-green, and gray where plant remains are abundant. Sandstone colors vary from grayish-red to gray-green to gray. Yellow-brown calcareous concretions commonly occur in well-defined layers, associated with *in situ* plant roots. They are irregular and globular, up to 5 cm. across, and similar to Type A of Allen (1974).

Primary sedimentary structures

Large-scale, trough cross-stratification, planar-lamination, and small-scale cross-stratification commonly occur in the sandstones. Large-scale cross-strata, typically in 0.05 to 0.2 m. thick sets, only occur in the thicker sandstone bedsets. Small-scale cross-strata are both planar and trough-type, and are dominated by climbing varieties (Type A of Jopling and Walker, 1968). Symmetrical and asymmetrical ripple forms are preserved on the tops of some sandstone and siltstone beds. In some instances, symmetrical ripples are superimposed on the troughs of asymmetrical ripples (Fig. 7C). Convolute lamination is common in sandstones interbedded with siltstone in fine-grained channel fills. Load casts and flow-parallel furrows (centimeters deep and wide) are locally associated with sandstone lying erosively above siltstone and mudstone. Desiccation cracks are ubiquitous, and raindrop imprints can rarely be seen in mudstone capping sandstone (Fig. 7A, D).

Biogenic structures

Primary sedimentary structures are commonly disrupted by plant roots and burrows. Drifted and *in situ* plant fragments, siltstone casts of stems and roots, are common, particularly in the siltstones immediately overlying major sandstone bodies. Silt-

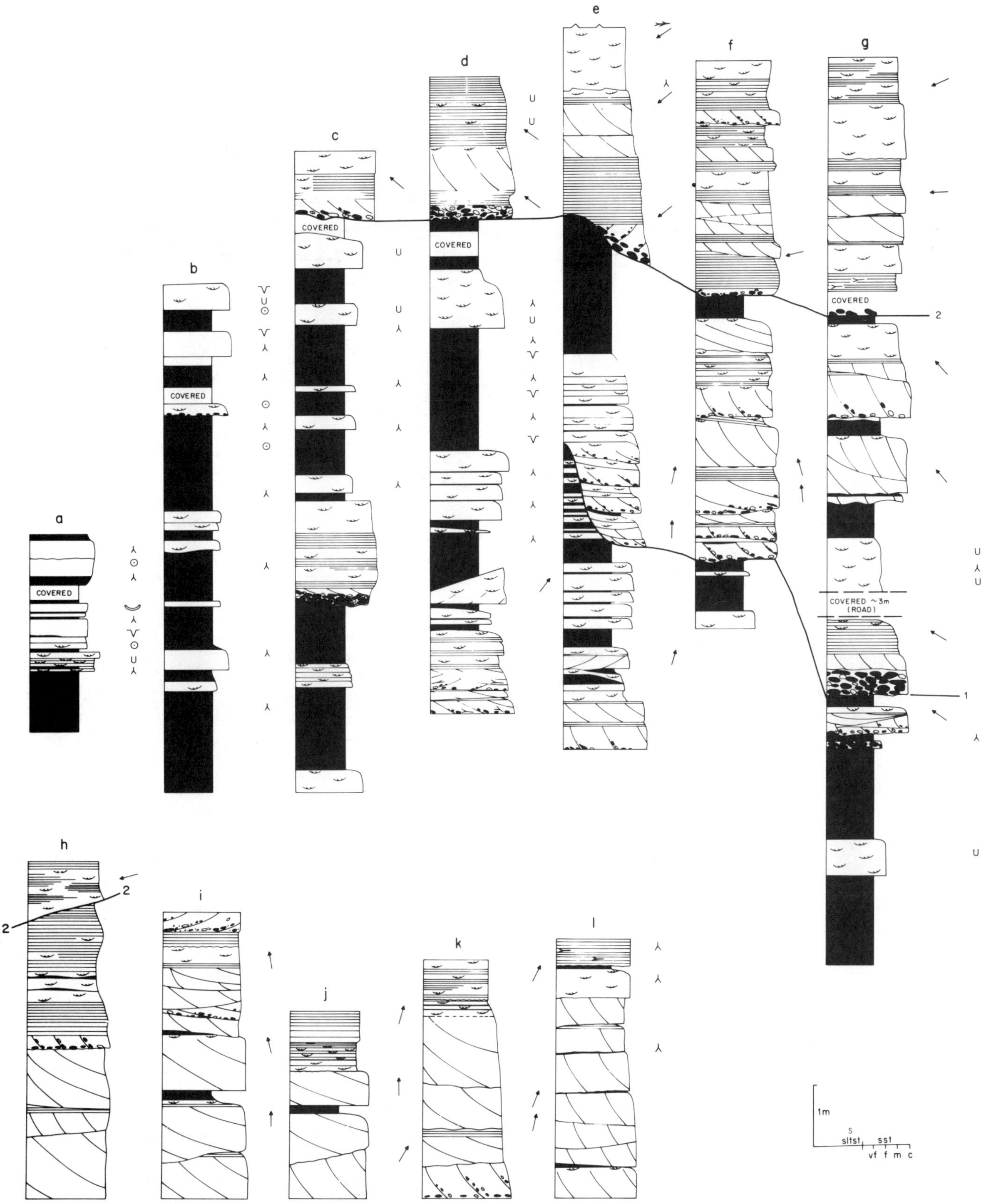

Figure 5B. Detailed vertical sections of the East Unadilla outcrop. Sections j, k, and l are from the east end of the outcrop, not shown in (A). The lower part of section g is from a lower roadcut also not shown in (A).

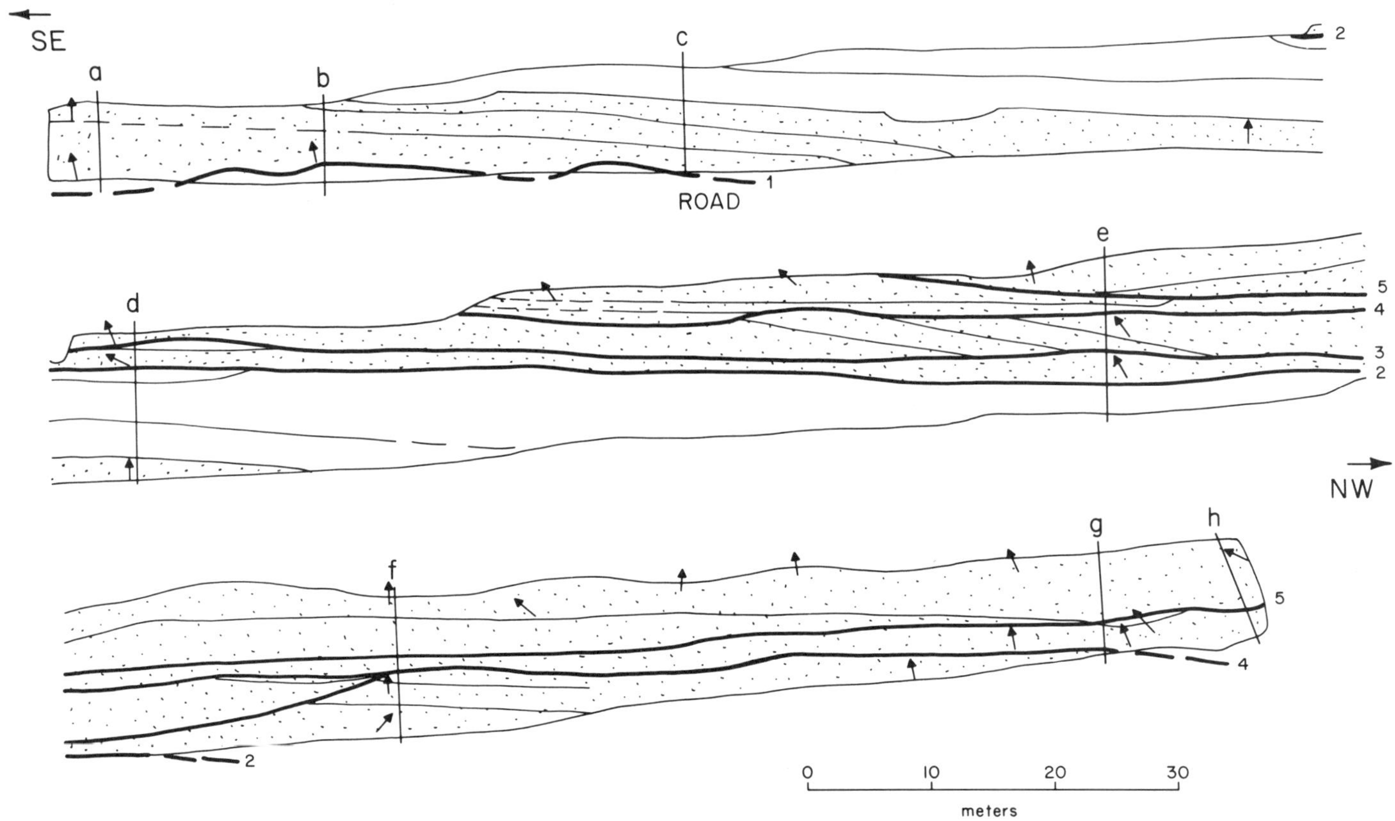

Figure 6A. Scale diagram of the East Guilford outcrop. The outcrop azimuth is N 134° E. See Figure 4 caption for explanation. Numbers refer to basal erosion surfaces of individual stories.

stone casts and immediately surrounding rock may display a local color change to greenish-gray.

Two burrow and four surface trail types have been recognized in this lithofacies association (Fig. 8). Small burrows (*Beaconites antarcticus*) occur in lithofacies immediately overlying the major sandstone bodies. Larger, shallow, dish-shaped burrows occur rarely in silty sandstones. They are circular to oval in plan, with longest diameter 10 to 17 cm. and depths of 2 to 5 cm. (Fig. 8D). The fill is distinctly coarser than the surrounding rock, consisting of fine-grained sandstone with angular mudstone chips up to 1 cm. long. Meniscus structure is not visible. The burrows are always associated with beds containing desiccation cracks and superimposed ripple marks. These features effectively distinguish this large burrow-type from *Beaconites barretti*.

Three of the surface trail types (Fig. 8E, F, G) are identified as *Arthropodichnus (Diplichnites) gouldi* (Gevers et al., 1971; Bradshaw, 1981; an arthropod walking trail), a small juliform millipede trace (Rolfe, 1980, his fig. 5A), and a larger myriapod trace resembling *Scaphiostreptus seychellarum* (Rolfe, 1980, his fig. 5F).

Lithofacies and their organization

The lithofacies of this association (Fig. 3) are gradational from those occurring in the channel fills and topmost parts of the sandstone bodies. The bedsets, centimeters to decimeters thick, always fine upward accompanied by a systematic change in sedimentary structures. Bedset bases are erosional, with locally developed intraformational breccias.

Many bedsets thin and change in lithofacies laterally. In some instances, these wedges thin away from the channel margin of a major sandstone body (e.g., Unadilla and East Unadilla outcrops). Other bedsets persist for hundreds of meters as horizontal or slightly inclined sheets, with little lateral facies change (Fig. 9F). Extensive bioturbation typically destroys most of the primary sedimentary structures in these bedsets.

Broad, shallow channel fills may form an overlapping complex or grade laterally into sheet-like bedsets. The channels are roughly symmetrical in section with maximum depths measured in decimeters and widths up to about 10 m. (Fig. 9G). Rarely these channels are several meters deep and tens of meters wide (Fig. 5). Systematic upward decrease in grain size is common to these channel fills, and filling bedsets are roughly conformable to channel margins. Isolated, relatively coarse-grained channel fills also occur, in some cases with an asymmetrical cross-section and some lateral-accretion bedding (Bridge and Dingman, 1981).

Large-scale lithofacies sequences on a scale of meters are also apparent. Upward-fining sequences involve upward change to finer-grained and thinner bedsets, whereas the converse is true for coarsening-upward sequences (Figs. 5, 6, and 9G). Upward-

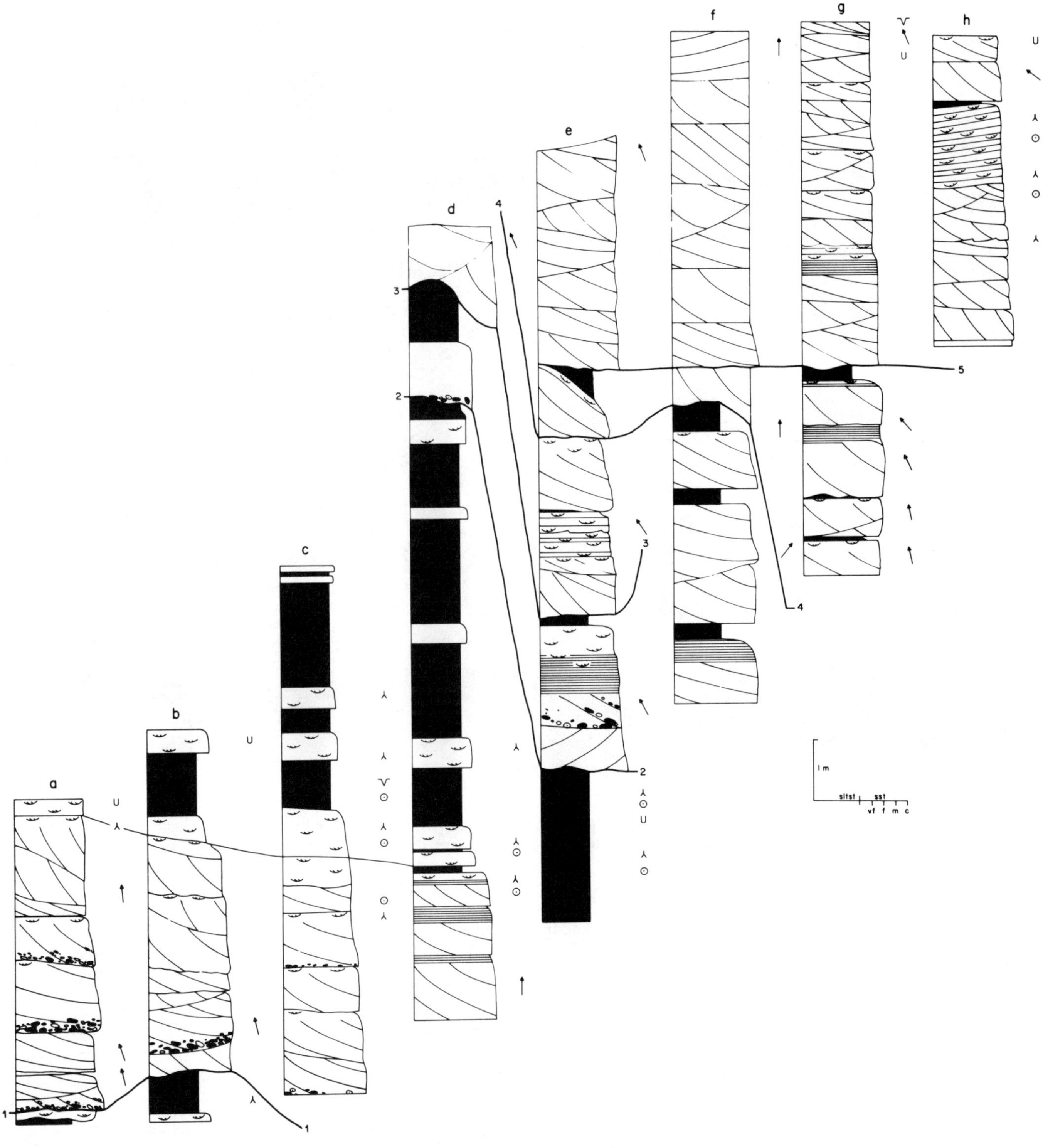

Figure 6B. Detailed vertical sections of the East Guilford outcrop.

Figure 7. (A) Desiccation cracks and symmetrical ripple marks. (B) Rill marks on siltstone draping local bedding surface. (C) Symmetrical ripple marks superimposed on troughs of asymmetrical ripple marks. (D) Raindrop imprints in siltstone draping asymmetrical ripple marks. Scale is 5 cm. long.

fining sequences are commonly developed immediately above major sandstone bodies, with calcrete horizons and abundant plant remains also concentrated in these locations.

Paleocurrents

Paleocurrents from large-scale cross-stratification, parting lineation, and channel axes indicate unidirectional flow, subparallel to paleocurrents from the sandstone bodies (Figs. 4, 5, and 6). Directional variation between bedsets of up to 30° is common, but small-scale cross-stratification and asymmetrical ripple marks give more variability. No systematic orientation of symmetrical ripple marks was observed.

SPATIAL DISTRIBUTION OF LITHOFACIES ASSOCIATIONS

In all outcrops studied, individual sandstone bodies are separated from each other by the interbedded sandstone-mudstone lithofacies association; therefore, connectedness ratios (Bridge and Leeder, 1979) must be less than about 0.1. The channel deposit density (relative proportion of sandstone bodies and associated channel fills) for each outcrop is shown in Table 1. The values range from about 40 percent to 70 percent; however, the higher values may well be overestimates due to limited exposure in road cuts which are mainly in sandstone. In fact, these channel deposit density values are remarkably close to Fletcher's (1963) values of sandstone/shale ratio for the Oneonta Formation.

QUALITATIVE INTERPRETATION OF SANDSTONE BODIES

The sandstone bodies are interpreted as deposits of aggrading, single-channel, sinuous rivers migrating laterally across alluvial plains. The laterally extensive basal erosion surfaces of each storey result from lateral migration of the deepest erosion-

Figure 8. (A) Bedding surface with *Beaconites barretti* burrows (scale is 5 cm.). (B) Cross-sectional view of *Beaconites barretti* burrow (scale is 5 cm.). (C) Bedding surface showing *Beaconites antarcticus* burrows (scale is 5 cm.). (D) Dipnoan aestivation (?) burrows (scale is 5 cm.) (E) Myriapod trail leading from small, shallow burrow; trace compares with *Scaphiostreptus seychellarum* (Rolfe, 1980, his fig. 5F; scale is 5 cm.). (F) Surface trail *Arthropodichnus (Diplichnites) gouldi* (Gevers et al., 1971; Bradshaw, 1981; an arthropod walking trail; cm. scales). (G) Small juliform millipede trace (Rolfe, 1980, his fig. 5A). (H) Unindentified trail.

A B C D

Figure 9 (this and facing page). (A) Upper sandstone body at East Guilford composed of four storeys. Note the inclined lateral-accretion surfaces in the second storey from the base. (B) Upper sandstone body at East Guilford. The uppermost storey exposes a section through part of a point bar parallel to flow direction. Note the lateral-accretion surfaces which rise from the basal erosion surface in the left part of the photograph, flatten out in the center of the picture, and dip back down to the basal erosion surface in the right part of the picture. (C) East end of the Otego outcrop showing abrupt thickening of the sandstone body, associated with transition from crossover to point bar. (D) Sandstone body at Otego which terminates in fine-grained channel fill. (E) Relatively coarse-grained channel fill in westernmost sandstone body at Unadilla. Note the distant channel margin cutting into the sandstone-mudstone lithofacies association. (F) Upper part of the lower sandstone body at East Guilford (north side of road) which grades into the sandstone/mudstone lithofacies association. Lateral-accretion surfaces within the sandstone body dip to the left in the photograph. Note the small channel form on the upper surface of the sandstone body. The sandstone-mudstone lithofacies association is dominated by laterally extensive sheet-like bedsets which grade upward into mudstone. These are interpreted as levee deposits. (G) Sandstone-mudstone lithofacies association beneath a sandstone body at Unadilla. Note the large-scale lithofacies sequence from fining-upward to coarsening-upward. Note channel-filling bedsets at base of outcrop and horizontal sheet-like bedsets in center of picture.

dominant areas of curved river channels, and the overlying breccias are formed from the material that slumped from the adjacent cut-banks. Overlying these basal erosion surfaces and lag breccias are the deposits of point bars that migrated laterally in association with the cut-banks, but also some channel fills. The inclined bedsets within storeys provide conclusive evidence for lateral accretion, each sandstone storey representing one point-bar deposit, or part of one. At the Otego and East Unadilla outcrops (Figs. 4 and 5), the consecutive lateral-accretion surfaces clearly "climb" upward indicating that the channel was aggrading as it migrated laterally; aggradation totalled 1.5 m. in the Otego outcrop.

The distribution of the lithofacies in each storey (point-bar deposit) has close parallels with the bar deposits of certain modern, single-channel, sinuous streams (McDowell, 1960; Frazier and Osanik, 1961; Harms and others, 1963; Davies, 1966; Sarkar and Basumallick, 1968; Bernard and others, 1970; Singh, 1972; Shelton and Noble, 1974; Schwartz, 1978). Each bedset lying between minor erosion surfaces was probably deposited during a single major flood event; the local decrease in grain size and change in internal structure in the uppermost parts of some of these bedsets record (minor) falling-stage deposition. The vertical and lateral changes in the lithofacies of bedsets in each storey record local spatial variation in the lithofacies deposited on the original point-bar surface, which is related mainly to flood-stage hydraulic conditions. Therefore, storeys that show an upward-

E

F

G

TABLE 1. ALLUVIAL ARCHITECTURE PARAMETERS AND NET AGGRADATION RATES

Outcrop	Otego	East Guilford	East Unadilla	Unadilla
Channel deposit density	0.64	0.69	0.58	0.51
Aggraded channel-belt thickness/ original thickness	≥1.3	2.5	2.5	≥1.3
Channel-belt aggradation rate, m/yr (from Bridge and Leeder, 1979, Fig. 5 with mean avulsion period ~1000 years)	0.004	0.003-0.004	0.004	0.007
Channel-belt aggradation rate, m/yr (from Bridge and Leeder, 1979, Fig. 9 with mean avulsion period ~1000 years)	0.0015	0.006	0.006	0.0015
Estimated average floodplain deposition rate, m/yr (see Bridge and Leeder, 1979, Figs. 2 and 4)	0.0015	0.0025	0.0025	0.0021

fining of successive bedsets were deposited in downstream parts of point bars, whereas those with little vertical grain-size variation, or coarsening-upward, were deposited in the upstream parts (Jackson, 1975, 1976; Bridge and Jarvis, 1976, 1982). The general dominance of upward-fining storeys suggests that lateral migration involved a downvalley component, resulting in preferential erosion of the upstream parts of point bars. Complete point-bar deposits should, however, be preserved after channel abandonment.

Lateral variation in the lithofacies and thickness of a given storey can be due to (1) stream-wise variability along a single point-bar deposit and/or (2) variability imparted by changing channel geometry and hydraulics as point-bar lateral accretion proceeded. The Otego outcrop (Figs. 4 and 9C, D) shows a storey that thickens to the west and changes from upward-coarsening to upward-fining in its lower parts. This is interpreted as a bend crossover deposit which passes westward into the upstream part of a point bar and then into the downstream part. This storey finally changes laterally into a fine-grained channel fill (analyzed in a later section). The topmost storey of the East Unadilla outcrop (Fig. 5) shows an increase in thickness (i.e., lowering of the basal erosion surface) and lithofacies change, possibly associated with the transition from a crossover to a point-bar deposit. At East Guilford (Figs. 6 and 9B), the topmost storey shows an unusual longitudinal section through a point bar. Bedsets and minor erosion surfaces rise from the basal erosion surface, flatten out where the local paleocurrent direction is roughly parallel to the outcrop face and then slope back to the basal erosion surface. Paleocurrent directions vary from about N 340° E at section e, parallel to the outcrop (N 314° E) at section f, to N 308° E at sections g and h. This section exposes only the bend apex region of the point bar (see later section). The lithofacies geometry of storeys in the Unadilla outcrop are interpreted in detail in Bridge and Dingman (1981) and Dingman (1982).

Most of the channel bars were apparently covered with sinuous-crested dunes (evidenced by large-scale, trough cross-strata) during flood-stage deposition and ripples were restricted mainly to shallow water near the inner bank. However, upper-stage plane beds were also common, particularly in the finer sands. The rare straight-crested dunes were apparently restricted to the upper parts of point bars (cf. chute bars), and their planar cross-strata deviated up to 15° from the local mean paleocurrent directions.

Falling-stage deposition is suggested by the decrease in grain size and changing sedimentary structures within some of the bedsets, particularly higher up in the storeys. At low stages, ripples replaced dunes or upper-stage plane beds on the lower bar surfaces, and suspended silt and clay were locally deposited on the upper bar surfaces. Subaerial exposure of the upper parts of point bars is recorded by desiccation cracks and raindrop imprints in the siltstone lenses, and the rill marks recording drainage off the bar into the adjacent channel. However, perennial river flow is suggested by the lack of the finest sediment sizes and *in situ* plant remains in the lower parts of point bars. The restriction of burrowing organisms and abundant plant remains to the upper point-bar deposits suggests land-dwelling animals that fed on vegetation and/or on herbivores.

The asymmetrical cross-sections of channel fills support a sinuous channel interpretation. The relatively fine-grained channel fills record relatively abrupt channel abandonment, by avulsion, so that the fill is deposited from slow-moving, overbank floodwaters. Relatively coarse-grained channel fills record gradual diversion of channel flow, such as during chute cut-off. Two adjacent coarse-grained channel fills in the westernmost sandstone body at the Unadilla outcrop show particularly complex channel behavior, involving chute cut-off, lateral migration and reoccupation of abandoned channels (Bridge and Dingman, 1981). Lateral-accretion bedding shows evidence of lateral migration during filling of these channels.

Relatively small channels which occur on the tops of some sandstone bodies, and having similar paleocurrent orientation to the underlying storey, are interpreted as chute channels. Spatially associated with these channels are sheets of sandstone that erosively truncate underlying lateral-accretion surfaces (e.g., Otego and Unadilla). Apparently, these are the result of sheet erosion and subsequent deposition on point-bar tops during major floods.

The multistorey nature of most of the sandstone bodies is expected when lateral channel migration within a channel belt is combined with net aggradation (Beerbower, 1964; Bluck, 1971; Bridge, 1975; Bridge and Leeder, 1979). Each storey with an eroded top is a single bar deposit that has been subsequently overridden by a different channel segment and associated bar. Channel belt abandonment (avulsion) will cause complete preservation of only the topmost point bar (storey). If the channels had any sinuosity, it is reasonable to expect differing mean paleocurrent directions from superimposed storeys. Such variation in paleocurrent directions within and between superimposed storeys can give an estimate of paleochannel sinuosity (discussed later).

INTERPRETATION OF SANDSTONE-MUDSTONE LITHOFACIES ASSOCIATION

Each bedset is interpreted as the deposit of a single flood on a floodplain. Lower bounding surfaces record some erosion of the floodplain surface prior to deposition. Large- and small-scale cross-stratification and planar-lamination record bedload deposition of sand moving as dunes, ripples, and upper-stage plane bed, respectively, but with appreciable suspended sediment concentrations in the floodwaters. The upward change in grain size (fining) and sedimentary structure within bedsets records waning overbank flow velocities, with the siltstone and mudstone representing suspended-load deposition only. Rare bedsets with upward-coarsening followed by fining result from deposition during rising and falling flood stages. Wind action on ponded water is evidenced by wave ripples, and the raindrop impressions and abundant desiccation cracks indicate subaerial exposure. The concretions may be the calcrete soil horizons common beneath floodplains in seasonally dry climates (Allen, 1974).

The thicknesses of these flood deposits are similar to those deposited on modern floodplains (summary in Bridge and Leeder, 1979). The coarser and thicker channel-filling lithofacies are similar to those in modern crevasse channels developed on splays (e.g., Kruit, 1955; Coleman, 1969; Singh, 1972). The isolated, relatively coarse-grained channel fills (e.g., Unadilla outcrop; Bridge and Dingman, 1981) are probably proximal crevasse channels. The lens- and wedge-shaped bedsets that thin and change lithofacies downcurrent are likely to be unchannelled crevasse-splay (or levee) deposits. The coarser-grained, sheet-like facies closely resemble levee deposits (e.g., Fisk, 1947; Singh, 1972, Ray, 1976), and in some cases (e.g., eastern part of Unadilla outcrop), the dip of the levee deposits away from the relatively elevated channel belt is apparent. The finest-grained bedsets are floodbasin deposits or fine-grained channel fills within sandstone bodies. Floodbasin bedsets are normally sheet-like in geometry; however, the fills of floodbasin drainage channels also occur.

The most abundant *in situ* and current-swept plant remains occur in lithofacies that are interpreted as upper point bar, levee, and crevasse splay: that is, channel-marginal, topographically elevated areas. Although plant remains commonly occur in floodbasin deposits, the roots are comparatively small. The restriction of the major flora to areas near channels may be due to a locally high water table near perennial streams in a semiarid climate and to enhanced preservation of their remains by rapid burial. In contrast, the dry hostile conditions in floodbasins would retard higher plant growth, and relatively slow deposition rates would not be conducive to preservation (Edwards, 1980).

The close association of calcretes with *in situ* plant roots suggests that concentration of carbonate in groundwater is associated with loss of water to the atmosphere through plants. The calcrete horizons (type A of Allen, 1974) are also associated with red, rubbly-weathering mudstone, which may increase in thickness away from contemporary channel margins (Retallack, personal communication). There is little doubt that these are calcareous paleosol horizons, and the implied seasonally dry climate is supported independently by palaeomagnetic and other evidence (Woodrow and others, 1973; Zeigler and others, 1979; Scotese and others, 1979; Heckel and Witzke, 1979).

The association of many ichnofossils with abundant plants on topographically high floodplain areas suggests the presence of a prolific ecosystem although not necessarily restricted to these areas. Arthropods were probably responsible for many of the traces, either through walking or by burrowing to feed on decaying plant debris and/or to avoid extreme temperatures (Rolfe, 1980; Bradshaw, 1981). The association of the large dish-shaped burrows with wave ripples and desiccation cracks (including on the top of the burrow) suggests an organism burrowing to escape the drying out of ponded water. They are similar to burrows interpreted elsewhere as dipnoan aestivation burrows (Woodrow, 1968).

The large-scale lithofacies sequences in this association reflect temporally changing environments on the floodplain, which may be caused by either (1) local crevasse-splay or levee progradation or abandonment while the channel belt remains essentially unchanged in position, or (2) environment and lithofacies changes due to channel-belt abandonment (avulsion) and relocation in a different area on the floodplain (Bridge, 1983). The scale of these sequences suggests that they reflect the avulsive behavior of the main channel belt (discussed below).

INTERPRETATION OF ALLUVIAL STRATIGRAPHY

The spatial organization of sandstone bodies within the fine-grained facies association can be explained by periodic avulsion of a channel belt while net aggradation proceeded (Allen, 1965, 1974, 1978, 1979; Bridge and Leeder, 1979). Table 1 shows estimated deposition rates using Bridge and Leeder's alluvial stratigraphy model. If avulsion periodicities of about 10^3 years and floodplain:channel-belt width ratios of about 10:1 are accepted as realistic estimates, the floodplain deposition rates (compacted) appear to be a few mm. per year, and appear not to vary appreciably between outcrops.

QUANTITATIVE INTERPRETATION OF PALEOCHANNEL GEOMETRY AND HYDRAULICS

Introduction

Paleochannel geometry and hydraulics were reconstructed by comparing each complete storey with a physical model that predicts the sedimentology of single point-bar deposits (Bridge, 1977, 1978, 1982). Given certain initial hydraulic and geometric conditions, the model predicts the local mean flow pattern, point-bar topography, mean grain size and bed configuration of bed-sediment at channel-forming (bankfull) discharge in bends of regular planform. As lateral deposition occurs mainly from bedload at channel-forming discharges, the model describes the nature of point-bar deposition as well as sediment transport. The model has been tested successfully against modern rivers (Bridge and Jarvis, 1982; Bridge, 1984) lending confidence to its application to interpreting ancient sediments.

Initial estimates of many of the input parameters for the model can be obtained from the deposits themselves. For instance, the geometry of lateral-accretion surfaces and channel fills, and lateral thickness variation in storeys can give an idea of the geometry of the paleochannels. Paleocurrent variability can be used to estimate channel sinuosity. Full details of initial parameter estimation are given in Bridge and Diemer (1983). Those hydraulic parameters that cannot be estimated from the exposed rocks are chosen by trial and error such that the model simulates the observed point-bar deposits as closely as possible. As successive simulations are compared with the observed deposits, the initial parameter estimates can be refined to improve the fit of theory to data. This technique differs significantly from those used

TABLE 2. INPUT PARAMETERS FOR MODEL SIMULATIONS

Location	Otego A	Otego B	Otego C	East Guilford Lower Sandstone Body	East Guilford Upper Sandstone Body	East Unadilla	Unadilla
Bankfull width, inner bank to talweg (m)	42.5	42	28	40	32-39	40-45	37.5
Estimated bankfull width (m)	64	63	42	60	48-58	60-68	56
Mean depth (m)	1.8	1.8	2.4	2.6	2.4-2.6	2.4-2.6	2.5
Centerline channel length in one bend (m)	540	540	540	750	730	750-800	628
Centerline radius of curvature at bend axis (m)	122	122	136	148	200-160	160-156	122
Centerline sinuosity	1.115	1.115	1.135	1.11	1.1-1.12	1.13-1.2	1.16
Mean Darcy-Weisbach-type friction coefficient	0.08	0.08	0.07	0.06	0.06	0.06	0.06
Dynamic friction coefficient	0.8	0.8	0.8	0.8	0.8	0.8	0.8
Mean friction slope	0.0001	0.000098	0.000085	0.000081	0.00011-0.000092	0.000086-0.000096	0.000095
Mean velocity ($m.s.^{-1}$)	0.37	0.37	0.39	0.45	0.57-0.55	0.53-0.65	0.45
Bankfull discharge ($m^3.s.^{-1}$)	43	43	39	70	66-83	76-115	45
Fluid density ($g.cm^{-3}$)	1.0						
Sediment density ($g.cm^{-3}$)	2.65						
Gravitational acceleration ($cm.s.^{-2}$)	981.0						

in most previous studies (e.g., Ethridge and Schumm, 1978; Bridge, 1978).

Results

Table 2 summarizes the reconstructed paleochannel geometry and hydraulics for the five point-bar deposits (complete storeys) that were simulated. The different storeys have similar paleochannel characteristics. Bankfull widths range from 42 to 68 m. and average 60 m.; mean depths range from 1.8 to 2.6 m., averaging about 2.5 m.; and sinuosities vary between 1.1 and 1.2. Mean velocities appear to average around 0.5 m./sec. and mean friction slopes are about 0.0001 (mean water surface and bed slopes may be up to twice this value [Bridge and Diemer, 1983]). The simulated channel geometries are also supported independently by the unsimulated storeys (that is, using complete upper storey thicknesses to estimate depths), channel fills, and mean paleocurrent variations. Within any sandstone body, the maximum mean paleocurrent variation is commonly between 80° and 120°. This range would be expected from channels with sinuosities of 1.1 to 1.3. This consistency in channel geometry and hydraulics suggests a stream (or streams) that did not change in character in time and space within this region. Variability in width, depth, and velocity can be explained easily by deposition at stages slightly different from bankfull, or different bend sinuosities, within one given stream.

Reconstructed bed topography, local mean velocity, and bedload mean grain size for several simulations are shown in Figure 10. Within any sandstone body, the preserved storeys all can be explained by stream-wise variation of the sedimentology of the simulated point bars. This supports the view that each sandstone body represents a channel belt deposited by a single, aggrading stream.

Reconstructions of upward-fining lateral-accretion surfaces at the Otego outcrop document changes in paleochannel geometry and hydraulics during lateral bend migration. Bankfull channel width decreased from about 63 m. to 42 m. while mean depth increased from 1.8 to 2.4 m.; mean transverse point-bar slope increased from 6° to 8°. These changes resulted from an increase in sinuosity during bend migration prior to avulsion. However, lateral changes in the geometry and sedimentology of lateral-accretion surfaces also occur when a point bar is viewed at different stream-wise positions. In this case, the lithofacies *and* paleocurrents should vary laterally, as seen in other storeys described here. There will also be lateral changes in lithofacies and geometry of bedsets as lateral-accretion deposits pass laterally into coarse-grained channel fills (e.g., at Unadilla; Bridge and Dingman, 1981).

CONCLUDING REMARKS

This study generally supports the prevalent view that the Oneonta Formation in this region was deposited by single, sinuous rivers flowing over a lowland alluvial plain. However, our detailed observations have uncovered important, previously undescribed, sedimentary features, resulting in a much more com-

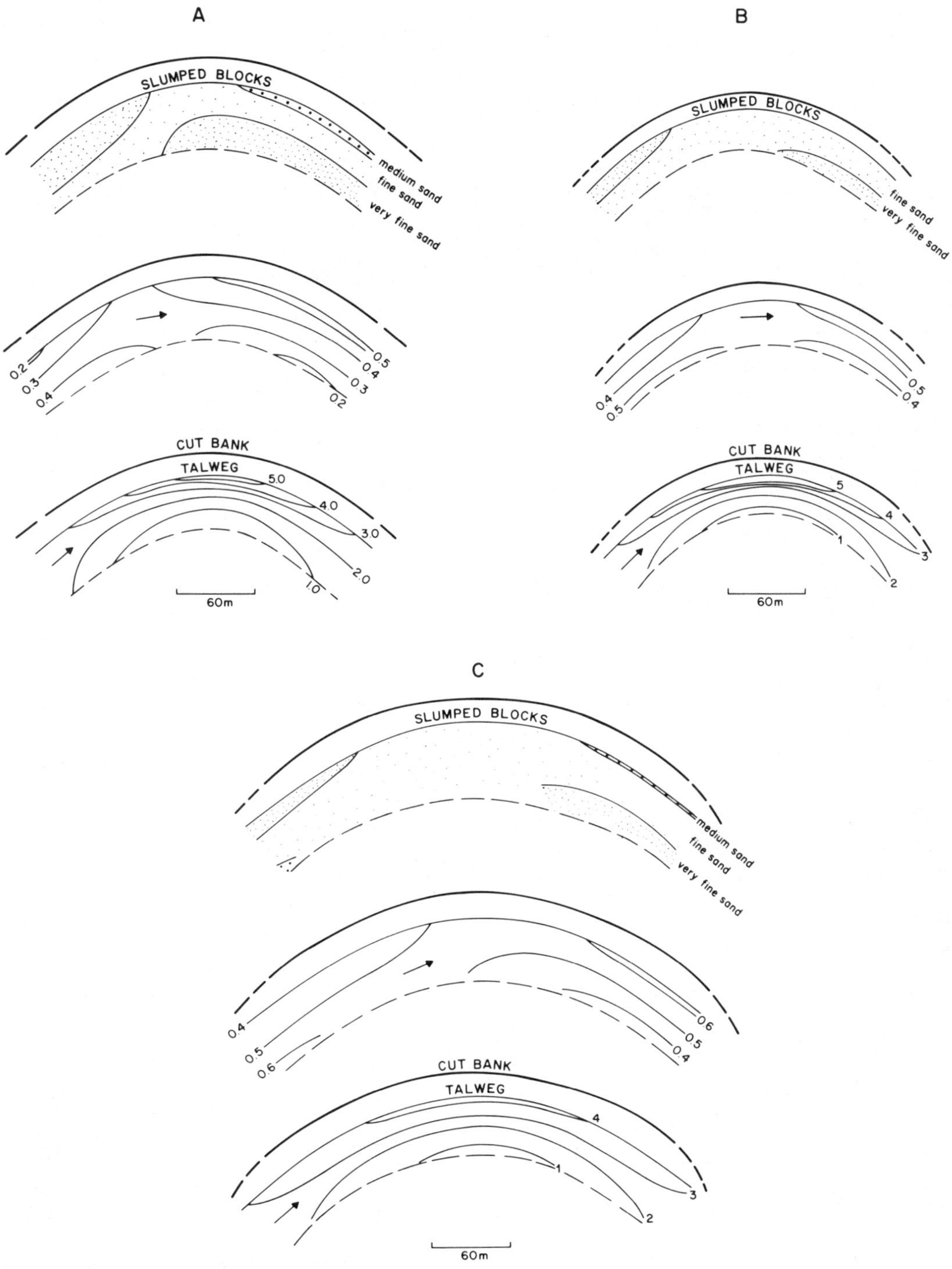

Figure 10. Reconstructions of bankfull channel geometries and hydraulics for Otego (A and B) and the lower sandstone body at East Guilford (C). Topmost maps show mean grain size of bedload sediment, middle maps show local mean flow velocity in $m \cdot s^{-1}$, and lower maps show bed elevation below bankfull level in m.

prehensive paleoenvironmental interpretation than possible hitherto. Similar features have been found in many other Catskill Magnafacies outcrops throughout the Catskill Mountains, but have not yet been formally documented. A channel sinuosity of 1.1 to 1.2 might be considered low (cf. Allen and Friend's [1968] Fig. 10), but it certainly does not preclude dominance of point bars and lateral channel migration. The presence of chute channels and chute cut-offs might impart somewhat of a braided appearance, lending partial credibility to Dunne and Bothroyd's (1981) reconstructions. The reconstructed channel slopes on the order of 10^{-4} are entirely consistent with a near-coastal plain; the lower-middle Frasnian shoreline was possibly less than 10 km. to the west of the study area (e.g., Rickard, 1975). It appears that the mean grain-size of the Catskill Magnafacies generally increases in coeval rocks to the east (up paleovalley) and in ascending through the sequence to the topmost Famennian (e.g., Fletcher, 1963; Allen and Friend, 1968). Explanation of these trends in terms of paleochannel and floodplain geometry and hydraulics, and ultimately in terms of extrabasinal controlling factors, must await further quantitative reconstructions which are underway.

ACKNOWLEDGMENTS

Mary Droser and Anne Gordon kindly assisted in the field, and David Tuttle generously helped with the photography. This research was supported by the Donors of the Petroleum Research Fund (administered by the American Chemical Society). We greatly appreciate comments on the initial manuscript by Bill Sevon, Don Woodrow, Mike Leeder, and Ed Cotter.

REFERENCES CITED

Allen, J.R.L., 1965, A review of the origin and characteristics of recent alluvial sediments: Sedimentology, v. 5, p. 89–191.

——1974, Studies in fluviatile sedimentation: implication of pedogenic carbonate units, Lower Old Red Sandstone, Anglo-Welsh outcrop: Geological Journal, v. 9, p. 181–208.

——1978, Studies in fluviatile sedimentation: an exploratory quantitative model for the architecture of avulsion-controlled alluvial suites: Sedimentary Geology, v. 21, p. 129–147.

——1979, Studies in fluviatile sedimentation: an elementary geometrical model for the connectedness of avulsion-related channel sand bodies: Sedimentary Geology, v. 24, p. 253–267.

Allen, J.R.L., and Friend, P. F., 1968, Deposition of the Catskill Facies, Appalachian Region: with notes on some other Old Red Sandstone Basins, *in* Klein, G. de V., ed., Late Paleozoic and Mesozoic continental sedimentation: Geological Society of America Special Paper 106, p. 21–73.

Allen, J.R.L., and Williams, B.P.J., 1981, *Beaconites antarcticus:* a giant channel-dwelling trace fossil from the Lower Old Red Sandstone of South Wales and the Welsh Borders: Geological Journal, v. 16, p. 255–269.

Beerbower, J. R., 1964, Cyclothems and cyclic depositional mechanisms in alluvial plain sedimentation: Bulletin of the Kansas Geological Survey, v. 169, p. 31–42.

Bernard, M. A., Major, C. F., Parrott, B. S., and LeBlanc, R. J., 1970, Recent sediments of southeast Texas: Bureau of Economic Geology, University of Texas at Austin, Guidebook 11.

Bluck, B. J., 1971, Sedimentation in the Meandering River Endrick: Scottish Journal of Geology, v. 7, p. 93–138.

Bradshaw, M. A., 1981, Paleoenvironmental interpretation and systematics of Devonian trace fossils from the Taylor Group (lower Beacon Supergroup), Antarctica: New Zealand Journal of Geology, v. 24, p. 615–652.

Bridge, J. S., 1975, Computer simulation of sedimentation in meandering streams: Sedimentology, v. 22, p. 3–43.

——1977, Flow, bed topography, grain size, and sedimentary structure in open channel bends: a three-dimensional model: Earth Surface Processes, v. 2, p. 401–416.

——1978, Palaeohydraulic interpretation using mathematical models of contemporary flow and sedimentation in meandering channels, *in* Miall, A. D., ed., Fluvial Sedimentology: Canadian Society of Petroleum Geologists Memoir 5, p. 723–742.

——1982, A revised mathematical model and FORTRAN IV program to predict flow, bed topography and grain size in open-channel bends: Computers and Geosciences, v. 8, p. 91–95.

——1983, Largescale facies sequences in alluvial overbank environments: Journal of Sedimentary Petrology, v. 54, p. 583–588.

——1984, Flow and sedimentary processes in river bends: comparison of field observations and theory *in* Elliott, M., ed., River Meandering American Society Civil Engineers, p. 857–872.

Bridge, J. S., and Diemer, J. A., 1983, Quantitative interpretation of an evolving ancient river system: Sedimentology, v. 30, p. 599–623.

Bridge, J. S., and Dingman, L. E., 1981, Depositional environment of the Oneonta Formation (Catskill Facies), near Unadilla, New York: New York State Geological Association Guidebook, 53rd Annual Meeting, Binghamton, p. 255–277.

Bridge, J. S., and Jarvis, J., 1976, Flow and sedimentary processes in the meandering River South Esk, Glen Clova, Scotland: Earth Surface Processes, v. 1, p. 303–336.

——1982, The dynamics of a river bend: a study of flow and sedimentary processes: Sedimentology, v. 29, p. 499–541.

Bridge, J. S., and Leeder, M. R., 1979, A simulation model of alluvial stratigraphy: Sedimentology, v. 26, p. 617–644.

Bridge, J. S., Van Veen, P. M., and Matten, L. C., 1980, Aspects of the sedimentology, palynology, and paleobotany of the Upper Devonian of southern Kerry Head, Co. Kerry, Ireland: Geological Journal, v. 15, p. 143–170.

Coleman, J. M., 1969, Brahmaputra River: channel processes and sedimentation: Sedimentary Geology, v. 3, p. 129–239.

Davies, D. K., 1966, Sedimentary structures and subfacies of a Mississippi River point bar: Journal of Geology, v. 74, p. 234–249.

Dingman, L. E., 1982, The sedimentology and paleohydraulics of the Upper Devonian Catskill Facies of the Unadilla-Hancock area, south central New York State: [unpublished Master's thesis], State University of New York at Binghamton, 99 p.

Dunne, L. A., 1980, Depositional environments of the Upper Devonian Oneonta Formation of south-central New York: [unpublished Master's thesis], University of Rhode Island, 104 p.

Dunne, L. A., and Boothroyd, J. C., 1981, The Upper Devonian Oneonta Formation, New York, and modern distal braided outwash streams: a comparison [abs.]: *in* Modern and Ancient Fluvial Systems: Sedimentology and Processes; University of Keele, U.K.

Edwards, D., 1980, Early land floras, *in* Panchen, A. L., ed., The Terrestrial Environment and the Origin of Land Vertebrates: Academic Press, London, p. 117–157.

Ethridge, F. G., and Schumm, S. A., 1978, Reconstructing paleochannel morphologic and flow characteristics: methodology, limitations, and assessment, *in* Miall, A. D., ed., Fluvial Sedimentology: Canadian Society of Petroleum Geologists Memoir 5, p. 703–721.

Fisk, H. N., 1947, Fine grained alluvial deposits and their effects on Mississippi River activity: Mississippi River Commission, Vicksburg, Mississippi, 82 p.

Fletcher, F. W., 1963, Regional stratigraphy of Middle and Upper Devonian non-marine rocks in southeastern New York, *in* Shepps, V. C., ed., Symposium on Middle and Upper Devonian Stratigraphy of Pennsylvania and Adjacent States: Pennsylvania Geological Survey Bulletin, G-39, p. 25–41.

Frazier, D. E., and Osanik, A., 1961, Point bar deposits, old river locksite, Louisiana: Transactions of the Gulf Coast Association Geological Society, v. 11, p. 121–137.

Gevers, T. W., Frakes, L. A., Edwards, D., and Marzolf, J. E., 1971, Trace fossils in the Lower Beacon sediments (Devonian), Darwin Mountains, Southern Victoria Land, Antarctica: Journal of Paleontology, v. 45, p. 81–94.

Graham, J. R., and Pollard, J. E., 1982, Occurrence of the trace fossil *Beaconites antarcticus* in the Lower Carboniferous fluviatile rocks of County Mayo, Ireland: Paleogeography, Paleoclimatology, and Paleoecology, v. 38, p. 257–268.

Harms, J. C., McKenzie, D. B., and McCubbin, D. G., 1963, Stratification in modern sands of the Red River, Louisiana: Journal of Geology, v. 71, p. 566–580.

Heckel, P. H., and Witzke, B. J., 1979, Devonian world paleogeography determined from distribution of carbonates and related lithic paleoclimatic indicators, *in* House, M., ed., The Devonian System: Special Papers in Paleontology, no. 23, p. 99–123.

Jackson, R. G., 1975, Velocity-bedform-texture patterns of meander bends in the Lower Wabash River of Illinois and Indiana: Geological Society of America Bulletin, v. 86, p. 1511–1522.

—— 1976, Depositional model of point bars in the Lower Wabash River: Journal of Sedimentary Petrology, v. 46, p. 579–594.

Johnson, K. G., and Friedman, G. M., 1969, The Tully clastic correlatives (Upper Devonian) of New York State: a model for recognition of alluvial, dune(?), tidal, nearshore (bar and lagoon), and off-shore sedimentary environments in a tectonic delta complex: Journal of Sedimentary Petrology, v. 39, p. 451–485.

Jopling, A. V., and Walker, R. G., 1968, Morphology and origins of ripple-drift cross-lamination, with examples from the Pleistocene of Massachusetts: Journal of Sedimentary Petrology, v. 38, p. 971–984.

Kruit, C., 1955, Sediments of the Rhone delta: grain size and microfauna: Verhandel. Koninkl. Ned. Geol. Mijnbouwk Genoot., Geology Series, v. 15, p. 357.

McDowell, J. P., 1960, Cross-bedding formed by sand waves in Mississippi River point bar deposits: Bulletin of the Geological Society of America, v. 71, p. 1925–1936.

Ray, P. K., 1976, Structure and sedimentological history of the overbank deposits of a Mississippi River point bar: Journal of Sedimentary Petrology, v. 46, p. 788–801.

Rickard, L. V., 1975, Correlation of the Silurian and Devonian rocks in New York State: New York State Museum and Science Service, Map and Chart Series no. 24, 16 p.

Rolfe, W.D.I., 1980, Early invertebrate terrestrial faunas, *in* Panchen, A. L., ed., The Terrestrial Environment and the Origin of Land Vertebrates: Academic Press, London, p. 117–157.

Sarkar, S. K., and Basumallick, S., 1968, Morphology, structure, and evolution of a channel island in the Barakar River, Barakar, West Bengal: Journal of Sedimentary Petrology, v. 38, p. 747–754.

Schwartz, D. E., 1978, Sedimentary facies, structures, and grain size distribution: The Red River in Oklahoma and Texas: Transactions of the Gulf Coast Association Geological Society, v. 28, p. 473–492.

Scotese, C. R., Bambach, R. K., Barton, C., Van der Voo, R., and Ziegler, A. M., 1979, Paleozoic base maps: Journal of Geology, v. 87, p. 217–277.

Sevon, W. D., 1981, The Middle and Upper Devonian clastic wedge in northeastern Pennsylvania: New York State Geological Association Guidebook for Fieldtrips, 53rd Annual Meeting, Binghamton, p. 31–63.

Shelton, J. W., and Noble, R. L., 1974, Depositional features of braided-meandering streams: American Association of Petroleum Geologists Bulletin, v. 58, p. 742–749.

Shepps, V. C., ed., 1963, Symposium on Middle and Upper Devonian stratigraphy of Pennsylvania and adjacent states: Pennsylvania Geological Survey Bulletin G–39.

Singh, I. B., 1972, On the bedding in the natural-levee and the point-bar deposits of the Gomti River, Uttar Pradesh, India: Sedimentary Geology, v. 7, p. 309–317.

Woodrow, D. L., and Fletcher, F. W., 1967, Late Devonian paleogeography in southern New York and northeastern Pennsylvania, *in* Oswald, D. H., ed., International Symposium on the Devonian System, Calgary, p. 1327–1334.

Woodrow, D. L., 1968, Devonian Dipnoan aestivation cylinders [abs.]: Geological Society of America Special Paper 121, p. 383–384.

Woodrow, D. L., Fletcher, F. W., and Ahrnsbrak, W. F., 1973, Paleogeography and paleoclimate at the deposition sites of the Devonian Catskill and Old Red facies: Geological Society of America Bulletin, v. 84, p. 3051–3064.

Ziegler, A. M., Scotese, C. R., McKerrow, W. S., Johnson, M. E., and Bambach, R. K., 1979, Paleozoic Paleogeography: Earth and Planetary Sciences Annual Review, v. 7, p. 473–502.

MANUSCRIPT ACCEPTED BY THE SOCIETY JULY 20, 1984

Printed in U.S.A.

Geological Society of America
Special Paper 201
1985

Relation of red-bed copper-uranium occurrences to the regional sedimentology of the Catskill Formation in Pennsylvania

Arthur T. Smith
Chevron USA, Inc. (Eastern)
935 Gravier St.
New Orleans, Louisiana 70112

Arthur W. Rose
Department of Geosciences
The Pennsylvania State University
University Park, Pennsylvania 16802

ABSTRACT

The Devonian Catskill Formation has been divided into four magnafacies (Mf) which have been correlated throughout Pennsylvania. Lithologies in 28 sections and wells were assigned to 10 facies on the basis of grain size, color, bed forms, fossils, and other sedimentary features. Repeating patterns of facies were used to identify magnafacies. Mf-A, the basal magnafacies, is composed dominantly of interbedded marine and non-marine shale and fine sandstone and is inferred to have been deposited mainly in a mud-rich tidal-flat environment. Mf-B is dominated by thick red shales accompanied by thin, fine-grained sandstones and is interpreted as a low-energy fluvial deposit on an inactive arid coastal plain. Mf-B is interrupted occasionally by thin, transgressive sandstones of tidal origin. Mf-C is composed of thick sandstones deposited by braided rivers and localized in three major northwest-trending zones of sediment input. Mf-D is composed of thick, fining-upward cycles with sub-equal amounts of sandstone and red shale. It is inferred to have been deposited by meandering rivers and is thickest in the same areas that Mf-C is thick. Mf-A and Mf-B are richer in sand near the sediment-input centers. In most sections, the Mf are present in the order A, B, C, and D upward, with C missing in many areas and more complex relations in extreme northeastern Pennsylvania. All Mf tend to thin northwestward. This pattern of distribution presumably results from northwestward progradation of the fluvial environment.

Numerous small non-economic Cu-U occurrences are concentrated in areas of thick, shale-rich Mf-B lying between the major sediment-input areas. Most are closely associated with marine transgressions. A few occurrences are in the upper part of Mf-A and in Mf-D, but none is in Mf-C. Most of the Cu-U occurrences are localized by small accumulations of plant trash in shallow fluvial or tidal channels. In contrast, Wyoming-type, roll-front uranium occurrences are localized near Jim Thorpe in large channel sandstones of thick Mf-D in the sediment-input areas. The Cu and U occurrences appear to have formed during diagenesis by migration of metalliferous pore fluids.

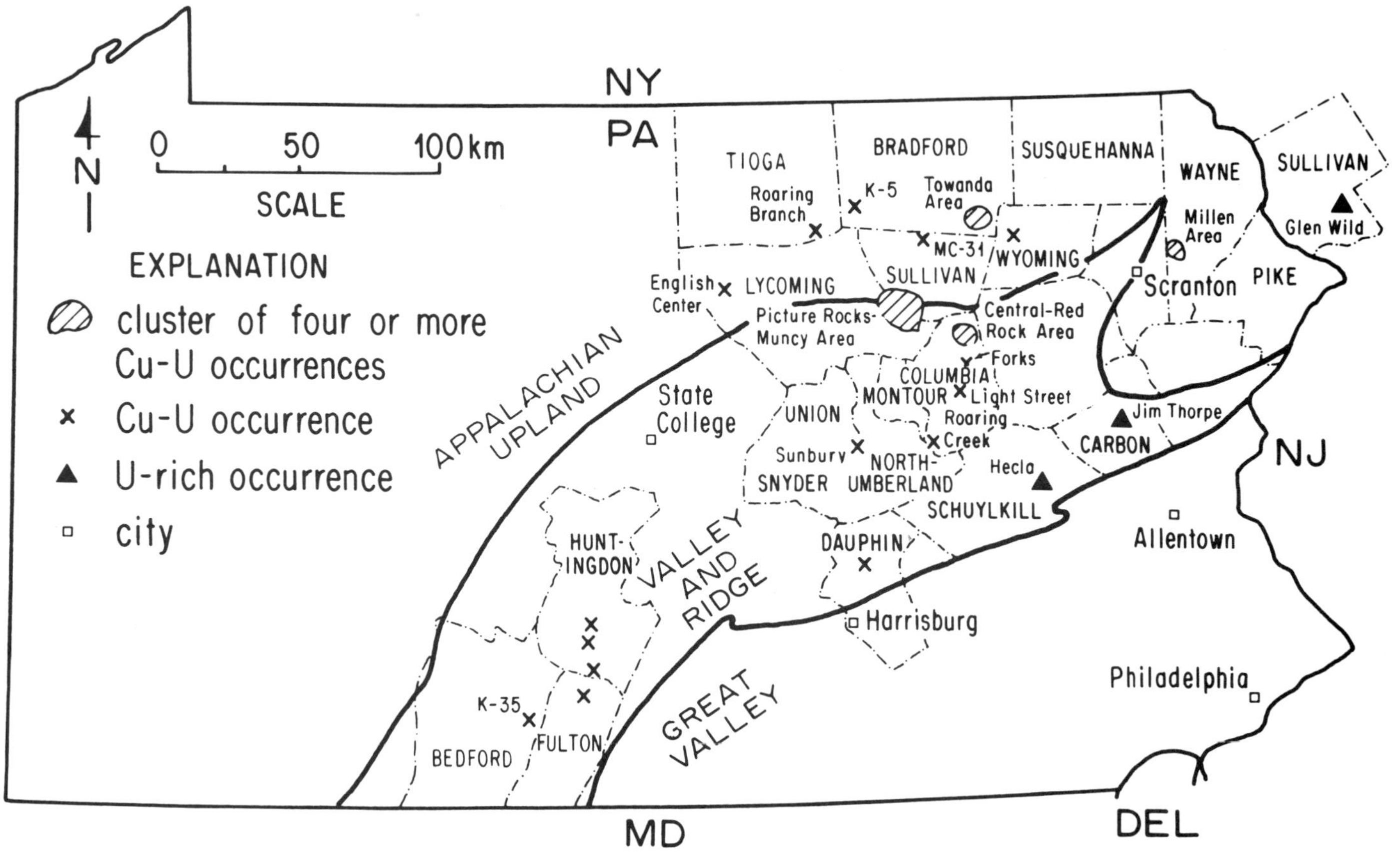

Figure 1. Index map of Pennsylvania showing Cu-U occurrences in the Catskill Formation, geological provinces, counties and cities.

INTRODUCTION

Numerous, small, sub-economic Cu-U occurrences of the red-bed type are located within the clastic red-beds of the Upper Devonian Catskill Formation in Pennsylvania (Figure 1). The origin of these occurrences appears to be related to the sedimentary and diagenetic environments. In this and earlier studies of these deposits (Smith, 1980, 1983), we have examined possible regional sedimentologic controls of their formation. Previous workers have correlated stratigraphy and sedimentary environments over only limited distances; consequently, a regional correlation of environments was developed by extending the work of Rahmanian (1979). This regional correlation and the distribution of the Cu-U occurrences within the regional sedimentologic framework are the major topics of this paper.

CATSKILL FORMATION

General Characteristics

The Catskill Formation is a dominantly fluvial clastic sequence derived from the eastern Acadian highlands and deposited during progradation into the Devonian epeiric sea to the northwest (Barrell, 1913, 1914; Willard, 1939; Rahmanian, 1979; and other papers in this volume). The Catskill Formation is over 2400 m thick in eastern Pennsylvania (Epstein and others, 1974) and thins to less than 1000 m in central and north-central Pennsylvania (Dyson, 1963, 1967; Walker and Harms, 1971; Rahmanian, 1979). The Catskill clastics interfinger with Upper Devonian marine deposits to the west.

The Catskill Formation is recognized by different criteria in eastern and central Pennsylvania, although the presence of red-beds and non-marine sediments are basic identifying criteria. In eastern Pennsylvania, the base of the Catskill is usually placed at the base of the first thick, gray sandstone overlying the thin-bedded sandstones and dark shales of the Trimmers Rock Formation (Epstein and others, 1974). The upper contact is at the top of the uppermost red unit underlying the yellowish-gray sandstones and polymictic diamictites of the Spechty Kopf Formation (Sevon, 1979). Throughout much of central and north-central Pennsylvania, the basal contact of the Catskill Formation is usually defined by the first red shale above a monotonous sequence of dark shales of the Trimmers Rock (Rahmanian, 1979) and Lock Haven (Faill and Wells, 1973) Formations. The upper contact is placed at the top of the last red shale unit underlying the thick, gray sandstones of the Burgoon or Pocono Formation (Rahmanian, 1979; Berg and others, 1981). Warg and Traverse (1973), Stolar (1977), and Streel and Traverse (1978)

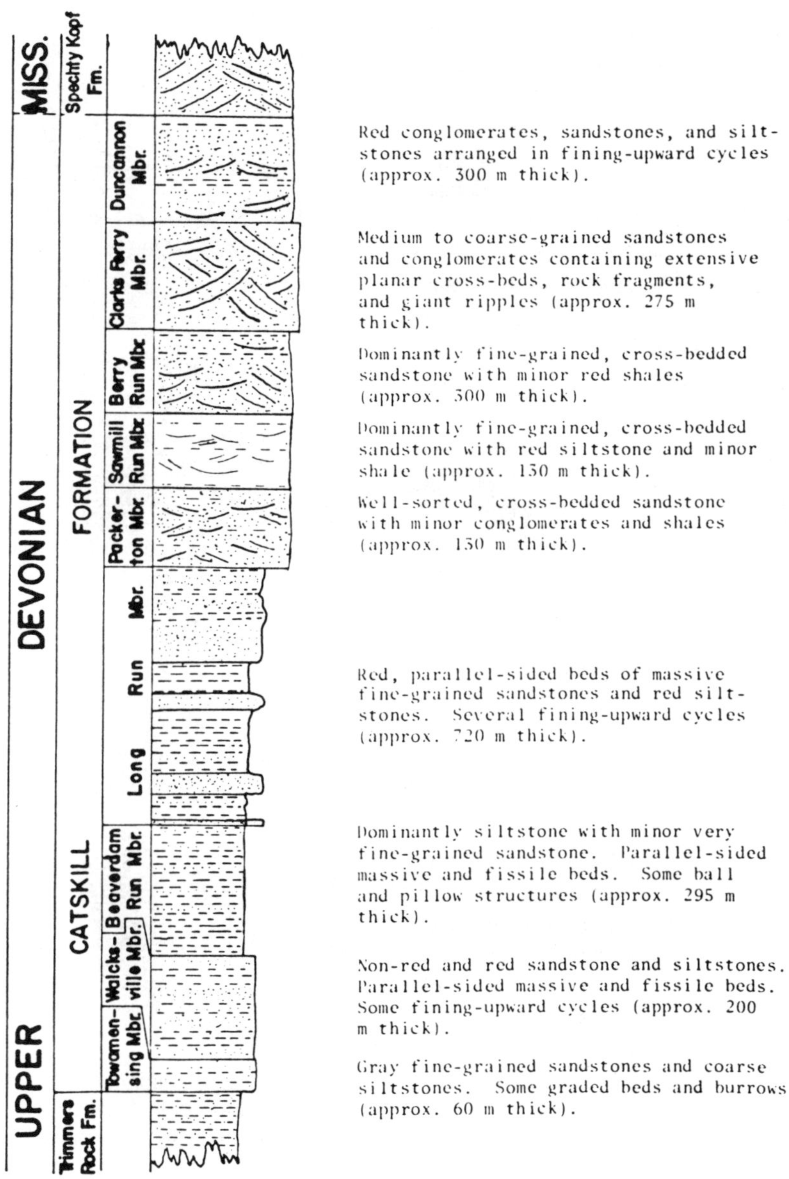

Figure 2. Idealized stratigraphic section for the Catskill Formation based on the Lehigh River section in eastern Pennsylvania (after Epstein and others, 1974). Not drawn to scale.

have shown that the Devonian-Mississippian boundary falls within the Pocono Formation based on palynological studies.

Near the Picture Rocks-Muncy, Central-Red Rock, and Towanda areas in north-central Pennsylvania (Fig. 1), the Catskill Formation is overlain by the recently defined Huntley Mountain Formation (Berg and Edmunds, 1979; Berg and others, 1981). Boundaries of the Catskill Formation used here do not completely agree with those of Berg and others (1981) who place the top of the Catskill at the highest red sandstone underlying the greenish sandstones of the Huntley Mountain Formation.

Our placement of the upper contact of the Catskill Formation in the Picture Rocks-Muncy and Central-Red Rock areas (Fig. 1) agrees with the placement of Berg and others (1981) for the lower contact of the Huntley Mountain Formation. However, in the Towanda area (Fig. 1), we have placed the boundary several tens of meters higher, within the Huntley Mountain Formation of Berg and others (1981). In this area, the lower part of the Huntley Mountain Formation of Berg and others (1981) and the upper part of the Catskill Formation are assigned to Magnafacies C and D of the Catskill Formation as defined herein, because the sediments appear to fit the regional distribution and character of these units.

In eastern Pennsylvania along the Lehigh River, the Catskill Formation has been divided into nine members (Epstein and others, 1974), summarized on Figure 2, and these members have been mapped in a number of adjacent quadrangles. Along the

Susquehanna and Juniata Rivers north of Harrisburg, Dyson (1967) divided the Catskill into four members (Irish Valley, Sherman Creek, Clarks Ferry, and Duncannon, Figure 3), and the sedimentary environment of these members has been extensively discussed by Walker and Harms (1971). Three of these members (Irish Valley, Sherman Creek, and Duncannon) have been carried northwestward to the Allegheny Front near Williamsport (Faill and Wells, 1977) and to central-southern Pennsylvania by Rahmanian (1979), who gives extensive detail on sedimentary environments. Glaeser (1974) correlated the Catskill Formation over a large area in northeastern Pennsylvania using genetically derived sedimentary units largely based on well data.

The correlation of the nine members in eastern Pennsylvania, the three or four members in central Pennsylvania, and the sedimentary environments of Glaeser (1974) is not clear, and the distribution of Catskill members and environments has not been mapped in north-central and many parts of northeastern Pennsylvania where most of the Cu-U occurrences are located. In order to understand the Cu-U occurrences, the concepts of previous workers, especially Rahmanian (1979), have been extended to develop a uniform classification system for all of central, northeastern, and north-central Pennsylvania.

Paleoenvironmental Models

A variety of paleoenvironmental models for the Upper Devonian Catskill Formation have been developed since Barrell (1913, 1914) first envisioned Catskill deposition to be the result of progradation of numerous streams flowing from a highland in the east. These models (Willard, 1939; Allen and Friend, 1968; Walker and Harms, 1971; Glaeser, 1974; Humphreys and Friedman, 1975; Sevon and others, 1978) are reviewed elsewhere in this volume and are not repeated here.

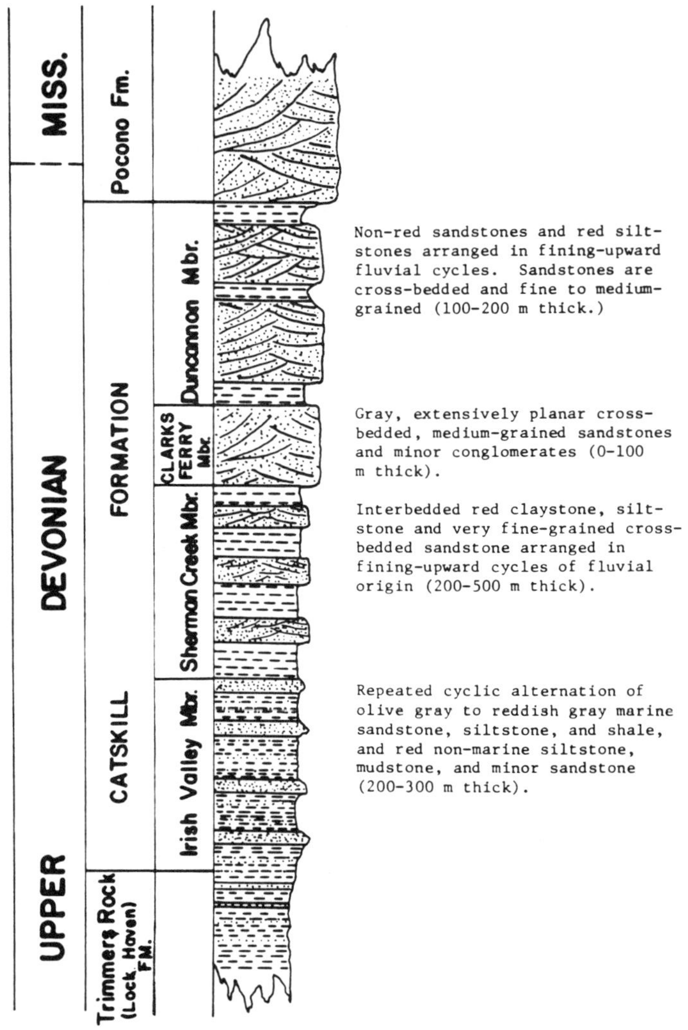

Figure 3. Idealized stratigraphic section for the Catskill Formation in central and north-central Pennsylvania (modified after Rahmanian, 1979, and Dyson, 1967). Not drawn to scale.

The model which seems most applicable in understanding regional facies relations and the distribution of the Cu-U occurrences in Pennsylvania is that of Rahmanian (1979). He proposed a paleoenvironmental model based on the stratigraphic and sedimentologic characteristics of the Catskill Formation in central and central-southern Pennsylvania (Fig. 4). In his model (Fig. 5a), drainage systems characterized by braided and meandering rivers developed in the eastern Acadian Highlands and coalesced to form several widely spaced, major river systems that transported sediment across a prograding coastal plain.

In most of the area described by Rahmanian (1979; Fig. 4), the Catskill Formation is composed of the Irish Valley, Sherman Creek, and Duncannon Members (Fig. 3). In his model the "motifs" (Walker and Harms, 1971) of the Irish Valley Member represent sedimentation on a low-energy, mud-rich tidal-flat and mudflat environment along a shoreline of repeated transgressions; the thin, fine- to very fine-grained, lenticular sandstones and thick shales and mudstones of the Sherman Creek Member represent deposition in a low-energy fluvial environment such as a broad, inactive coastal plain; and the fining-upward sequences of the Duncannon Member represent deposition by meandering rivers higher on the coastal plain (Fig. 5).

Near Port Matilda in an area estimated to be 30-35 km wide (Fig. 4), the Irish Valley, Sherman Creek, and Duncannon Members tend to be richer in sand and are interpreted to have been deposited on an active alluvial plain and tidally-dominated delta fed by a major river system (Fig. 4; Rahmanian, 1979, p. 286–290). At this locality, the Irish Valley Member is not composed of "motifs" but is sand-rich and contains evidence for rapid deposition and a greater importance of fluvial processes than farther south; it exhibits only a few marine transgressions. Rahmanian (1979) interpreted the Sherman Creek Member near Port Matilda to be composed of delta plain and flood plain deposits. The Duncannon Member at Port Matilda is also sand-rich and contains thicker sandbodies and little shale. It is interpreted to be typical of distributary and fluvial environments landward of the delta plain.

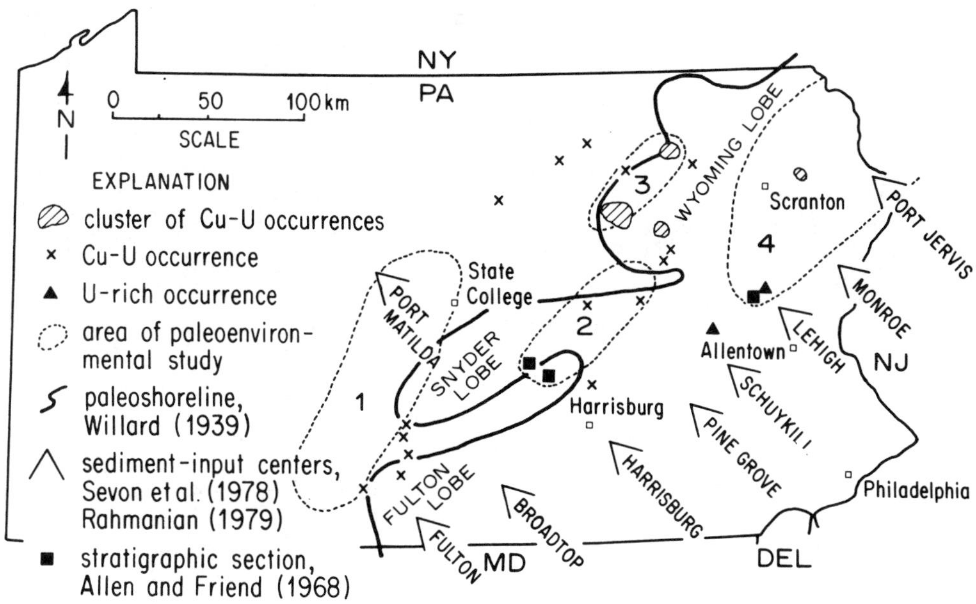

Figure 4. Location of proposed sediment-input centers and areas for which paleoenvironmental models have been proposed. 1 - Rahmanian (1979); 2 - Walker and Harms (1971); 3 - Humphreys and Friedman (1975); 4 - Glaeser (1974).

Rahmanian (1979) also found evidence, including reversals in paleocurrent indicators, suggesting that the sediment-input center at Port Matilda was moderately affected by tidal and/or long shore currents. Recent deltas with similarities to the Port Matilda center are those of the Brahmaputra and Ganges Rivers (Scott and Fisher, 1969; Morgan, 1970) and the Irrawaddy River (Scott and Fisher, 1969).

COPPER AND URANIUM OCCURRENCES

Forty Cu-U occurrences in the Catskill Formation have been examined in detail by Smith (1980). Characteristics common to all occurrences (McCauley, 1961; Mahar, 1978; Smith, 1980) are as follows:

1. Mineralization is associated with and in part replaces fossil plant material.
2. Mineralization occurs in or adjacent to a sandstone or conglomerate.
3. Host rocks for mineralization are most commonly shales but also sandstones and rarely calcareous conglomerates.
4. Copper mineralization is discontinuous and has dimensions on the order of 1-10 cm thick and 1-10 m long depending on the extent of beds containing the reducing agent (fossil plant material). The grade of mineralization is typically 0.5 percent Cu or less.
5. Primary Cu-sulfides include chalcocite, digenite, djurleite, tennantite(?), chalcopyrite, and covellite (Sevon and others, 1978). Uranium is generally amorphous, but uraninite has been collected from several localities (R. C. Smith, personal communication, 1983). Secondary minerals include malachite, azurite, chrysocolla, brochantite, metazunerite, and metatorbenite. Minor amounts of barite have been found. Authigenic pyrite occurs in reduced shales and fossil plant material.
6. The presence of orthorhombic chalcocite (T $<100°C$), exsolution of bornite from chalcocite on heating above 75°C, a wide range in $\delta^{34}S$ of sulfides, an apparent pre-compaction age of sulfides, fine grain-sizes, and a lack of igneous rocks or other evidence for elevated temperatures all indicate that the Cu and U were deposited at relatively low temperatures.

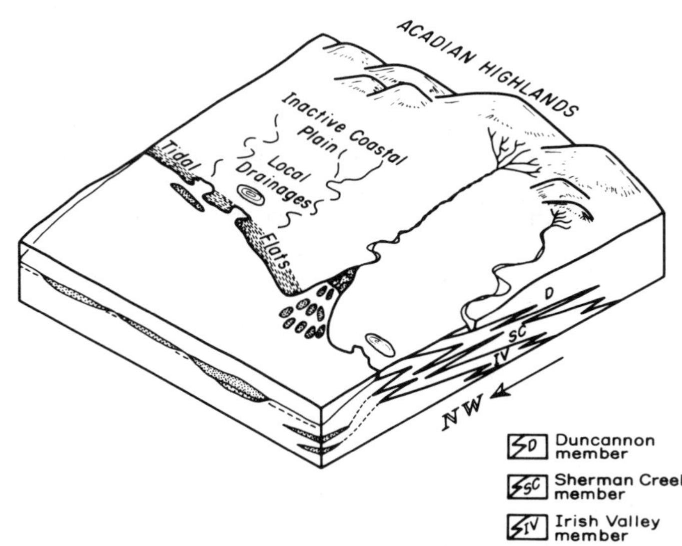

Figure 5. Paleoenvironmental model for the Catskill Formation (modified after Rahmanian, 1979).

Several U occurrences lacking Cu are present in the Catskill Formation near Jim Thorpe along the Lehigh River (McCauley, 1961; Sevon and others, 1978). These U occurrences are in thick,

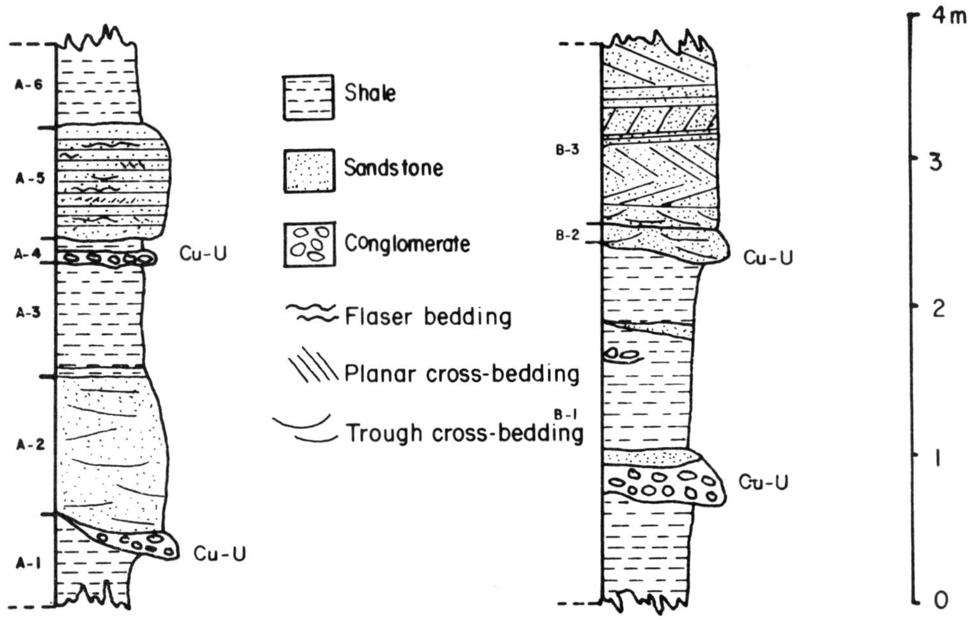

Figure 6. Two idealized lithofacies models for Cu-U occurrences in the Catskill Formation (after Smith, 1980).

medium- to coarse-grained, channel sandstone bodies at boundaries between oxidized (hematite-stained) and reduced sandstone. Redox boundaries and ore zones form "rolls" curving across the bedding in a fashion similar to the roll-type U ores of Wyoming (Rackley, 1976). These Wyoming occurrences are believed to form by down-slope flow of oxidizing groundwater through aquifers in large alluvial fans. A similar origin is indicated for the Catskill uranium occurrences near Jim Thorpe (Sevon and others 1978; Schmiermund, 1977).

Facies Associations of Cu-U Occurrences

Smith (1980) described a series of lithofacies models for the Cu-U occurrences and found certain common stratigraphic relationships. Two idealized models derived from the series are summarized in Figure 6.

Most of the Cu-U occurrences are associated with lithofacies presented in Model A (Fig. 6a). The model is characterized by a basal siltstone or mudstone (Bed A-1) that is disconformably overlain by a trough-crossbedded, fining-upward sandstone or calcareous intraformational conglomerate (Bed A-2). Bed A-2 contains abundant fossil plant material near its base and grades upward into siltstones and mudstones (Bed A-3) that may contain root casts and fossil caliche horizons. The sequence has the characteristics typical of a meandering stream deposit (Allen, 1965; Rust, 1972; Campbell, 1976; Collinson, 1978; Rahmanian, 1979).

Overlying the fining-upward cycle is a semi-conformable sandstone (Bed A-5) that is generally less than 1.5 m thick and up to 100 m wide. This sandstone is thin-bedded and contains various combinations of small trough crossbeds, bipolar planar crossbeds, climbing and mud-draped ripples, flaser beds, marine fossils or casts, and fossil plant material. It is overlain by shales (Bed A-6) that contain root casts and are interpreted to be fluvial. In some areas, the semi-conformable sandstone is underlain by a continuous, undulatory calcareous conglomerate (Bed A-4), which contains fossil plant material and marine fossils and is interbedded with flaser-bedded siltstones. The structures and facies relationships of Beds A-4 and A-5 are described in detail in Smith (1980) and are similar to marine sand bars described by Kenyon (1970) and Johnson (1977) in the shallow marine and inter-tidal environments in the North Sea. These sandstones (Bed A-5) are referred to as "marine bars" throughout the rest of this paper.

Several Cu-U occurrences are associated with facies typical of Model B (Fig. 6b). This stratigraphic model is characterized by basal red shales (Bed B-1) of variable thickness. The shales are interpreted to be fluvial, because they contain vertical root casts and desiccation cracks. The shales are sometimes interbedded with thin, lenticular, trough-crossbedded sandstones and calcareous conglomerates.

Overlying these red shales are several discrete quartz-rich, medium-grained, sandstone units. The basal unit (Bed B-2) is lenticular and contains bipolar, trough crossbeds, and marine fossils. This unit is conformably overlain by thick-bedded sandstones (Bed B-3), which are up to 10 m thick, coarsen-upward, and contain multi-directional planar crossbeds. Based on their internal structure and similarities with other recent and ancient deposits (e.g., Bernard and others, 1962; Land, 1972; Klein, 1977), they are interpreted to be small barrier island deposits

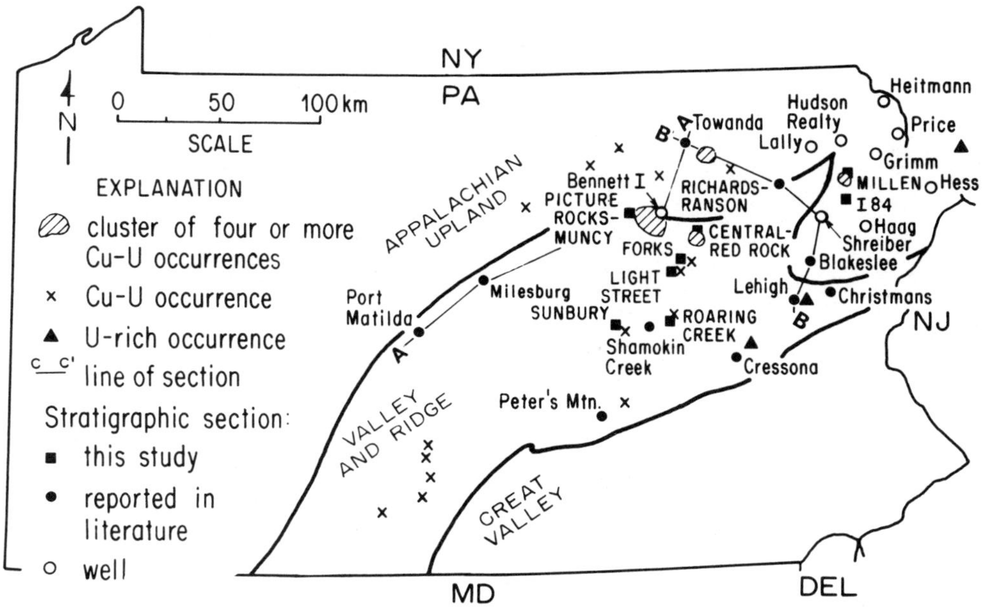

Figure 7. Index map of stratigraphic sections and wells used in regional correlations of the Catskill Formation.

formed in intertidal zones (Fig. 5). Such deposits are rare and only weakly mineralized (Smith, 1980). They are referred to as the "barrier" sandstones throughout the rest of the paper.

Of 30 Cu-U occurrences described in detail by Smith (1980), 23 are overlain within 10 m stratigraphically, by a "marine bar" or "barrier" sandstone. In some localities, the marine sandstone lies directly over the mineralization and may be mineralized. At other localities, the marine sandstone is separated stratigraphically from the mineralization by fluvial deposits (Fig. 6a). In the latter case a fluvial sandstone (Bed A-2) lies directly over the mineralized zone. Of the seven occurrences not associated with recognizable marine units, six are within the Towanda and Picture Rocks clusters and are therefore close to occurrences with marine facies; at several of these occurrences the exposed stratigraphic section is less than 2 m thick, and the facies analysis may be incomplete. The remaining occurrence, English Center in Lycoming County (Fig. 1), is located west of the other occurrences and closer to the dominantly marine Devonian section.

REGIONAL CORRELATION AND THE DISTRIBUTION OF CU-U OCCURENCES

As noted earlier, stratigraphic correlation of subdivisions of the Catskill Formation is complex and has not been accomplished everywhere in Pennsylvania. The regional correlations discussed here are based on a modified version of the facies modeling method described by Walker (1976). In this method correlations are done by using regionally defined facies instead of stratigraphically defined units and thus facies changes are incorporated into the correlations.

Methodology

The data for the regional correlations were collected from 18 stratigraphic sections and eight well logs (Fig. 7; Table 1; Smith, 1983). Eight of the stratigraphic sections were described and measured by Smith (1980, 1983) in central, north-central, and northeastern Pennsylvania. Several other stratigraphic sections described by other workers were re-examined. In areas where stratigraphic sections were incomplete or poorly exposed, well logs were utilized. The thickness of the measured Catskill Formation within each of these sections is listed in Table 1. The total thickness of the Catskill Formation was either measured, estimated from the isopach map for the total Catskill Formation (Fig. 8), or estimated from the dip, outcrop width, and elevation of contacts in the vicinity.

Although there are no marker beds within the Catskill Formation to provide a datum for the regional correlations, the Middle Devonian Tioga Bentonite is a persistent marker bed that underlies the Catskill Formation throughout Pennsylvania and can be used as a datum. The depth of the Tioga below the Upper Devonian Tully Limestone and the total thickness of the Upper Devonian were estimated from published isopach maps (Ayerton, 1964; Oliver and others, 1967) and are listed in Table 1.

Ten basic facies (Table 2) were necessary to describe all rock units found between the lower and upper contacts of the Catskill Formation in the 29 stratigraphic sections and wells. The facies are defined according to their color, geometry, lithology, grain size, sedimentary structures, and associated features. The definitions are sufficiently general to incorporate the differences in descriptions of similar rock types by different workers.

TABLE 1. DATA AND SOURCES FOR REGIONAL FENCE DIAGRAM AND ISOPACH AND SANDSTONE/SHALE MAPS. ALL VALUES ARE IN METERS

Site (Abbreviation)	County	Type[1]	Measured Catskill[2]	Thickness U. Dev.[3]	Thickness Tully-Tioga[3]	Total Catskill[4]	Comments	References[5]
Towanda (T)	Bradford	O	939	2141	534	939	These sources are	4,5
Christmans (C)	Monroe	O	2465	2752	728	2465	the basis of the	3
Cressona (CR)	Schuylkill	O	1861	2905	642	1861	isopach map of	11
Hudson Realty (HR)	Wayne	W	792	2269[d]	719	792	the total Catskill	7
Lehigh (L)	Carbon	O	2403	2753	917	2403	Fm. (Fig. 8).	8
Milesburg (MI)	Centre	O	856	2141	385	856	"	6
Muncy[a] (PRM)	Sullivan	O	294	2416	489	926	"	1
Peter's Mtn. (P)	Dauphin	O	2086	2875	529	2086	"	9,11
Picture Rocks[a] (PRM)	Lycoming	O	417	2416	489	926	"	1
Port Matilda (PM)	Centre	O	803	2049	336	803	"	6
Richards-Ransom (RR)	Lackawanna	O/W	1058	2431	607	1058	"	7,10
Shamokin Creek (SC)	Northumberland	O	1959	2752	642	1959	"	11
Bennett #1 (BE)	Sullivan	W	743	2416	520	934	Total thickness	2
Blakeslee (BL)	Carbon	O	703	2690	750	1865	of the Catskill Fm.	3
Central[b] (CRR)	Columbia	O	83	2447	550	1100	was approximated	1
Forks (F)	Columbia	O	274	2752	633	1406	from isopach map	1
I84[c] (I84)	Wayne	O	337	2446	765	1700	of total Catskill	1
Light Street (LS)	Columbia	O	802	2753	627	1450	FM. (Fig. 8).	1
Millen (M)	Wayne	O	223	2294	734	1500	"	1
Red Rock[b] (CRR)	Columbia	O	124	2447	550	1100	"	1
Roaring Creek (RC)	Columbia	O	181	2752	642	1600	"	1
Sunbury (S)	Snyder	O	727	2722	627	1600	"	1,12
Grimm (G)	Wayne	W	1052	1472	734	1843	These sources are	7
Haag[e] (H)	Pike	W	1391	1737	743	2155[e]	described in Glaeser	7
Heitmann[e] (HE)	Wayne	W	964	1421	781	1806[e]	(1974). The thick-	7
Hess (HS)	Pike	W	392	1240	751	1574	ness of the U. Dev.	7
Lally[c] (LA)	Lackawanna	W	349	1651	688	991	is the preserved	7
Price (PR)	Wayne	W	1080	1511	777	1832[e]	thickness of the Catskill	7
Schreiber (SCH)	Lackawanna	W	1635	2232	765	1895[e]	approximated from Fig. 8 (see e below).	7

1. O = outcrop; W = well.
2. Refers to Catskill measured in outcrop and identified in well cuttings based on the criteria discussed in text. The measured Catskill is the same as the total Catskill at the first 11 sites (see a. below).
3. These values were taken from isopach maps (Ayerton, 1963; Oliver and others, 1967) for the Upper Devonian (Top of the Catskill Fm. to the Tully limestone), and the stratigraphic interval between the Tully ls., and the datum for this project, the Tioga bentonite. The sum of both columns is the total interval between the datum and the top of the Catskill Fm. Only for the wells described by Glaeser (1974) does the thickness of U. Dev. represent preserved thickness.
4. These values either equal the measured thickness or are approximate values taken from the isopach map of total Catskill (Fig. 8). They are the thicknesses plotted on the fence diagram (except in northeastern PA, where preserved thicknesses are plotted).
5. (1) This study; (2) Sevon and others (1978); (3) Sevon (1975); (4) Woodrow (1968); (5) Berg and others (1981); (6) Rahmanian (1979); (7) Glaeser (1974); (8) Epstein and others (1974); (9) Dyson (1963, 1967); (10) Kehn and others (1966); Hoskins and others (1963); (12) Walker and Harms (1971).

a. The Picture Rocks and Muncy sections are combined to form a complete section with 215 m missing.
b. Central and Red Rock sections are also from the same area and are combined. The section is not complete, but the upper boundary of the Catskill is the top of the Red Rock section.
c. I84 and Roaring Creek sections and the Lally well are not plotted on the fence diagram because they do not add significant information.
d. Like the other wells described by Glaeser (1974), this value represents preserved thickness.
e. These values are also plotted on Figure 8. The base of the Catskill Fm. was encountered in each and the total thickness was estimated by Glaeser (1974).

All units in each stratigraphic section were assigned a number from 1 to 10 corresponding to the 10 recognized facies (Table 2). Examination of the resulting sequences of numbers led to the identification of four repeated patterns of facies illustrated in Figure 9.

The term "magnafacies" appears appropriate for the four sequences of facies. Magnafacies was originally defined (Caster, 1934; Glossary of Geology, 1980) for facies belts or zones of major size, similar lithology, and paleontology that cut across time planes in the Upper Devonian of northwestern Pennsylvania. Caster's seven zones (magnafacies) included a "Pocono" magnafacies comprised of gray sandstone and conglomerate, and a Catskill magnafacies of red and green sandstone and shale giving way eastward to coarse conglomeratic red or green sandstone, and locally including thin marine tongues. Other magnafacies were marine units underlying the Catskill. The zones being called magnafacies in this paper are smaller, but seem similar in concept. The interpretation derived from the paleoenvironmental model (Fig. 5) as well as the generally progradational nature of the Catskill indicates that the zones cross time planes. The use of magnafacies rather than members emphasizes their complex nature and avoids confusion with previously defined stratigraphic members.

The four magnafacies have been given letter designations Mf-A, Mf-B, Mf-C, and Mf-D. These magnafacies are defined as follows:

1. Mf-A consists of gray shales, sandstones, and conglomerates that may contain marine fossils (Facies 1, 2, 3, 7) and may be interbedded with red siltstones, mudstones, and minor fine- to very fine-grained sandstones (Facies 9, 8, 4). The red units may contain fossil root casts and desiccation cracks.

2. Mf-B is composed of red siltstones, mudstones (Facies 9), and fine-grained to very fine-grained sandstones (Facies 4) and minor interbedded gray sandstones that commonly contain fossil plant material (Facies 4, 8). Gray shales (Facies 10) are minor.

3. Mf-C is characterized by abundant fine- to coarse-

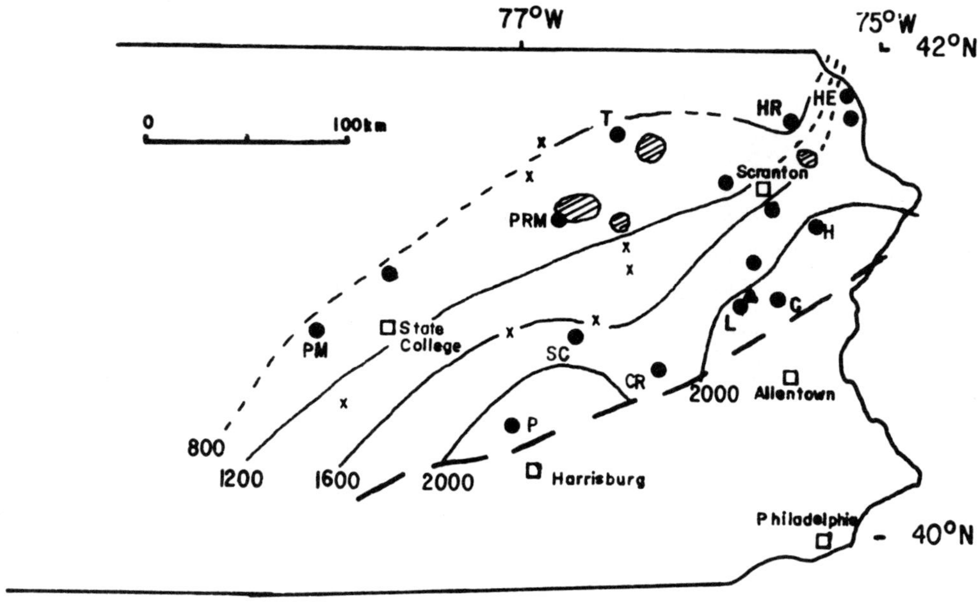

Figure 8. Isopach map for total Catskill Formation in meters. Clusters of 4 or more Cu-U occurrences are shown as lined areas, single Cu-U occurrences shown as "x," U-rich occurrences as solid triangles, and stratigraphic sections used for isopachs as solid circles. See Tables 1 and 3 for abbreviations of localities.

grained sandstone (Facies 5) and conglomerate (Facies 6) with minor interbeds of shale (Facies 10). The conglomerates occur as lag deposits at the base of sandstones or within sandstones as lags in intrasandstone troughs and channels.

4. Mf-D comprises repeated sequences of conglomerate-sandstone-shale (Facies 6, 5, 4, 9) or sandstone-shale (Facies 5, 4, 9). The sandstones and shales are similar in thickness.

The thickness and sandstone/shale ratios (ss/sh) for each magnafacies for each stratigraphic section and well log are listed in Table 3. Several sections are incomplete so that some thicknesses of magnafacies are minima, and total thicknesses could not be calculated. For other sections, the descriptions of individual units were ambiguous or incomplete and ss/sh values could not be calculated.

Two cross-sections showing the magnafacies are presented in Figure 10. Isopach and ss/sh ratio maps of the total Catskill Formation (Fig. 8) and each magnafacies (Fig. 11) are also derived from data in Table 3.

It should be emphasized that the magnafacies distinguished here have no time significance, but are lithostratigraphic in nature. The boundaries are clearly intertongued in some areas, as indicated by multiple occurrences of some units in certain sec-

TABLE 2. GENERALIZED CRITERIA WHICH DEFINE 10 FACIES ASSOCIATIONS IN THE CATSKILL FORMATION OF PENNSYLVANIA

Facies	Color[1]	Fossils[2]	Geometry	Grain Size and Lithology[3]	Sedimentary[4] Structures	Commonly Associated Features
1	g	M and/or P	tabular	f-vf ss	massive beds, flasers, ripples, trough x-beds	-
2	g	M and/or P	lenticular	f-c ss $\pm$ quartz pebbles	" "	-
3	g	M and/or P	tabular	" "	" "	-
4	r or g	P	lenticular	f-vf ss	trough and planar x-beds, erosive basal contacts	fines up, basal cgl. lag
5	g or r	P	"	m-c ss	" "	cgl. lag
6	g	P	"	cgl	massive	facies 4,5
7	g	M and/or P	lent./tabular	sst, mst	bioturbated	thin, fss and facies 1,2,3
8	g	P	"	f-c sst	bioturbated, root casts	" "
9	r	-	"	sst, mst	root casts	fines up and underlain by facies 4
10	g	P	"	sst, mst	-	facies 5

1. g = gray; r = red (first letter is most common).
2. M = marine; P = fossil plant material.
3. f = fine; vf = very fine; c = coarse; m = medium; cgl = conglomerate; ss = sandstone; sst = siltstone; mst = mudstone.
4. x-beds = cross-beds.

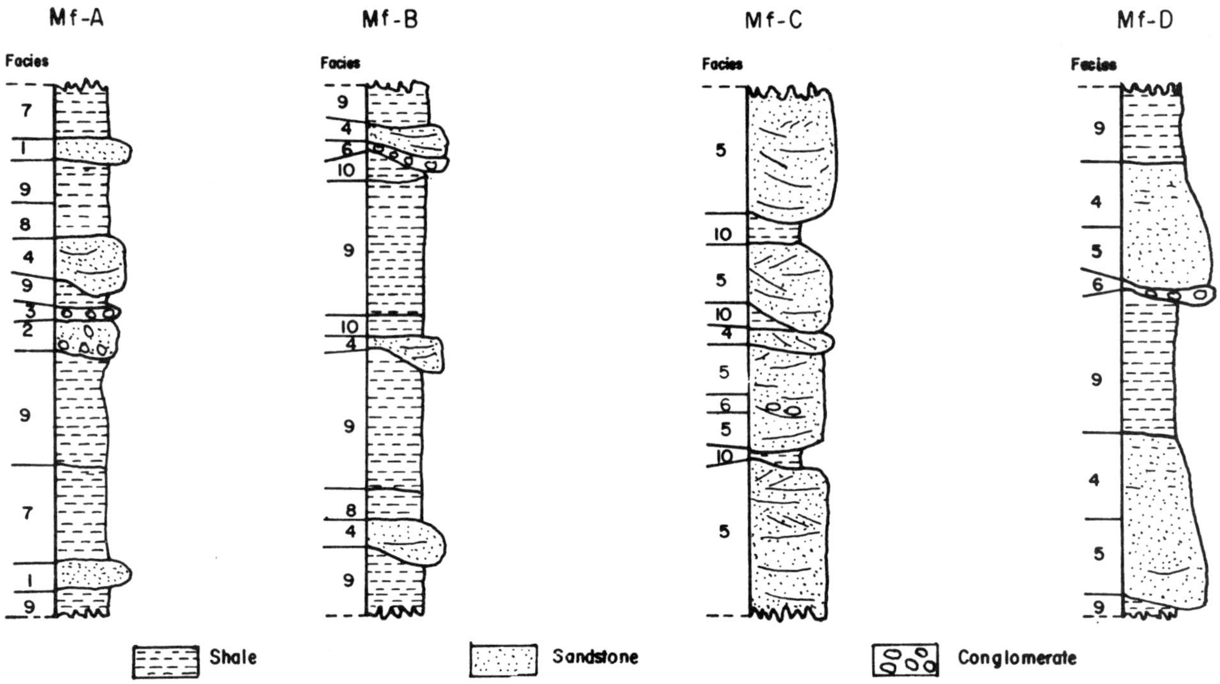

Figure 9. Idealized stratigraphic sections for 4 magnafacies used in regional correlations of the Catskill Formation. See Table 2 for descriptions of component facies. Not drawn to scale.

TABLE 3. MAGNAFACIES THICKNESS AND SANDSTONE/SHALE VALUES CALCULATED FOR THE STRATIGRAPHIC SECTIONS AND WELLS USED FOR THIS REPORT

		Magnafacies[3]							
		Thickness (m)				ss/sh			
Site[1]	Abbreviation[2]	A	B	C	D	A	B	C	D
Bennett #1	BE	254	278	96	115	---	---	---	---
Blakeslee	BL	---	> 36	486	183	---	---	---	---
Towanda	T	520	198	74	147	0.6	0.4	1.7	1.0
Central[a]	CRR	---	> 83	---	---	---	1.2	---	---
Christmans	C	355	1193	642	275	---	---	---	---
Cressona	CR	278	1232	127	223	---	---	---	---
Forks	F	237	> 37	---	---	0.4	0.3	---	---
Grimm	G	401	223	168	>260	1.6	1.1	4.0	1.5
Haag	H	661	207	352	>171	1.3	2.8	2.9	1.2
Heitmann	HE	>189	0	323	452	1.6	0.0	4.7	1.7
Hess	HS	63	175	154	0	6.4	---	3.4	2.8
Hudson Realty	HR	599	193	0	0	---	---	0	0
I84	I84	---	---	---	>337	---	---	---	1.6
Lally	LA	>349	---	---	---	1.6	---	---	---
Lehigh	L	363	920	824	296	0.8	0.3	>15.0	1.2
Light Street	LS	128	>674	---	---	1.1	0.1	---	---
Milesburg	MI	191	367	0	298	---	---	---	---
Millen	M	---	>370	0	110	---	0.4	0.0	0.6
Muncy[b]	PRM	---	>158	31	106	---	0.4	12.0	0.6
Peter's Mtn.	P	612	1070	416	358	---	---	---	---
Picture Rocks[b]	PRM	319	> 97	---	---	---	0.5	---	---
Port Matilda	PM	237	260	46	260	---	---	---	---
Price	PR	297	116	124	543	1.9	>3.0	>15.0	2.5
Red Rock[a]	CRR	---	---	---	>124	---	---	---	0.9
Richards-Ransom	RR	571	204	30	253	2.1	1.4	8.0	1.4
Roaring Creek	RC	>112	> 69	---	---	0.6	0.2	---	---
Schreiber	SCH	448	484	360	343	1.6	1.0	---	7.0
Shamokin Creek	SC	611	1348	0	0	---	---	---	---
Sunbury	S	595	>152	---	---	1.7	0.4	---	---

1. All underlined sites are complete sections of Catskill Formation.
2. Abbreviations identify control points in Figs. 8 and 11.
3. 96 = Complete; --- = Unknown or eroded.
a. Central and Red Rock are from the same area.
b. Picture Rocks-Muncy is a complete section

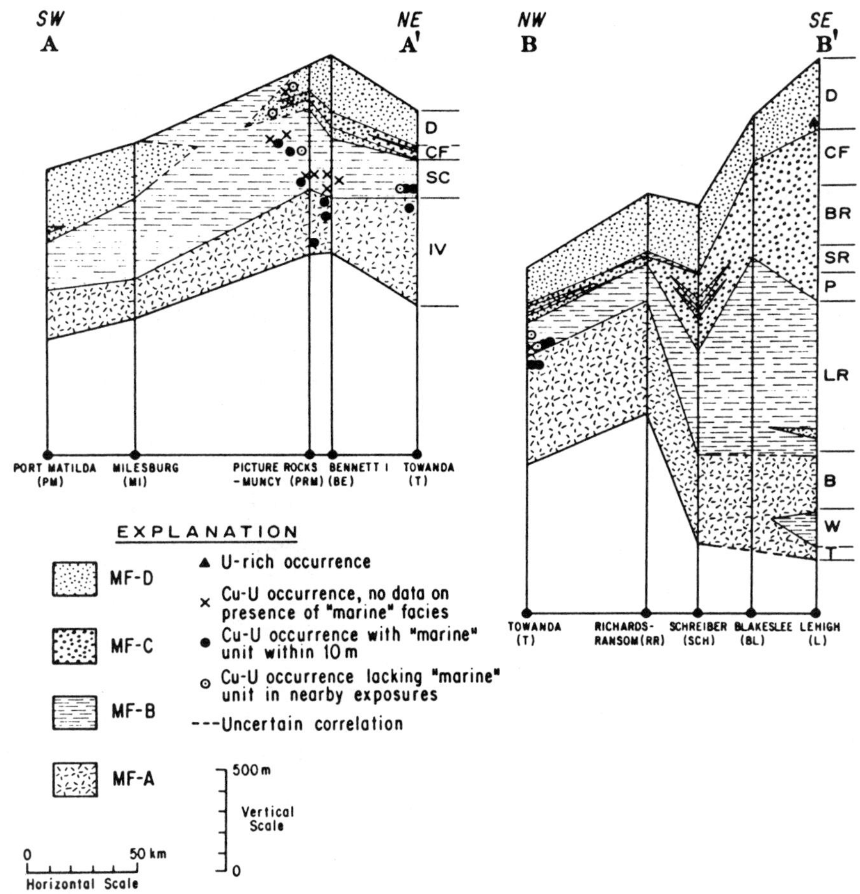

Figure 10. Regional cross-sections, for sections shown on Figure 7. Datum is Tioga Bentonite + 1500 m. Area lacking pattern is marine Devonian. a) Panel A-A'. b) Panel B-B'. Abbreviations: IV, Irish Valley Member; SC, Sherman Creek Member; CF, Clarks Ferry Member; D, Duncannon Member, approximately as identified by Rahmanian (1979) at Port Matilda and Milesburg, and Faill and Wells (1977) near the Picture Rocks-Muncy section; T, Towamensing Member; W, Walcksville Member; B, Beaverdam Run Member; LR, Long Run Member; P, Packerton Member; SR, Sawmill Run Member; BR, Berry Run Member; CF, Clarks Ferry Member; and D, Duncannon Member of Epstein and others (1974) along the Lehigh River.

tions, and the paleoenvironmental interpretations indicate lateral transgressions must occur. The classification is nevertheless useful in correlating lithologic and environmental characteristics of widely separated parts of the Catskill Formation.

Palinspastic maps for both isopach and ss/sh ratios have been prepared using the reconstruction of Pedlow (1976) but do not markedly change the shape or location of the contours.

The stratigraphic position of most Cu-U occurrences has been located within the magnafacies using structural cross-sections and field observations (Smith, 1980, 1983) and has been projected onto the fence diagram panels (Fig. 10).

Discussion

The patterns of magnafacies thickness and ss/sh shown in Figure 11 and their relationship to the Cu-U occurrences in the Catskill Formation can be used along with the paleoenvironmental interpretation of Rahmanian (1979; Fig. 5) to derive a number of conclusions about Catskill deposition and Cu-U concentration in Pennsylvania.

Rahmanian (1979) proposed that centers of sediment-input and upper alluvial plain sedimentation are characterized by thick, extensively crossbedded sandstones of braided stream origin or by thick, fining-upward cycles deposited by meandering streams. By the definitions presented previously and the idealized models (Fig. 9), Mf-C corresponds to the braided stream environment and Mf-D to the meandering stream environment. The isopach and ss/sh ratio maps for these two magnafacies define the position of three sediment-input centers (Fig. 12), which have also been recognized by previous workers in the area (Barrell, 1913, 1914; Rahmanian, 1979; Sevon and others, 1978), and are discussed below. No Cu-U deposits occur in the zones of thick, coarse sediment defining the input centers. However, a few Cu-U occurrences are present in Mf-D between input centers where ss/sh is less than 1 and the facies is thinner than 200 m. The two Cu-U occurrences studied in Mf-D lack evidence for marine

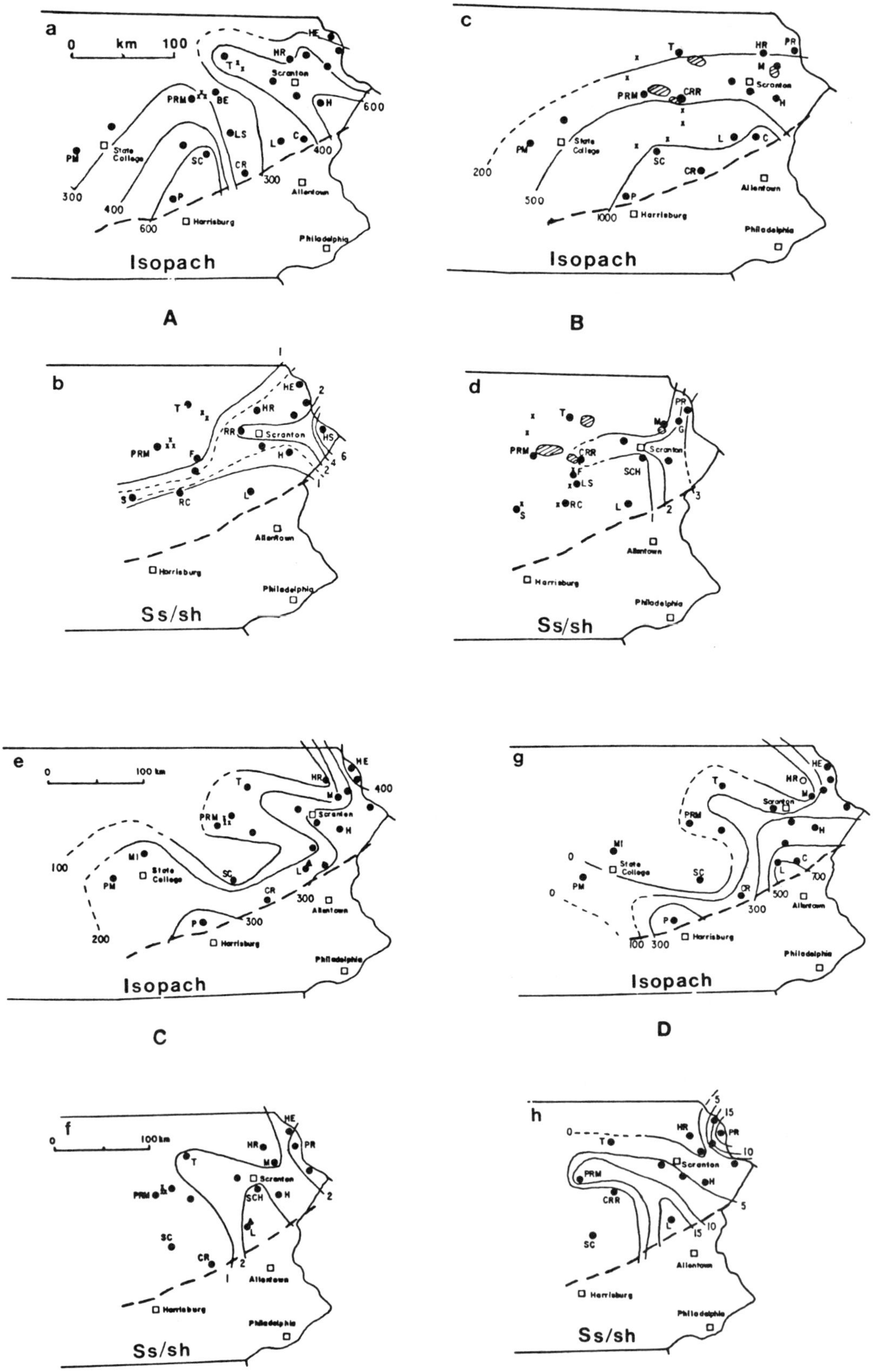

Figure 11. Isopach and ss/sh ratio maps for magnafacies. a, b) Magnafacies A. c, d) Magnafacies B. e, f) Magnafacies C. g, h) Magnafacies D. Isopach contours are in meters. x = Cu-U occurrences within magnafacies; o = control point for thickness or ss/sh. See Tables 1 and 3 for abbreviations of localities.

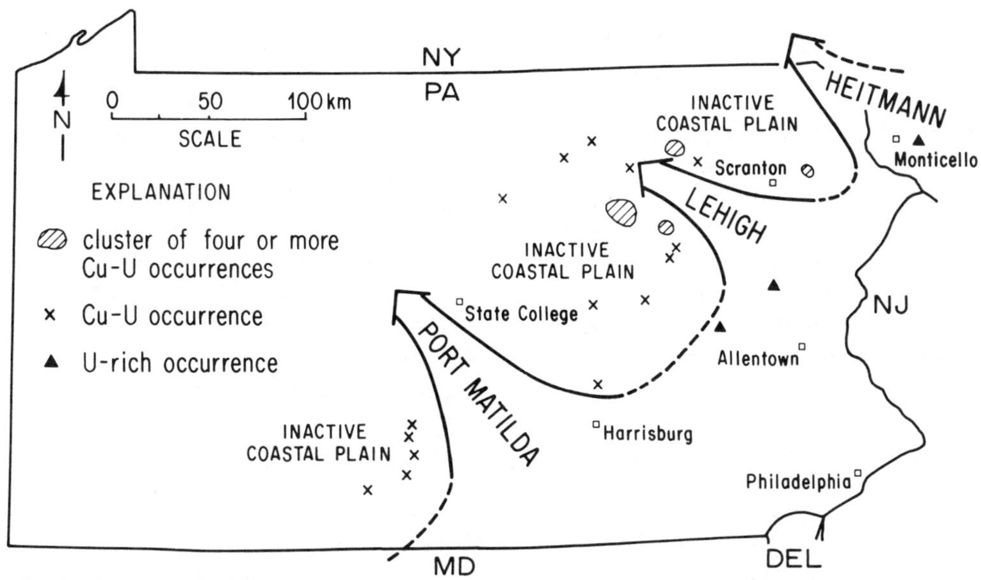

Figure 12. Schematic diagram of proposed sediment-input centers and areas of coastal plain deposits based on the results of the regional correlations.

facies, but are very close to the boundary with Mf-B (Fig. 10a). In addition to the Cu-U occurrences, roll-type U occurrences lacking Cu are present in the proximal sections of the input centers, near Jim Thorpe (Fig. 10b).

The model of Rahmanian (1979) indicates that deposits of the interdeltaic environment are composed of thick red shales and thin, fine-grained, lenticular sandstones deposited by low-energy meandering streams. Such interdeltaic areas are also characterized by occasional marine transgressions during which gray sediments bearing marine fossils are deposited. Mf-B contains all of these features and was presumably deposited in a similar situation. The isopach map of this magnafacies (Fig. 11c) indicates a relatively simple pattern of northwestward decrease in thickness for this facies in Pennsylvania. The ss/sh ratio map (Fig. 11d) suggests a localized coarser-grained phase possibly associated with a sediment-input center (Fig. 12). Most of the known Cu-U occurrences are located within Mf-B. The occurrences are all in areas with ss/sh ratios of less than 1 and most are in areas where Mf-B is less than 500 m thick. The occurrences are most abundant in the lower third of the magnafacies, but occur higher in the sequence in the Picture Rocks-Muncy and Millen areas. Facies evaluations of Cu-U occurrences, based on data in Smith (1980) and summarized previously in this paper, indicate that several occurrences in the middle third of Mf-B and one in upper Mf-B (Jordan Pits, Sevon and others, 1978) are associated with marine transgressions into Mf-B.

The patterns of thickness and ss/sh ratio values for Mf-A differ considerably from those for Mf-B (Fig. 11). These patterns are closely related to sediment-input centers. Mf-A is sand-rich in the sediment-input center areas and shale-rich in the interdeltaic areas. Mf-A includes completely marine zones (Towamensing and Beaverdam Run Members along the Lehigh River) as well as the "motifs" of Walker and Harms (1971). Five Cu-U occurrences clustered in two areas are located within Mf-A. In both areas, Mf-A is 300-450 m thick and has an ss/sh ratio of less than 1.

We recognize three major sediment-input centers within the Catskill Formation in Pennsylvania (Fig. 12) based on characteristics of the various magnafacies. The most southerly is a northwest-trending zone in which Mf-D is thick and Mf-C is present. Both magnafacies thin towards Port Matilda (Fig. 7). This northwest-trending zone correlates with the Harrisburg sediment-input center (Fig. 4; Sevon and others, 1978) and with the Port Matilda lobe (Fig. 4; Rahmanian, 1979). It is hence called the Port Matilda center. The second sediment-input center, the Lehigh center, is characterized by thick, sand-rich Mf-C and Mf-D. Both magnafacies thin to the northwest towards the Bradford and Picture Rocks-Muncy areas (Fig. 10b). The Lehigh center also correlates with the westward-trending zone of sand-rich Mf-B. The third sediment-input center, the Heitmann center, is located in extreme northeastern Pennsylvania where Mf-C and Mf-D are sand-rich and thick. The trend and extent of this center is not fully defined because its northern edge is located to the north of the area investigated.

The Lehigh and Port Matilda centers cover broad areas in eastern Pennsylvania, but are restricted to much narrower areas in central Pennsylvania. This distribution probably reflects the coalescing of many sediment-input centers (Fig. 4) near the source in the east into several widely spaced drainage systems to the northwest.

Two areas considered to be inactive coastal plain environments occur between the sediment-input centers (Fig. 12). These inactive coastal plain areas are characterized by thick, shale-rich Mf-B, and by either an absence or deficiency of Mf-C and Mf-D. Rahmanian (1979) indicates that another inactive coastal plain environment is located south of the Port Matilda center (Fig. 12).

The distribution of Mf-A indicates that two thick Catskill sequences occur in Pennsylvania. One corresponds to the Lehigh center and the other occurs in the inactive coastal plain area south of the Lehigh center (Fig. 11a). The two areas are separated by a thin zone of Mf-A, which parallels the trend of the Lehigh center and forms its approximate southern limit.

This thin zone in Mf-A is of interest because many Cu-U occurrences are along or near it. The thin zone may represent a tectonic arch that was active during the deposition of Mf-A. Diment and others (1979), using geophysical evidence, describe a northwest-trending, basement-controlled lineament that occurs in the same area as the thin zone of Mf-A. Vertical movement along this lineament could have produced the zone of thin Mf-A. A second explanation is that rapid fluvial sedimentation in the Lehigh center inundated the typical interbeds of root-bearing shales and sandstones and fossiliferous sediments of Mf-A with sand-rich fluvial deposits. This explanation is supported by the trend of the ss/sh values in Mf-A (Fig. 11b). These values cross the thin zone and are not perturbed or broken. This pattern suggests that deposition of Mf-A, as defined in this study, was not influenced by rapid sedimentation in the Lehigh center and that the thin zone reflects limited deposition of Mf-A. A third explanation is that marine conditions persisted longer in the area of the thin zone than in the areas to the north and south. This possibility is indicated by the thick Tioga-Bentonite-Catskill Formation interval in the Cressona (CR), Light Street (LS), Picture Rocks-Muncy (PRM), and the Bennett #1 (BE) sections, all of which are characterized by relatively thin Mf-A.

In summary, several statements can be made about the relationship between the Cu-U mineralization and the host rocks of the Catskill Formation. First, the occurrences are located in stratigraphic intervals that are shale-rich; i.e., ss/sh values of the host magnafacies are less than 1.0. Secondly, most of the occurrences are located within the lower third of Mf-B. The occurrences within the upper parts of Mf-B and in Mf-D are few and are restricted to areas where Mf-C is thin and interbedded with Mf-D or where Mf-C is absent. Thirdly, no occurrences were found or described within the sand-dominated Mf-C. Only the U-rich occurrences in eastern Pennsylvania occur within the sand-rich sequences of Mf-D and Mf-C.

As discussed by Rose (1976) and Smith (1983), the presence of marine Cl-bearing pore waters in the relatively oxidizing environment of the red-beds allows mobilization of Cu to levels of 100 ppm in solution, whereas in fresh waters, Cu solubility is at best a few ppm. The marine incursions into Mf-B are believed to have furnished this Cl to the red-beds. Uranium is also soluble under these oxidizing conditions. On the other hand, the marine rocks are generally too reducing for Cu or U mobility. Shales and impure sandstones contain 10-50 ppm Cu and 2-5 ppm U, probably concentrated initially in the Fe-oxides and able to be mobilized by oxidizing Cl-rich pore solutions.

The accumulation of Cu and U is inferred to have occurred during diagenesis, based on interpreted low temperatures of deposition and textural features indicating early emplacement (Smith, 1983). Causes for flow of the Cu- and U-bearing pore fluids include compaction, changes in sea level which would cause reversal of flow in the inundated areas, and flow of fresh waters down the alluvial fan sediments of the input centers. The greatest amount of flow would be in the most permeable sandstones. The copper and uranium would be precipitated whenever solutions moving through the permeable units encountered reducing plant trash at the base of fining-upward cycles.

ACKOWLEDGMENTS

The research reported in this paper was supported by the Mineral Conservation Section, Pennsylvania State University, and by National Science Foundation Grant EAR-8036454 to A. W. Rose, H. Ohmoto, and L. M. Cathles. The senior author was also supported during part of the research by fellowships from the Pennsylvania Mining and Mineral Resources Institute and an MMMFC Fellowship of the U.S. Department of Education. We are greatly indebted to Victor Rahmanian, E. G. Williams, and R. L. Slingerland for initial concepts and discussions during the research.

REFERENCES CITED

Allen, J.R.L., 1965, A review of the origin and characteristics of recent alluvial sediments: Sedimentology, v. 5, p. 89–191.

Allen, J.R.L., and Friend, P.F., 1968, Deposition of the Catskill facies, Appalachian regions: With notes on some other Old Red Sandstone Basins: Geological Society of America Special Paper 106, p. 21–74.

Ayerton, W. G., 1963, Isopach and lithofacies map of the Upper Devonian of northeastern United States: Pennsylvania Geological Survey, 4th series, General Geology Report G-39, p. 3–6.

Barrell, J., 1913, The Upper Devonian delta of the Appalachian geosyncline: Part I. The delta and its relations to the interior areas: American Journal of Science, 4th series, v. 36, p. 429–472.

—— 1914, The Upper Devonian delta of the Appalachian geosyncline: Part II. Factors controlling the present limits of strata: American Journal of Science, 4th series, v. 37, p. 87–109.

Berg, T. M., and Edmunds, W. E., 1979, The Huntley Mountain Formation: Catskill-to-Burgoon Transition in north-central Pennsylvania: Pennsylvania Geological Survey, 4th series, Information Circular 83, 80 p.

Berg, T. M., Crowl, G. H., Edmunds, W. E., Luce, P. B., Sevon, W. D., Wilshusen, J. P., Woodrow, D. L., and Pohn, D. L., 1981, Geology of Tioga and Bradford Counties, Pennsylvania: Guidebook to 46th Annual Field Conference of Pennsylvania Geologists, Pennsylvania Geological Survey, 162 p.

Bernard, H. A., LeBlanc, R. J., and Major, C. F., Jr., 1962, Recent and Pleistocene geology of southeast Texas, *in* Geology of the Gulf Coast and Central Texas and guidebook of excursions: Houston Geological Society, Houston, Texas, p. 175–225.

Campbell, C. V., 1976, Reservoir geometry of a fluvial sheet sandstone: American Association of Petroleum Geologists Bulletin, v. 60, p. 1009–1020.

Caster, K. E., 1934, The stratigraphy and paleontology of northwestern Pennsylvania, Part I: Stratigraphy: Bulletins of American Paleontology, v. 21, no. 71, p. 9–29.

Collinson, J. D., 1978, Alluvial sediments, *in* Reading, H. G., ed., Sedimentary environments and facies: Elsevier Publishing Co., New York, p. 15–60.

Diment, W. J., Muller, O. H., and Lavin, P. M., 1979, Basement tectonics of New York and Pennsylvania as revealed by gravity and magnetic studies, *in* The Caledonides in the U.S.A., Wones, D. R., ed.: Memoir 2, Virginia Polytechnical Institute, Department of Geological Sciences, p. 221–227.

Dyson, J. L., 1963, Geology and mineral resources of the northern half of the New Bloomfield quadrangle, Pennsylvania: Pennsylvania Geological Survey, 4th series, Atlas 137ab, 63 p.

——1967, Geology and mineral resources of the southern half of the New Bloomfield quadrangle, Pennsylvania: Pennsylvania Geological Survey, 4th series, Atlas 137cd, 86 p.

Epstein, J. B., Sevon, W. D., and Glaeser, J. D., 1974, Geology and mineral resources of the Lehighton and Palmerton quadrangles, Carbon and Northampton Counties, Pennsylvania: Pennsylvania Geological Survey, 4th series, Atlas 195 ed., 460 p.

Faill, R. G., 1977, Bedrock geology and mineral resources of the Salladasburg and Cogan Station quadrangles, Lycoming County, Pennsylvania: Pennsylvania Geological Survey, Atlas 133 ed, 44 p.

Faill, R. T., and Wells, R. B., 1973, Structure and Silurian and Devonian stratigraphy of the Valley and Ridge province in central Pennsylvania: Guidebook to 38th Annual Field Conference of PennsylvaniaGeologists, Pennsylvania Geological Survey, 168 p.

Glaeser, J. D., 1974, Upper Devonian stratigraphy and sedimentary environments in northeastern Pennsylvania: Pennsylvania Geological Survey, 4th series, General Geology Report G-63, 89 p.

Glossary of Geology, 1980, Bates, R. L., and Jackson, J. A., eds.: American Geological Institute, Falls Church, Va., 749 p.

Hoskins, D. M., Arndt, H. H., Wood, G. H., Jr., Conlin, R. R., Dyson, J. L., and Trexler, J. P., 1963, Lithology, subdivision and correlation of the Catskill Formation in east-central Pennsylvania: Pennsylvania Geological Survey, 4th series, General Geology Report G-39, p. 147–164.

Humphreys, M., and Friedman, G., 1975, Late Devonian Catskill deltaic complex in north-central Pennsylvania, *in* Broussard, M. L., ed.: Deltas: Models for exploration: Houston Geological Survey, p. 369–379.

Johnson, H. D., 1977, Shallow marine sand bar sequences: An example from the late Precambrian of North Norway: Sedimentology, v. 24, p. 245–270.

Kehn, T. M., Glick, E. E., and Culbertson, W. C., 1966, Geology of the Ransom quadrangle, Lackawanna, Luzerne and Wyoming Counties, Pennsylvania: U.S. Geological Survey Bulletin 1213, 81 p.

Kenyon, N. H., 1970, Sand ribbons of European tidal seas: Marine Geology, v. 9, p. 25–39.

Klein, G. deV., 1977, Tidal circulation model for deposition of clastic sediment in epeiric and mioclinal shelf seas: Sedimentary Geology, v. 18, p. 1–12.

Land, C. B., 1972, Stratigraphy of Fox Hills sandstone and associated formations, Rock Spring Uplift and Wamsutter Arch area, Sweetwater County, Wyoming: A shoreline-estuary sandstone model for the late Cretaceous: Colorado School of Mines Quarterly, v. 67, no. 2, 69 p.

Mahar, D. L., 1978, Geology and geochemistry of uranium deposits near Beaver Lake, Sullivan County, Pennsylvania [M.S. Thesis]: Pennsylvania State University, 142 p.

McCauley, J. F., 1961, Uranium in Pennsylvania: Pennsylvania Geological Survey, 4th series, Bulletin M43, 71 p.

Morgan, J. P., 1970, Depositional processes and products in the deltaic environment, *in* Morgan, J. P., ed.: Deltaic sedimentation, modern and ancient: Society of Economic Paleontologists and Mineralogists, Special Publication 15, p. 31–47.

Oliver, W. A., Jr., DeWitt, W., Jr., Dennison, J. M., Hoskins, D. M., and Huddle, J. W., 1967, Devonian of the Appalachian Basin, United States, *in* International symposium of Devonian System, v. 1, Oswald, D. H., ed.: Alberta Society of Petroleum Geologists, Calgary, p. 1001–1040.

Pedlow, G. W., III, 1976, Palinspastic base map: Central and southern Appalachians: Geological Society of America Bulletin, v. 87, p. 133–136.

Rackley, R. I., 1976, Origin of western-states type uranium mineralization, *in* Handbook of stratabound and stratiform ore bodies, v. 7, Wolf, K. H., ed.: Elsevier, Amsterdam, p. 89–156.

Rahmanian, V. D. Kh., 1979, Stratigraphy and sedimentology of the Upper Devonian Catskill and Uppermost Trimmers Rock Formations in central Pennsylvania [Ph.D. Thesis]: Pennsylvania State University, 329 p.

Rose, A. W., 1976, The effect of cuprous chloride complexes in the origin of red-bed copper and related deposits: Economic Geology, v. 71, p. 1036–1048.

Rust, B. R., 1972, Structure and process in a braided river: Sedimentology, v. 18, p. 221–245.

Schmiermund, R. L., 1977, Geology and geochemistry of uranium deposits near Penn Haven Junction, Carbon County, Pennsylvania [M.S. Thesis]: Pennsylvania State University, 153 p.

Scott, A. J., and Fisher, W. L., 1969, Delta systems and deltaic deposition, *in* Delta systems in the exploration for oil and gas: A research colloquium, Fisher, W. L., Brown, L. F., Scott, A. J., and McGowan, J. H., eds.: Bureau of Economic Geology, University of Texas, p. 10–29.

Sevon, W. D., 1975, Geology and mineral resources of the Christmans and Pohopoco Mountain quadrangles, Carbon and Monroe Counties, Pennsylvania: Pennsylvania Geological Survey, 4th series, Atlas 195ab.

——1979, Polymictic diamictites in the Spechty Kopf and Rockwell Formations, *in* Devonian shales in south-central Pennsylvania and Maryland, Dennison, J. M. and others, eds.: Guidebook for 44th Annual Field Conference of Pennsylvania Geologists, Pennsylvania Geological Survey, p. 61–66, 107–110.

Sevon, W. D., Rose, A. W., Smith, R. C., and Hoff, D. T., 1978, Uranium in Carbon, Lycoming, Sullivan, and Columbia Counties, Pennsylvania: Guidebook for 43rd Annual Field Conference of Pennsylvania Geologists, Pennsylvania Geological Survey, 99 p.

Smith, A. T., 1980, Stratigraphic and sedimentologic controls for copper and uranium in red-beds of the Upper Devonian Catskill Formation in Pennsylvania [M.S. Thesis]: Pennsylvania State University, 216 p.

——1983, Geology of the red-bed Cu-U occurrences in the Upper Devonian Catskill Formation, Pennsylvania [Ph.D. Thesis]: Pennsylvania State University, 240 p.

Stolar, J., 1977, Megaspores and the Devonian-Mississippian boundary along Route 322, Centre County, Pennsylvania [D.Ed. Thesis]: Pennsylvania State University, 102 p.

Streel, M., and Traverse, A., 1978, Spores from the Devonian-Mississippian transition near the Horseshoe Curve section, Altoona, Pennsylvania, USA: Review of Paleobotany and Palynology, v. 26, p. 21–39.

Walker, R. G., 1976, Facies models: 1. General introduction: Geoscience Canada, v. 3, p. 21–24.

Walker, R. G., and Harms, J. C., 1971, The 'Catskill delta,' a prograding muddy shoreline in central Pennsylvania: Journal of Geology, v. 79, p. 381–399.

Warg, J. B., and Traverse, A., 1973, A palynological study of shales and coals of a Devonian-Mississippian transition zone, central Pennsylvania: Geoscience and Man, v. 7, p. 39–46.

Willard, B., 1939, The Devonian of Pennsylvania: Pennsylvania Geological Survey, 4th series, General Geology Report G-19, 481 p.

Woodrow, D. L., 1968, Stratigraphy, structure and sedimentary patterns in the Upper Devonian of Bradford County, Pennsylvania: Pennsylvania Geological Survey, 4th series, General Geology Report G54, 78 p.

MANUSCRIPT ACCEPTED BY THE SOCIETY JULY 20, 1984

Printed in U.S.A.

Geological Society of America
Special Paper 201
1985

Late Devonian marine ecosystems of the lower West Falls Group in New York

George R. McGhee, Jr.
Department of Geological Sciences
Wright Geological Laboratory
Rutgers University
New Brunswick, New Jersey 08903

Robert G. Sutton
Department of Geological Sciences
University of Rochester
Rochester, New York 14627

ABSTRACT

The shallow-water strata of the lower West Falls Group (Upper Devonian) in New York can be subdivided into three units on the basis of extensive black shale tongues from the deep-water Rhinestreet Shale. These units are the "Millport equivalent," "Beers Hill equivalent," and "Meads Creek equivalent" in ascending stratigraphic order. Two distinctive community-types developed during lower West Falls deposition: a shallower water spiriferacean-rhynchonellide-bivalve species ensemble, and a deeper water assemblage of unattached epibenthic and sessile semi-infaunal brachiopods. During deposition of the "Beers Hill equivalent," an additional diverse assemblage—the *Nervostrophia-Devonochonetes* Community—developed in a unique array of muddy, delta platform environments flanked by seaward, delta front sand bars. The evolution of lower West Falls marine ecosystems represents a complex interaction of biological and physical patterns, of which regional differences in substratum type, and the evolution of those differences through time, played a major role.

INTRODUCTION

The present study is the third in a series of comparative analyses of the ecology, environments, and zoogeography of the New York region during the Late Devonian (McGhee and Sutton, 1981, 1983). The goal of this research is the complete paleoecological characterization of the marine Catskill delta stratigraphic units within the Frasnian stage. Such environmental and biological control is crucial to the examination of evolutionary changes in marine ecosystems over geologic time (McGhee, 1981).

The stratigraphic and sedimentologic framework analyzed here is restricted to the lower West Falls Group in south-central New York state (Fig. 1). The relatively unfossiliferous, thin, black and dark gray shales that provide the stratigraphic framework are not part of this study. The upper West Falls has been considered earlier by the writers (McGhee and Sutton, 1983), and the underlying Sonyea Group has been previously studied by R. G. Sutton, Z. P. Bowen, and A. L. McAlester (Sutton and others, 1970). The Sonyea study describes in some detail the shelf subenvironments used in this article as well as the faunal and sedimentological criteria used to define them. The same criteria were used by Bowen, McAlester, and Sutton to identify the shelf environments in over 400 post-Sonyea outcrops, including those designated here in the lower West Falls Group. The evolution of the shallow-marine environments and the resultant impact upon the marine communities (during the entire Frasnian) are subjects of a separate article (Sutton and McGhee, this volume).

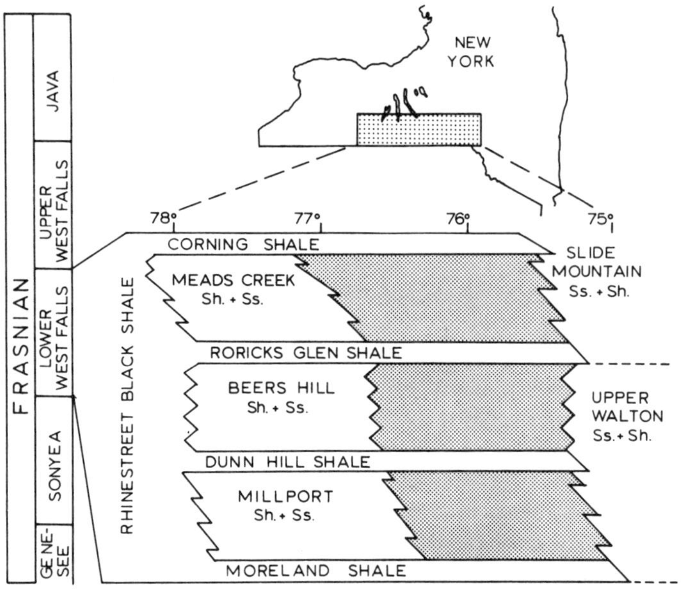

Figure 1. Stratigraphic units within the Frasnian stage in New York, and within the lower West Falls Group in detail. Modified from Rickard (1975). Shaded portions represent the fossiliferous shelf equivalents of the slope facies to the west.

LOWER WEST FALLS STRATIGRAPHY AND PALEOENVIRONMENTS

The West Falls Group is the second oldest of four groups that comprise the Frasnian in New York; they are, in ascending order: Genesee (in part), Sonyea, West Falls, and Java (Fig. 1). The lower West Falls consists of a wedge of clastics that thickens eastward from Lake Erie in western New York to the Catskill Mountains in the eastern part of the state. In the west, it is composed of the Rhinestreet Black Shale and is less than 27 m thick. In west-central New York, the black shale splits into four tongues separated by gray shale and siltstones. These black-shale tongues are called the Moreland (lower), Dunn Hill, Roricks Glen, and Corning (upper). The intervening gray shales and siltstones are, in ascending order: Millport, Beers Hill, and Meads Creek. Eastward, facies changes occur and the tongues of black shale change to dark gray shales and the intervening strata become more arenaceous and fossiliferous. The fossiliferous sequence, including the dark gray shales, measures more than 300 m thick in the study area. In east-central New York, the entire sequence passes into non-marine red and green shales and sandstones. The lower part is equivalent to the upper Walton Formation, whereas the upper part correlates with the lower part of the Slide Mountain Formation (Sutton, 1963).

The undivided Rhinestreet represents basinal facies, whereas the Millport, Beers Hill, Meads Creek, and their boundary black shales in west-central New York represent slope facies. These stratigraphic names, which have been applied only to the slope facies, are extended informally to the east. For example, the fossiliferous shelf equivalent of the Beers Hill shale will be referred to as the "Beers Hill equivalent" (see Fig. 1). Unless designated otherwise, usage of the terms Millport, Beers Hill, or Meads Creek in this article implies "equivalent."

The stratigraphic framework used here was constructed essentially in the same manner as those used in the studies of the Java Group and upper West Falls (McGhee and Sutton, 1981, 1983), and builds upon the work of Ehrets (1981) in the nearshore facies. Because of the extensive east-west trend and narrow north-south width of the outcrop belt, measured sections were projected into an east-west stratigraphic profile (Figs. 2 and 3). The location of the sections (Fig. 2) and the environmental identification (based upon physical and biological criteria) are shown in Figure 3. Twenty-five Millport, 21 Beers Hill, and 19 Meads Creek sections comprise the stratigraphic profile. A generalized profile of the paleoenvironments is shown in Figure 3.

LOWER WEST FALLS MARINE COMMUNITIES

Sampling methodology and the determination of species relative abundances at each field locality have been previously described in detail (Sutton and others, 1970; McGhee and Sutton, 1981), and are only briefly reviewed here. At each locality, the

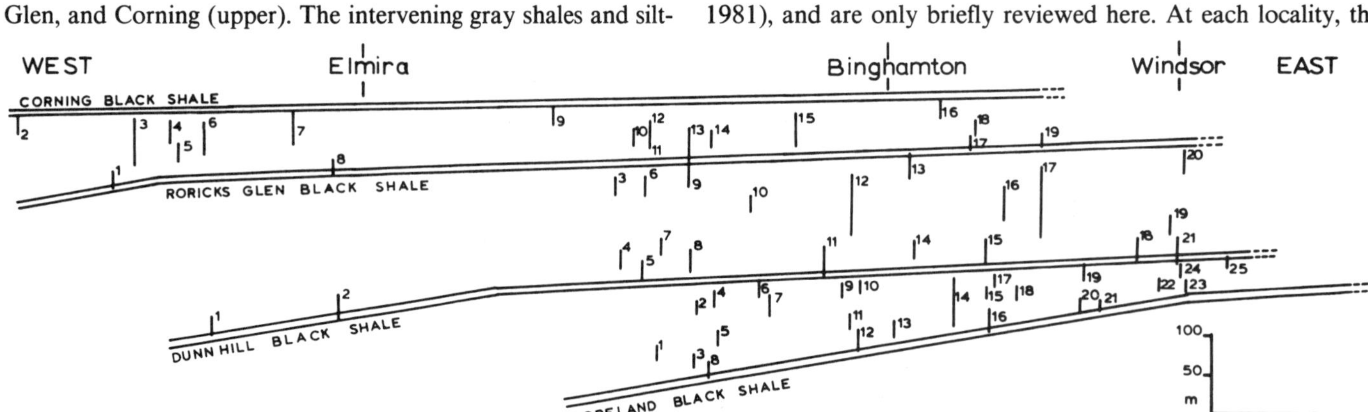

Figure 2. Stratigraphic cross section of lower West Falls strata in the study area showing stratigraphic and geographic relationships of the field localities. Stratigraphic framework modified and expanded from Ehrets (1981).

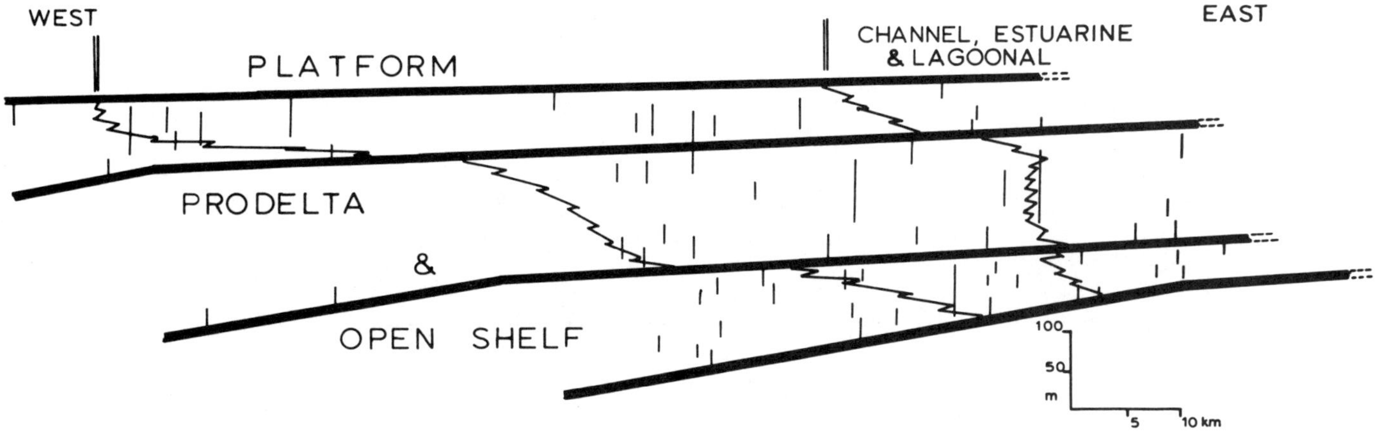

Figure 3. Generalized stratigraphic distribution of marine sedimentary environments relative to the field locations.

most abundant species is given a rank of four, the second most abundant a rank of three, and so on, on a logarithmic scale. Usage of rank abundances helps to dampen large fluctuations in absolute numbers of fossils from locality to locality. Variation in absolute numbers of fossils from outcrop to outcrop may be considerable, although the rank abundance of species at the same outcrops remains nearly constant. Species rank abundances were then plotted along the general east-west (Figs. 2 and 3) onshore-offshore environmental gradient (Figs. 4-6). In Figures 4-6, the field localities have been ordered from furthest offshore to nearest onshore, and the running average of the rank abundance of each species plotted for each locality. Running averages of rank abundance were used in order to smooth the species gradient curves for graphical presentation. Communities are defined by the distribution of dominant species along an environmental gradient (Whittaker, 1975). Congruent associations of species abundance gradients are recognized along the onshore-offshore gradient, and these associations are named by two of the characteristic species in the species ensemble.

"Millport Equivalent"

Two distinct benthic communities are recognized in the Millport interval (Fig. 4). The *AMBOCOELIA-TYLOTHYRIS* Community is a low-diversity, deep-water assemblage, and is best developed in prodelta and open shelf environments west of the general position of Binghamton, N.Y. East of Binghamton, the *ORTHOSPIRIFER-CYPRICARDELLA* Community obtains its maximum development in the narrow platform region (cf. Fig. 3), and the abundance of the fauna then declines in the estuarine zones to the east towards the present position of Windsor (Figs. 3 and 4).

The *AMBOCOELIA-TYLOTHYRIS* Community is characterized by the co-occurrence of the species *Ambocoelia umbonata, Tylothyris mesacostalis, Productella speciosa, P. lachrymosa, Douvillina mucronata, D. cayuta, Spinatrypa spinosa,* and *S. hystrix* (Fig. 4), all of which are generally deep-water species throughout the Frasnian (Sutton and others, 1970; Bowen and others, 1974; Thayer, 1974; McGhee, 1976; McGhee and Sutton, 1981, 1983). *Tylothyris mesacostalis* is a particularly wide-ranging species, however, and increasingly ventures into shallower water platform and nearshore habitats with time (McGhee, 1976; McGhee and Sutton, 1981). Other characteristic moderate- to deep-water species not shown in Figure 4 are *Devonochonetes lepidus, D. deflectus, D. setigerus,* and a holdover from the older, deeper water communities of the Sonyea Group, *D. vanuxemi* (Table 1). *Devonochonetes,* however, while abundant throughout prodelta and shelf habitats, actually obtains its maximum abundance in outer platform environments during Millport time (Fig. 4), and constitutes a sizeable fraction of the individuals within the *ORTHOSPIRIFER-CYPRICARDELLA* Community (Table 2).

The diverse *ORTHOSPIRIFER-CYPRICARDELLA* Community is characterized by the co-occurrence of the large, robust spiriferacean *Orthospirifer mesastrialis,* the morphologically similar *Cyrtospirifer chemungensis,* the mobile infaunal bivalve species *Cypricardella gregaria* and *C. bellastriata,* and sizeable populations of species of *"Camarotoechia"* and *Devonochonetes* (Fig. 4, Table 1). The spiriferaceans, rhynchonellids, and infaunal burrowing bivalves are particularly characteristic of Frasnian shallower water communities. *Orthospirifer mesastrialis* and *Cypricardella gregaria* are also dominant elements in the *CYPRICARDELLA* Community of the older Glen Aubrey Formation of the Sonyea Group, which is separated from the Millport by the Moreland Shale (Sutton and others, 1970; Bowen and others, 1974), and it would appear that nearshore communities change very little in their organization between the two periods of deposition. One major change of note is the appearance for the first time in the Appalachian Sea of the spiriferacean *Cyrtospirifer chemungensis.* Species of this genus become dominant elements within the late Frasnian communities, simultaneously with the demise of *Orthospirifer mesastrialis* (McGhee, 1981).

In general, the offshore *AMBOCOELIA-TYLOTHYRIS* Community is dominated by epifaunal free-living animals (*Am-*

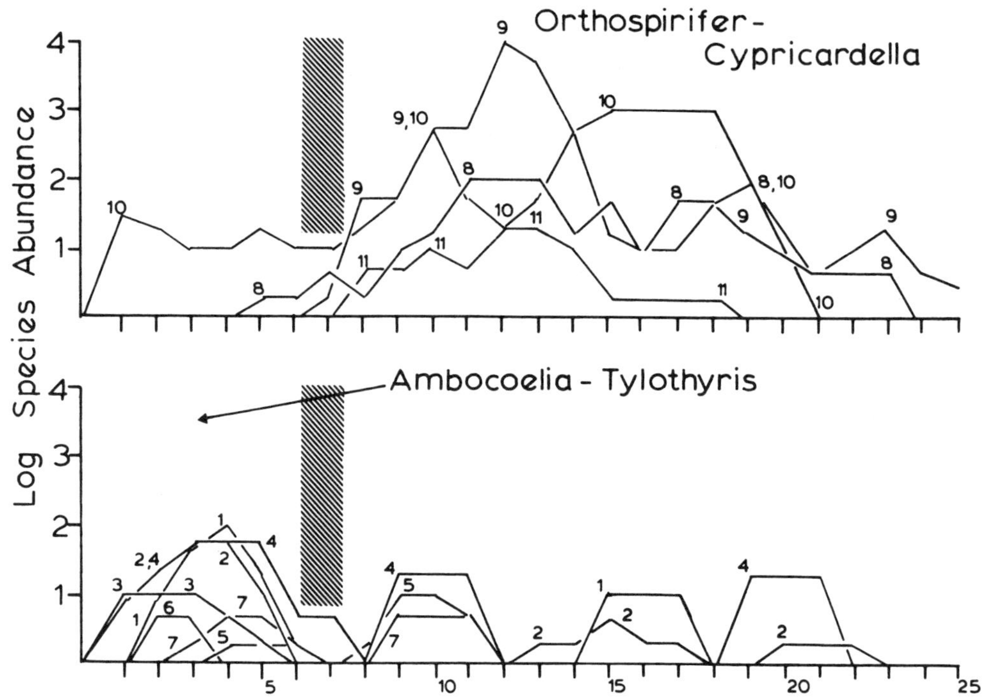

Figure 4. Characteristic species of the *ORTHOSPIRIFER-CYPRICARDELLA* and *AMBOCOELIA-TYLOTHYRIS* Communities within the "Millport equivalent." Shaded bars represent boundaries assigned to the areal distribution of each community. See Figure 2 for stratigraphic positions. Species identifications: 1, *Ambocoelia umbonata*; 2, *Tylothyris mesacostalis*; 3, *Productella* spp.; 4, *"Camarotoechia"* spp.; 5, *"Cornellites" chemungensis*; 6, *Douvillina* spp.; 7, *Spinatrypa* spp.; 8, *Cypricardella* spp.; 9, *Orthospirifer mesastrialis*; 10, *Devonochonetes* spp.; 11, *Cyrtospirifer chemungensis*.

bocoelia, Spinatrypa) or semi-infaunal sessile brachiopods (*Productella, Douvillina*). The shallower water *ORTHOSPIRIFER-CYPRICARDELLA* Community is characterized by populations of pediculately attached epibenthic brachiopods (*Orthospirifer, Cyrtospirifer,* and also *"Camarotoechia"*), and larger numbers of bivalve molluscs. The bivalve species are mainly infaunal mobile burrowers (*Cypricardella, Palaeoneilo constricta, Nuculoidea corbuliformis, Eoschizodus? chemungensis*) or semi-infaunal less mobile species (*Goniophora chemungensis, Leiopteria nitida*). In proportion, unattached epibenthic or semi-infaunal brachiopods are much less common than in deeper water habitats.

For further discussion concerning the characteristic habitats and mode-of-life of the species listed above, and in following sections of the paper, the reader is referred to the previous works of Bowen and others (1974), Thayer (1974), McGhee (1976), and McGhee and Sutton (1981).

"Beers Hill Equivalent"

In contrast to the Millport interval, a distinct three-fold pattern of community development emerges during deposition of the Beers Hill (Fig. 5). From the time of the deposition of the Millport to that of Beers Hill, there occurred a broadening of the delta platform to the west (Fig. 3), although the nearshore channel, lagoonal, and estuarine region remained virtually stationary geographically. This virtual still-stand in the progradation of nearshore facies into the basin is further documented in the study of the eastern strata of the lower West Falls Group by Ehrets (1981).

With the broadening of the delta platform, the characteristic deeper water community is displaced somewhat to the west of its position during Millport deposition. During deposition of the Beers Hill, the *AMBOCOELIA-SPINATRYPA* Community developed in prodelta and shelf habitats (Figs. 3 and 5). This low-diversity community is similar to the older *AMBOCOELIA-TYLOTHYRIS* Community, but significant dominance rearrangements occur in the species populations of offshore regions between the Millport and the Beers Hill (Table 2). *Tylothyris mesacostalis* and *"Camarotoechia"* spp. decline markedly in abundance, whereas the species populations of *Spinatrypa* increase three-fold in numbers. While the spiriferacean *Tylothyris mesacostalis* declines in abundance (shifting to shallower water habitats to the east; see Table 2), the spiriferaceans *Cyrtospirifer chemungensis* and *Orthospirifer mesastrialis* expand their environmental range and occupy deeper water habitats, where they previously are not present in this deeper water setting. However, the community as a whole is still dominated by sessile brachiopods, either unattached epibenthos (with the introduction of the small orthide *Thiemella danbyi*) or semi-infaunal species.

Reorganization in species dominance also occurs in near-

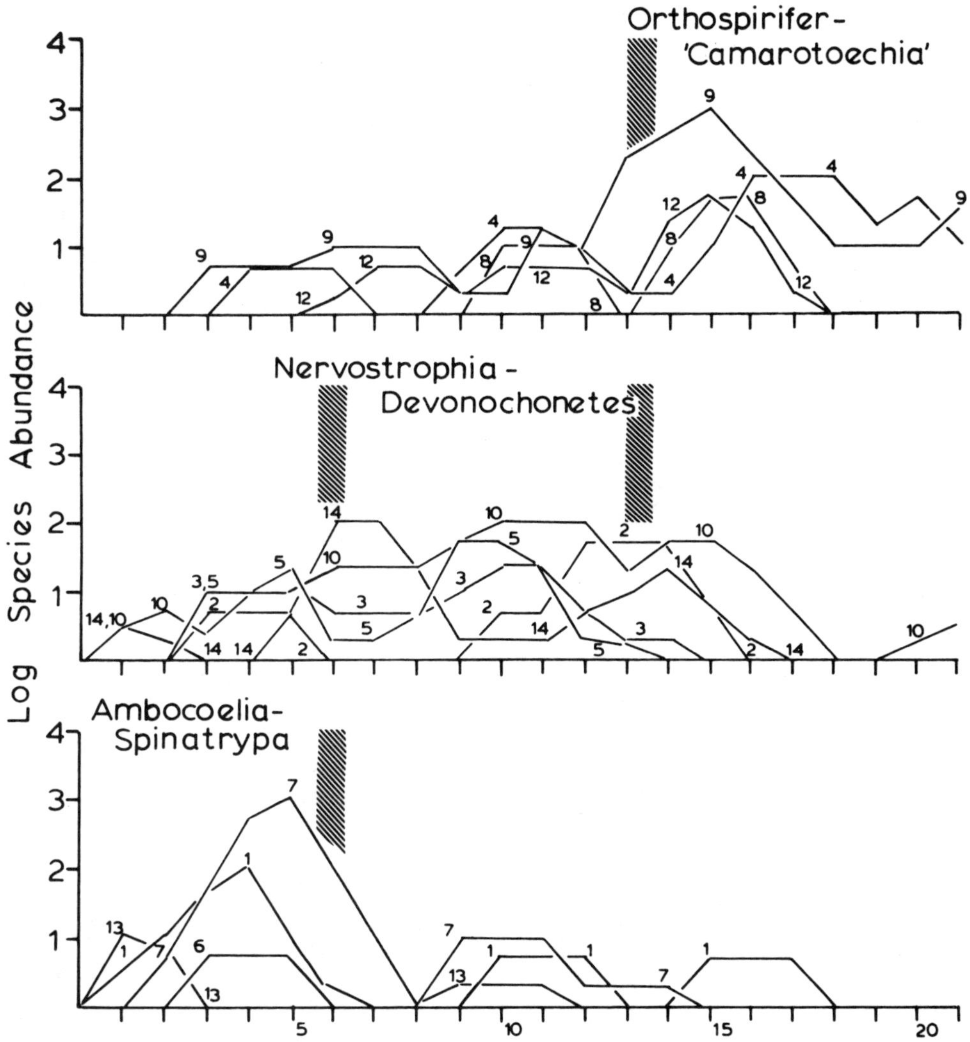

Figure 5. Characteristic species of the *ORTHOSPIRIFER-"CAMAROTOECHIA," NERVOSTROPHIA-DEVONOCHONETES,* and *AMBOCOELIA-SPINATRYPA* Communities within the "Beers Hill equivalent." Identities of the species not previously listed in Figure 4: 12, *Palaeoneilo constricta;* 13, *Thiemella danbyi*; 14, *Nervostrophia nervosa.*

shore, shallower water populations with the development of the *ORTHOSPIRIFER-"CAMAROTOECHIA"* Community (Fig. 5). Characteristic species of this assemblage include *Orthospirifer mesastrialis, "Camarotoechia" congregata* and *"C." eximia.* The infaunal bivalve *Palaeoneilo constricta* increases in abundance, while simultaneously the infaunal species of *Cypricardella* slip in overall numbers (Fig. 5, Table 2). The temporal decline in abundance of *Cypricardella* is consistent from the Sonyea Group—where it is very abundant—throughout the successive strata of the lower West Falls Group. Structurally, however, the *ORTHOSPIRIFER-"CAMAROTOECHIA"* Community is very similar to the older *ORTHOSPIRIFER-CYPRICARDELLA* Community, being dominated by attached epifauna (*Orthospirifer, "Camarotoechia," Cyrtospirifer, Tylothyris*) and mobile infauna (*Palaeoneilo, Cypricardella, Eoschizodus?, "Lingula"*).

The broad outer platform regions, newly developed to the west of Binghamton, are inhabited by a diverse species ensemble not present in the narrower platform habitats of the Millport. Characteristic members of the *NERVOSTROPHIA-DEVONOCHONETES* Community include *Nervostrophia nervosa, Devonochonetes lipidus, D. deflectus, D. setigerus, "Cornellites" chemungensis, Productella speciosa, P. lachrymosa,* and *Tylothyris mesacostalis* (Fig. 5). The community is also characterized by a high diversity of adaptations for living in benthic habitats and in this regard resembles other outer platform communities, which are only developed in the later Frasnian (the *ORTHOSPIRIFER-CYPRICARDELLA* Community in the central Appalachians (McGhee, 1976), and the *TYLOTHYRIS-SCHIZOPHORIA* Community in New York (McGhee and Sutton, 1981)). Life habits represented include unattached free-living epibenthos (*Nervostrophia, Devonochonetes, Spinatrypa, Ambocoelia*) and

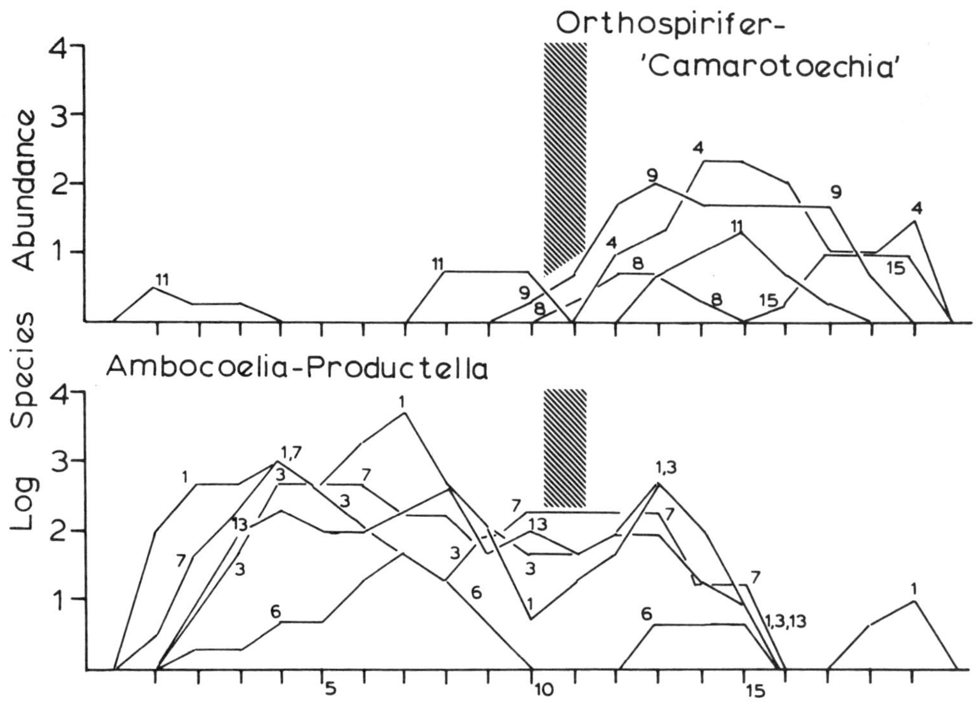

Figure 6. Characteristic species of the *ORTHOSPIRIFER-"CAMAROTOECHIA"* and *AMBOCOELIA-PRODUCTELLA* Communities within the "Meads Creek equivalent." Species not previously identified in Figure 4 and 5: 15, *Mucrospirifer mucronatus.*

attached epifauna, either by pedicle (*Orthospirifer, Cyrtospirifer, Tylothyris, Mucrospirifer*), by byssus (*"Cornellites," Leiopteria*), or by pedicular fibrils (*Schuchertella chemungensis*). Semi-infaunal brachiopods (*Productella*) and infaunal molluscs (*Cypricardella, Palaeoneilo*) are also present (Table 3), as are colonies of bryozoans. Interestingly, bellerophontid molluscs are also present in modest numbers. These animals—probably grazers (Bowen and others, 1974)—are usually found only in shallow-water, nearshore regions in the Sonyea Group (Sutton and others, 1970) and in the Millport equivalent (Table 1). Their presence argues for the presence of benthic plants, which in turn indicates the shallow-water nature of outer platform habitats of the Beers Hill.

"Meads Creek Equivalent"

The diverse, intermediate community-type is not developed in the Meads Creek, and marine ecosystems return to the two-fold pattern of a characteristic deeper water and shallower water species ensemble (Fig. 6). The shallower water *ORTHOSPIRIFER-"CAMAROTOECHIA"* Community now, however, shifts and expands considerably to the west of the Binghamton area (cf. sections locations in Figs. 6 and 2). During both Millport and Beers Hill deposition, the spiriferacean-rhynchonellide-bivalve community-type is geographically stationary in the area from Binghamton to the east.

There is considerable blurring and overlap of species gradients in the Meads Creek (Fig. 6, Table 3). The division of the two community-types—a shallower water attached epifaunal and a mobile infaunal assemblage—is less distinct than anywhere else in the lower West Falls.

The characteristic species of the *ORTHOSPIRIFER-"CAMAROTOECHIA"* Community retain their co-occurrences from the Beers Hill to the Meads Creek (cf. Figs. 5 and 6), and thus it is considered that the recognition of a new community in nearshore regions of the Meads Creek is not justified. There is, however, an increase in species within the community which are usually more abundant in deeper water regions (Table 3). Although still dominated by attached epibenthic brachiopods and infaunal bivalves, the community has become "diluted" by populations of free-lying and semi-infaunal brachiopods.

Species of the brachiopod *Spinatrypa* expand their geographic distribution considerably from the Beers Hill to the Meads Creek, and lose their ecological distinctiveness as being predominantly confined to deeper water habitats (Table 3). Characteristic deeper water species include *Ambocoelia umbonata, Productella speciosa, P. lachrymosa, Douvillina mucronata, D. cayuta,* and *Thiemella danbyi* of the *AMBOCOELIA-PRODUCTELLA* Community (Fig. 6). Although still dominated by unattached epibenthic brachiopods, there is an increase in numbers of sessile semi-infauna (*Productella, Douvillina*) in this community. As with the *ORTHOSPIRIFER-"CAMAROTOECHIA"* Community, the ecological distinctiveness of the community is blurred by the addition of elements such as the byssally attached bivalve *"Cornellites" chemungensis* and the ecologically

TABLE 1. SPECIES COMPOSITION OF MARINE COMMUNITIES WITHIN THE MILLPORT*

Ambocoelia-Tylothyris Community (Offshore)	*Orthospirifer-Cypricardella* Community (Nearshore)
1. Major elements:	1. Major elements:
'Camarotoechia' spp.	*Orthospirifer mesastrialis*
Devonochonetes spp.	*'Camarotoechia'* spp.
Tylothyris mesacostalis	*Devonochonetes* spp.
Ambocoelia umbonata	cricoconarids
'Spirifer'? posterus	*Cypricardella* spp.
Productella spp.	*Goniophora chemungensis*
Leptodesma spp.	*Leiopteria nitida*
Palaeoneilo constricta	*Ambocoelia umbonata*
Spinatrypa spp.	*Palaeoneilo constricta*
Rhipidomella vanuxemi	*Nervostrophia nervosa*
Goniophora chemungensis	bryozoans
	Cyrtospirifer chemungensis
	Eoschizodus? chemungensis
2. Minor elements:	*'Cornellites' chemungensis*
Schizophoria spp.	bellerophontid
Douvillina spp.	*Spinatrypa* spp.
Leiopteria nitida	*Leptodesma* spp.
'Cornellites' chemungensis	*Nuculoidea corbuliformis*
Cyricardella spp.	
Grammysia elliptica	2. Minor Elements:
Grammysioidea subarcuata	*Tylothyris mesacostalis*
Lyriopecten spp.	*Mucrospirifer mucronatus*
	Grammysia elliptica
	Cyrtina hamiltonensis
	Rhipidomella vanuxemi
	Schizophoria spp.
	Edmondia? philipi
	Lyriopecten spp.
	Myalina? newelli
	Paracyclas? sp.
	Spathella typica

*Species are listed in order of rank abundance; species listed as "minor elements" occur in less than 10 percent of the field locations. Pelmatozoan echinoderm fragments are abundant throughout lower West Falls environments, but population sizes are very difficult to estimate and their abundance was not ranked.

similar pedicular-fibril attached brachiopod *Schuchertella chemungensis.*

Although the absolute diversity of deeper water communities declines from the Beers Hill to the Meads Creek (cf. Tables 2 and 3), the abundances of individuals within species populations increase dramatically (cf. Figs. 5 and 6). The *AMBOCOELIA-SPINATRYPA* Community is richer in species numbers than the *AMBOCOELIA-PRODUCTELLA* Community, yet average population sizes in the latter community are an order of magnitude larger than the former.

Summary

The geographic distribution pattern of marine communities during deposition of the lower West Falls is given in Figure 7 and will be discussed here in relation to the distribution of major environmental zones given in Figure 3.

Throughout the lower West Falls, shallower water, inner platform, lagoonal, and estuarine environments are inhabited by communities of the spiriferacean-rhynchonellide-bivalve composition. These communities are characteristically composed of an attached epifauna of brachiopods and bivalve molluscs, and a mobile infauna of shallow burrowing bivalves. Deeper water shelf, prodelta, and outer platform habitats are inhabited by lower diversity communities of unattached epibenthic brachiopods and sessile semi-infaunal brachiopods, with very few molluscs.

During deposition of the Millport, the *AMBOCOELIA-TYLOTHYRIS* Community was confined to deeper water prodelta and open shelf regions, whereas the *ORTHOSPIRIFER-CYPRICARDELLA* Community developed in narrow platform regions in the area of Binghamton, extending eastward into estuarine regions in the vicinity of Windsor (Figs. 2, 3, and 7).

During deposition of the Beers Hill, the *ORTHOSPIRIFER-"CAMAROTOECHIA"* Community developed in approximately the same geographic area as the older *ORTHOSPIRIFER-CYPRICARDELLA* Community. During this period of time, there was a virtual still-stand of progradation of nearshore facies to the west (Ehrets, 1981). The basin shallowed overall, however, with the expansion of a broad platform region not present during Millport deposition (Fig. 3). In this region, the highly diverse *NERVOSTROPHIA-DEVONOCHONETES* Community flourished (Fig. 7), while the *AMBOCOELIA-SPINATRYPA* Community developed in outer platform, platform edge, and prodelta regions.

TABLE 2. SPECIES COMPOSITION OF MARINE COMMUNITIES WITHIN THE BEERS HILL*

Ambocoelia-Spinatrypa Community (Offshore)	Nervostrophia-Devonochonetes Community	Orthospirifer-'Camarotoeshia' Community (Nearshore)
1. Major elements:	1. Major elements:	1. Major elements:
Spinatrypa spp.	Orthospirifer mesastrialis	Orthospirifer mesastrialis
Cyrtospirifer chemungensis	bryozoans	Tropidoleptus carinatus
Ambocoelia umbonata	Nervostrophia nervosa	'Camarotoechia' spp.
bryozoans	Cyrtospirifer chemungensis	cricoconarids
'Cornellites' chemungensis	Spinatrypa spp.	Cyrtospirifer chemungensis
auloporid cnidarians	Devonochonetes spp.	Tylothyris mesacostalis
Productella spp.	Tylothyris mesacostalis	Cypricardella spp.
Devonochonetes spp.	bellerophontid	Palaeoneilo constricta
Nevrostrophia nervosa	Productella spp.	'Lingula' spp.
Schuchertella chemungensis	'Cornellites' chemungensis	Devonochonetes spp.
'Camarotoechia' spp.	Schuchertella chemungensis	Nervostrophia nervosa
Douvillina spp.	Cypricardella spp.	Ambocoelia umbonata
Mucrospirifer mucronatus	cricoconarids	Schuchertella chemungensis
Orthospirifer mesastrialis	'Camarotoechia' spp.	Eoschizodus? chemungensis
'Spirifer' posterus	Palaeoneilo constricta	Leioptoria nitida
Thiemella danbyi	Leiopteria nitida	
Tylothyris mesacostalis	Ambocoelia umbonata	
	Mucrospirifer mucronatus	
2. Minor elements:	2. Minor elements:	2. Minor elements:
rugosid cnidarians	'Spirifer'? posterus	Goniophora chemungensis
Vertumnia? reversa	Goniophora chemungensis	Spinatrypa spp.
	Sphenotus tiogensis	Productella spp.
	Cyrtina hamiltonensis	Rhipidomella vanuxemi
	Leiorhynchus spp.	'Spirifer' posterus
	Thiemella danbyi	bellerophontid
	Pseudoaviculopecten fasciculatus	bryozoans
	Edmondia? philipi	
	Grammysia elliptica	
	Myalina? newelli	
	auloporid cnidarians	

*Table format identical to Table 1.

During deposition of the Meads Creek, nearshore facies prograded kilometers to the west, and the platform expanded by tens of kilometers (Fig. 3). The diverse mid-platform *NERVOSTROPHIA-DEVONOCHONETES* Community vanished, and a two-fold pattern of community development was re-established along the onshore-offshore environmental gradient. The *ORTHOSPIRIFER-"CAMAROTOECHIA"* Community expanded considerably to the west, while the *AMBOCOELIA-PRODUCTELLA* Community advanced across outer platform environments to the east (Figs. 7 and 3). The ecological distinctiveness of the two community-types is considerably blurred during this interval of time. Species diversity dropped in offshore regions, although overall within species population sizes increased markedly.

EVOLUTION OF LOWER WEST FALLS ECOSYSTEMS

Role of Sand Amount and Distribution on the Shelf

Two variables must be considered here: The total amount of sand found in the platform area, and the distribution pattern of

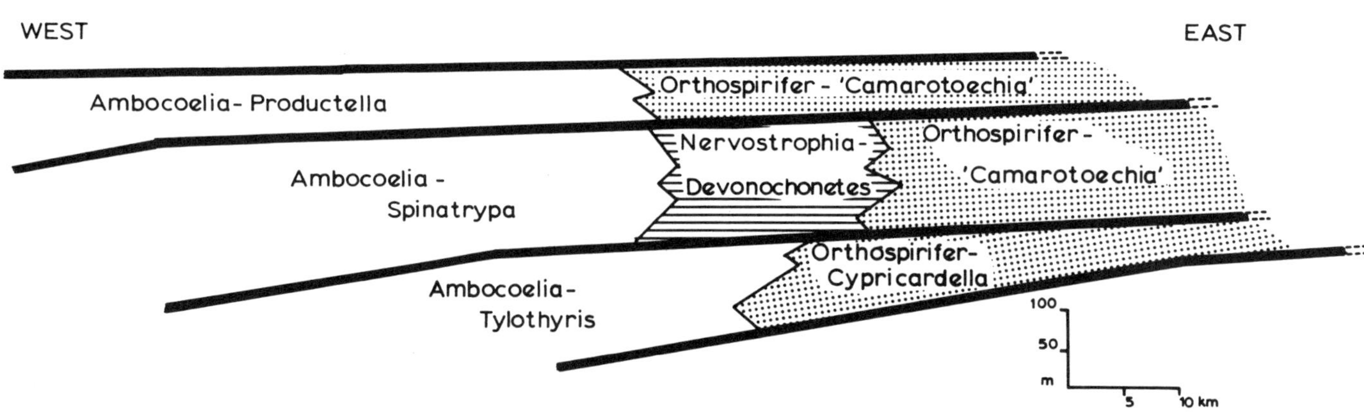

Figure 7. East-west cross section showing marine benthic communities in the lower West Falls.

TABLE 3. SPECIES COMPOSITION OF MARINE COMMUNITIES WITHIN THE MEADS CREEK STRATIGRAPHIC INTERVAL*

Ambocoelia-Productella Community (Offshore)	Orthospirifer-'Camarotoechia' Community (Nearshore)
1. Major elements:	1. Major elements:
Ambocoelia umbonata	Spinatrypa spp.
Spinatrypa spp.	Nervostrophia nervosa
Nervostrophia nervosa	Tylothyris mesacostalis
Thiemella danbyi	'Camarotoechia' spp.
Productella spp.	auloporid cnidarians
Schuchertella chemungensis	Ambocoelia umbonata
'Cornellites' chemungensis	Devonochonetes spp.
auloporid cnidarians	Productella spp.
Tylothyris mesacostalis	Orthospirifer mesastrialis
Douvillina spp.	Thiemella danbyi
Leptodesma spp.	Leiorhynchus spp.
Devonochonetes spp.	Palaeoneilo constricta
Cyrtospirifer chemungensis	'Spirifer'? posterus
Palaeoneilo constricta	Tropidoleptus carinatus
	Cyrtospirifer chemungensis
2. Minor elements:	Mucrospirifer mucronatus
Orthospirifer mesastrialis	Leptodesma spp.
bellerophontid	rugosid cnidarians
rugosid cnidarians	Douvillina spp.
Edmondia? philipi	
Pseudoaviculopecten fasciculatus	2. Minor elements:
Spathella typica	Goniophora chemungensis
Vertumnia? reversa	Cypricardella spp.
	'Cornellites' chemungensis
	Actinopteria boydi
	Schuchertella chemungensis
	Glossites? depressus
	Leiopteria nitida
	Myalina? newelli
	Paracyclas? sp.
	Pseudaviculopecten fasciculatus
	Eoschizodus? chemungensis
	Spathella typica
	Vertumnia? reversa
	cricoconarids
	bellerophontid

*Table format identical to Table 1.

TABLE 4. LITHOLOGIC COMPOSITION OF 65 SHELF SECTIONS IN THE LOWER WEST FALLS GROUP

	Sandstone (%)	Mudstone (%)	Shale (%)	Siltstone (%)	Coquinite (%)
Meads Creek	37	39	14	10	tr
Beers Hill	23	48	24	5	tr
Millport	5	71	15	9	tr
Average for the Lower West Falls Group	25	49	19	7	tr

sand within that area. Table 4 displays a substantial increase in total sand content from 5 percent in the Millport to 37 percent in the Meads Creek. Mudstones show a corresponding decrease from 71 percent to 39 percent in that same interval. Shales vary by 10 percent but show no trend. Siltstones vary even less.

The second factor is the distribution of the sand within the shelf subenvironments. Typically, sand is concentrated at the shoreline as channel deposits and distributary mouth bars, or at the western edge of the platform as bars or delta front sands. The intervening platform is dominated by mudstones and other fine-grained clastics. There are two ways in which the actual setting can depart from this pattern. One is in the uniformity of sand distribution. Sand may be distributed uniformly from the shoreline to the platform, or it may be concentrated in one or several positions. In order to understand the changes in the sediment distribution on the shelf, several profiles were constructed showing the percentages of each rock type at each section. The profile illustrating percent-sandstone utilized not only the data from the sections shown in Figure 2 but was supplemented by data from 20 others. The additional 20 sections had been measured during the course of the post-Sonyea field studies but lacked fossil data for use here. The profile for sand distribution, where sand is shown as percentage of the measured section, is shown in Figure 8.

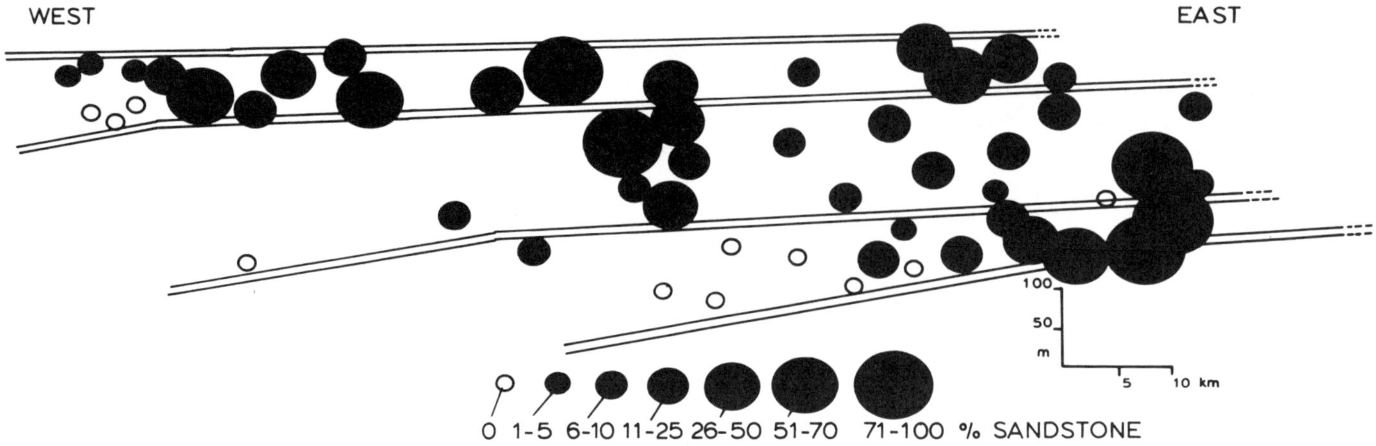

Figure 8. Distribution of sandstone in the lower West Falls Group. See text for discussion. Only 11 of the 43 sections without sand shown here in the prodelta region; sections without sand in platform and nearshore regions are left blank.

Temporal Patterns of Physical and Biological Change

Variations in the amount of sand as well as its position on the shelf determined the nature of the substrate and are crucial paleoecological factors. During Millport deposition, most of the sand was concentrated in the channels or distributary mouth bars. Little reached the open shelf (Figs. 8 and 9). The platform was very narrow and composed mostly of mud, with a wide shelf covered by somewhat deeper water than the narrow platform to the east. Some of this mud was carried beyond the shelf edge and was added to the slope, broadening the shelf at the same time.

During Beers Hill deposition, a new pattern was established (Fig. 9). Sand bars developed in the mid-shelf area effectively separating a much broader platform of shallow water with muds trapped behind the sands. These muds represented the environment inhabited by the *NERVOSTROPHIA-DEVONOCHONETES* Community. This community extended eastward to a point where sand content increased—which, in conjunction with possible salinity changes, resulted in a change to the *ORTHOSPIFIFER-"CAMAROTOECHIA"* fauna. The outer platform sands were, in turn, occupied by the *AMBOCOELIA-SPINATRYPA* fauna. West of the sand bars were the fine-grained sediments of the prodelta and open shelf in yet deeper waters (Figs. 8 and 9).

The reason for this dramatic change in sand distribution lies in the increase in sand available to the shelf. The sand content, having been only 5 percent in Millport equivalent, jumped to 23 percent. The *NERVOSTROPHIA-DEVONOCHONETES* Community apparently failed to develop earlier during Millport deposition because of the virtual absence of a well-developed sand barrier on the platform's seaward edge. It is possible that the platform environments were particularly shallow during deposition of the Beers Hill.

During deposition of Meads Creek sediments, a still different pattern was established on the shelf (Fig. 9). Following deposition of the dark gray muds of the Roricks Glen, the sand barrier was once again established in the same position as in the Beers Hill, but it enlarged rapidly both to the west and to the east. Some mud was trapped behind the sands but the total volume was considerably reduced from that of the Beers Hill, and with it, the species ensemble characteristic of the *NERVOSTROPHIA-DEVONOCHONETES* Community. Part of the reduction was caused by the expansion of nearshore channel and distributary mouth bar sands that were inhabited by the *ORTHOSPIRIFER-"CAMAROTOECHIA"* Community.

Except for the muds, the other fine-grained sediments (shales and siltstones) may be found in two concentrations: in the nearshore platform and the prodelta-open shelf. Both environmental zones are viewed as low-energy, with weak and intermittent currents forming plane-beds or small ripples.

The dark gray shales and siltstones on the shelf and identified as the Moreland, Dunn Hill, Roricks Glen, and Corning represent an environment unfavorable to the fauna found stratigraphically above or below. They record no evidence of active bottom currents. Fossils are rare and the sediments show little bioturbation. From this, it is concluded that few organisms inhabited the bottom or the waters above even temporarily. It is not known if all of the shelf was covered by such sediment, but both subsurface data and field observations strongly suggest that a drape of dark muds blanketed it. It is possible that some offshore sand bars and distributary mouth bars provided projections into less anoxic waters close to the surface. These would have served as temporary havens for remnants of the previously extensive platform faunas. The stratigraphic control currently being employed is not sufficient to permit a test of that hypothesis.

CONCLUSIONS

The distribution and development of distinctive animal communities during lower West Falls time were strongly con-

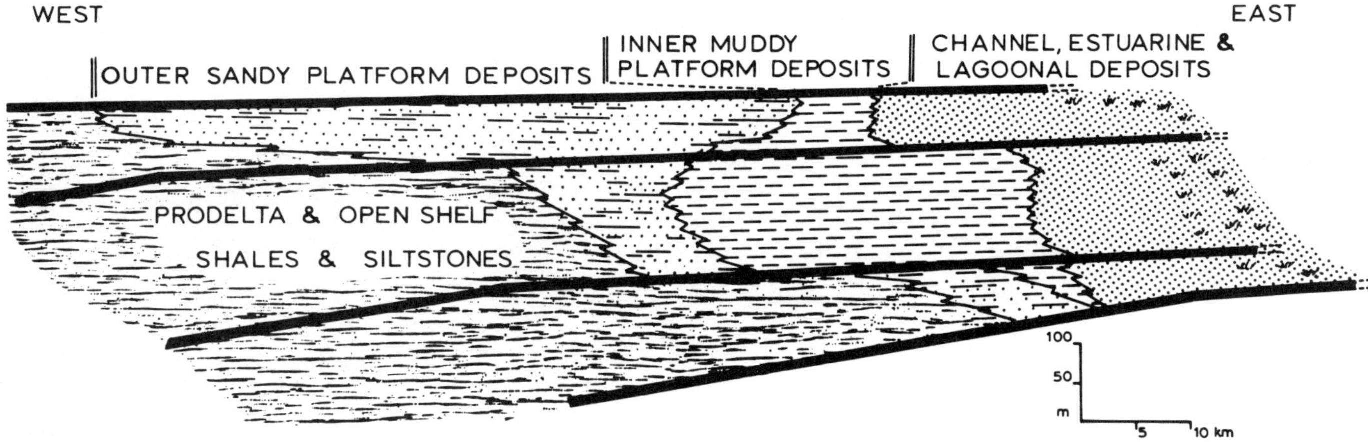

Figure 9. Changes in sediment distribution patterns on the shelf during deposition of the lower West Falls.

trolled by the sedimentary character of the substratum (Figs. 7 and 9), which in itself was related to differences in water depth and the rate of input of sediments into the basin. Initially, the environments consisted of a narrow platform and an extensive deep-water prodelta and open shelf (Millport equivalent). Coarse sediments were largely confined to shallow-water nearshore regions. Two distinctive benthic community-types developed within this environmental context—a shallow-water spiriferacean-rhynchonellide-bivalve assemblage and a deeper water assemblage of unattached epibenthic and sessile semi-infaunal brachiopods. These two community-types persisted throughout lower West Falls deposition in one form or another.

With time, the delta platform broadened considerably (Beers Hill equivalent). Coarse sediments were now transported across the shallow platform and formed delta front sand bars near the platform edge. Landward of these bars, finer sediments accumulated, and a new community of both high species and ecological diversity in modes-of-life developed in mid-platform regions (the *NERVOSTROPHIA-DEVONOCHONETES* Community).

During deposition of the Meads Creek equivalent, the western edge of the platform migrated rapidly to the west (Fig. 3). There is no evidence to suggest that the subsidence rate of the shelf area during this time was significantly different from that which existed during the deposition of the underlying Beers Hill equivalent. Therefore, the most logical explanation for the enlargement of the platform and the rapid prograding of its western edge would be an increase in the rate of sediment input. Meads Creek platform sediments do consist of a relatively large amount of sand (Table 4). These sands destroyed the previous regional differences in the substratum type; the Meads Creek Communities intergraded and became less distinct than at any other time during West Falls deposition.

ACKNOWLEDGMENTS

We thank Drs. Z. P. Bowen and A. L. McAlester for the use of some of their unpublished paleontological data, collected in the 1968 and 1969 field seasons. We also thank Drs. J. J. Sepkoski and R. K. Bambach for their helpful reviews of the manuscript. The Rutgers Research Council provided travel funds which made further examination of the University of Rochester paleontological collections possible. The research was supported in part by a grant from the Earth Sciences Division, National Science Foundation, NSF Grant EAR 81-15518 to G. R. McGhee. Also, acknowledgment is made to the donors of the Petroleum Research Fund, administered by the American Chemical Society, for partial support of this research (ACS-PRF 13481-AC2).

REFERENCES CITED

Bowen, Z. P., Rhoads, D. C., and McAlester, A. L., 1974, Marine benthic communities in the Upper Devonian of New York: Lethaia, v. 7, p. 93–120.

Ehrets, J. R., 1981, The West Falls Group (Upper Devonian) Catskill Delta Complex: stratigraphy, environments, and sedimentation: New York Geological Association, Field Trip Guidebook 53, p. 3–25.

McGhee, G. R., Jr., 1976, Late Devonian benthic marine communities of the central Appalachian Allegheny Front: Lethaia, v. 9, p. 111–136.

—— 1981, Evolutionary replacement of ecological equivalents in Late Devonian benthic marine communities: Palaeogeography, Palaeoclimatology, Palaeoecology, v. 34, p. 267–283.

McGhee, G. R., Jr., and Sutton, R. G., 1981, Late Devonian marine ecology and zoogeography of the central Appalachians and New York: Lethaia, v. 14, p. 27–43.

—— 1983, Evolution of late Frasnian (Late Devonian) marine environments in New York and the central Appalachians: Alcheringa, v. 7, (1-2), 9–21.

Rickard, L. V., 1975, Correlations of the Silurian and Devonian rocks in New York State: New York State Museum and Science Service, Map and Chart Series, no. 24, 16 p.

Sutton, R. G., 1963, Correlation of the Upper Devonian strata in south-central New York: Pennsylvania Geological Survey General Report, G39, p. 87–101.

Sutton, R. G., Bowen, Z. P., and McAlester, A. L., 1970, Marine shelf environments of the Upper Devonian Sonyea Group of New York: Geological Society of America Bulletin, v. 81, p. 2975–2992.

Thayer, C. W., 1974, Marine paleoecology in the Upper Devonian of New York: Lethaia, v. 7, p. 121–155.

Whittaker, R. H., 1975, Communities and Ecosystems: Macmillan Pub. Co., Inc., New York, 385 p.

Manuscript Accepted by the Society July 20, 1984

Printed in U.S.A.

Geological Society of America
Special Paper 201
1985

The evolution of Frasnian marine "community-types" of south-central New York

Robert G. Sutton
Department of Geological Sciences
University of Rochester
Rochester, New York 14627

George R. Mc Ghee, Jr.
Department of Geological Sciences
Wright Geological Laboratory
Rutgers University
New Brunswick, New Jersey 08903

ABSTRACT

The evolution of Frasnian marine "community-types" in south-central New York represents a complex interaction of biological and physical processes. The two most important physical factors were shelf profile (bathymetric configuration) and regional differences in substratum type, and temporal changes in both. Offshore marine communities were more sensitive to changes in bathymetric configuration than were the nearshore communities; the latter retained an essentially unchanging ecological structure throughout Frasnian time. Three major biome phases can be discerned in the Frasnian strata of south-central New York. These reflect the progressive shallowing on the shelf, the increase in overall sand content with time, and the geographic expansion of sand-grade substrata in shallow marine habitats.

INTRODUCTION

During Frasnian time, the bulk of the clastic deposits known as the Catskill delta were deposited across New York state. During the subsequent Famennian time, marine deposition was confined to the western part of the state. This deltaic sequence has played a historic role in the development of geologic thinking and served as a basis for models constructed by Hall, Chadwick, Caster, Cooper and others during the 19th and first half of the 20th centuries. Space does not permit us to review these contributions nor to credit the hundreds of contributors whose labors have led to the advances in our understanding of these strata and their faunas. The results of detailed studies outlined below permits a re-examination of the existing concepts and the construction of a new model that helps to explain the faunal changes that occurred in shallow marine environments during Frasnian time.

Three important advances have occurred in the past 40 years that have made possible our revision of the clastic facies models of Chadwick (1933) and Caster (1934). They emphasized the facies relationships that coexisted during the progradation of the delta toward the west. Both emphasized the striking similarities of the faunas and sediments that represented the same paleoenvironments. The first advance occurred when the basinal sediments were accurately correlated to the nearshore marine deposits and non-marine clastics to the east. The tongues of black shales in the basin pass eastward into fossiliferous strata and subdivide the faunas. This subdivision made possible the second advance, a detailed examination of these faunas and the enclosing sediments and population studies of coexisting faunas. Finally, multivariate techniques permitted the analysis of masses of population data so that overall patterns emerged from otherwise confusing numbers. The purpose of this paper is to examine the overall faunal patterns during Frasnian time and to account for the changes observed.

STRATIGRAPHY OF THE FRASNIAN STAGE IN NEW YORK

The Frasnian consists of four groups of strata in New York state. They are in ascending order: Genesee (in part), Sonyea,

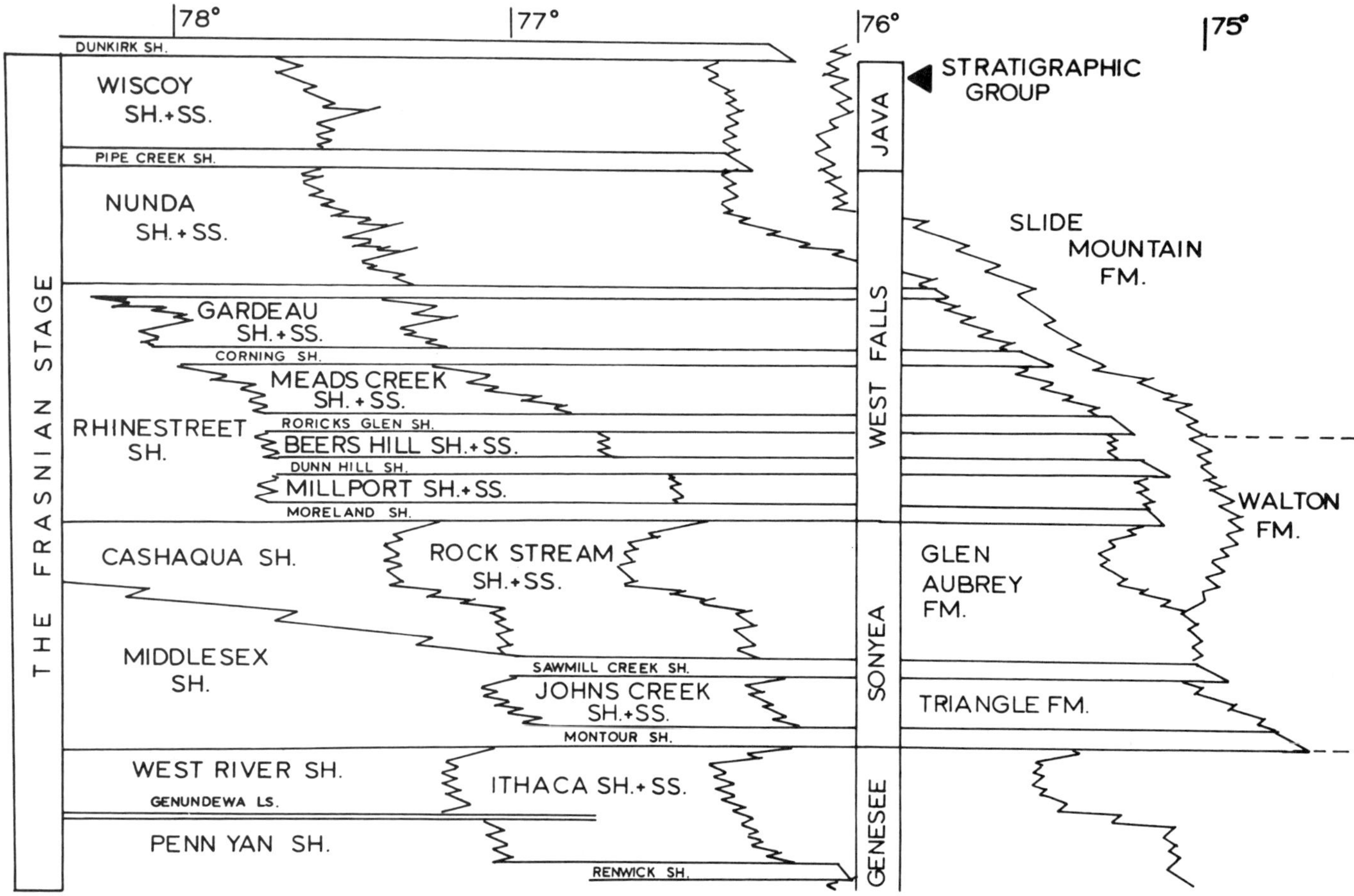

Figure 1. The Frasnian stratigraphic record of south-central New York state. Modified from Rickard (1975).

West Falls, and Java (Fig. 1). The International Subcommission for Devonian Stratigraphy has recently formally recommended that the base of the Frasnian Stage be placed at the base of the lower *Polygnathus asymmetricus* conodont Zone, with the result that the lower two-thirds of the Genesee Group now properly belongs in the Givetian Stage (Middle Devonian).

In western New York, the base of each group mentioned above is the base of a black shale (Geneseo, Middlesex, Rhinestreet, and Pipe Creek). Each black shale is overlain by gray shales (West River, Cashaqua, Angola, and Hanover). Eastward, the black shales are split into tongues by wedges of eastward-coarsening gray shales, siltstones and sandstones. A few examples of these wedges are the Sherburne, Rock Stream, Beers Hill, and Nunda. In east-central New York, the wedges become fossiliferous and in some places have been named in order to differentiate them from their relatively unfossiliferous equivalents to the west. Some examples of these wedges are the Unadilla, Otselic, Cincinnatus, Triangle, and Glen Aubrey. Eastward, these strata pass into nonmarine conglomerates, sandstones and shales. Examples are the Oneonta, Walton, Slide Mountain, and Sunfish. Facies terms traditionally applied to these strata are the "Catskill" for the nonmarine portion, "Chemung" for the fossiliferous part, "Portage" and "Cleveland" for the relatively unfossiliferous gray and black shales in the west (Fig. 2).

BASIN EVOLUTION DURING FRASNIAN TIME

In order to understand the changes in the Frasnian shallow marine faunas, it is necessary to identify the changes that took place in the marine basin to the west as well as in the non-marine areas to the east. The configuration of the marine basin changed considerably from its beginning during Genesee deposition to the end of Java deposition. At the beginning, the marine areas were shallow and extended from east-central New York to Lake Erie. In the west, the Geneseo lies unconformably upon the Hamilton Group and upon the Tully limestone in the east-central part. Thayer (1974) found that the earliest Genesee deposition was followed by a rapid westward progradation of the shoreline. He concluded that extensive subsidence occurred when Genesee time drew to a close. Clearly, at its inception, the water in the Frasnian basin was relatively shallow.

The source of the Frasnian clastics lay to the east in the Acadian orogen. This conclusion is based upon two lines of evidence: the petrology of the clastics and directional data. Burtner's

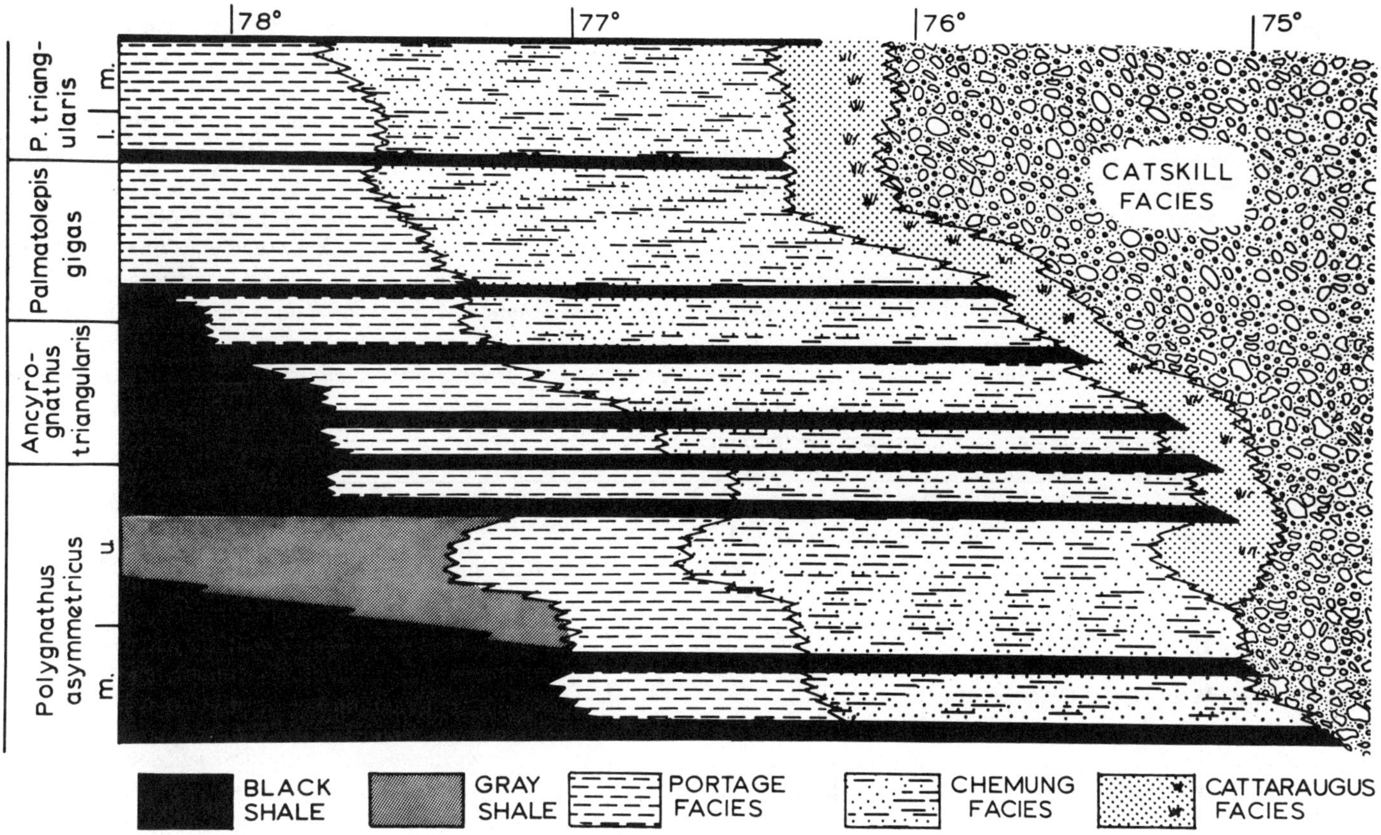

Figure 2. Stratigraphic and geographic distribution of the major Frasnian sedimentary facies in New York, with reference to conodont zonations. The upper part of the Genesee Group has here been omitted, though it belongs in the lowermost Frasnian. See text for discussion.

(1963) analysis of directional data in the entire 2,400 m (8,000-foot) sequence of the Catskill facies indicated sediment transport to the west. Lucier (1966) studied the non-marine facies equivalent to the Hamilton and Genesee Groups, a 600 m (2,000-foot) section at the Catskill front. He found not only directional data consistent with Burtner's but a petrographic assemblage that matched the rocks in the present-day Taconics, a few kilometers to the east. Chellman (1983) analyzed the nearshore marine sandstones from the base of the Sonyea Group into the Famennian strata above, a composite section of 1,800 m (6,000 feet). Her compositional data at the base matched that at the top of Lucier's section and showed a gradual but steady increase in metamorphic rock fragments, a continuation of the Lucier pattern, a trend which can be explained only by the continued unroofing of the region of the present-day Taconics.

During Frasnian time the marine basin in central and western New York received clastic sediments supplied by westward-flowing streams which had their origins in the Acadian Moutains. In the region of the present-day Catskills, in eastern New York, the streams had a high gradient and formed wedges of braided stream deposits. To the west, the streams had a much lower gradient and crossed a broad floodplain to the marine basin to the west. The bulk of the sediments transported to the shoreline by the streams was deposited directly in the nearshore marine area while relatively little found its way into the deeper parts of the marine basin to the west. The result was a pattern of progradation of nearshore marine and non-marine facies toward the west during Frasnian time.

During the Frasnian, the progradation rate was far from uniform. Sediments of most of the lower West Falls Group display no progradation at all and the lower part of the Middlesex Black Shale in the Sonyea Group transgresses eastward over non-marine strata of the Genesee Group (see Fig. 1). Times of non-progradation occur when dark muds and little sand accumulated in the nearshore marine areas (see McGhee and Sutton, this volume). These muds, now shales, have been correlated with black shales to the west. The marine strata of the Middlesex shale were traced eastward to a wedge of sandstones in the Catskill Mountains by Chadwick (1935), and the Geneseo Black Shale is correlated with the Twilight Park Conglomerate according to L. V. Rickard (personal communication). Similarly, the tongues of the Rhinestreet Black Shale in the lower West Falls Group were each correlated with distinct tongues of the Slide Mountain Conglomerate by Nugent (1960). Whatever explanation offered for non-progradation must also account for accumulation of black muds in the marine basin to the west, a paucity of sands and presence of muds in the nearshore marine area, and an accumulation of gravel deposits on the flanks of the Acadian Mountains.

One explanation for this sediment pattern lies in the changes of the sediment supply rate and its relation to rates of subsidence of the Appalachian Basin during Frasnian time. The Frasnian non-marine clastics form a wedge that thins westward across New York state, almost all deposited close to sea level. Subsidence must have been greater in the east than the west. The tongues of the Twilight Park and Slide Mountain Conglomerates suggest that the uplift of the Acadians may have been episodic. With these episodes of uplift, a corresponding subsidence of the Appalachian Basin is postulated. The maximum subsidence would occur in the area of the floodplain, between the shoreline of the marine basin to the west and the flanks of the Acadian Mountains to the east. Subsidence in this area would explain the marine transgressions during deposition of the Geneseo and Middlesex. Also, it would result in a decrease in stream gradients near the strand line resulting in a diminished sediment supply to the marine basin during deposition of the Rhinestreet. Each tongue of black shale in the basin is explained by a diminished sediment supply caused by a pulse of tectonic activity.

These times of diminished supply are visualized as temporary. The uplifted Acadians would begin to provide sediment at an increased rate; streams and their floodplains would become readjusted to their previous gradients and levels. As a result, the volume of sediment supplied to the marine basin proper would return to the pre-pulse rates. The marine basin reflects these changes in rate of sediment supply. During a high rate of supply, the marine muds in the basin are gray, the result of greater dilution of organic matter. At the same time, wedges of clastics build along the basins eastern margin and the facies display marked progradation. During times of low sediment supply, the marine muds in the basin are black because of a higher organic content due to less dilution. Relatively little clastic sediment is contributed to the nearshore marine area and virtually no progradation occurs.

Ettensohn and Barron (1981) provide a somewhat different explanation for the tongues of black shale. They propose that the Acadians were periodically uplifted and blocked the easterly moisture-laden winds, thereby reducing the erosion on the lee side. Erosion eventually reduced the mountains, more rain fell on the western side, and the clastic removal increased.

In New York state, the marine basin displayed marked changes in its configuration during Frasnian time. As stated before, the water in the basin was not deep at its inception. However, due to the differential subsidence, greater in the east than the west, and the accumulation of the bulk of the sediments along the eastern margin, the water in the basin deepened and a definite shelf and slope developed there. The basin was well established by the end of the Genesee deposition, continued to deepen during Sonyea and West Falls deposition, but Java sediments were emplaced in a marine basin that was shallower and progradation had resulted in the displacement of its deepest point into western New York. During Famennian time, deposition occurred in a relatively shallow basin with the marked distinctions of the shelf, slope and basin facies becoming blurred. By then, progradation was rapid and marine deposition quickly gave way to non-marine over much of the area (McGhee and Sutton, 1981, 1983, this volume).

The significance of the marine basin evolution lies in its development and demise as a sink for sediments transported from the shelf area. The winnowing of the fine-grained sediments and their transportation into deeper water resulted in more arenaceous shelf subenvironments. The changes in the sediment substrate and the configuration of the shelf help explain the changes in the distribution of the benthic marine communities during Frasnian time.

MARINE BIOMES DURING THE FRASNIAN STAGE

The temporal and geographic distribution of benthic marine communities throughout the Frasnian Stage in New York is summarized in Figure 3. In this figure are compiled the results of a series of paleoecological studies which have spanned the past decade (Sutton and others, 1970; Bowen and others, 1974; McGhee and Sutton, 1981, 1983, this volume).

Figure 3 shows the successive marine shellfish communities, and their geographic distributions, which inhabited the shallow water regions of the Catskill Sea during the deposition of each of the stratigraphic units given in Figure 1—excepting only the uppermost part of the Genesee Group—for which the reader is referred to Thayer (1974). Detailed ecological descriptions of each community's structure—taxonomic composition, species diversity, dominance structure, and overall trophic structure—as well as the community's characteristic habitats and discussions of the functional morphology of many of the community's constituent species, can be found in the series of papers listed above. Space limitations do not permit us to reiterate or summarize here the detailed and specific information given for each community in these previous papers.

Of primary interest to us in the present paper are large-scale ecosystem changes (or the lack thereof) in the New York shallow marine realm during the major span of Frasnian time. An ecosystem is here defined as "a community and its environment treated together as a functional system of complimentary relationships" (Whittaker, 1975). Discussions of these "complimentary relationships" of physical and biological factors can be found for each stratigraphic unit in the papers listed above. What we wish to examine here are larger groupings of ecosystems which are similar in community structure or physiognomy, which can alternately be designated as "community-types" or biomes" (Whittaker, 1975).

Can larger scale "community-types" or "biomes" be recognized in Frasnian fauna? If so, did these biomes change with time, or remain essentially static within their characteristic physical environments? The reader can easily recognize a rough correspondence between facies distributions and community distribu-

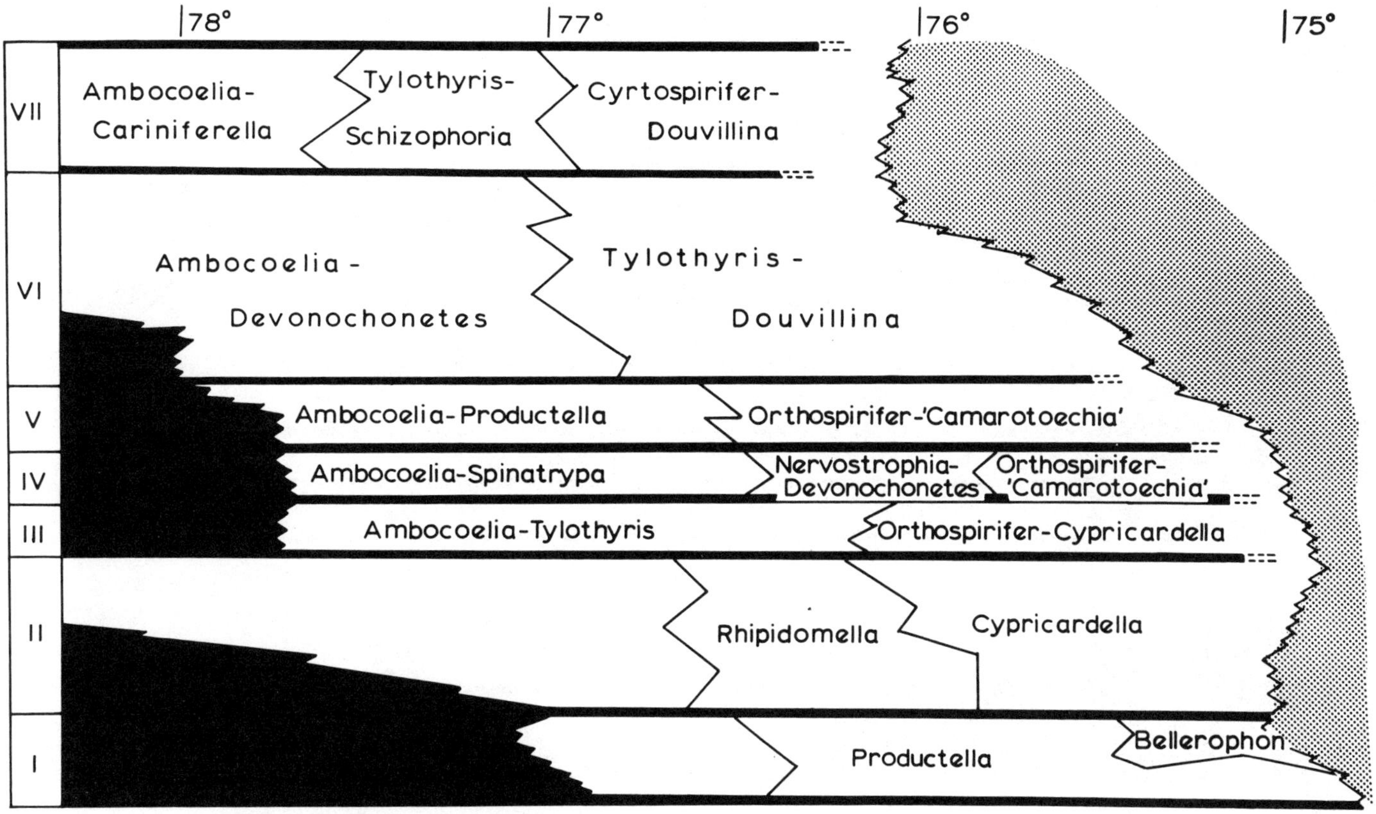

Figure 3. Stratigraphic and geographic distribution of Frasnian benthic marine communities. Roman numerals correspond to individual paleoecological studies in each stratigraphic unit as cited in text. Explanation: *VII*, Wiscoy Formation; *VI*, Nunda Formation; *V*, Meads Creek equivalent; *IV*, Beers Hill equivalent; *III*, Millport equivalent; *II*, Glen Aubrey Formation; *I*, Triangle Formation (see Fig. 1 for nomenclature).

tions by comparing Figure 2 with Figure 3. What we might roughly designate as the "*Ambocoelia*-type" grouping of communities (Fig. 3) largely, and consistently, occurs in the Portage facies (Fig. 2). A "spiriferacean-type" grouping of communities (those dominated by *Cyrtospirifer, Tylothyris,* and *Orthospirifer;* see Fig. 3) is largely and consistently found in the Chemung facies (Fig. 2). Do changes occur within these rough, but visually recognizable, "community-types" with time—or do they remain essentially unchanging throughout the span of the Frasnian?

The recognition of large-scale biome patterns within the Frasnian strata was undertaken through Q-Mode factor analysis. Taxonomic abundance data from a total of 224 fossiliferous field sections (Fig. 4) were analyzed. These sections are the same as those used in the recognition of the distinctive animal associations given in Figure 3. Factoring of the data was accomplished with the use of the Biomedical Series Program BMDP4M, designed at the University of California, Los Angeles. For a brief and clear explanation of the technique of factor analysis, the reader is referred to Davis (1973).

Five orthogonal factors were chosen, based on the pattern of decay of the eigenvalues (Fig. 5A) and the change in proportion of total variance exhibited by the successive factors (Fig. 5B). In the scree graph (Fig. 5A), the slope of the initial rapid decay of eigenvalues breaks after the fifth factor, the first five together accounting for 48 percent of the total variance within the data. Proportional variance of factors beyond the five chosen trail off in magnitude (Fig. 5B) such that to account for an additional 42 percent in variance (or 90 percent of the total variance in the data), an increase to 23 factor axes would be necessary. While these additional 18 axes contain information, the first five, which account for roughly half of the total variance, are sufficient to pick up the major patterns of geographic and temporal variation within the data. These factors were then rotated to maximize case loadings.

Factor analysis is a continuum technique, and can portray gradients and overlapping trends in data (unlike discrete association techniques such as cluster analysis, which cannot). Positive loadings of field sections on each axis were plotted on a series of stratigraphic diagrams, using the section locations given in Figure 4, and are shown in Figure 6-10. Within each figure, the magnitude of section loadings on the particular axis is indicated by ellipses of varying sizes, and the overall distribution of sections with positive loadings for a given factor is given by the shaded pattern. The five factors reveal two nearshore biome patterns (Figs. 6 and 7), and three offshore biome patterns (Figs. 8-10) within strata of the Frasnian Stage.

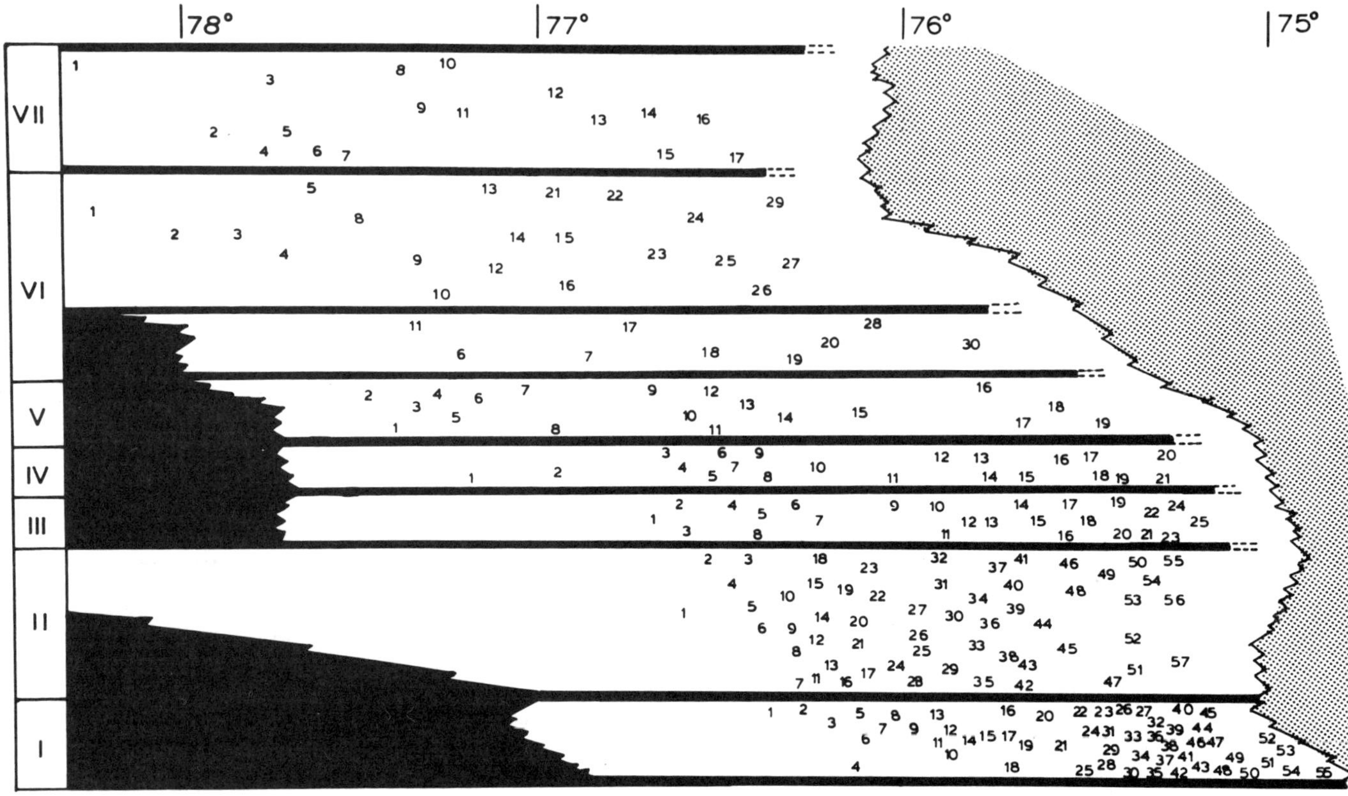

Figure 4. Stratigraphic and geographic distribution of the 224 field sections analyzed in the present study. Explanation of the stratigraphic code (given in the left margin of the figure) and sources of data are given in the caption to Figure 3.

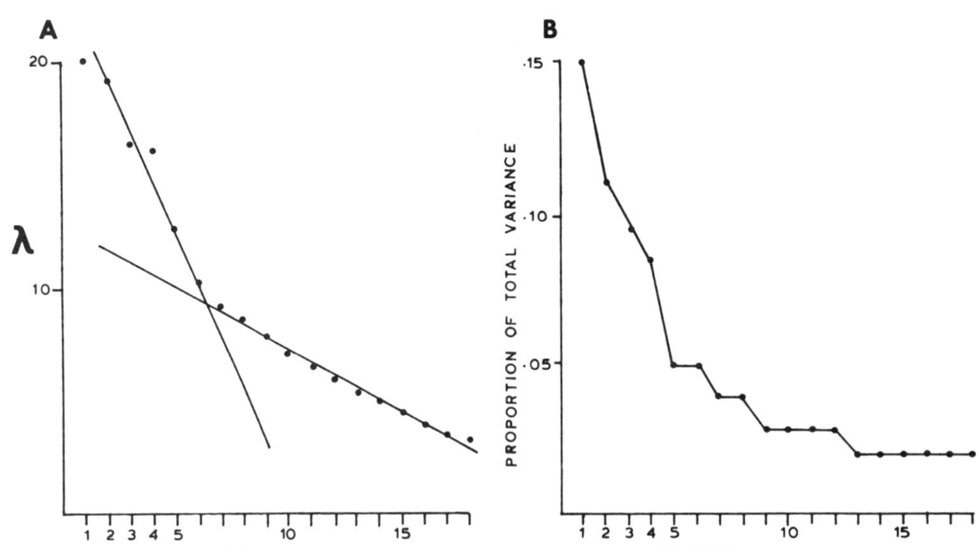

Figure 5. A. A scree graph showing the decay of eigenvalues (λ) for the Q-Mode factor analysis of Frasnian strata. In scree graphs, a break in slope of the decay of eigenvalues reflects a change in the nature of variation associated with the corresponding eigenvectors (i.e., factor axes). B. A graph of the relative proportion of the total variance for each factor, to accompany the scree graph in Figure 5A. The change in proportion for each factor beyond the fifth is small in relation to the change in proportion for the first five factors.

TABLE 1. FACTOR SCORES FOR TAXA WITH POSITIVE SCORES ON NEARSHORE BIOME FACTOR AXES, DIAGRAMMED STRATIGRAPHICALLY IN FIGS. 6 AND 7 (See text for discussion)

Factor 1		Factor 4	
Camarotoechnia	(7.67)	Orthospirifer	(7.61)
Grammysia	(0.58)	Cypricardella	(1.14)
Lyriopecten	(0.45)	Palaeoneilo	(0.43)
Leiopteria	(0.28)	Cyrtospirifer	(0.38)
Goniophora	(0.23)	Devonochonetes	(0.28)
Productella	(0.17)	Goniophora	(0.20)
Sphenotus	(0.11)	Nautiloid	(0.09)
Spinatrypa	(0.01)	Nervostrophia	(0.05)
		Bryozoans	(0.02)
		Tylothyris	(0.01)

TABLE 2. FACTOR SCORES FOR TAXA WITH POSITIVE SCORES ON OFFSHORE BIOME FACTOR AXES, DIAGRAMMED STRATIGRAPHICALLY IN FIGS. 8 - 10 (See text for discussion)

Factor 2		Factor 5		Factor 3	
Mucrospirifer	(5.31)	Ambocoelia	(7.43)	Spinatrypa	(5.20)
Productella	(5.02)	Nervostrophia	(0.44)	Thiemella	(3.88)
Schizophoria	(1.25)	Cyrtospirifer	(0.34)	Productella	(2.34)
Leiorhynchus	(0.82)	Bryozoans	(0.28)	Ambocoelia	(1.68)
Eoschizodus?	(0.74)	Devonochonetes	(0.26)	Tylothyris	(1.09)
Stropheodonta	(0.46)	Leptodesma	(0.25)	Auloporid	(1.73)
Pugnoides?	(0.31)	Rhipidomella	(0.24)	'Cornellites'	(0.73)
Lyriopecten	(0.15)	Stropheodonta	(0.20)	Douvillina	(0.70)
Cyrtina	(0.12)	Mucrospirifer	(0.16)	Schuchertella	(0.65)
Gastropod	(0.12)	Vertumnia	(0.10)	Leptodesma	(0.23)
Warrenella	(0.07)	Productella	(0.09)	Orthospirifer	(0.16)
Orthospirifer	(0.07)	Schizophoria	(0.08)		
Nautiloid	(0.06)	Palaeoneilo	(0.07)		
		Tentaculites	(0.05)		
		Leiopteria	(0.04)		
		'Camarotoechia'	(0.04)		

To assist in the interpretation of the factor axes, a factor-scores matrix was computed. Factor scores yield the relative contribution of each variable (in this case, taxon) in the determination of the factor axis. Taxa having positive scores for the two nearshore biome axes are listed in Table 1, and for the three offshore biome axes in Table 2. These tables will be discussed relative to Figures 6-10.

Factor 1 delimits a nearshore community-type zone which persisted throughout the Frasnian in New York (Fig. 6). High loadings (large ellipses) are distributed throughout the strata, and differences in the density of ellipses are largely due to stratigraphic differences in density of field sections (cf. Fig. 4). Taxa which have high scores for the first factor are overwhelmingly bivalve molluscs (*Grammysia, Lyriopecten, Leiopteria, Goniophora, Sphenotus*; see Table 1). The rhynchonellid *"Camarotoechia"* has the highest score; other brachiopods include *Productella* and *Spinatrypa*.

While the first factor reveals a fairly uniform pattern throughout the nearshore realm during the Frasnian, a subordinant nearshore biome is differentiated by factor 4 (Fig. 7). This community-type has its highest concentration in the lower West Falls Group, expanding from the Sonyea Group below and tapering into the upper West Falls Group above (Fig. 7). Like factor 1, bivalve molluscs have positive scores for this axis (*Crypricardella, Palaeoneilo, Goniophora*; see Table 1). *Orthospirifer* by far has the highest score, and in general the biome corresponds to the *Orthospirifer*-dominated communities of the Millport, Beers Hill, and Meads Creek equivalents (Fig. 3).

In addition to the spiriferacean *Orthospirifer,* the spiriferaceans *Cyrtospirifer* and *Tylothyris* also have positive scores for factor 4 (Table 2). These three spiriferaceans replace one another in importance in nearshore communities throughout the Frasnian—*Orthospirifer* is more common in the Sonyea and lower West Falls Groups, *Tylothyris* in the upper West Falls, and *Cyrtospirifer* in the Java (Fig. 3; see also Bowen and others (1974) and McGhee and Sutton (1981, 1983, this volume).

Taxonomic turnover in the spiriferacean elements of nearshore faunas is the only major biological change which affected nearshore communities during the Frasnian (McGhee, 1981). Combining the taxa which have positive scores for factors 1 and 4 (Table 1) yields the basic elements of the spiriferacean-rhynchonellid-bivalve "community-type" or biome, which persisted in shallower water regions throughout the span of the Frasnian stage.

In contrast to the nearshore communities, offshore communities were more variable in structure during the Frasnian. Three

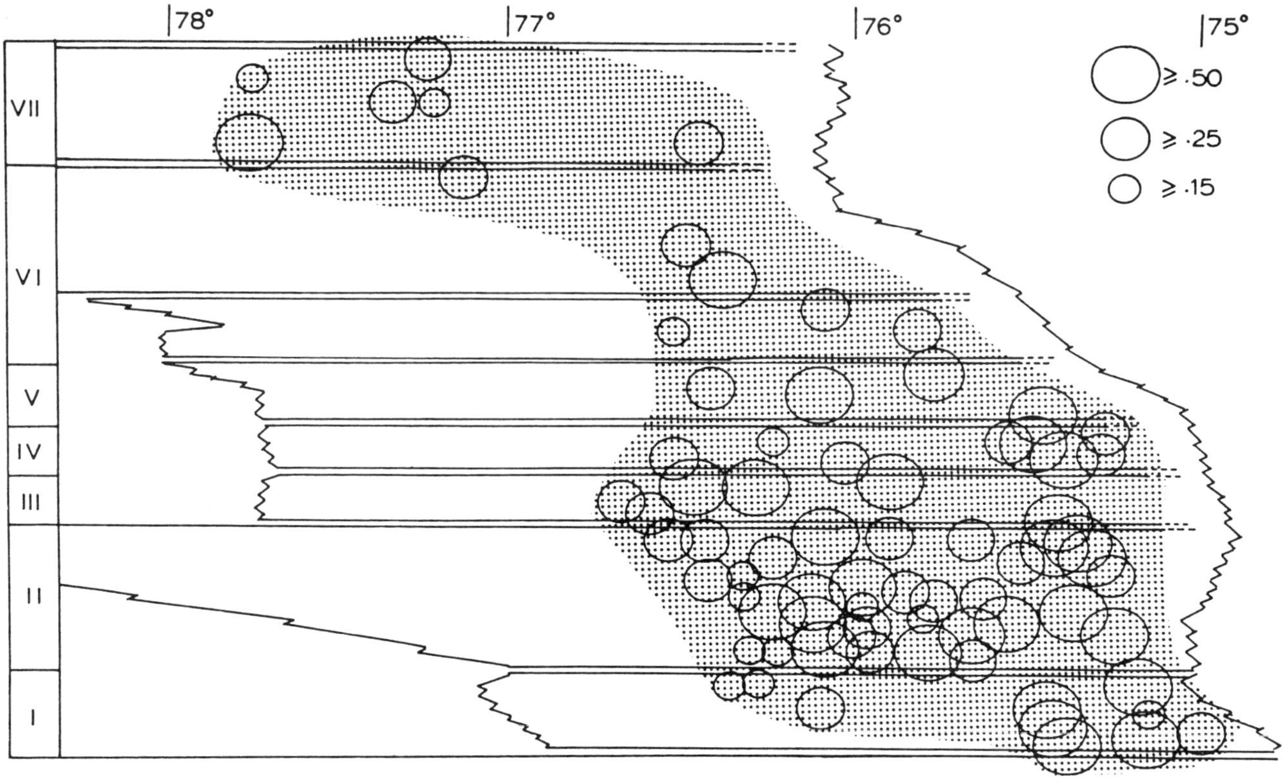

Figure 6. Distribution of field sections having positive loadings on factor axis 1. Magnitudes of factor loadings indicated by the sizes of the ellipses given in the upper right-hand corner of the figure. Field section format identical to Figure 4.

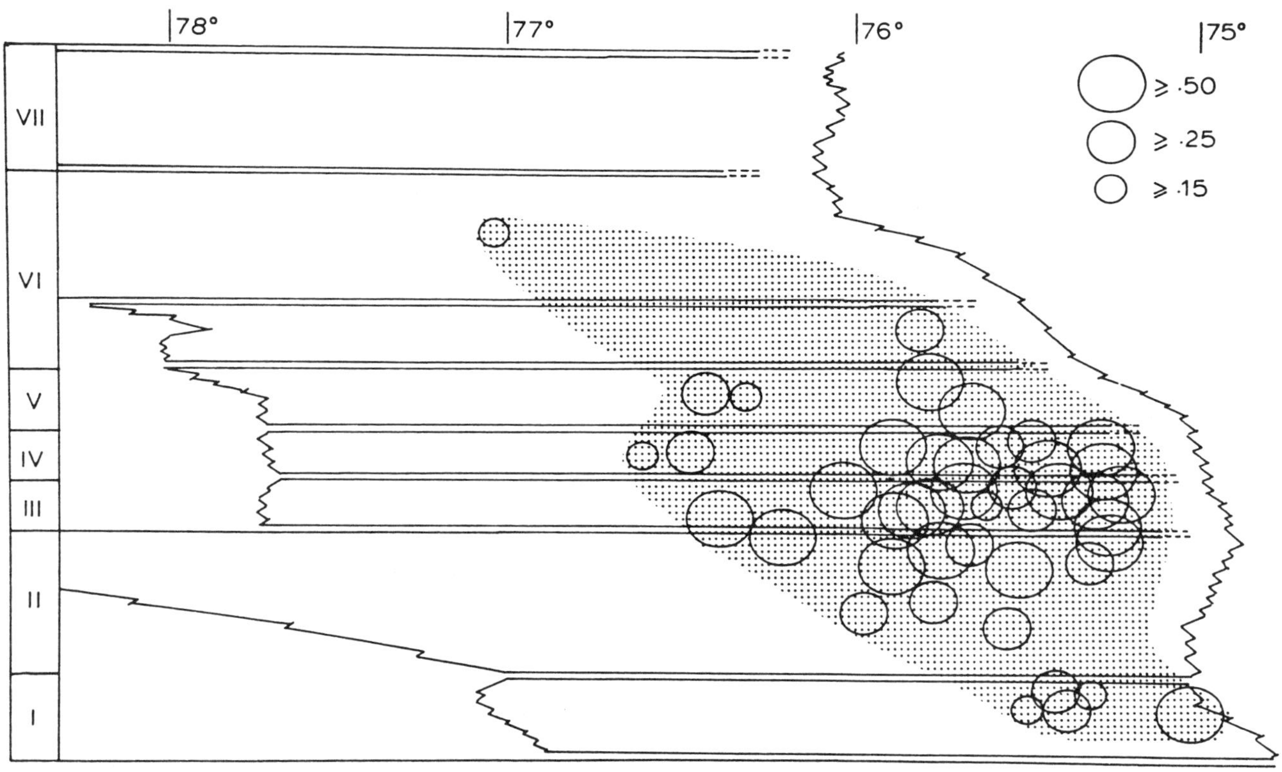

Figure 7. Distribution of field sections having positive loadings on factor axis 4. Format identical to Figure 6.

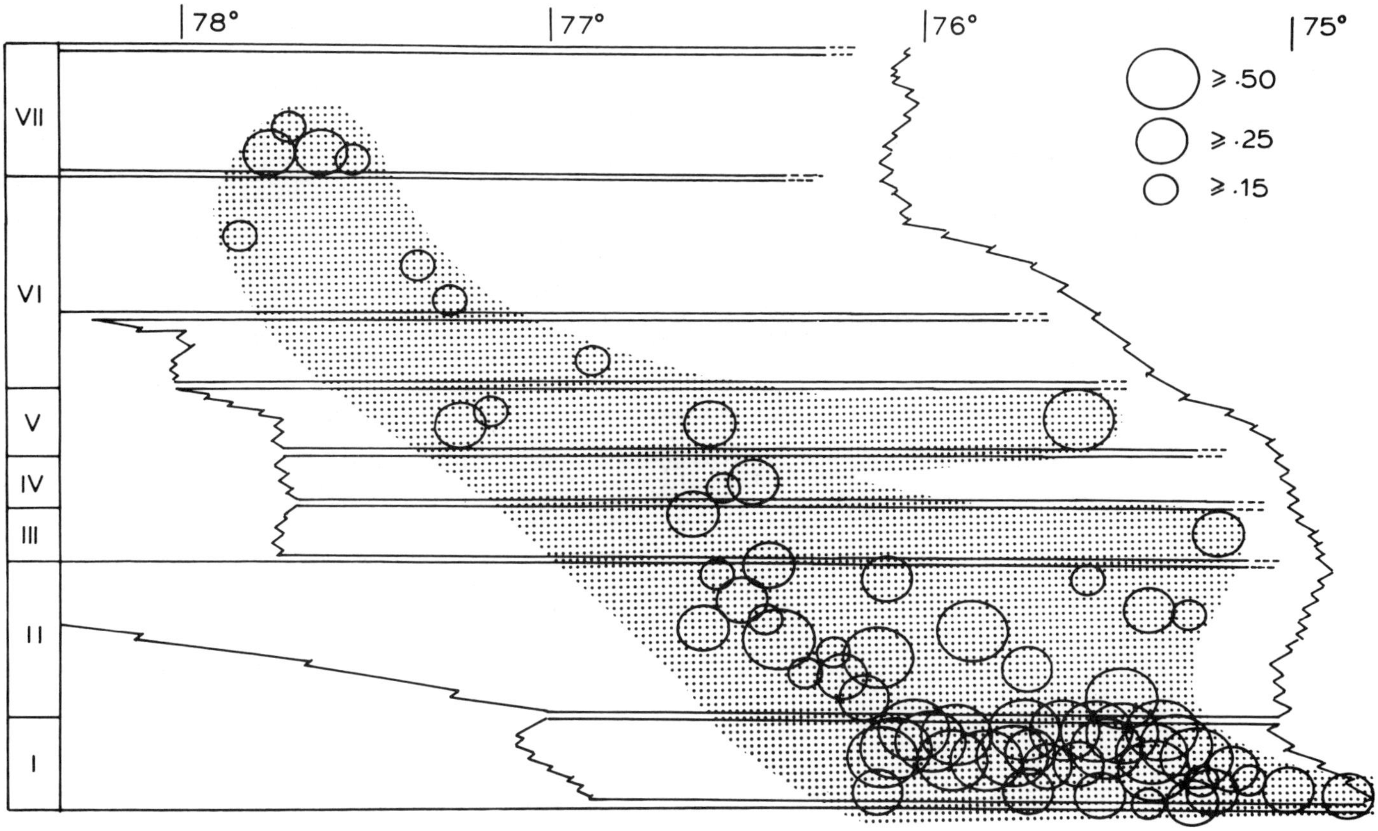

Figure 8. Distribution of field sections having positive loadings on factor axis 2. Format identical to Figure 6.

broad biome patterns, which overlapped and successively replaced one another, occurred in offshore regions during the Frasnian (Figs. 8-10).

The first biome phase is delimited by factor 2 (Fig. 8). Both density and magnitude of factor loadings are highest in the field sections of the lower Sonyea Group (Triangle Formation). The biome pattern then dwindles and tapers into deeper water strata with time (Fig. 8). Many of the taxa which have high scores for this factor (Table 2) are more characteristic late Givetian (Middle Devonian) forms (*Mucrospirifer, Warrenella, Cyrtina, Stropheodonta*; see Thayer (1974).

The second biome phase is broadly distributed (Fig. 9) but is more concentrated in the upper Sonyea Group and the lower West Falls Group. Factor scores indicate that the newly introduced *Ambocoelia* dominated the deeper water fauna (Table 2) of this biome (cf. Fig. 3). New elements include the characteristic deeper water taxa *Ambocoelia, Nervostrophia, Devonochonetes, Rhipidomella* plus a mixture of the earlier phase 1 fauna (*Mucrospirifer, Stropheodonta,* etc.).

The third phase in offshore community-type development is shown in Figure 10. This biome contains a few field sections in the lower Frasnian but progressively geographically expands with time—in a pattern which is virtually the reverse of biome phase 1 (Figs. 8 and 10). Deeper water taxa are still present in the factor-scores matrix (Table 2), but there has been an increase in attached epifauna (*"Cornellites," Schuchertella*) and semi-infaunal sessile brachiopods (*Productella, Douvillina*). The highest score is contributed by the brachiopod *Spinatrypa*—very common in the middle and upper Frasnian. This third biome phase is terminated at the close of the Frasnian, as the atrypide brachiopods all perish in the Frasnian-Famennian extinction event.

Offshore biomes throughout the Frasnian are characterized by unattached epibenthic and sessile semi-infaunal brachiopods. Bivalves, with the exception of a few pterioids, are rare in offshore biofacies (Table 2).

In summary, many distinct animal associations existed in the shallow marine environments of New York during the Frasnian (Fig. 3). Each community represents the local adaptation of a particular fauna to a particular environment in time. Through factor analysis, however, larger scale biome patterns are revealed. Nearshore communities are more stable and less variable in structure than offshore communities. Changes in biome pattern through time reflect both evolutionary changes in the fauna (extinctions and origination of new species) and large-scale changes in the physical marine environment.

It is difficult to separate large-scale changes in the physical environment, which may cause a reorganization in animal associ-

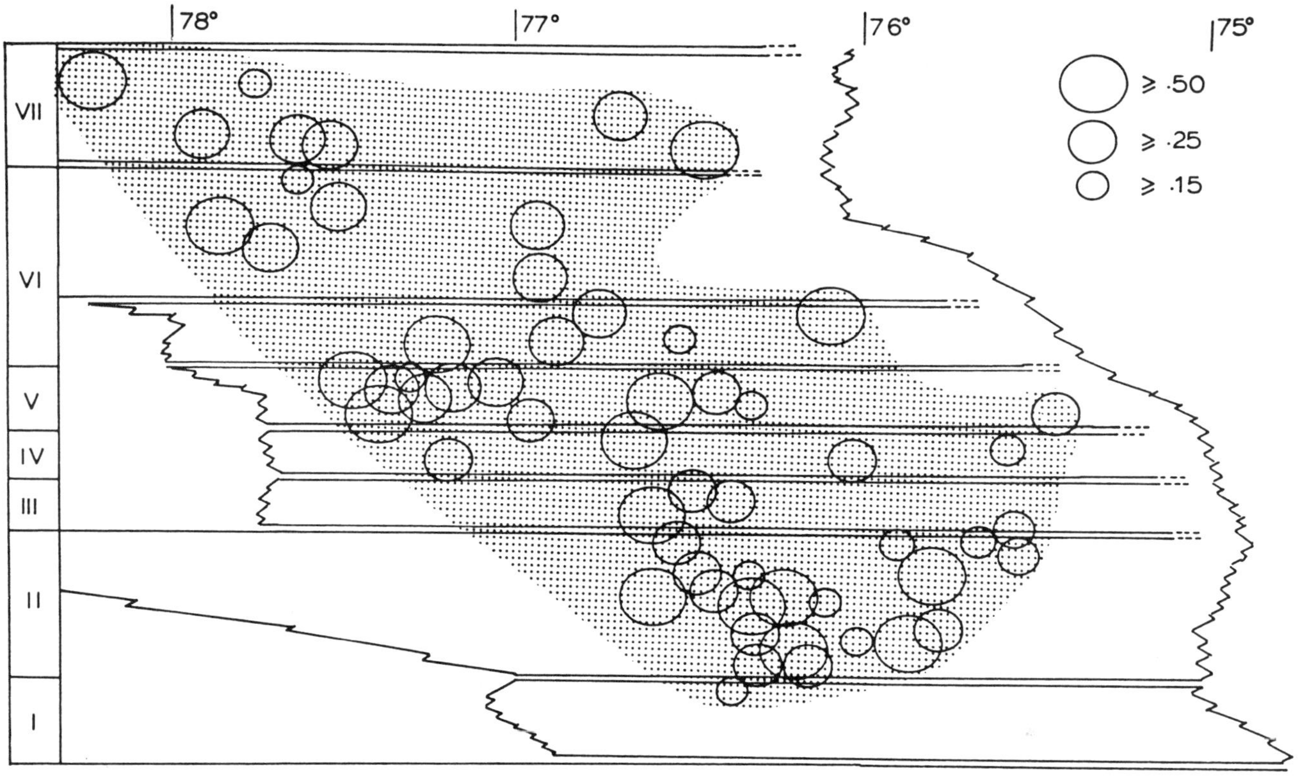

Figure 9. Distribution of field sections having positive loadings on factor axis 5. Format identical to Figure 6.

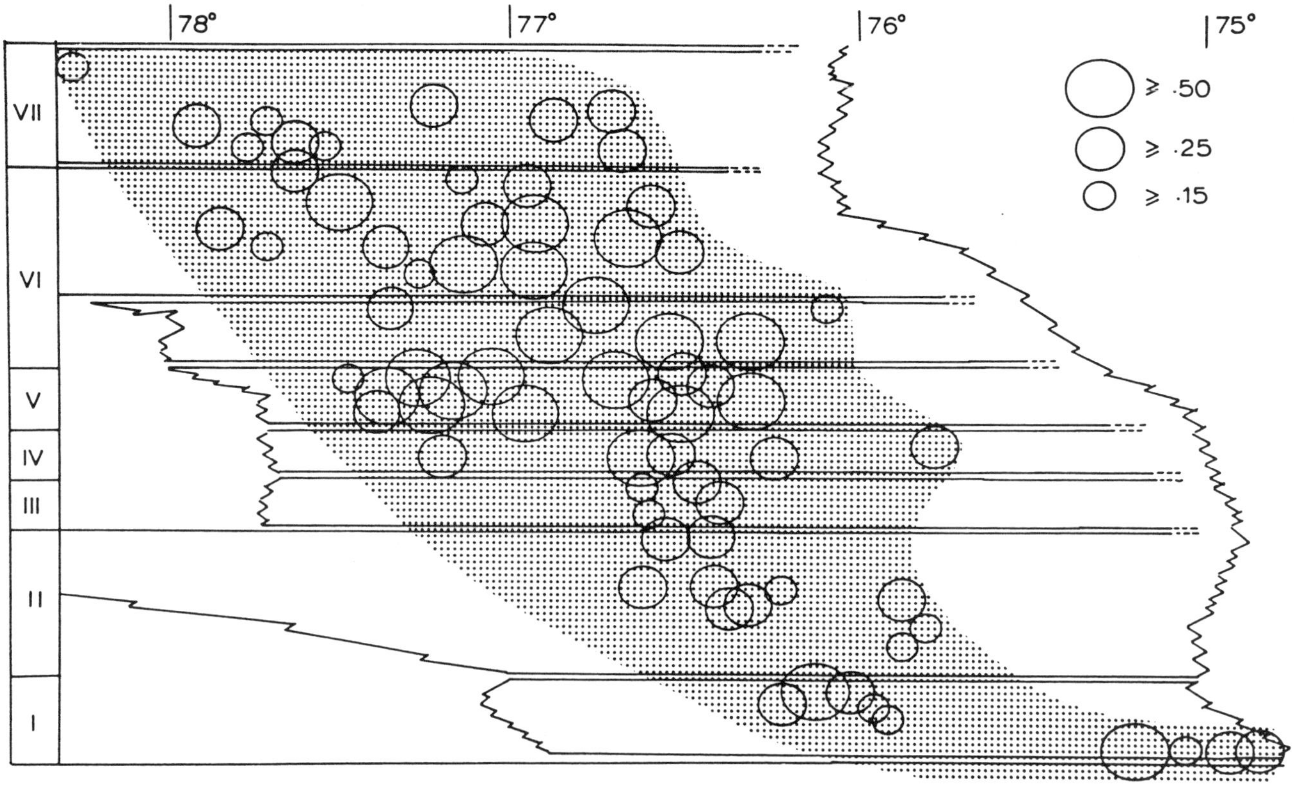

Figure 10. Distribution of field sections having positive loadings on factor axis 3. Format identical to Figure 6.

ations and produce a new community-type, from biologically induced changes in community structure (driven by species migration and evolution), which may also produce new community types. This is only to be expected, as evolution does not proceed in an ecological vacuum. As demonstrated above, three distinct biome phases occurred in offshore regions during the span of the Frasnian; yet in Figure 2 the major Portage facies is shown as simply prograding to the west during the same span of time, with no facies change.

Local adaptation of each community's fauna to local environments, for each stratigraphic unit, has been described in the series of papers listed previously. Yet, these adaptations themselves must have changed with time, particularly in the more variable offshore communities. A "Portage fauna" and a "Chemung fauna" can be said to exist in only a crude sense, as significant ecological reorganization (at least in the offshore regions) occurred during the Frasnian.

We will attempt in the next section of the paper to derive some independent physical (non-biological) parameters to search for major physical-environmental changes in the Catskill basin during the span of the Frasnian. Changes in these parameters will then be compared with the changes in Frasnian shallow marine community-types which have been outlined in the present section.

SHELF ENVIRONMENTS DURING FRASNIAN TIME

The traditional view of the shallow portion of the marine basin was one of an area of fossiliferous sands and muds that uniformly prograded westward to Ohio from the beginning to the end of late Devonian time. This simplistic view became untenable when the subdividing dark-shale tongues were traced through the shelf sediments demonstrating that the rates of progradation were highly variable as were the contemporaneous faunas and lithologies between them. Some studies emphasized the benthic marine communities (Bowen and others, 1974; Thayer, 1974), other studies the pattern of sedimentary structures and sediment distribution (Sutton and Ramsayer, 1974), and still others a combination of the two (Sutton and others, 1970). The result is an array of intricate patterns so complex that no two shelf sequences can be said to be exactly the same, either paleontologically or lithologically. Some major trends do emerge (as discussed previously) and the inorganic aspects will be considered here.

The shelf subenvironments used here have been described elsewhere (Sutton and others, 1970). They consist of the outer shelf, prodelta, and a platform. The latter, in turn, may be divided into outer, middle, and inner parts and may also contain delta front or bar deposits, channel, lagoonal, or distributary-mouth bar sediments. To the west of the shelf occurred the slope and deeper parts of the marine basin.

In this study three variables are considered most significant. They are the shelf width, water depth over various subenvironments of the shelf, and the lithology of the shelf sediments.

SHELF WIDTH

As a result of the previously described studies, the identification of the shelf environments permitted a much more accurate determination of the width of the shelf and the changes in that width than had been possible previously. During earliest Genesee sedimentation, the marine waters were shallow with no distinct slope or deeper water basin discernable. As a result, shelf-type sediments were broadly distributed across the entire marine area with marine benthic populations occupying that same area. Sands were concentrated in the eastern part, while muds were deposited in the west. Thayer (1974) describes this early phase as one of rapid progradation. During deposition of the upper Genesee Group, the area of shallow water deposition had narrowed with deepening water to the west and a prograding shoreline to the east. Thayer called this a time of "slow progradation."

By the time Sonyea sediments were being deposited, the shelf had narrowed to less than 95 km (60 miles), less than one-half the values calculated for the Genesee. The width remained fairly constant during deposition of the lower West Falls, broadened somewhat during deposition of the upper West Falls, and doubled to more than 195 km (120 miles) during deposition of the Java Group.

The shelf width is one of the factors that determines the residence time of the sediment on the shelf. For example, a narrow shelf might be cut by numerous channels that could transport stream sediments directly to the slope and basin, effectively bypassing the shelf proper.

WATER DEPTH

Water depth over the shelf was controlled mainly by two factors: subsidence rate and the rate of sediment contribution to the shelf by streams. These opposing factors were nearly in balance most of the time, but the imbalances resulted in differing shelf profiles (Fig. 11). In one example an open shelf existed with relatively deep water over it (Fig. 11A). A very narrow platform with shallow water existed at the shoreline. In another example, sands accumulated in about the mid-shelf position. These delta front sands served as a trap for muds behind them to the east while the prodelta and open shelf environments were covered with deeper water over the balance of shelf to the west (Fig. 11B). If sediment contribution exceeded subsidence rate, the platform extended seaward so that the prodelta merged with the slope environment. The platform was then separated from the latter by bar or delta front sand accumulations. To the east of these sands, muds were trapped (Fig. 11C) or sands concentrated (Fig. 11D). In the last example, the sands were concentrated by a winnowing of the fine-grained sediments from the platform and their transportation into deeper water to the west. Naturally, variations between these examples existed owing to evolution from one form to another. The examples in Figure 11 display a very important factor in sediment distribution on the shelf. The outer shelf environment receives relatively little sediment compared with the

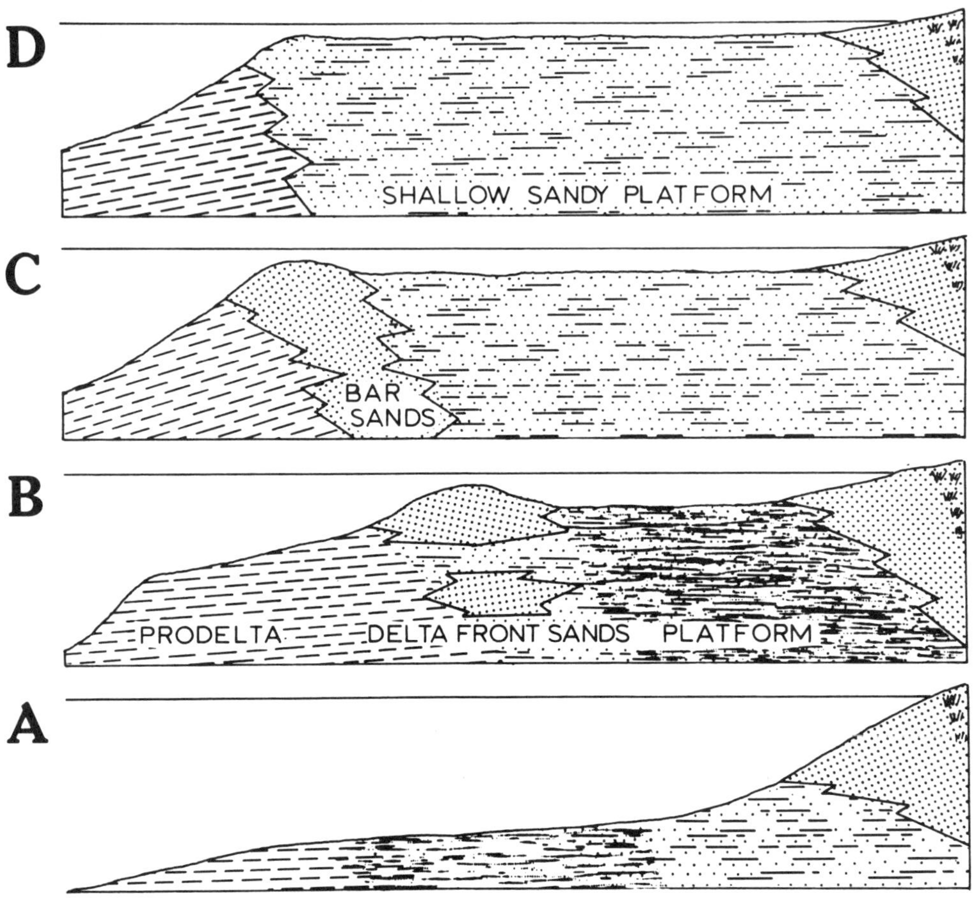

Figure 11. Temporal variation in shelf profile, bathymetric gradients, and distribution of sedimentary environments during the span of the Frasnian in New York. Variation in these physical parameters are further outlined in Table 3. See text for discussion.

platform. However, it subsides at a rate comparable to the platform. Where the open shelf sediments rest directly above older platform sediments, the subsidence distance is approximately that of the thickness of the open shelf sediments as such. These thicknesses range from 90-150 m (300-500 feet). The open shelf sediments are statistically deficient in sand, crossbedding, wave and current ripples while the opposite is true for most of the platform subenvironments (Sutton and Ramsayer, 1975). These differences are explained by the subsidence of the open shelf from the aerobic zone above to the dysaerobic zone below. The relative abundance of the wave and current ripples, having already been recorded for each of the shelf subenvironments is used here as a rough index of water depth (Table 3).

A second depth indicator is provided by the lithology of the subenvironments. Because wave- and current-action winnows the sediments in shallow water carrying the silts and clays into deeper water, a ratio of their proportions could serve as a depth indicator:

$$\text{Wave-Current ``Depth Ratio''} = \frac{\text{shale + siltstone}}{\text{sandstone}}$$

Bioturbation destroys some of the stratification normally displayed by the shales and siltstones. The result is a mudstone. However, bioturbation decreases in amount with increasing water depth, due to a less favorable environment. Therefore a biogenic depth ratio may be obtained:

$$\text{Biogenic ``Depth Ratio''} = \frac{\text{shale + siltstone}}{\text{mudstone}}$$

The two depth ratios may be combined as a lithologic depth ratio:

$$\text{Lithologic ``Depth Ratio''} = \frac{2\ \text{(shale + siltstone)}}{\text{sandstone + mudstone}}$$

Lower values indicate shallower water, higher values deeper water. The average depth ratio has been calculated from all of the measured sections identified by each of the benthic communities (Table 3). There is a considerable amount of agreement between the ripple depth indicator and the lithologic depth ratio. However, these depth indicators should be used only as a general guide. For example, a protected lagoon may have a deficiency of current ripples even though it is very shallow. Or a shallow, nearshore platform setting may have an abnormally high salinity, thereby greatly reducing biogenic activity. In both examples, er-

TABLE 3. FRASNIAN BENTHIC COMMUNITIES IN RELATION TO SELECTED PHYSICAL PARAMETERS OF THE MARINE ENVIRONMENT, AND THE DISTRIBUTION OF THOSE COMMUNITIES AND PARAMETERS THROUGH TIME

Stratigraphic Position	Benthic Community	Shelf Position	Lithology (%) SS.	Mdst.	Shale	Sitst.	Cong.	Ripples Current	Wave	Inferred Depth: Ripples	"Depth Ratio"	Shelf Profile
VII. Java	*Ambocoelia-Cariniferella*	Outer	15	18	59	8	-	None	None	Deep	Deep (4.1)	
	Tylothyris-Schizophoria	Middle	52	24	14	4	6	Some	None	Moderate	Shallow? (0.5)	D
	Cyrtospirifer-Douvillina	Inner	43	18	51	3	5	Some	None	Moderate	Moderate (1.1)	
VI. Upper West Falls	*Ambocoelia-Devonochonetes*	Outer	30	35	30	5	Trace	Some	Some	Moderate	Moderate (1.1)	C
	Tylothyris-Douvillina	Inner	40	30	24	5	1	Many	Many	Shallow	Shallow (0.8)	
V. Meads Creek	*Ambocoelia-Productella*	Outer	13	34	36	17	-	Some	None	Moderate	Deep (2.8)	C-D
	Orthospirifer-'Camarotoechia'	Inner	40	28	18	9	1	Some	Some	Shallow	Shallow (0.8)	
IV. Beers Hill	*Ambocoelia-Spinatrypa*	Outer	26	46	24	4	-	Some	Some	Shallow?	Shallow (0.8)	
	Nervostrophia-Devonochonetes	Middle	11	34	46	8	-	None	None	Deep?	Deep? (3.2)	B
	Orthospirifer-'Camarotoechia"	Inner	33	32	26	9	-	Some	Some	Shallow	Moderate (1.1)	
III. Millport	*Ambocoelia-Tylothyris*	Outer	2	75	12	11	-	None	None	Deep	Deep (5.9)	A
	Orthospirifer-Cypricardella	Inner	27	38	28	6	1	Few	None	Shallow	Moderate (1.1)	
II. Glen Aubrey	*Rhipidomella*	Outer	26	69	1	4	-	Some	Few	Moderate	Shallow (0.1)	B-C
	Cypricardella	Inner	22	51	18	8	1	Rare	None	Shallow	Shallow (0.8)	
I. Triangle	*Productella*	Outer	9	45	27	18	1	Many	None	Shallow?	Deep (3.0	A
	Bellerophon	Inner	19	53	12	15	1	Few	None	Shallow	Moderate (1.0)	

roneously high depth values would result. However, the writers are convinced by the overall results that depth played an important role in determining the inhabitants of the various benthic marine communities and that changes in depth resulted in shifts of those communities on the shelf during Frasnian time.

LITHOLOGY OF THE SHELF SEDIMENTS

Knowledge of the character of the shelf sediments is crucial in reconstructing the Frasnian shelf environments. It determines the nature of the bottom conditions and the substrate and is a guide to the turbulence of the waters above the interface. Data on sediment composition and structures is available from the Sonyea shelf environments (Sutton and others, 1970) and from their subsequent field studies in the post-Sonyea Frasnian strata. These studies resulted in more than 300 measured sections (the 224 most fossiliferous of which are given in Fig. 4). An environment was assigned to each section by two independent means: inorganic and organic. Subsequent work carried the analysis further by plotting species rank abundances along an onshore-offshore environmental gradient, thereby recognizing distinct communities in the Java Group (McGhee and Sutton, 1981), in the upper West Falls (McGhee and Sutton, 1983), and finally the lower West Falls (McGhee and Sutton, this volume). These communities are identified in Figure 3. Once the communities were established, it was a simple matter to isolate the appropriate stratigraphic sections, determine the relative amounts of sediment type, and combine the data to provide a sediment profile for each community (Table 3).

Although too few data are available for the Genesee Group, there appears to be a gradual increase in sand content of the shelf throughout the rest of Frasnian time. One possible explanation for this increase may be an increase in the relative amount of sand being contributed from the source area. A second explanation may lie with the evolution of the marine basin itself. Seaward of the shelf, the slope and deeper part of the marine basin provided a sink into which fine-grained sediments could be transported, leaving a higher volume of sand on the shelf as a residue.

A second trend is the changes in the shelf profiles. During early Frasnian time, the profiles exhibited a greater irregularity in depth and where deeper, served as traps for fines that accumulated behind delta front sands (Fig. 11A and B). By the late Frasnian, the platforms were more extensive, more sandy, waters were of more uniform depth and generally shallower (Fig. 11C and D).

Finally, it must be remembered that variations in shelf width can influence the nature of the substrate on the shelf. During deposition of the Genesee sediments, the shelves were much broader. Muds dominated the Genesee environments due, in part, to a lack of a deeper basin offshore to serve as a sink. Thayer (1974) points out that sand bars were so few that the sand-loving *Rhipidomella* was rare in Genesee sediments. By latest Frasnian time, the shelf had broadened once again and the Java sediments were being deposited in a shallow platform setting with a much higher sand content. By Famennian time, the marine profile defined by the terms "shelf," "slope," and "basin" had blurred and soon vanished from New York state.

SUMMARY AND CONCLUSIONS

The evolution of Frasnian marine ecosystems in New York represents a complex interaction of biological and physical processes. On the local ecosystem level, physically induced changes in community composition and distributions were variously due to changes in the amount and distribution of sand-grade substrate over the shelf (McGhee and Sutton, this volume), rates of delta progradation (Thayer, 1974), steepness of onshore-offshore submarine gradients (McGhee and Sutton, 1983), and so on. Biologically induced changes in community composition include ecological replacement of dominant species, which appear to be ecological replacement of dominant species independent of spatial environmental gradients (McGhee, 1981).

On the larger scale biome or community-type level, major changes can also be seen. On this level, however, the two most important physical factors were shelf profile (bathymetric gradient) and regional differences in substratum type. Offshore communities were more sensitive to changes in shelf configuration than nearshore communities, which retained an essentially unchanging ecological structure throughout Frasnian time. Considering the Frasnian strata of New York as a whole, three major biotic-physical phases can be discerned in the evolution of the shelf and its marine fauna.

The first phase occurred during deposition of the Sonyea Group. Bathymetric gradients were initially steep during deposition of the lower Sonyea (Triangle) sediments with only a very narrow platform developed to the east (Fig. 11A and Table 3). Following this, the deposition of the upper Sonyea (Glen Aubrey) shows an evolution to a broader platform with a more gentle bathymetric gradient (Fig. 11B, C, and Table 3). Delta front sands developed and spread geographically, resulting in the replacement of the *Productella* Community by the shallower water *Rhipidomella* Community (Sutton and others, 1970; Bowen and others, 1974).

During the deposition of the lowermost West Falls sediments, the waters deepened over the shelf, resulting in a steeper bathymetric gradient once again. The shelf sediments of the Millport equivalent conformed to a type A profile with the sands confined to a narrow platform to the east (Fig. 11A; Table 3). An *Ambocoelia* Community (offshore biome 2) dominated the fine-grained, deeper substrates of the prodelta and open shelf to the west. Sediments of the Beers Hill equivalent display a type B profile (Fig. 11) with the diverse *Nervostrophia-Devonochonetes* Community inhabiting the platform muds, landward of the delta front sands. During deposition of the sediments of the Meads Creek equivalent, sands spread across a broadening platform to form a type C profile evolving to type D (Fig. 11).

Initially, the sedimentation rate of the upper West Falls was low (Corning Shale). The water on the shelf deepened temporarily but the sedimentation rate increased once again providing a type C profile (Table 3). Java sands spread far to the west and developed a broad, sandy platform over 195 km (120 miles) wide (McGhee and Sutton, 1983).

The progressive shallowing of the shelf, the increase in overall sand content, and the geographic expansion of sand-grade substrate are reflected in the reciprocal replacement of offshore biome 1 (Fig. 8) with offshore biome 3 (Fig. 10) in the Frasnian strata with time.

ACKNOWLEDGMENTS

We thank Drs. T. Lewis and L. V. Rickard for their critical review and constructive comments on the manuscript.

The research was supported in part by a grant from the Earth Sciences Division, National Science Foundation, NSF Grant EAR 81-15518 to G. R. McGhee. Also, acknowledgment is made to the donors of the Petroleum Research Fund, administered by the American Chemical Society, for partial support of this research (ACS-PRF 13481-Ac2).

REFERENCES CITED

Bowen, Z. P., Rhoads, D. C., and McAlester, A. L., 1974, Marine benthic communities in the Upper Devonian of New York: Lethaia, v. 7, p. 93–120.

Burtner, R. L., 1963, Sediment Dispersal Patterns within the Catskill Facies of Southeastern New York and Northeastern Pennsylvania, in Symposium on Middle and Upper Devonian Stratigraphy of Pennsylvania and Adjacent States: Bull. G39, Penn. Geol. Survey, 4th Ser. p. 7–23.

Caster, K. E., 1934, The stratigraphy and paleontology of northwestern Pennsylvania; Part 1, Stratigraphy: Bull. Am. Paleont., v. 21, no. 71, 185 p.

Chadwick, G. H., 1933, Great Catskill Delta, and revision of late Devonic succession: Pan-American Geol. v. 60, p. 91–107, 189–204, 275–286, 348–360.

—— 1935, Summary of Upper Devonian Stratigraphy: Am. Midland Naturalist, v. 16, no. 1, p. 857–862.

Chellman, K. K., 1983, Provenance of some nearshore marine sandstones in the Upper Devonian of south-central New York [M.S. dissertation]: Univ. of Rochester, Rochester, N.Y., 55 p.

Davis, J. C., 1973, Statistics and Data Analysis in Geology: John Wiley and Sons, Inc., New York, 550 p.

Ettensohn, F. R., and Barron, L. S., 1981, Depositional Model for the Devonian-Mississippian black shales of North America: A paleoclimatic-paleogeographic approach: Geol. Soc. Amer. Field Trip Guidebook No. 3, v. 2, p. 344–361.

Lucier, W. A., 1966, The petrology of the Middle and Upper Devonian Kiskatom and Kaaterskill Sandstones: A vertical profile [Ph.D. dissertation]: Univ. of Rochester, Rochester, N.Y., 92 p.

McGhee, G. R., Jr., 1981, Evolutionary replacement of ecological equivalents in Late Devonian benthic marine communities: Palaeogeog., Palaeoclimat., Palaeoecol., v. 34, p. 267–283.

McGhee, G. R., Jr., and Sutton, R. G., 1981, Late Devonian marine ecology and zoogeography of the central Appalachians and New York: Lethaia, v. 14, p. 27–43.

—— 1983, Evolution of Late Frasnian (Late Devonian) marine environments in New York and the central Appalachians: Alcheringa, v. 7 (1-2), p. 9–21.

—— in press, Late Devonian marine ecosystems of the lower West Falls Group in New York; this volume.

Nugent, R. C., 1960, Geology of the Elmira, Waverly, and Owego Quadrangles [M.S. dissertation]: Univ. of Rochester, Rochester, N.Y., 103 p.

Rickard, L. V., 1975, Correlation of the Devonian Rocks in New York: New York State Mus. and Science Service, Map and Chart Series No. 24, plate 3.

Sutton, R. G., Bowen, Z. P., and McAlester, A. L., 1970, Marine Shelf Environments of the Upper Devonian Sonyea Group of New York: Geol. Soc. Amer. Bull. 81, p. 2975–2992.

Sutton, R. G., and Ramsayer, G. R., 1975, Association of lithologies and sedimentary structures in marine deltaic paleoenvironments: Journal of Sedimentary Petrology, v. 45, p. 799–807.

Thayer, C. W., 1974, Marine paleoecology in the Upper Devonian of New York: Lethaia, v. 7, p. 121–155.

Whittaker, R. H., 1975, Communities and Ecosystems: Macmillan Pub. Co., New York, 385 p.

MANUSCRIPT ACCEPTED BY THE SOCIETY JULY 20, 1984

Printed in U.S.A.

Geological Society of America
Special Paper 201
1985

Ammonoid horizons in the Upper Devonian Genesee Formation of New York: Legacy of the Genesee, Portage, and Chemung

William T. Kirchgasser
Department of Geology
State University College of Arts and Science
Potsdam, New York 13676

ABSTRACT

The names Genesee, Portage and Chemung are indelibly linked to the early Upper Devonian stratigraphy of the Catskill delta in western New York. For over a century generations of stratigraphers struggled in the attempt to correlate these groups eastward from the Genesee Valley into sections around the delta-front at Cayuga Lake. Rapid thickening and intertonguing of facies were the chief difficulties, but the misinterpretation of ammonoids also played a major role in obscuring the facies relationships. *Manticoceras sinuosum* (Hall, 1843), the index fossil of the Portage Group, was mistaken by James Hall and J. M. Clarke and later workers for species of *Ponticeras* and *Koenenites* from horizons in the older Genesee Formation east of Canandaigua Lake.

Some of these Genesee ammonoid horizons have been traced from the pelagic facies in the basin into the benthic facies of the delta-front. One of these horizons in the Upper Penn Yan and Middle Ithaca Members may record an interval of transgression or standstill in the progradation of the delta. The ammonoid horizons in the Genesee Formation at Cayuga Lake are stratigraphic markers which should aid in the taxonomic and biostratigraphic revision of the benthic groups and in the refinement of the paleoecological associations in the delta-front facies.

> **In commencing investigations of any kind, where the objects are numerous, lines must be occasionally drawn; they are to the mind, like steps or rounds of a ladder: we have then a place to stand and look around, and to ascertain fully our position. Vanuxem (1842)**

INTRODUCTION

Almost 25 years have elapsed since the first edition of the now familiar correlation chart of Rickard (1964) summarized the results of a century and a quarter of revision of the early Upper Devonian stratigraphy of the Catskill delta in New York. It seems appropriate on the occasion of this publication to follow Vanuxem's advice and review again our position. The complicated history of stratigraphic revision of the marine part of the succession has been thoroughly recounted in major summaries, most notably Hall (1843), Williams (1884), Clarke (1904), Chadwick (1933, 1935a), Cooper and others (1942), deWitt and Colton (1959, 1978), Sutton (1963), and Rickard (1964, 1975, 1981). There is general agreement that the confusion of usage and miscorrelations of the classical groups Genesee, Portage and Chemung in sections shoreward of their type section in the Genesee Valley were the result of complex facies and thickness changes in these clastic units near the delta-front between Seneca and Cayuga Lakes (Fig. 1). Not until the 1930s was it fully realized that the dark shales of the Genesee, the lighter colored shales and siltstones of the Portage and the sandstones of the Chemung not only followed one another in stratigraphic sequence but also intertongued with one another between the basin in the west (Genesee facies) and the delta-front to the southeast (Chemung facies). Still, the precise alignment of the facies tongues, particularly within the intermediate Portage facies, was not established until the late 1950s when key black shales were traced within the Genesee and Sonyea Formations of present terminology from the basin sections in the west to the delta-front sections at Cayuga Lake.

Recently, several ammonoid cephalopod horizons within

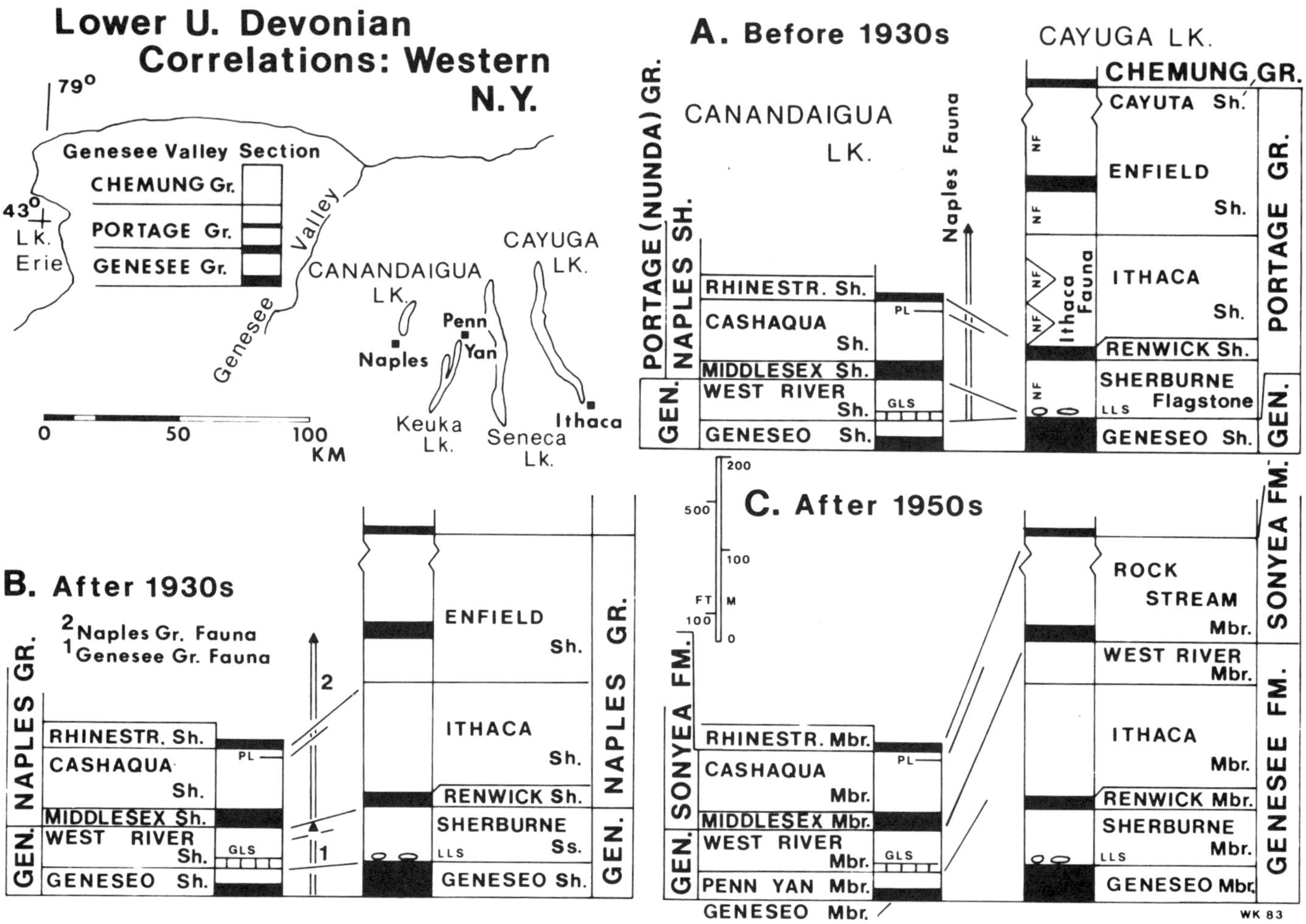

Figure 1. Diagrammatic history of changes in lower Upper Devonian correlations across western New York. A. and B. based on Chadwick (1933, 1935, 1935a). C. based on deWitt and Colton (1959, 1978), Sutton and others (1962), Sutton (1963), and Rickard (1964, 1975). Abbreviations: GEN—Genesee; LLS—Lodi Limestone; GLS—Genundewa Limestone; PL—Parrish Limestone; NF—Naples Fauna.

the Genesee Formation have been shown, like the black shales, to be traceable from the basin sections into the sections at the delta-front (Kirchgasser and House, 1981). Particular ammonoid horizons are recognized in the field by their unique lithologic character and association of ammonoid species; some of the horizons also mark changes in other groups such as conodonts and brachiopods. These ammonoid horizons confirm the accuracy of the black shale correlations and further refine the intraformational correlations. They also indicate that the misinterpretation of ammonoid species by early workers was a major factor in delaying the discovery of the facies relations and correlations as currently understood.

It is hoped that the ammonoid horizons which have been found in the delta-front facies can provide useful stratigraphic markers for the long overdue revision of the taxonomy and stratigraphic ranges of the benthic species. Unlike the black shale bands which are thought to record intervals of transgression and the siltstones which mark intervals of delta-building (progradation), the most widespread of the ammonoid horizons may record intervals of hiatus or standstill in the construction of the delta.

The age of the Genesee Formation is regarded here as early Upper Devonian, following Rickard (1975, 1981). In 1982, the Subcommission on Devonian Stratigraphy of the International Union of Geological Sciences voted to define the Middle/Upper Devonian boundary at the base of the Lower *asymmetricus* conodont Zone (entry of *Ancyrodella rotundiloba*); at this writing, a decision by the Stratigraphic Commission of the IUGC is pending. The conodont data at hand indicate that the proposed 1982 SDS boundary in New York is in the lower part of the Penn Yan Member.

EARLY CORRELATIONS

When Vanuxem offered his metaphor, he was struggling with the drawing of lines of subdivision and correlation within the intertonguing shales and sandstones exposed in the ravines around Cayuga Lake in central New York. In early late Devonian time, Cayuga Lake was the line of transition between the outer-shelf and slope of the Catskill delta and the basin farther to the west. The problem facing Vanuxem, and generations of stratigraphers to follow, was to correlate the Cayuga Lake section westward to the standard succession in the Genesee Valley, from base upwards, the Genesee, Portage (or Nunda) and Chemung Groups. The correlation of these units and their subdivisions between the Genesee Valley and Canandaigua Lake was straightforward. The difficulties in correlation came farther southeast between Canandaigua and Cayuga Lakes where the thickness of the succession increases rapidly as the shales of the basin tongue into the wedges of siltstone around the delta-front. The principal stages of the exceedingly complicated history of changes in correlation and stratigraphic nomenclature are illustrated in Figure 1.

Before Chadwick's revision in the 1930s, the Sherburne flagstone or siltstone at Cayuga Lake was correlated westward with the Cashaqua Shale at the base of the Portage Group (Fig. 1A). The key to the correlation was the belief that the Geneseo black shale at Cayuga Lake was the sole representative of the Genesee Group in that section. Another factor was the belief that the pelagic fauna of the Sherburne was an eastern representative of the pelagic fauna of the Portage Beds, called the Naples Fauna. More specifically, the pelagic fauna of the Sherburne and Lodi Limestone at its base was thought to be the same as the "prenuncial" or early Naples Fauna of the mid-Genesee Genundewa Limestone. The western Naples Fauna was also identified as intertonguing with the benthic Ithaca Fauna within the siltstones and shales comprising the Ithaca Shale. Above the Ithaca Shale or beds, the Naples Fauna reappeared in the Enfield Shale. The succeeding Cayuta shale marked the return of benthic faunas and the introduction of the Chemung Fauna which defined the base of the Chemung Group. There were many problems concerning this higher boundary, but the problems in correlation began lower down, at the Genesee/Portage boundary.

CHADWICK'S REVISION

By the 1930s the facies relations of the Genesee, Portage and Chemung had been discovered, and Chadwick (1933, 1935a) overhauled the classical stratigraphy, Figure 1B. The name "Portage" (and several other geographically restricted facies names) was abandoned as a rock-stratigraphic unit. In revising the correlations, Chadwick found that the units in the western sections correlated higher in the eastern sections than had been previously thought. The Genesee Group now correlated with the Geneseo and Sherburne at Cayuga Lake, and the Cashaqua correlated with the Ithaca beds. The key discovery was the tracing (between Keuka and Seneca Lakes) of the West River Shale of the upper Genesee beds into the Sherburne equivalents and the Parrish Limestone of the upper Cashaqua into the Ithaca equivalents (Chadwick, 1935a, p. 348–349). In support of the new correlations, Chadwick (1935) compiled faunal lists to differentiate between the fauna of the Genesee Group and the fauna of the succeeding Naples Group, but unfortunately, his lists were based on range data from the same classical units whose correlations he was revising. Nearly 25 years elapsed before new discoveries showed that Chadwick had not gone far enough in shifting the correlation lines upward from west to east.

BLACK SHALE CORRELATIONS

In the late 1950s, Sutton (1959, 1963), Sutton and others (1962), and deWitt and Colton (1959, 1978) succeeded in tracing the Middlesex black shale at the boundary between the Genesee and Naples Groups and the Rhinestreet black shale above the Cashaqua into the section at Cayuga Lake (Figure 1C). The Middlesex shale was found in the uppermost Ithaca beds, which meant that the Geneseo, Sherburne and Ithaca at Cayuga Lake belonged entirely to the Genesee Group and were thus older than the Naples Group and Portage Group of classical usage. The major black shales were interpreted as transgressive horizons (time-lines) that punctuated the major depositional cycles of the delta. In the current classification, the Geneseo, Middlesex and Rhinestreet are the basal members of (in sequence) the Genesee, Sonyea and West Falls Formations. In the basin, the black shales are overlain by gray shales which intertongue toward the delta-front with eastward-thickening wedges of siltstone and shale, represented by the Sherburne and Ithaca Members of the Genesee Formation and the Rock Stream Member of the Sonyea Formation (Rickard, 1964, 1975).

AMMONOID FAUNAS AND HORIZONS

Since its establishment, the modern classification outlined above has provided the stratigraphic framework for the study of many aspects of Catskill geology. The paleoecology of the major benthic communities of the shelf and delta-front facies of the Genesee and Sonyea have been described in modern quantitative studies (Thayer, 1973, 1974; Macdonald, 1975; Bowen and others, 1974; and McGhee and Sutton, 1981). A resurvey of the biostratigraphy of the ammonoid cephalopods was begun by M. R. House to determine the horizons and ranges of the type specimens in terms of the newer correlations. Several reports have been published (House, 1962, 1968, 1978; Kirchgasser, 1975, 1982; Kirchgasser and House, 1981), and a forthcoming monograph on the project will document the taxonomic revisions and locality details.

Figure 2 ilustrates the sequence of ammonoid faunas for the early Upper Devonian. In the course of work, it was found that the ammonoids are mostly concentrated in thin bands or horizons (often thin limestones or concretionary beds) within thicker intervals of shale or siltstone in which cephalopods are rare or absent. A horizon usually contains two or three genera and species (rarely more) with one or more species dominant (in numbers of individuals) and/or restricted to the particular horizon; the ammonoid associations are numbered in stratigraphic sequence following the scheme in House (1978).

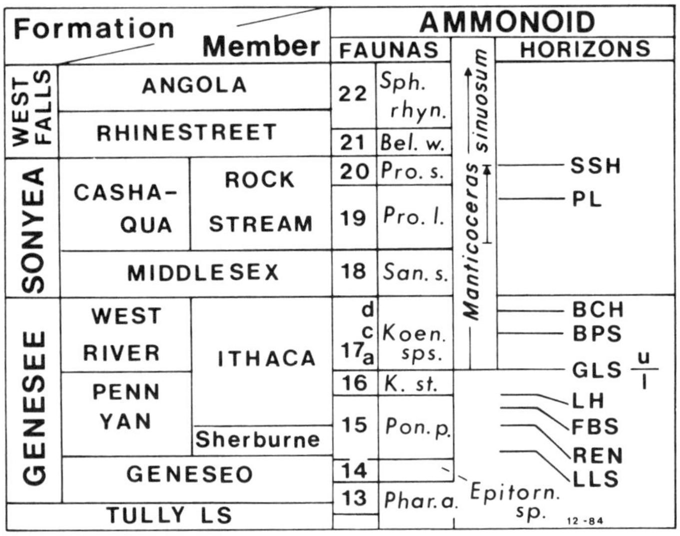

Figure 2. Sequence of ammonoid faunas and horizons in the lower Upper Devonian of New York. Numbers for faunas follow scheme in House, 1978 and Kirchgasser and House, 1981. Abbreviations (faunas): *Pharciceras amplexum, Epitornoceras* species, *Ponticeras perlatum, Koenenites styliophilum, Koenenites* species, *Sandbergeroceras syngonum, Probeloceras lutheri, Beloceras williamsi, Sphaeromanticoceras rhynchostomum.* Abbreviations (horizons): LLS—Lodi Limestone, REN—Renwick Shale, FBS—Firestone Beds, LH—Linden Horizon, GLS—Genundewa Limestone (lower, upper), BPS—Bluff Point Siltstone, BCH—Beards Creek Horizon, PL—Parrish Limestone, SSH—Shurtleff Septarian Horizon. Note that the range of *Manticoceras sinuosum,* the index fossil of the classical Portage Group, is restricted to the Cashaqua Member of the Sonyea Formation. The misidentification of this species by early workers in horizons in the shoreward sections of the older Genesee Formation was a major factor in the history of miscorrelations across western New York.

In outcrop, individual ammonoid horizons either are lithologically distinctive or can be readily located with reference to other key beds such as a thin black shale horizon. Most important of all for correlation within the Catskill delta complex, some of the horizons cross facies boundaries and have been traced tens of kilometers from the basin into the benthic environments of the delta-front.

Four such horizons have been located in the Cayuga Lake section of the Genesee Formation (Fig. 3: solid diamond symbols) along with a fifth, but geographically restricted horizon (open diamond). The significant point about these horizons, with respect to the history of stratigraphic revision sketched above, is that while their occurrence in the Cayuga Lake section (excepting possibly the uppermost one) was known to the early stratigraphers, particularly Hall, Clarke, and H. S. Williams, the misinterpretation of the ammonoid species in them played a central role in reinforcing the miscorrelations from the time of the first survey in the middle of the last century through the period of Chadwick's revision. There is irony in the fact that the ammonoid cephalopods, which now provide one of the best means of faunal correlation between the basin and delta-front, were once the chief culprits in obscuring the facies relations and correlations.

MANTICOCERAS SINUOSUM: INDEX FOSSIL OF THE PORTAGE GROUP

The misinterpretation of one species in particular, *Manticoceras sinuosum,* was at the heart of the failure to recognize the Genesee rather than Portage age of the Sherburne and Ithaca beds at Cayuga Lake. This species was named by Hall (1843) as *Goniatites sinuosus* for specimens from the lower Portage Cashaqua Shale in the Genesee Valley. Farther east around Naples, N.Y., at the south end of Canandaigua Lake, Clarke (1885) found the pelagic fauna of the lower Portage to be better developed than in the Genesee Valley and introduced the name Naples Shale or Beds for the more fossiliferous lower Portage shales (Fig. 1A). The pelagic molluscan fauna of the Portage shales was described in a monograph by Clarke (1889, 1904) under the name "Naples Fauna." *Goniatites sinuosus* (now assigned to the genus *Manticoceras*) is a prominent species in the Cashaqua Shale, and in time the species was taken as the index fossil for the Naples Fauna. Its identification (with a companion species *Probeloceras lutheri*) served as a guide to the recognition of the Portage Group east of Canadaigua Lake and in sections farther south in the Appalachian Basin (Kirchgasser, 1975, 1982).

The problem was that Hall and subsequent workers applied the name *Goniatites sinuosus* (and its synonym *Goniatites pattersoni* Hall, 1860) to specimens collected or noted at horizons within what is now the Genesee Formation, which were not the same species or genus as *Manticoceras sinuosum* of the Cashaqua Shale (Figure 4). In the section at Cayuga Lake, the species and horizons are (in sequence): *Ponticeras perlatum* (Hall, 1879): Lodi Limestone and base of Renwick black shale member at the bottom and top of the Sherburne Member; *Ponticeras* cf. *regale* (Holzapfel, 1899): Firestone Beds of Williams (1884) in the middle of the Ithaca Member; and *Koenenites styliophilum* (Clarke, 1898): Linden Horizon (Kirchgasser and House, 1981) and equivalents higher in the Ithaca Member.

LODI LIMESTONE

At its type area between Seneca and Cayuga Lakes, the Lodi Limestone (Lincoln, 1897) is a concretionary limestone in the lower Sherburne (or Penn Yan) Member of the Genesee Formation, near the contact with the underlying black shales of the Geneseo Member (deWitt and Colton, 1978; Kirchgasser and House, 1981) (Fig. 3, column G). Lincoln (1897) believed the Lodi horizon at the Geneseo/Sherburne contact correlated westward to the boundary between the Genesee and Portage Groups. The correlation was based on Clarke's interpretation of the Lodi fauna (J. M. Clarke in Lincoln, 1897, p. 100–101). At the top of Clarke's faunal list were the Portage ammonoids *Goniatites pattersoni* and *Gon. sinuosus* along with several other molluscan species characteristic of the Portage Group. Also noted were the brachiopods *Atrypa reticularis, Leiorhynchus mesacostalis* and *Spirifer* cf. *subumbona* which indicated to him the "influences

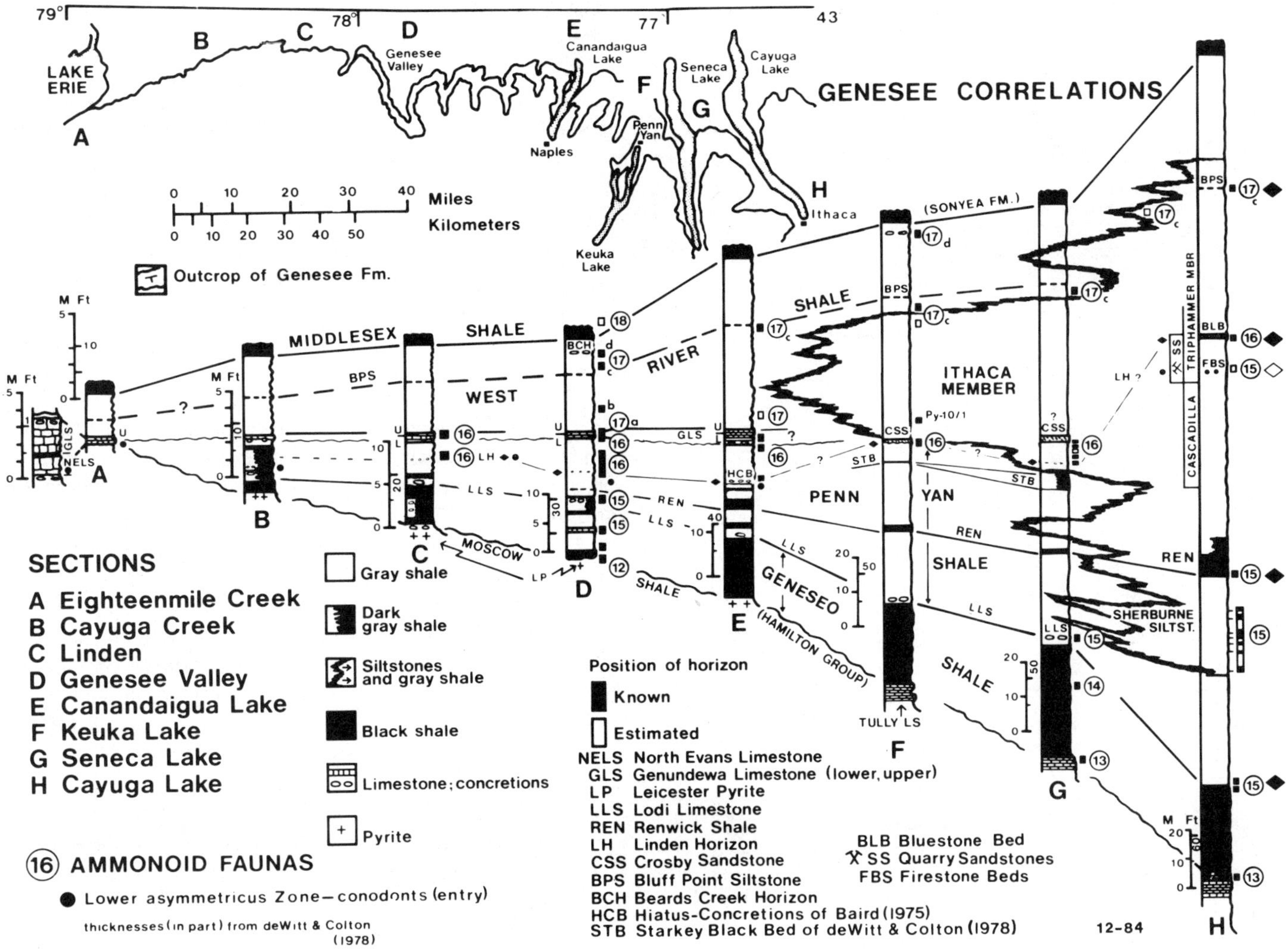

Figure 3. Correlations within the Genesee Formation between Lake Erie and Cayuga Lake. Numbering of ammonoid faunas as in Figure 2. Solid diamonds—principal horizons which have been traced from the basin sections in the west into the delta-front section to the southeast; open diamond-horizon not recognized west of Cayuga Lake. Modified from Kirchgasser and House (1981, Fig. 6) and based on stratigraphic framework of deWitt and Colton (1978).

and presence of the Ithaca fauna or eastern representative of the Portage group."

In Clarke's view, the Lodi ammonoids marked the entry into the central New York section of the western Naples Fauna (Fig. 1A). The Lodi fauna was compared to the fauna of the Genundewa Limestone of the middle Genesee Group at Canandaigua Lake, the horizon which in that section marked the appearance of the Naples Fauna. The Lodi-Genundewa correlation had significant ramifications for mapping and correlation higher in the Seneca and Cayuga Lake sections (Luther, 1909; Williams and others, 1909) because it forced the Sherburne and Ithaca above the Lodi (or "false Genundewa") to be correlated westward with post-Genundewa strata and lent support to the assignment of these units to the Portage Group (deWitt and Colton, 1978, p. A10). The correlation was corrected by deWitt and Colton (1959, p. 2824) when the Lodi horizon was found far below the Genundewa in sections west of Seneca Lake. DeWitt and Colton (1978) did not recognize the Lodi Limestone outside the area of Seneca and Cayuga Lakes (Fig. 3, columns G and H) but recently the horizon has been traced as far west as Cayuga Creek (column B) (Baird and Brett, 1984, personal communication). At Fall Brook, in the Genesee Valley, the Lodi is a nodular horizon about 33 ft (10 m) above the base of the Genesee Formation; it is beneath a pair of thin black shales which define the Geneseo/Penn Yan contact (Fig. 3, column D). Previously, the Lodi was thought to occur much lower in the Genesee Valley section (Kirchgasser, 1975; Kirchgasser and House, 1981) and the Geneseo/Penn Yan contact was drawn at the top of the first tongue of black shale at the base of the Genesee Formation (deWitt and Colton, 1978).

The prominent goniatite in the Lodi horizon is not *Goniatites pattersoni/sinuosus* of the Portage fauna but is *Ponticeras*

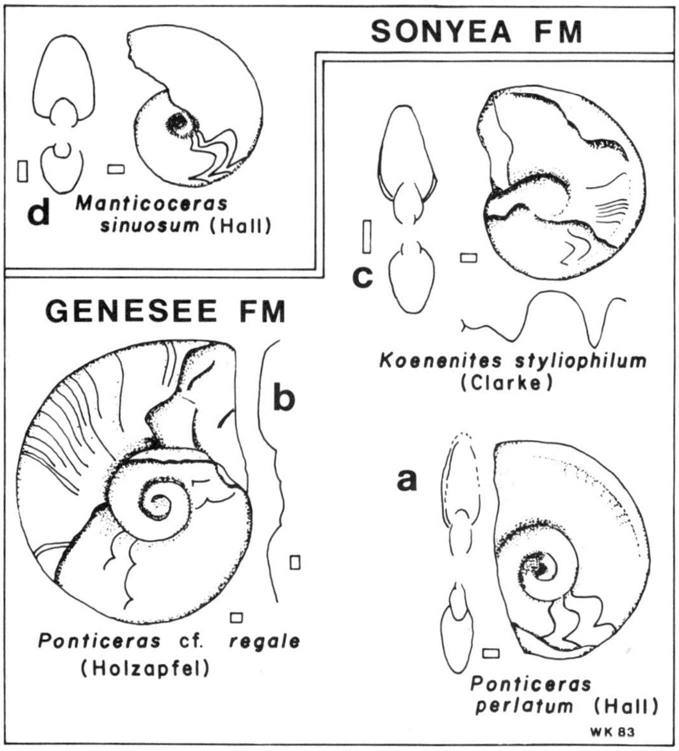

Figure 4. Diagrammatic representation of ammonoid species from the Genesee Formation (a-c) that were mistaken by Hall, Clarke and others for *Manticoceras sinuosum* of the Sonyea Formation (d), the index fossils of the Naples Fauna and classical Portage Group. a. *Ponticeras perlatum* (Hall, 1879), NYSM 13173, Lodi Limestone (base of Sherburne/Penn Yan Member), Ovid, N.Y. (Seneca Lake). b. *Ponticeras* cf. *regale* (Holzapfel, 1899), NYSM 3753 (illustrated by Hall, 1888), Firestone Beds, University Quarry, Middle Ithaca Member, Ithaca, N.Y. (Cayuga Lake). c. *Koenenites styliophilum* (Clarke, 1898), Linden Horizon and equivalents of the Upper Penn Yan Member: cross-section—specimen from Linden Horizon, Taunton Gully, near Leicester, N.Y. (Genesee Valley); lateral view—specimen from Crosby Sandstone, Sunset Point Creek, Keuka Lake; suture—specimen collected by D. D. Luther from Crosby Sandstone, Sartwell Ravine, Penn Yan, N.Y. (Keuka Lake). d. *Manticoceras sinuosum* (Hall, 1843), Cashaqua Member, Sonyea Formation: cross-section specimen from Parrish Limestone, Cashaqua Member, Conklin Gully, near Naples, N.Y. (Canandaigua Lake); lateral view—specimen from upper Cashaqua at Pine Tavern, N.Y. (Genesee Valley). Scale equals 5 mm.

perlatum (Hall, 1879), a Genesee species (Wells, 1959; House, 1962; Kirchgasser, 1975) (Fig. 4a). *Ponticeras* (Fauna 15) enters the section in the middle of the Geneseo Member and ranges to the middle of the Penn Yan and Ithaca Members. The appearance of *Ponticeras perlatum* in the uppermost Geneseo Member and overlying Lodi Limestone at Seneca and Cayuga Lakes corresponds to the beginning of the *Leiorhynchus quadracostatus* brachiopod zone (Wells, 1959; Johnson, 1970) and is probably within the Upper *disparilis* conodont zone (Klapper, 1984, personal communication). In the Lodi horizon in the Genesee Valley, *Ponticeras perlatum*? and *Leiorhynchus quadracostatus*? occur with Lowermost *asymmetricus* Zone conodonts.

RENWICK BLACK SHALE

At Cayuga Lake, the Renwick Member of the Genesee Formation (deWitt and Colton, 1978) is a black shale between the Sherburne and Ithaca Members (Fig. 3, column H). Locally at its basal contact with the Sherburne is the *Warrenella laevis* faunal horizon of Williams (1884) and Kindle (1896, 1906) from which Williams reported the Portage *Goniatites sinuosus* (along with *Goniatites perlatus*). The significance of Williams' record is that it was accepted by Kindle (1896) and later workers and thus reinforced the belief that Portage-age fossils were present in the Sherburne at Cayuga Lake. Aside from tornoceratids, *Ponticeras perlatum* appears to be the only ammonoid in the *laevis* fauna.

Ammonoids have not been reported from the Renwick outside of the *laevis* horizon at Cayuga Lake, but the Renwick Member itself is widespread. Thayer (1974) recognized its position in the separation of the facies equivalents of the Sherburne and Ithaca on the delta platform southeast of Cayuga Lake. Toward the basin, deWitt and Colton (1978) traced the Renwick in outcrop westward to Keuka Lake but were unable to find it with certainty among the black shale bands in the Penn Yan Member farther west. The Renwick horizon has now been identified at Canandaigua Lake and traced, by gamma-ray correlation, into the Genesee Valley (Fig. 3, columns E, D). In the Fall Brook section, the base of the Renwick horizon is about 4 ft (1.2 m) above the Geneseo/Penn Yan contact (Rickard, 1984, personal communication).

Although the Renwick Member does not appear to record a significant break in the ammonoid sequence (no new species are added), it does mark a major change in the benthic faunas around the delta-front (Thayer, 1974). According to Thayer, the geographic distribution of the benthic faunal associations ("faunal facies") was primarily controlled by the rate of delta-progradation, and the Renwick black shale records a transgressive or standstill interval, separating an interval of rapid progradation in the early Genesee time (Sherburne) from an interval of comparatively slow progradation in late Genesee time (Ithaca). In the basin and open shelf environments around Cayuga Lake and farther southeast, the *Cladochonus* facies (*Cladochonus* and *Warrenella*) of the Sherburne is replaced at the Renwick level by the *Plumalina* facies (*Productella* and *Plumalina*) of the Renwick and lower Ithaca (Thayer, 1974, fig. 17). Farther shoreward on the delta platform, the Renwick separates the *Rhipidomella-Leiorhynchus* facies from the succeeding *Schizophoria-Strophomenid* facies.

FIRESTONE BEDS

In his description of the section at Ithaca, at the south end of Cayuga Lake, H. S. Williams (1884, p. 17–18) referred to a bed of calcareous siltstones consisting of a mass of broken fossils in the middle of the Ithaca beds (Fig. 3, column H). This distinctive horizon, which is apparently a series of beds, was quarried at several localities around Ithaca and was called the "firestone bed"

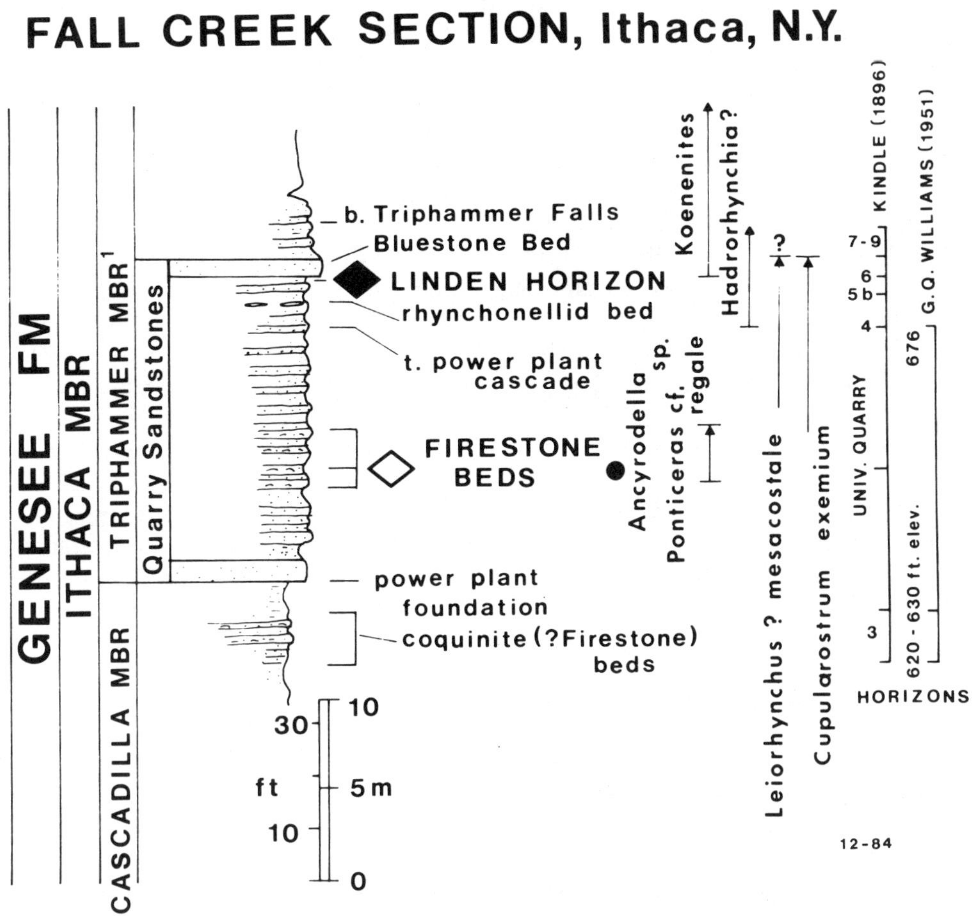

Figure 5. Location of ammonoid horizons in middle part of the Ithaca Member at Fall Creek. Position of main Firestone Beds within Quarry Sandstone interval based on Kindle's (1896, p. 23) reference to the University Quarry about 235 ft (72 m) above the base of the Renwick Shale; coquinites with firestone lithology and fauna also occur in Fall Creek in the cascade below the power plant foundation (Kindle's Horizon 3). The Linden Horizon (Kindle's Horizon 6) lies beneath the prominent siltstone bed (herein called the Bluestone Bed) just above the top of the power plant cascade. Thicknesses and ranges based on Kindle (1896) and G. Q. Williams (1951). 1: Members in classification of Caster (1933).

because of its resistance to breakage when heated (Williams, 1884; Kindle, 1896). In the section at Fall Creek, the Firestone horizons were located by Kindle (1896, p. 23) in the cascade behind the electric power plant; the main horizons correspond to the level of the University Quarry (Fig. 5). The benthic fauna in the Firestone Beds and surrounding siltstones was referred to the Ithaca Fauna by Williams (1884, 1906, 1907; Williams and others, 1909; Kindle, 1896; and G. Q. Williams, 1951). The association of *Spirifer (?Spinocyrtia/?Platyrachella) mesastrialis, Cryptonella eudora* and the brevicone cephalopod *Gomphoceras tumidum* is characteristic of the horizon (Williams, 1884).

The significance of the Firestone Beds to history of Catskill stratigraphy is that James Hall (1879, 1888) identified his Portage ammonoid *Goniatites sinuosus* in it and thereby established the Portage age of the Ithaca beds at Cayuga Lake. The ammonoid in the Firestone Beds is not *Manticoceras sinuosum* of the Cashaqua but belongs to the earlier genus *Ponticeras* (Fig. 4b). The species, which is rare in the horizon (only Hall's specimens are known), will be described in a forthcoming monograph with M. R. House, as *Ponticeras* cf. *regale* (Holzapfel).

Although he made no reference to the better preserved of Hall's specimens (Hall, 1888), Clarke accepted Hall's identification in his monograph of the Naples Fauna (1898, 1904). Although H. S. Williams does not refer to specific occurrences of *Goniatites sinuosus* in the middle Ithaca, his papers on the Ithaca section (Williams, 1884, 1906, 1907, 1913) make frequent references to the intermingling of the Portage and Ithaca faunas in the Ithaca beds. Kindle (1896) and later workers through the period of Chadwick's revision followed in the belief of the Portage age of the Ithaca beds.

Ponticeras cf. *regale* is apparently restricted to the middle part of the Ithaca Member of the Genesee Formation; the species and the Firestone horizon are unknown west of Cayuga Lake. A few fragments of conodonts, tentatively assigned to *Ancyrodella,* have been recovered from the matrix of Hall's University Quarry specimen, and they indicate that the Firestone Beds are in the Lower *asymmetricus* conodont Zone.

LINDEN HORIZON

At its type locality at Linden, New York, the Linden Horizon (Kirchgasser and House, 1981) is a layer of small, irregular, calcareous nodules with grains of pyrite and immature but well-preserved specimens of the ammonoid *Koenenites styliophilum* (Clarke) (Fig. 3, column C). The horizon is in the Upper Penn Yan Shale Member of the Genesee Formation, 7 ft (2 m) below the Genundewa limestone member. At Cayuga Creek (column B) and between Linden and the Genesee Valley (column D), the horizon is at or near the base of a series of thin styliolinid limestones also containing *Koenenites styliophilum,* above which is the contact with the Genundewa Limestone.

New species of *Koenenites* are introduced in the lower Genundewa Limestone. The *Koenenites* fauna with *K. styliophilum* in the Upper Penn Yan and the *Koenenites* fauna of the lower Genundewa comprise ammonoid Fauna 16 (Fig. 2; Fig. 3, column D). Ammonoid Fauna 17 begins in the upper Genundewa Limestone with the entry of species of *Manticoceras.* Fauna 17, which also includes species of *Koenenites,* ranges to the top of the West River Member and is informally subdivided into faunas 17a, b, c, d.

When first noted in 1970, the common goniatite in the Linden Horizon was thought to belong to *Manticoceras,* although none of the specimens then available showed the diagnostic adult suture (Kirchgasser, 1973). Specimens of the Linden ammonoid showing the diagnostic suture of the genus *Koenenites* were discovered by accident in the reserve collection of the Naples Fauna in the New York State Museum. There a collection made by D. D. Luther in 1904 from near Penn Yan on Keuka Lake was found to contain specimens with the unmistakable suture pattern of *Koenenites,* a genus then only known in New York from the lower West River Shale (Wells, 1956). The Linden species is assigned to *Koenenites styliophilum* (Clarke, 1898) (Figure 4c).

It is clear from the text with his map of the area and from the labels with the collection that Luther thought his Keuka Lake fauna was the typical Naples Fauna, and that the large ammonoid was *Manticoceras pattersoni.* He provided precise locality data and on his map (Luther, 1906), the horizon is mapped within the Cashaqua Shale (the Parrish Limestone at the top of the Cashaqua and the overlying Rhinestreet black shale are shown upsection from the locality). In his discussion of the Cashaqua Shale around Keuka Lake, Luther (1906, p. 45) cited the standard fauna list of the Naples Fauna (including *Manticoceras pattersoni*) of the type that appears in the texts that accompanied the Museum Bulletin maps of the region from Canandaigua to Seneca Lakes published in the early part of the century. Luther's misidentification of Genesee rocks as belonging to the Cashaqua and higher units was carried east of Keuka Lake to Seneca Lake (Clarke and Luther, 1905; Luther, 1909) and was not corrected until the Parrish Limestone in the upper Cashaqua was found hundreds of feet higher in the section than previously thought (Chadwick, 1935a, p. 348). Luther's Keuka Lake fauna actually came from nodules of the Linden Horizon at the base of the Crosby Sandstone (Torrey and others, 1932) at the contact of the Penn Yan and Ithaca Members of the Genesee Formation (Fig. 3, column F).

To the west of Keuka Lake, the Linden Horizon may be the same level as the Hiatus Concretion Bed of Baird (1975), a horizon of reworked concretions in the Upper Penn Yan at Canandaigua Lake (Fig. 3, column E). There are lithologic similarities to the Linden nodules, but thus far the ammonoid evidence from Baird's horizon is inconclusive. The details of the stratigraphic and sedimentological relations of the Linden Horizon, the Hiatus Concretion Bed, the Crosby Sandstone, and the Genundewa Limestone need further investigating. deWitt and Colton (1959, 1978) reasoned on the basis of stratigraphic position that the Crosby at Keuka Lake correlated westward with the easternmost exposures of the Genundewa Limestone at Canandaigua Lake. However, the discovery of the Linden ammonoid fauna and a distinctive conodont fauna (Klapper, 1984, personal communication) in the Crosby suggest that Genundewa and Crosby are not correlatives and that the Crosby is a somewhat older horizon (Fig. 4, columns E, F).

Following upon the work of Baird (1975), it would appear that the Linden Horizon is a reworked level and marks a hiatus during the deposition of the Upper Penn Yan Shale. Around the north end of Keuka Lake, the Upper Penn Yan Shale appears to be condensed relative to sections farther south along the lake and sections to the east and west (deWitt and Colton, 1978, charts) (Fig. 3, column F). These stratigraphic relations suggest that the nodules of the Linden Horizon were reworked by submarine erosion and later incorporated into the base of the Crosby Sandstone, as the Ithaca sediments (turbidites?) first reached the Keuka Lake area. Farther east at Seneca Lake, the Linden Horizon occurs in a tongue of Penn Yan Shale several meters below the horizon which deWitt and Colton (1978) identified as the Crosby Sandstone.

At Cayuga Lake, the Linden Horizon occurs in the middle of the Ithaca Member above the level of the Firestone Beds (Kirchgasser and House, 1981). In the section at Fall Creek, Ithaca, the horizon is the one from which Kindle (1896, his Station 6) reported Hall's *Goniatites sinuosus* (Fig. 5). Kindle's specimens, found in the Paleontological Research Institution at Ithaca, are *Koenenites styliophilum.* The ammonoid horizon is a bedding plane of dark shale beneath a prominent bed of siltstone, informally named the Bluestone Bed. The benthic faunas surrounding the Linden Horizon belong to the classical Ithaca Fauna. It is important to note that the ammonoids occur only in a separate bedding plane; they are not mixed with the benthic species. The bedding plane occurrence of the Linden ammonoids within the delta-front facies further supports the view that this widespread horizon records a hiatus or standstill in the progradation of the delta. The location of the Linden Horizon in the middle of the Ithaca Member indicates that the horizon of the Genundewa Limestone must be higher in the Ithaca section than suggested by lithostratigraphic projections of the Crosby horizon (deWitt and Colton, 1978, plate 3). The Linden Horizon at Cayuga Lake may prove to record a change in the brachiopod faunas. Underneath the Linden Horizon at the base of the Bluestone Bed is Kindle's Station 5b, a calcareous siltstone with abundant rhynchonellids, identified by Kindle as *Rhynchonella pugnus.* This may be the species Harrington (1972) referred to as *Hadrorhynchia*? aff. *H. solon.* Based on the faunal records of Kindle (1896) and G. Q. Williams (1951, plate 7), this species enters the Ithaca section at about the Linden Horizon, which also is near the top of the range of the rhynchonellids, *Cupularostrum exemium* and *Leiorhynchus? mesacostale.*

BLUFF POINT SILTSTONE

The Bluff Point Siltstone (Bergin, 1964) is a distinctive and widespread horizon of convolutely bedded siltstone in the upper West River Member. It was named for exposures around Keuka Lake (Fig. 3, column F). deWitt and Colton (1978) recognized the horizon as far west as Cayuga Creek (column B) but were unable to find it with certainty east of Keuka Lake. A faunal horizon with species of *Koenenites* and *Manticoceras* (Fauna 17c) occurs just beneath the Bluff Point Siltstone between Canandaigua and Keuka Lakes, and this horizon has been traced eastward into the upper Ithaca Member at Seneca and Cayuga Lakes (Kirchgasser and House, 1981) (Fig. 3, columns G and H). The discovery of the ammonoid horizon associated with the Bluff Point Siltstone in the section at Cayuga Lake provides faunal confirmation of the accuracy of the present correlation of the Genesee and lower Sonyea Formations across central New York. Like the Linden Horizon below it, the Bluff Point faunal horizon at Seneca and Cayuga Lakes is a bedding plane of dark gray shale within a sequence of siltstones and silty shale with benthic faunas. As such it may also record an interval of hiatus or standstill in delta-progradation.

SUMMARY AND POSTSCRIPT

The sequence of ammonoid faunal horizons in the basin facies of the Genesee Formation in western New York is now reasonably well understood. A few of the most widespread horizons have been traced shoreward into the delta-front facies around Ithaca at the south end of Cayuga Lake. The principal ammonoid horizons in the Genesee Formation in the Ithaca section are (in sequence) 1. the Lodi Limestone (with *Ponticeras perlatum*) at the base of the Sherburne/Penn Yan Member, 2. the base of the Renwick Member (with *Ponticeras perlatum*), 3. the Firestone Beds (with *Ponticeras* cf. *regale*) in the Middle Ithaca Member, 4. the Linden Horizon (with *Koenenites styliophilum*) in the Middle Ithaca Member, and 5. the Bluff Point Siltstone (with species of *Koenenites* and *Manticoceras*) in the Upper Ithaca Member. Some of the Genesee ammonoid horizons in the Ithaca section were known to James Hall, J. M. Clarke, H. S. Williams, and E. M. Kindle in the last century, but it was their mistaken belief that those horizons contained *Goniatites sinuosus/pattersoni* Hall, the index fossil of the Portage Group, that was apparently at the root of the long-held miscorrelation of the Sherburne siltstones and the Ithaca beds at Ithaca with the post-Middlesex Portage Group to the west.

The legacy of the faunal lists and collections from the classical units Genesee, Portage and Chemung provided critical clues in the rediscovery and revision of the Genesee ammonoids, and it is clear that attention to that legacy will be required if we are to make progress in the revision of other faunal groups. It is hoped that the stratigraphic control provided by these marker beds will encourage specialists on other faunal groups to begin the long overdue taxonomic and biostratigraphic revision of the benthic faunas around the delta-front. The well-documented faunal horizons reported by Williams (1884), Kindle (1896), and G. Q. Williams (1951) from the Ithaca Member should be re-examined and located in the present scheme of correlations.

As a start to formal biostratigraphic revision, the Finger Lakes and Chemung Stages should be abandoned or redefined. As currently defined, they are nothing more than the old group names elevated to the rank of stage. As Cooper et al. (1942, p. 1734) noted, the Finger Lakes Stage "included the Genesee and Naples groups as conceived by Chadwick" and the Chemung Stage "has the same limits that Chadwick gave to the group." Reliance on these stage names, rather than faunal horizons, as biostratigraphical boundaries has unfortunately limited the value of attempts to refine the biostratigraphy of the brachiopods (Greiner, 1957; Dutro, 1981) and bivalves (MacAlester, 1963).

There appears to be some correspondence in the first occurrences of ammonoid, rhynchonellid and conodont taxa in the ammonoid horizons, but more collecting, taxonomic work, and cross-checking of old faunal lists remains to be done to establish species sequences in these and other groups. The association of *Leiorhynchus quadracostatus* and *Ponticeras perlatum* along with probable Upper *disparilis* Zone conodonts in the Lodi Limestone at Seneca and Cayuga Lakes and the association of *Warrenella laevis* and *Ponticeras perlatum* in the basal Renwick Member at Cayuga Lake have already been documented (Rickard, 1975; Klapper, 1981, 1984, personal communication). In the Lodi horizon in the Genesee Valley, *Ponticeras perlatum*? and *Leiorhynchus quadracostatus*? are found with Lowermost *asymmetricus* Zone conodonts. The Firestone Beds in the Middle Ithaca Member with *Ponticeras* cf. *regale* mark the first occurrence of *Ancyrodella* in the Ithaca section, indicating the Lower *asymmetricus* Zone. The Linden Horizon, which marks the entry of the genera *Koenenites (K. styliophilum)* and *Probeloceras* in the New York section, may correspond to the entry of the genus *Hadrorhynchia* (*H.* ?aff. *solon*) in the rhynchonellid sequence (Harrington, 1972), but the biostratigraphic value of this and the other associations for even local correlation remains to be established.

There are still important gaps in our understanding of the

ammonoid sequence and correlations, particularly in the more shoreward facies. More collecting is needed in the upper parts of the Sherburne, West River and Ithaca Members of the Genesee Formation and in the Middlesex Member of the Sonyea Formation between Keuka and Cayuga Lakes. While this report was in preparation, the Genundewa horizon was found east of Canandaigua Lake. Gordon Baird discovered what appears to be the lower Genundewa ammonoid fauna in the lower Ithaca Member at Keuka Lake (Fig. 3, column F, horizon PY 10/1), and Baird and Carlton Brett have located the Genundewa horizon about 50 ft (15 m) above the Crosby horizon on the west side of Seneca Lake.

With regard to paleoecology, the new ammonoid marker horizons may make it possible to refine and place biostratigraphic boundaries on the sequence of benthic "faunal facies" of the Genesee Formation defined by Thayer (1974). As currently understood, the major change or break in these interfingering paleoecological associations around the delta-front occur at the level of the Renwick Shale. The next step is to look for evolutionary changes within these faunal associations, and the possibility of major breaks at the level of some of the other ammonoid horizons.

ACKNOWLEDGMENTS

I especially wish to thank Michael House (University of Hull), Wallace deWitt, Jr. (U.S.G.S., Reston), Lawrence Rickard (New York State Museum), and John Wells (Cornell University) for their generous assistance and support of my work on the Genesee Formation. I also greatly appreciate the exchange of information on Genesee stratigraphy with Gordon Baird (S.U.N.Y., Fredonia) and Carlton Brett (University of Rochester) and on Genesee conodonts with Gilbert Klapper (University of Iowa). The late Katherine Palmer (Paleontological Research Institution, Ithaca) made the collections of the P.R.I. available for study and graciously assisted in tracking down some of E. M. Kindle's localities. Ed Landing (New York State Museum) arranged for the loan of specimens from the New York State Museum. At Potsdam College, Suzanne Hammill drafted an early version of Figure 3, and Dick Bitely, Roy Madison, and Ron Riley helped prepare the illustrations. Laura Tharrett typed the manuscript. Support for fieldwork was provided by grants from the Research Foundation of the State University of New York.

REFERENCES CITED

Baird, G. C., 1975, Coral encrusted concretions: A key to recognition of a "shale on shale" erosion surface, Lethaia, v. 9, p. 293–302.

Bergin, M. J., 1964, Bedrock geology of the Penn Yan and Keuka Lake quadrangles, New York: U.S. Geological Survey Bulletin, 1161–G: 34 p.

Bowen, Z. P., Rhoads, D. C., and McAlester, A. L., 1974, Marine benthic communities in the Upper Devonian of New York: Lethaia, v. 7, p. 93 120.

Caster, K. E., 1933, Upper Devonian rocks at Ithaca, New York [abs.]: Geological Society of America Bulletin, v. 44, p. 201–202.

Chadwick, G. H., 1933, Great Catskill delta and revision of late Devonic succession: Pan-American Geologist, v. 60, p. 91–360.

——1935, Faunal differentiation in the Upper Devonian: Geological Society of America Bulletin, v. 46, p. 305–342.

——1935a, Chemung is Portage: Geological Society of America Bulletin, v. 46, p. 343–354.

Clarke, J. M., 1885, On higher Devonian faunas of Ontario County, N.Y.: U.S. Geological Survey Bulletin, v. 16, 80 p.

——1898, The Naples Fauna (fauna with *Manticoceras intumescens*) in Western New York: N.Y. State Geologist, Ann. Rept. 16, p. 29–161.

——1904, Naples Fauna in Western New York: Part 2: N.Y. State Museum Memoire 6, p. 199–454.

Clarke, J. M., and Luther, D. D., 1904, Stratigraphic and paleontologic map of the Canandaigua and Naples Quadrangles: N.Y. State Museum Bulletin, v. 63, 76 p.

——1905, Geology of the Watkins and Elmira Quadrangles: N.Y. State Museum Bulletin, v. 81, 29 p.

Cooper, G. A. and others, 1942, Correlation of the Devonian sedimentary formation of North America: Geological Society of America Bulletin, v. 53, p. 1729–1794.

deWitt, Wallace, Jr., and Colton, G. W., 1959, Revised correlations of lower Upper Devonian rocks in western and central New York: American Association of Petroleum Geologists Bulletin, v. 43, p. 2810–2828.

——1978, Physical stratigraphy of the Genesee Formation (Devonian) in western and central New York: U.S. Geological Survey Professional Paper, 1032–A, p. 1–22.

Dutro, J. T., 1981, Devonian brachiopod biostratigraphy: in Oliver, W. A., Jr., and Klapper, Gilbert, eds., Devonian biostratigraphy of New York, Part 1, Text: International Union of Geological Sciences, Subcommission on Devonian Stratigraphy, Washington, D.C., p. 67–82.

Greiner, Hugo, 1957, *"Spirifer disjunctus"*: Its evolution and paleoecology in the Catskill Delta: Peabody Museum of Natural History (Yale Univ.) Bulletin II, 75 p.

Hall, James, 1843, Geology of New York, Part IV, Comprising the survey of the Fourth Geological District: Albany, N.Y., 683 p.

——1879, Descriptions of the Gasteropoda, Pteropoda and Cephalopoda of the Upper Helderberg, Hamilton, Portage and Chemung groups: New York Geological Survey, Palaeontology of New York, v. 5 (2), 492 p.

——1888, Descriptions of the Gasteropoda, Pteropoda and Cephalopoda of the Upper Helderberg, Hamilton, Portage and Chemung groups: New York Geological Survey, Palaeontology of New York, v. 5 (2), Supplement (with Palaeontology of New York, v. 7), 42 p.

Harrington, J. W., 1972, Rhynchonellid brachiopod zonation of the New York Senecan (Early Upper Devonian): Proceedings of 24th International Geological Congress, Sec. 6, p. 278–284.

House, M. R., 1962, Observations on the ammonoid succession of the North American Devonian: Journal of Paleontology, v. 36, p. 247–284.

——1968, Devonian ammonoid zonation and correlation between North America and Europe: in Oswald, D. H., ed., International Symposium on the Devonian System, Calgary, Alberta, 1967: Alberta Society of Petroleum Geologists, v. 11, p. 1061–1068.

——1978, Devonian ammonoids from the Appalachians and their bearing on international zonation and correlation: Palaeontological Association, Special Paper 21, 70 p.

Johnson, J. G., 1970, Taghanic onlap and the end of North American provinciality: Geological Society of America Bulletin, v. 81, p. 2077–2105.

Kindle, E. M., 1896, The relation of the Ithaca Group to the faunas of the Portage and Chemung: Bulletins of American Paleontology, v. 2, 56 p.

——1906, Range and distribution of *Reticularia laevis*: Journal of Geology, v. 14, p. 188–193.

Kirchgasser, W. T., 1973, Lower Upper Devonian stratigraphy from the Batavia-Warsaw meridian to the Genesee Valley: goniatite sequence and correlation: N.Y. State Geological Association, 45th Annual Meeting, Guidebook, State University of New York at Brockport and Monroe Community College, p. C1–C21.

——1975, Revision of *Probeloceras* Clarke, 1898 and related ammonoids from the Upper Devonian of western New York: Journal of Paleontology, v. 49, p. 58–90.

——1982, Ammonoid correlations near the delta-front: Late Devonian Genesee Formation, New York: NE/SE Sections Geological Society of America, Washington, D.C., Abstracts with Program, p. 31.

——and House, 1981, Upper Devonian goniatite biostratigraphy: in Oliver, W. A., Jr., and Klapper, Gilbert, eds., Devonian biostratigraphy of New York. Part 1, Text: International Union of Geological Sciences, Subcommission on Devonian Stratigraphy, July 1981, Washington, D.C., p. 39–55.

Klapper, Gilbert, 1981, Review of New York Devonian conodont biostratigraphy: in Oliver, W. A., Jr., and Klapper, Gilbert, eds., Devonian biostratigraphy of New York, Part 1, Text: International Union of Geological Sciences, Subcommission on Devonian Stratigraphy, July 1981, Washington, D.C., p. 57–66.

Lincoln, D. F., 1897, Report on the structural and economic geology of Seneca County: New York State Geologist, Annual Report, v. 14, p. 57–125.

Luther, D. D., 1906, Geology of the Penn Yan-Hammondsport Quadrangles: New York State Museum Bulletin 101, 58 p.

——1909, Geology of the Geneva-Ovid quadrangles: New York State Museum Bulletin 128, 41 p.

Macdonald, K. B., 1975, Quantitative community analysis: recurrent and cluster techniques applied to the fauna of the Upper Devonian Sonyea Group, New York: Journal of Geology, v. 82, p. 473–499.

McAlester, A. L., 1963, Pelecypods as stratigraphic guides in the Appalachian Upper Devonian: Geological Society of America Bulletin, v. 74, p. 1209–1224.

McGhee, G. R., Jr., and Sutton, R. G., 1981, Late Devonian marine ecology and zoogeography of the central Appalachians and New York: Lethaia, v. 14, p. 27–43.

Rickard, L. V., 1964, Correlation of the Devonian rocks in New York State: New York State Museum and Science Service, Map and Chart Series 4.

——1975, Correlation of the Silurian and Devonian rocks in New York State: New York State Museum and Science Service, Map and Chart Series 24, 16 p.

——1981, The Devonian System of New York: in Oliver, W. A., Jr., and Klapper, Gilbert, eds., Devonian biostratigraphy of New York. Part 1, Text: International Union of Geological Sciences, Subcommission on Devonian Stratigraphy, July 1981, Washington, D.C., p. 5–22.

Sutton, R. G., 1959, Use of flute casts in stratigraphic correlation: American Association of Petroleum Geologists Bulletin, v. 43, p. 230–237.

——1963, Correlation of the Upper Devonian strata in south-central New York, in Shepps, V. C., ed., Middle and Upper Devonian stratigraphy of Pennsylvania and adjacent states—a symposium: Pennsylvania Geological Survey, 4th Series, Bulletin G-39, p. 87–101.

——Humes, E. C., Jr., Nugent, R. C., and Woodrow, D. L., 1962: New stratigraphic nomenclature for Upper Devonian of south-central New York: American Association of Petroleum Geologists Bulletin, v. 46, p. 390–393.

——Bowen, Z. P., and McAlester, A. L., 1970, Marine shelf environments of the Upper Devonian Sonyea Group of New York: Geological Society of America Bulletin, v. 81, p. 2975–2992.

Thayer, C. W., 1973, Marine paleoecology of the Upper Devonian Genesee Group of New York [Ph.D. Thesis]: Yale University, 226 p.

——1974, Marine paleoecology in the Upper Devonian of New York: Lethaia, v. 7, p. 121–155.

Torrey, P. D., Fralich, G. H., Young, W. H., Brewer, Charles, Jr., and Phillipi, P. M., 1932, The geology of New York and northern Pennsylvania: American Petroleum Institute, Div. Production, Paper 826-4A, 19 p.

Vanuxem, Lardner, 1842, Geology of New York, Part III, Comprising the survey of the Third Geological District: Albany, N.Y., 306 p.

Wells, J. W., 1956, *Koenenites* and *Beloceras* from the Upper Devonian of New York: Journal of Paleontology, v. 30, p. 749–751.

——1959, Geology of the Cayuga Lake Basin: *in* Cole, W. S., ed., New York State Geological Association, 31st Annual Meeting (Guidebook): Cornell University, Ithaca, N.Y., 36 p.

Williams, G. Q., 1951, The stratigraphy of the type Ithaca Formation [Ph.D. Thesis]: Cornell University, Ithaca, N.Y., 118 p.

Williams, H. S., 1884, On the fossil faunas of the Upper Devonian: U.S. Geological Survey Bulletin, v. 3, p. 55–86 (5–36).

——1906, The Devonian section of Ithaca, N.Y., Part I: Journal of Geology, v. 14, p. 579–598.

——1907, The Devonian section of Ithaca, N.Y., Part II: Journal of Geology, v 15, p. 93–112.

——1913, Recurrent *Tropidoleptus* zones of the Upper Devonian of New York: U.S. Geological Survey Professional Paper 79, 103 p.

——Tarr, R. S., and Kindle, E. M., 1909, Watkins Glen-Catatonk Folio: U.S. Geological Survey Folio 169, 242 p.

Manuscript Accepted by the Society July 20, 1984

Printed in U.S.A.

Geological Society of America
Special Paper 201
1985

Destruction of a Late Devonian ophiuroid assemblage: A victim of changing ecology at the Catskill delta front

Daniel B. Sass
Department of Geology
Alfred University
Alfred, New York 14802

Robert A. Condrate
State University of New York
College of Ceramics
Alfred, New York 14802

ABSTRACT

The fossilized remains of an isolated but once-thriving assemblage of the brittle starfish *Furcaster* cf. *F. leptosoma*(?) (Salter, 1857) record a singular event in the succession of late Devonian (Famennian?) marine life. In similar cases of fossil starfish mortality reported elsewhere, the cause of death was cited as suffocation due to rapid burial. In order to construct a paradigm which reflects the circumstances before and after the destruction of the assemblage in question, trace elements in the rock matrix were analyzed and correlated with neontologic data relative to important elements of the community and its environment.

The circumstantial evidence points to burial as the ultimate factor in the death of the starfish. They were probably doomed by a series of subtle changes, of which they were not aware, taking place at the front of the delta they inhabited. The changes, hydrologic in nature, were instrumental in dulling the instincts of the starfish and preventing their escape from burial.

INTRODUCTION

Although the Alfred Shale is best known as a ceramic raw material, knowledge of its geologic attributes antedates its commercial usage by almost a century. James Hall, New York's second and most durable paleontologist, visited outcrops in the vicinity of Alfred Station (Figure 1) circa 1857. That he knew of the rock sequence is documented by unpublished correspondence between Hall and the Jonathan Allen family of Alfred University, whose son Alfred was a fossil collector for Hall. The Alfred Shale was primarily of interest to local fossil collectors prior to its formal designation as a distinct facies (Figure 2 a-d) by Rudy Eller (1935, p. 263):

> The shale in which the fossils are found is about five feet thick, very fine grained, and was probably deposited in quiet and fairly deep water. Above the shale is a barren, silicious shale about forty feet thick. . . .The name Alfred Shale is proposed here for the shale facies with its large unique fossil assemblage as well as for the thicker silicious shale lentil above. This Alfred Shale is overlaid by a heavy calcareous sandstone [coquinite] about six feet thick.

He (Eller) established the two-part nature of the Alfred sequence and noted that his contemporary G. H. Chadwick considered the calcareous sandstone (Figure 2d) to be a part of the Rushford Sandstone of the Canadaway Group.

Woodruff (1942, Section 2) regarded the Alfred Shale as a "fossil zone near Alfred Station"—at the base of the Caneadea Shale. This position agreed with that of Cooper et al., (1942) placing the Alfred Shale as a correlative of the lower part of the European Famennian Stage. Manspeizer (1963) did not regard the Alfred Shale as a separate entity, but instead designated the "Alfred Station Coquinite" (= Chadwick's Rushford Sandstone?) as the stratigraphic equivalent of the Rushford Sandstone.

Our efforts to confirm the Famennian age of the Alfred

Dedicated to the memory of Eugene R. Eller, 1904–1976.

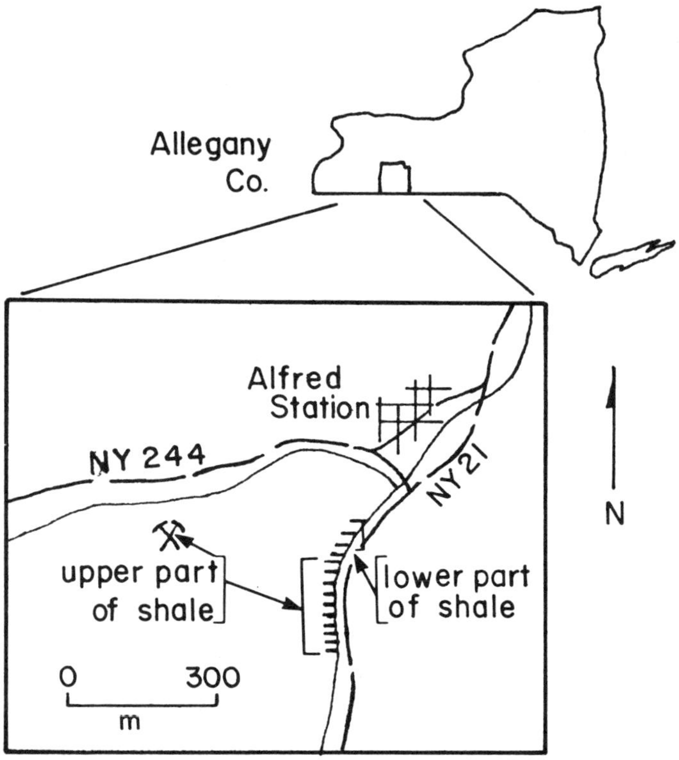

Figure 1. Location map of principal Alfred Shale outcrops in western New York. The exposure along the creek west of N.Y. Rt. 21 is the "type section."

Shale have not been rewarding. The abundant remains of the rhynchonellid brachiopod genus *Camarotoechia* were rarely well enough preserved to make the serial sections requisite for comparison with species determined to be Famennian in Europe. The dissolution of random samples of the Alfred Shale, totalling one kilogram in weight, produced exactly four microfossils: two scolecodonts and two conodonts. Eller (personal communication, 1975) identified one of the scolecodonts as *Nereidavus* sp. *"A"* without determining its geologic range; the second specimen was unidentifiable. The conodonts were sent to T. T. Uyeno of the Geological Survey of Canada. He (Uyeno, personal communication, 1975) identified one as *Icriodus alternatus* Branson and Mehl (I Element) which ranges in age from the European *Ancyrognathus triangularis* s. f. Zone (*do* Iδ) to the Upper *Palmatodella quadrantinodus* Zone (*do* III α)—a time spanning the mid-Frasnian through mid-Famennian. The question of precise correlation of the Alfred Shale with the European section will have to await a more thorough analysis of its flora and fauna.

THE LITHOFACIES

Eller's perception of the bipartite physical and chemical nature of the Alfred Shale is reasonably accurate. The lower contact has no known surface exposure nor is it recorded in available drilling logs. The exposed lower 5 ft. (1.5 m) grade upward into the caprock of a small falls (Figure 2c). The caprock is distinguishable from the dense, dark gray shale below by its superior resistance to erosion and a detectable increase in free quartz. The dark gray shale is thinly bedded and has a conchoidal fracture. Occasional limonite-stained burrows (?) occur along or parallel to the bedding surfaces and less commonly, in the vertical dimension. Also, on occasion, in the lowermost 2 ft. (0.6 m) of the basal sequence, smooth, flattened, and rounded green shale pebbles (clasts) are found imbedded in the gray shale matrix. They may represent physical perturbation temporarily affecting sedimentation rates; but their source is not apparent, nor do green shales occur commonly in this part of the geologic sequence.

The upper member, about 40 ft. (12.0 m) thick, is well exposed in a quarry along N.Y. 244, approximately 2,000 ft. (600 m) SW of Alfred Station. The rocks are light to medium gray siliceous shales (on fresh surface) with intercalations of thin sandstone lenses exhibiting "varve-like" regularity. Weathered sequences are heavily limonite-stained and decrepitate rapidly into a fine mud. R. Snyder (personal communication, 1980), citing X-ray study, has confirmed that the upper unit contains appreciably more free quartz (up to 35 percent total) than the lower unit. The upper member terminates abruptly with a marked disconformity against the base of the coquinite.

The coquinite is a gray, massive, coarse sandsatone (Figure 2d) which weathers to a golden brown. It contains up to 40 percent calcium carbonate (by weight) in the form of cross-bedded accumulations of disarticulated brachiopod shells. The layer can be traced southwest to the quarry on N.Y. 244 (Figure 1) where recent excavations have exposed a similar massive sandstone 10 ft. higher in the section.

THE BIOFACIES

Eller (date unknown) compiled the list of 80 taxa distributed through nine phyla and 10 classes of organisms from the Alfred Shale. He stated that the fauna came from the lowermost 6 inches (0.15 m) of the basal member and that the remainder of the lower, and all of the upper member, were barren. He devoted his attention to the scolecodonts and the malacostracans (Figure 3a) and left the crinoids (Figure 3b) to Winifred Goldring (1935); the eurupterids to Rudolph Ruedemann (1935); and the remainder of the taxa to posterity. This benign neglect changed in 1975 with the interest generated by the discovery of a fossil brittle star (ophiuroid) in the float at Alfred Station (Figure 3d).

Study of the exposed basal 6 inches (0.15 m) of the lower member revealed some interesting relationships. The earliest occupants of the environment may have been malacostracans (e.g., *Echinocaris* sp.) and diverse-infaunal burrowers. A complex community subsequently developed at this site with exotic elements, ophiuroids, crinoids; a sessile benthos, brachiopods, mollusca, and sponges; an infauna, worms; and plants? (Figure 3c). The bivalves frequently show orientations with long axes parallel to the bedding as is presumed for some feeding positions in life. There is no evidence for extensive anoxic conditions.

This complex Devonian marine community flourished only

Figure 2. Vistas of the "type section" showing the discrete units described by Eller in 1935. a. The lowest exposed portion of the sequence. The 512 ft. (155 m) of section from the N.Y. Rt. 244 bridge in the background to the position of the student in the foreground is covered with glacial till and stream debris. The student is standing on the ophiuroid surface. b. The arrow in the foreground marks the position of the ophiuroid layer. Eller (1935) considered the succeeding 0.5 ft. (0.15 m) as the only fossiliferous part of the Alfred Shale. c. The exposure in the stream bed south of the ophiuroid bed. The lip of the falls marks the top of Eller's basal unit and is 5 ft. (1.5 m) above the ophiuroid bed. d. The 40 ft. (14 m) thick upper unit of the Alfred Shale is separated from the overlying 6 ft. (1.8 m) massive calcareous sandstone by a marked disconformity (upper arrow). The sandstone is more properly designated as a coquinite. The lower arrow marks the position of one of many large blocks of conquinite float, one of many which break away from the exposure in the cliff above.

during the deposition of the lowest 6 inch (0.15 m) interval at Alfred Station; but contrary to Eller, the succeeding sequences were not barren. Except for the ophiuroids, subsequent colonization appeared sporadically in small lenses in the lower member. The same regimen pervaded the upper member—except at less frequent intervals, with the most common colonizer being the brachiopod *Camarotoechia contracta.* A hitherto unreported find in the upper member was an occasional remnant of the land plant *Calamites.*

The Ophiuroid Occurrence

The ophiuroids (Figure 3d) were found in place on a single surface (Figure 2b) at the lowest accessible exposure of the Alfred Shale—1561.06 ft. AT along Rt. 21. They do not subsequently appear in the exposed section beyond the time interval represented by the surface on which they occur. The possibility of an earlier occurrence cannot be determined without a suitable outcrop. F.H.C. Hotchkiss (personal communication, 1974) deter-

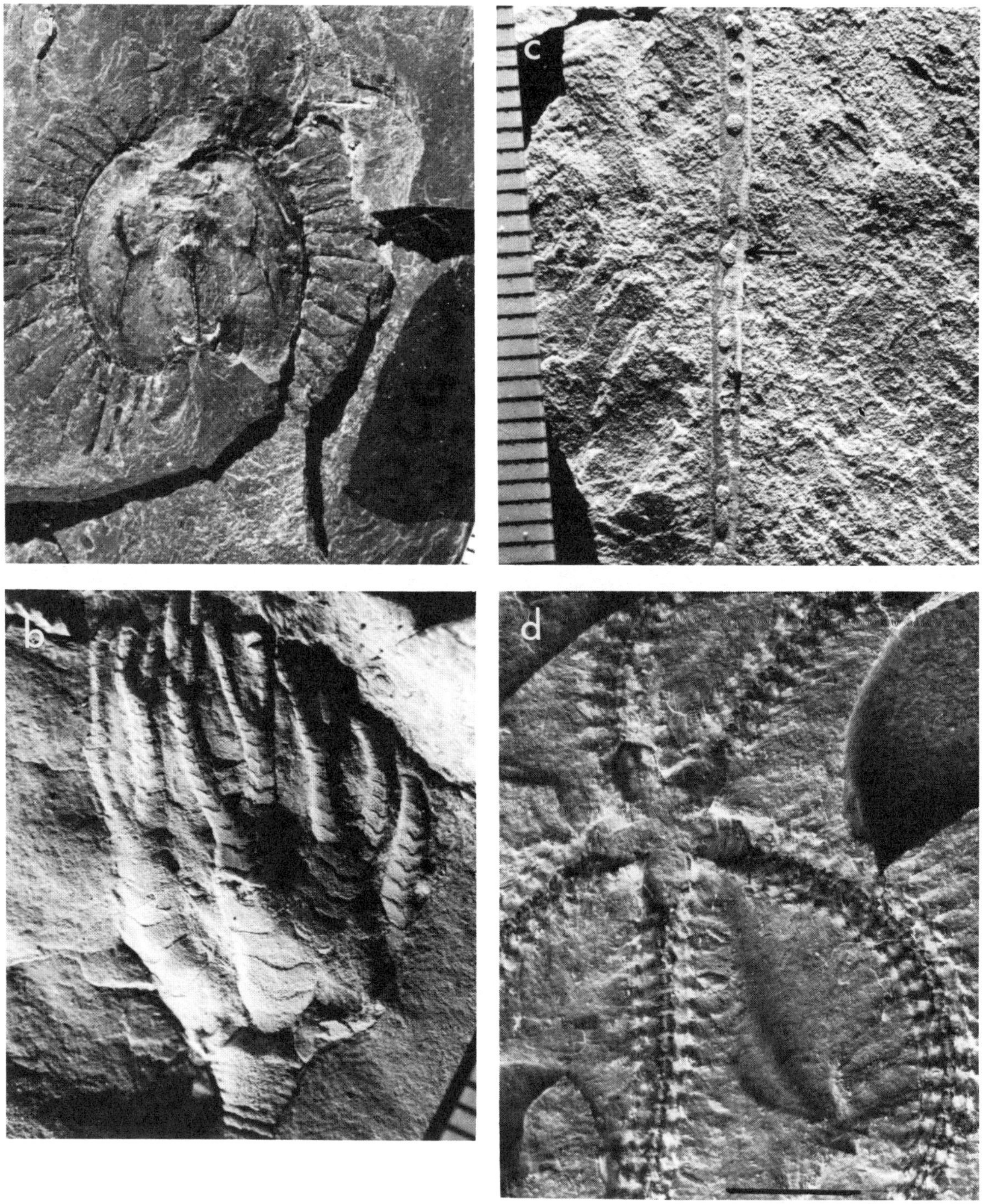

Figure 3. Elements of the biota of the Alfred Shale. a. The carapace and carbonized appendages of an unusually well-preserved phyllocarid malacostracan *Pephricaris* cf. *P. horripalata* (Clarke). Basic unit on scale to lower right is 1 mm. University of Cincinnati Geological Museum accession number 42615. b. An unaccessioned topotype specimen of the flexible crinoid *Eutaxocrinus alfredi* Goldring, 1935 in the Alfred collection. The basic unit on the scale to the lower right is 1 mm. Holotype is Carnegie Museum accession number 7229. c. An unaccessioned specimen of a carbonized impression of a brown algae (?). The arrow points to the row of nodes which may have had a reproductive or flotational function. The basic unit of the scale on the left is 1 mm. d. The aboral surface and partially exposed arms of the zeugophiurinid ophiuroid *Furcaster leptosoma*? (Salter, 1857). Note the total articulation of the arms, the tips of which are concealed in the rock matrix. The bar measures 1 cm. United States National Museum accession number 210249.

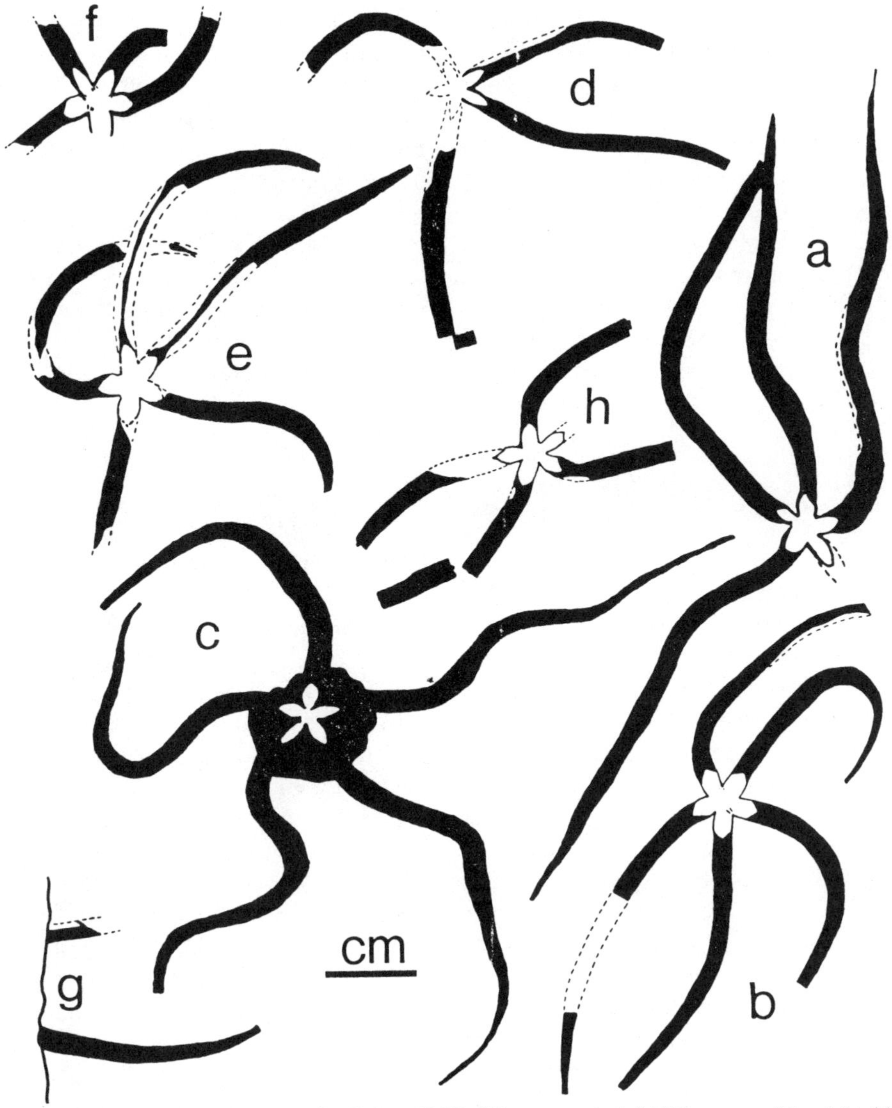

Figure 4. *Furcaster* cf. *F. leptosoma* (?) (Salter, 1857) Silhouettes. /a-h/ Silhouettes of the eight Alfred Shale specimens. Catalogue numbers for specimens b through g, in sequence, are United States National Museum accession numbers 210249 through 210524. (After H.F.C. Hotchkiss, 1974).

mined the assemblage to consist of the single species *Furcaster* cf. *F. leptosoma* (?) (Salter, 1857). The horizon is only the fifth known from the New York State Devonian (Hotchkiss, op. cit.). Analogous starfish interments from three European Paleozoic horizons, reported by Goldring and Stephenson (1972), were attributed to smothering by rapid accumulation of sediments in paleoenvironments somewhat different from that of the Alfred Shale.

The 12 (±) ophiuroid specimens obtained by the authors and others thus far comprise a number of size classes (Figure 4) typical of a biocoenosis. Rokop (1974) confirmed that extant ophiuroids breed throughout the year unfettered by constraints which govern the reproductive activities of other organisms and, therefore, are likely to exhibit considerable size variation. Further, most of the ophiuroids in the assemblage are still essentially articulated, a condition which suggests that their burial took place immediately after death or that interment was rapid, almost instantaneous.

Postmortem articulation implies burial sufficiently rapid to protect the corpse from predation. Meyer (1971) determined experimentally that the isopod *Cirolana parva* would disarticulate Recent ophuiroids in two to six days unless they are insulated by burial. Secondly, the apparent failure of the majority of assemblage members to exhibit autotomy under the stress of rapid sedimentation or some other environmental anomaly requires explanation; this subject will be addressed, with alternatives, in the conclusions reached by the authors.

According to Hotchkiss (personal communication, 1974), the preservation of the ophiuroids renders them difficult to interpret. The solid calcite parts are lacking in surface detail due to secondary deposition. The ossicles are small, thin, and discontinuous due to rock parting (= shrinking?) However, sufficient detail exists to identify them as to genus and suggest specific affinities. The orientation of all specimens suggests normal-unstressed movement. No evidence of dorsal arm flexure (used for vertical motion) exists.

Because of the fragmentary and sporadic exposure of the ophiuroid layer, it is difficult to be too specific about contemporary elements of the fauna. The crinoids may have occurred with the starfish, but we have no specimen with both on the same surface. Goldring's (1935) descriptions do not include detailed analyses of their state of preservation or its environmental significance. Our slab samples frequently show complete calices attached to long columnals—a circumstance which might suggest rapid burial rather than normal expiration and slow decomposition.

ANALYSES AND TECHNIQUES

In order to have the broadest data base possible for interpretations of the paleoenvironment represented by the Alfred Shale, a decision was made to analyze its trace element content. The first selection involved the basal 5 ft. (1.5 m) of the section along Rt. 21 beginning at the ophiuroid horizon 1561.06 ft. (454 m) AT. The 5 ft. (1.5 m) sequence was measured with a plane table and divided in 1 ft. (0.31 m) increments from each of which a channel sample approximating 10 lbs. (4.6 kg) was cut. Each sample was pulverized to -200 standard ASTM mesh and subdivided into 100 gm lots with a sample splitter. Subsequently, after consulting authoritative works in the field (see Keith and Degens, 1959; Potter, Shimp and Witters, 1963) and personally consulting with geochemists actively engaged in trace-element studies, some 16 elements were selected as potentially relevent to this study.

Using standard atomic absorption techniques, three different splits of each interval, two different machines, and three operators, virtually the same results (within acceptable variation) were obtained for each element tested as recorded in Table 1. Significant changes in the concentration of the elements Be, Co, Ga, Mn, Pb and V were noted between the first and second sample horizons, providing pause for reflection on cause and effect apropos of environmental change.

To investigate the possible effect trace-element changes might have on the life processes of the ophiuroids, data were sought on the trace-element content of living brittle stars. Published information was difficult to obtain. The problem was tken to Dr. Jerry D. Reber of the Department of Physics of S.U.N.Y. Geneseo. He applied proton activation analysis to samples of dried specimens of *Ophiocoma echilata* from Grahams Harbor, San Salvador Island, Bahamas.

According to Dr. Reber (personal communication, 1982), elements definitely detected were Ca, Zn, and Li; missing were Cd, Co, Ga, Mn, Pb, Ni, and As. Although this is admittedly from a small sample, it would seem that the elements in which there were noticeable changes in concentration (Co, Ga, Mn) in the Alfred Shale probably do not affect the *in vitro* processes of brittle stars.

Finally, Dr. Robert Snyder of S.U.N.Y. College of Ceramics of Alfred (personal communication, 1979) reported that X-ray analysis of a random sample of the lower "Alfred Shale" unit disclosed a mineral composition of 50 percent illite, 25 percent quartz, 6 percent feldspar, and 4 percent kaolin, and 8 percent chlorite. These values correspond closely to those obtained for an analysis of the bottom sediment of Keuka Lake, New York, possibly suggesting the influence of less-than-normal salinity of the water during part of "Alfred Shale time."

ENVIRONMENTAL ANALYSIS AND SYNTHESIS

The relationship between the Alfred Shale and its biota must

TABLE 1. CONCENTRATIONS (ppm) OF SELECTED TRACE ELEMENTS OF THE LOWERMOST FIVE FEET OF THE ALFRED SHALE

Trace Element	Interval (Ft) *0-1	1-2	2-3	3-4	4-5
B	81.00	72.00	91.00	58.00	83.00
Be	00.38	00.17	00.15	00.16	00.23
Cd	00.06	00.10	00.11	00.19	00.10
Co	21.00	12.00	14.00	08.00	14.00
Cr	79.70	79.10	85.50	74.00	72.40
Cu	28.00	26.60	26.50	20.20	25.70
Ga	21.20	07.20	06.90	06.30	06.80
Li	35.30	32.50	35.20	30.60	37.00
Mn	949.00	522.00	548.00	491.00	582.00
Ni	57.00	48.00	58.00	47.00	62.00
Pb	05.20	08.80	07.00	08.20	05.60
Rb	224.00	202.00	221.00	193.00	251.00
Sr	22.30	23.20	24.40	22.90	19.50
V	160.00	171.00	165.00	162.00	138.00
Zn	83.00	92.00	86.00	77.00	89.00
S_{Tot}(wt%)	00.19	00.24	00.28	00.35	00.39

*0 - ophiuroid layer @ 1561.06 AT

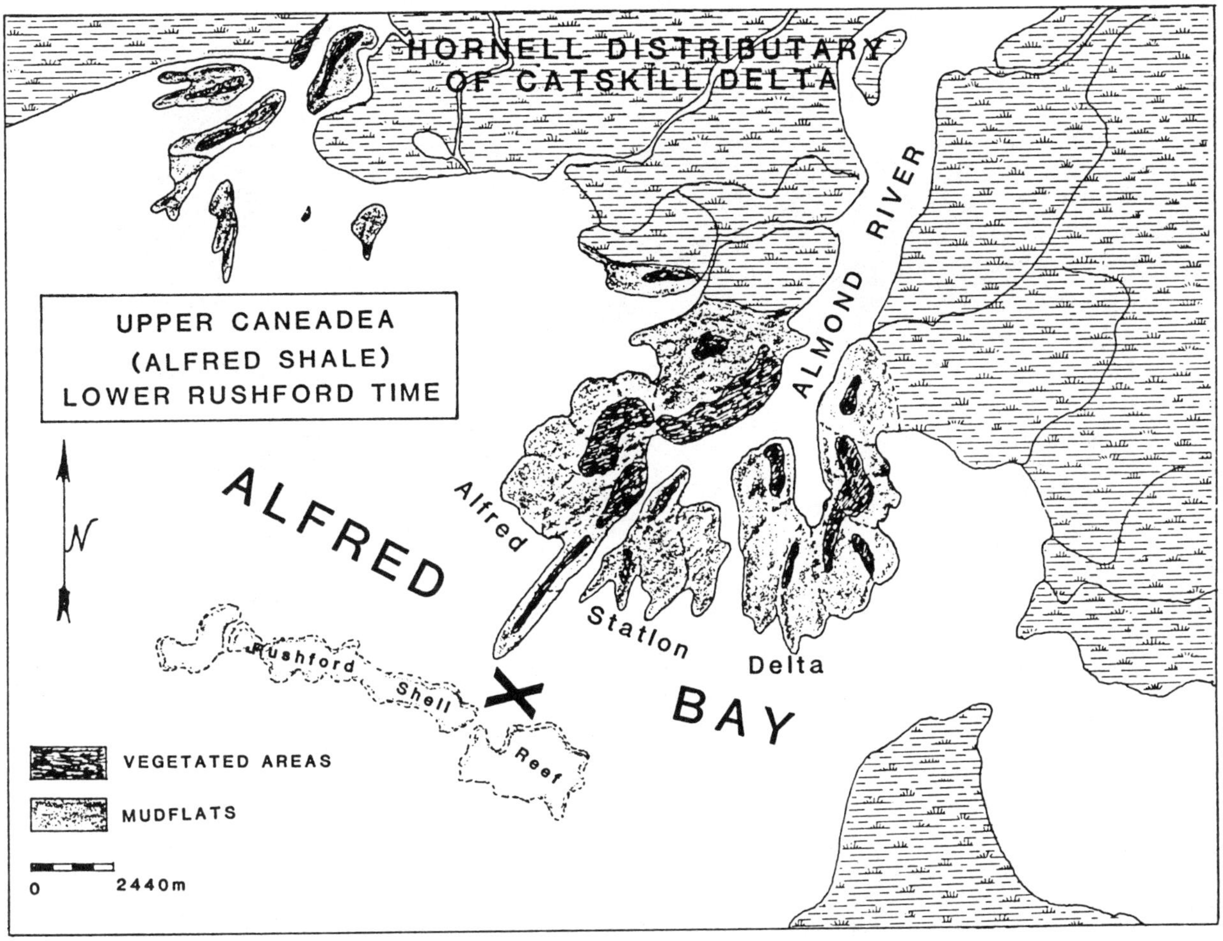

Figure 5. A suggested reconstruction of the paleoenvironment during upper Alfred Shale-lower Rushford time. The "X" marks the position of the ophiuroid horizon now buried beneath the subsequent load of 45+ ft. (14 m) of sediment on the seaward side of the advancing distributary. The Rushford shell reef is beginning to develop and will be one of a number of such features in the Upper Devonian of New York. (With the permission of the editor of Aquanotes, a Sea Grant Publication of Louisiana State University—after a diagram in Vol. 12, Issue 2, p. 3, June 1983.)

be examined in the context of the larger event of which they were but a part—the growth and development of the Catskill delta. By "Alfred Shale time," the marine phase of the delta had retreated westward from its easternmost extent to a point near the current site of Alfred Station (Figure 5). The Alfred Shale cannot be traced laterally for any distance; it appears to have been a unique embayment at the delta front. From the assumptions made above, the following environmental conditions are interpreted as reflecting the factors at work in the embayment.

According to Epstein and others (1977, p. 1), the recovery of as few as two conodonts (as herein) per kilogram of sample would indicate rapid deposition of sediment. The green shale clasts in the beds above the ophiuroids reinforce this impression and suggest periodic energy changes, probably generated by storms, which might challenge the stability of the resident community as well. The thin sandstone layers may suggest an alternatively transgressive or regressive interface between saline and fresh water on a sloping bottom (R. D. Matthews, personal communication, 1982). That the net effect was progradation may be interpreted from the increased percentage of free quartz and the fossil land plants in the upper member. Finally, with regard to the front *per se,* Goldring's (1978, p. 139–143; Figures 36 and 37) reconstructions of the Upper Devonian Baggy beds coastal deltaic and clastic shelf communities match the ecology suggested by the upper and lower members of the Alfred Shale. Goldring (1978, p. 139), for an analogous circumstance, suggests that "the sands and muds were probably deposited just off a distributory and subjected to repeated scour by storms and floods." He (ibid. p. 140) adds:

> The Upper Devonian Baggy Formation . . . is a suite of shallow water sediments deposited in a variety of near shore environments but always under some degree of influence from a delta or distributary bringing in fresh water, sediment and plant debris. An offshore benthic fauna, . . . occurs at scattered horizons, . . .

A distributary would explain the land plant debris in the

Figure 6. Reconstruction of the Devonian sea bottom during early "Alfred Station time." The biota is not shown in its entirety. The crinoids are, from left to right, *Eutaxocrinus alfredi, Eutaxocrinus tenuiramosus,* and *Anamesocrinus lutheri.* Note the spirifer brachiopods in the living-feeding position. Drawing by Ms. Pam Lea Miller, AU 1976.

Alfred Shale and the possible source of the diluent for the normally saline waters. The presence of brown algae (H. Banks, personal communication, 1980 and Figure 3c) would suggest a water depth of less than 200 m at the Alfred Station site.

The trace element residuum within the lower fossiliferous member of the Alfred Shale was studied, as noted above, to determine the nature of the geochemical environment of "Alfred Shale time."

The resulting data (Table 1) revealed significant changes in some trace-element quantities between the first (0-1 ft.) and second (1-2 ft.) intervals. Of these, the 200-percent decrease in the concentration of gallium between the first and second intervals was selected by the writers to be the most significant. Gallium has been used for estimating the salinity of paleooceans, but there is a disagreement as to its interpretation. Keith and Degens (1959) concluded that increases in gallium concentration accompany decreases in salinity. Potter and others (1963) reached an opposite conclusion. Either interpretation fits the proposed scenario of the incursion of fresh water for "Alfred Shale time." In the first case (Keith and Degens, op. cit.) the change from 21.20 ppm gallium for interval one to 07.20 ppm for interval two can be interpreted as a change from what had been a low to a higher salinity. In the second case (Potter and others, op. cit.), the change of the same magnitude might be interpreted as a circumstance where salinity changed from a high to lower value. Either way, it can be assumed that the brittle stars would be subjected to water of less-than-normal salinity.

DISCUSSION

In-life positions of the brachiopods and pelecypods and the random ophiuroid-size classes support the thesis that the lowermost portion of the Alfred Shale represents the remains of a biocoenosis. It is probable that its demise was rapid, taking hours or a few days. The observed articulation and arm postures of the brittle stars as a function of rapid burial are well covered by Schäfer (1972, p. 97–99) and Hotchkiss (herein). Periodic changes in environmental stress are suggested by the occasional green shale clasts and sandstone lenses. The marine setting for the above was limited in areal extent and to shallow-water depths in which critical changes in salinity are suggested by fluctuations in gallium concentration.

The task remaining includes 1) an explanation for the proliferation of the benthic community (Figure 6) in early "Alfred Station time," and 2) an understanding of the failure of the ophiuroid assemblage to either exhibit serious stress behavior or flee in the face of traumatic environmental change. A paradigm relating these circumstantive anomalies to the deltaic environment in which they occurred follows.

Prior to "Alfred Shale time," a distributary of the Catskill delta formed to the east, perhaps with a bird's-foot extension eventually combining settings analogous to those of the Upper Devonian Baggy Formation of north Devon, England and Atchafalaya Bay, La., characterized by crustaceans, malacostracans and infaunal burrowing organisms. As the distributary lengthened westward, the infusion of terrestrial nutrients enriched an area of limited extent but capable of supporting a large, diverse population (see Figure 6.). The penetration of sunlight in the shallow water and establishment of algae gardens produced oxygen, adding to the attractiveness of the environment. Initially, the fresh water carried by the distributary, was well mixed with sea water before it reached the Alfred Shale. But with continued westward extension of the distributary, the fresh water was no longer thoroughly mixing with the saline offshore waters. The diminishing salinity could have "anesthetized" the ophiuroids, dulling their normal instincts. A duplication of this natural process is employed by purveyors of ophiuroids who collect them alive for biological studies and wish to prevent the moribund organisms from committing autotomy. The salt water in the confining receptacle is slowly diluted until the starfish expire peacefully (see Steinbeck, 1954, p. 142).

Shortly thereafter, the slope of the leading edge of the distributary assumed an unstable habit and failed, producing a turbidity current which innundated the Alfred Shale community. The sessile benthos, which could not escape, became victims of their own immobility. Some elements of the vagrant benthos survived, but the brittle stars, their senses dulled by fresh water anesthesia, surrendered to their fate: entombment and death.

Dr. Carlton Brett (personal communication, 1983) suggests an alternative explanation of the starfish demise which is also consistent with the data. He suggests that during major storms fresh water charged with particulate matter might move as a "slug" toward the starfish as a density flow. The first slurry might have anesthetized the brittle stars with later mudflows burying the entire community—brittle stars included.

Other scenarios or interpretations are eminently possible and should be explored, particularly as our appreciation of the role of trace elements is enhanced and better understood.

ACKNOWLEDGMENTS

The authors are indebted to many friends and colleagues for their interest and contributions during the lengthy investigations which culminated in the conclusions put forth in this paper. Field work and collections were enriched by the efforts of Messrs. Jay Simpson, Merrill Frank, Craig Samuel, Gerold Bastedo, Richard Hamell, Herbert Ehrig, and members of the classes in Geology 322 (Invertebrate Palaeontology) at Alfred University from 1973-79. Faunal analysis and identification were enhanced by the attention of Drs. Chris Barnes, Paul Sartenaer, Irving Tesmer, Tom Uyeno, Fred H. C. Hotchkiss, and the late Rudy Eller. Analyses of sedimentary characteristics were supplemented by Drs. Robert Snyder and Wayne E. Brownell of S.U.N.Y. College of Ceramics at Alfred.

Mr. Paul Gignac provided the photographic expertise. Ms. Pam Millter created the diorama of the Devonian sea bottom. Ms. Carolyn Dannheim typed and helped edit the manuscript.

The Alfred University Research Foundation awarded the senior author a stipend to initiate the project in the summer of 1973.

REFERENCES CITED

Cooper, G. A. and others, 1942, Correlation of the Devonian sedimentary formations of North America: Geological Society of America Bulletin, v. 53, p. 1729–1794.

Eller, E. R., 1935, New species of *Echinocaris* from the Upper Devonian, of Alfred Station, New York: Carnegie Museum Annals, v. 24, p. 263–275.

Epstein, A. G., Epstein, J. B., and Harris, L. D., 1977, Conodont color alteration—An index to organic metamorphism: United States Geologic Survey Professional Paper No. 995, p. 27.

Goldring, R., 1978, Devonian: *in* McKerrow, W. S., ed., The Ecology of Fossils: Massachusetts, The MIT Press, p. 125–145.

Goldring, R., and Stephenson, D. G., 1972, The depositional environment of three starfish beds: Neues Jarhbuch für Geologie und Paläeontologie Monatschefte, H. 10, p. 611–624.

Goldring, W., 1935, Some Upper Devonian crinoids from New York: Carnegie Museum Annals, v. 24, p. 337–348.

Keith, M. L., and Degens, F. T., 1959, Geochemical indicators of marine and fresh water sediments: *in* Albeson, P. H., ed., Researches in Geochemistry, New York, John Wiley, p. 38–61.

Manspeizer, W., 1963, A restudy of the Chautauquan Series of Allegany County, New York: *in* Shepps, V. C., ed., Symposium on Middle and Upper Devonian Stratigraphy of Pennsylvania and Adjacent States: Pennsylvania Geological Survey, Fourth Series, General Geology Report G 39, p. 259–277.

Meyer, D. L., 1971, Post mortem disarticulation of Recent crinoids and ophiuroids under natural conditions: Geological Society of America Abstracts with Programs, p. 645–646.

Potter, P. E., Shimp, N. F., and Witters, J., 1963, Trace elements in marine and fresh water argillaceous sediments: Geochemica et Geochemica Acta, v. 27, p. 669–694.

Rokop, F. J., 1974, Reproductive patterns in the deep sea benthos: Science, v. 186, p. 743–745.

Ruedemann, R., 1935, A review of the genus *Pterygotus* with descriptions of two new Devonian species: Carnegie Museum Annals, v. 24, p. 69–73.

Schäfer, W., 1972, Ecology and Palaeocology of Marine Environments, Craig, G. Y., ed.: Chicago, The University of Chicago Press, p. 586.

Steinbeck, J., 1954, Sweet Thursday: Viking Press, New York, New York, p. 142.

Woodruff, J. G., 1942, Geology of the Wellsville Quadrangle, New York: New York State Museum Bulletin, no. 326, p. 1–135.

MANUSCRIPT ACCEPTED BY THE SOCIETY JULY 20, 1984

Typeset by WESType Publishing Services, Inc., Boulder, Colorado
Printed in U.S.A. by Malloy Lithographing, Inc., Ann Arbor, Michigan